ELECTROCHEMISTRY OF METAL COMPLEXES

Electrochemistry of Metal Complexes

Editor:

Maria Emilova Velinova, Ph.D.

ArclerPress

www.arclerpress.com

Electrochemistry of Metal Complexes

Editor: Maria Emilova Velinova, Ph.D.

708 3rd Avenue, 6th Floor
New York
NY 10017
United States of America
www.arclerpress.com

ISBN: 978-1-68094-417-4

Library of Congress Control Number: 2016951125

About the Editor

Maria Emilova Velinova, Ph.D.

Maria Velinova is Ph.D. holder in Quantum chemistry at the University of Sofia since April 2012. Her major research experience is in the field of Computational Chemistry, especially in statistical mechanics methods applied to different sorts of biomolecules. Member of the Laboratory of Quantum and Computational Chemistry at the University of Sofia.

Contents

List of Contributors

Sarmed Abdalrasoul Salih
Electrochemical Eng. Department, University of Babylon, Hillah, Iraq

Abdalrasoul Salihh Mehdi
Electrochemical Eng. Department, University of Babylon, Hillah, Iraq

Ali Safa Nouri Alsaegh
Electrochemical Eng. Department, University of Babylon, Hillah, Iraq

P. N. Bartlett
Chemistry, The University of Southampton, Southampton SO17 1BJ, UK

D. A. Cook
Chemistry, The University of Southampton, Southampton SO17 1BJ, UK

M. W. George
School of Chemistry, The University of Nottingham, Nottingham NG7 2RD, UK

A. L. Hector
Chemistry, The University of Southampton, Southampton SO17 1BJ, UK

J. Ke
School of Chemistry, The University of Nottingham, Nottingham NG7 2RD, UK

W. Levason
Chemistry, The University of Southampton, Southampton SO17 1BJ, UK

G. Reid
Chemistry, The University of Southampton, Southampton SO17 1BJ, UK

D. C. Smith
Physics & Astronomy, The University of Southampton, Southampton SO17 1BJ, UK

W. Zhang
Chemistry, The University of Southampton, Southampton SO17 1BJ, UK

Claudia A. Berger
Institute of Electrochemistry, Ulm University, Albert-Einstein-Allee 47, 89081 Ulm, Germany

Maria Arkhipova
Institute of Organic Chemistry I, Ulm University, Ulm, Germany

Gerhard Maas
Institute of Organic Chemistry I, Ulm University, Ulm, Germany

Timo Jacob
Institute of Electrochemistry, Ulm University, Albert-Einstein-Allee 47, 89081 Ulm, Germany
Helmholtz-Institute-Ulm (HIU), Ulm, Germany

Liyan Zhao
WATLab and Department of Chemistry, University of Waterloo, Waterloo, Ontario, Canada N2L 3G1

Joseph P. Thomas
WATLab and Department of Chemistry, University of Waterloo, Waterloo, Ontario, Canada N2L 3G1

Nina F. Heinig
WATLab and Department of Chemistry, University of Waterloo, Waterloo, Ontario, Canada N2L 3G1

Marwa Abd-Ellah
WATLab and Department of Chemistry, University of Waterloo, Waterloo, Ontario, Canada N2L 3G1

Xiongyao Wang
WATLab and Department of Chemistry, University of Waterloo, Waterloo, Ontario, Canada N2L 3G1

K. T. Leung
WATLab and Department of Chemistry, University of Waterloo, Waterloo, Ontario, Canada N2L 3G1

J. Porcayo-Calderon
CIICAp, Universidad Autónoma del Estado de Morelos, Avenida Universidad 1001, 62209 Cuernavaca, MOR, Mexico
Instituto de Ciencias Físicas, Universidad Nacional Autónoma de México, Avenida Universidad s/n, 62210 Cuernavaca, MOR, Mexico

M. Casales-Diaz
Instituto de Ciencias Físicas, Universidad Nacional Autónoma de México, Avenida Universidad s/n, 62210 Cuernavaca, MOR, Mexico

V. M. Salinas-Bravo
Instituto de Investigaciones Eléctricas, Avenida Reforma 113, Colonia Palmira, 62490 Cuernavaca, MOR, Mexico

L. Martinez-Gomez
Instituto de Ciencias Físicas, Universidad Nacional Autónoma de México, Avenida Universidad s/n, 62210 Cuernavaca, MOR, Mexico
Corrosion y Protección (CyP), Buffon 46, 11590 México City, DF, Mexico

Burçin Bozal
Department of Analytical Chemistry, Faculty of Pharmacy, Ankara University, Tandogan, 06100 Ankara, Turkey

Bengi Uslu
Department of Analytical Chemistry, Faculty of Pharmacy, Ankara University, Tandogan, 06100 Ankara, Turkey

Sibel A. Özkan
Department of Analytical Chemistry, Faculty of Pharmacy, Ankara University, Tandogan, 06100 Ankara, Turkey

Thomas Dann
Wolfson Materials and Catalysis Centre, School of Chemistry, University of East Anglia, Norwich, NR4 7TJ, UK

Dragoş-Adrian Roşca
Wolfson Materials and Catalysis Centre, School of Chemistry, University of East Anglia, Norwich, NR4 7TJ, UK

Joseph A. Wright
Wolfson Materials and Catalysis Centre, School of Chemistry, University of East Anglia, Norwich, NR4 7TJ, UK

Gregory G. Wildgoose
Wolfson Materials and Catalysis Centre, School of Chemistry, University of East Anglia, Norwich, NR4 7TJ, UK

Manfred Bochmann
Wolfson Materials and Catalysis Centre, School of Chemistry, University of East Anglia, Norwich, NR4 7TJ, UK

Fangquan Xia
School of Chemistry and Chemical Engineering, University of Jinan, Jinan, 250022 Shandong, China

Xin Zhang
School of Chemistry and Chemical Engineering, University of Jinan, Jinan, 250022 Shandong, China

Changli Zhou
School of Chemistry and Chemical Engineering, University of Jinan, Jinan, 250022 Shandong, China

Danzi Sun
School of Chemistry and Chemical Engineering, University of Jinan, Jinan, 250022 Shandong, China

Yanmin Dong
School of Chemistry and Chemical Engineering, University of Jinan, Jinan, 250022 Shandong, China

Zhen Liu
School of Chemistry and Chemical Engineering, University of Jinan, Jinan, 250022 Shandong, China

Bao N. Nguyen
Department of Chemistry, Imperial College London, Exhibition Road, South Kensington, London SW7 2AZ, UK

Luis A. Adrio
Department of Chemistry, Imperial College London, Exhibition Road, South Kensington, London SW7 2AZ, UK

Tim Albrecht
Department of Chemistry, Imperial College London, Exhibition Road, South Kensington, London SW7 2AZ, UK

Andrew J. P. White
Department of Chemistry, Imperial College London, Exhibition Road, South Kensington, London SW7 2AZ, UK

Mark A. Newton
European Synchrotron Radiation Facility (ESRF), 6 rue Jules Horowitz, 38043, Grenoble, France

Maarten Nachtegaal
Paul Scherer Institut, 5232 Villigen PSI, Switzerland

Santiago J. A. Figueroa
European Synchrotron Radiation Facility (ESRF), 6 rue Jules Horowitz, 38043, Grenoble, France

King Kuok (Mimi) Hii
Department of Chemistry, Imperial College London, Exhibition Road, South Kensington, London SW7 2AZ, UK

Hui Xu
Key Laboratory of Functional Inorganic Material Chemistry, Ministry of Education and School of Chemistry and Materials, Heilongjiang University, 74 XueFu Road, Harbin 150080, China
Department of Chemistry, Faculty of Science, National University of Singapore, 3 Science Drive 3, Singapore 117543, Singapore

Runfeng Chen
Department of Chemistry, Faculty of Science, National University of Singapore, 3 Science Drive 3, Singapore 117543, Singapore
Key Laboratory for Organic Electronics & Information Displays, Institute of Advanced Materials, Nanjing University of Posts & Telecommunications, Nanjing 210023, China

Qiang Sun
Department of Chemistry, Faculty of Science, National University of Singapore, 3 Science Drive 3, Singapore 117543, Singapore

Wenyong Lai
Key Laboratory for Organic Electronics & Information Displays, Institute of Advanced Materials, Nanjing University of Posts & Telecommunications, Nanjing 210023, China

Qianqian Su
Department of Chemistry, Faculty of Science, National University of Singapore, 3 Science Drive 3, Singapore 117543, Singapore

Wei Huang
Key Laboratory for Organic Electronics & Information Displays, Institute of Advanced Materials, Nanjing University of Posts & Telecommunications, Nanjing 210023, China
Singapore-Jiangsu Joint Research Center for Organic/Bio-electronics and Information Displays and Institute of Advanced Materials, Nanjing University of Technology, Nanjing 211816, China

Xiaogang Liu
Department of Chemistry, Faculty of Science, National University of Singapore, 3 Science Drive 3, Singapore 117543, Singapore
Institute of Materials Research and Engineering, Agency for Science, Technology and Research, 3 Research Link, Singapore 117602, Singapore

I. Alcón
Institut de Ciència de Materials de Barcelona (ICMAB-CSIC), Networking Re Center on Bioengineering, Biomaterials and Nanomedicine (CIBER-BBN), Campus de la UAB, 08193, Bellaterra, Spain

M. Gonidec
Institut de Ciència de Materials de Barcelona (ICMAB-CSIC), Networking Re Center on Bioengineering, Biomaterials and Nanomedicine (CIBER-BBN), Campus de la UAB, 08193, Bellaterra, Spain

M. R. Ajayakumar
Institut de Ciència de Materials de Barcelona (ICMAB-CSIC), Networking Re Center on Bioengineering, Biomaterials and Nanomedicine (CIBER-BBN), Campus de la UAB, 08193, Bellaterra, Spain

M. Mas-Torrent
Institut de Ciència de Materials de Barcelona (ICMAB-CSIC), Networking Re Center on Bioengineering, Biomaterials and Nanomedicine (CIBER-BBN), Campus de la UAB, 08193, Bellaterra, Spain

J. Veciana
Institut de Ciència de Materials de Barcelona (ICMAB-CSIC), Networking Re Center on Bioengineering, Biomaterials and Nanomedicine (CIBER-BBN), Campus de la UAB, 08193, Bellaterra, Spain

Chen Yang
State Key Laboratory of Synthetic Chemistry, Institute of Molecular Functional Materials, HKU- Joint Laboratory on New Materials and Department of Chemistry, The University of Hong Kong, Pokfulam Road, Hong Kong, China
HKU Shenzhen Institute of Re and Innovation, Shenzhen, China

Faisal Mehmood
State Key Laboratory of Synthetic Chemistry, Institute of Molecular Functional Materials, HKU- Joint Laboratory on New Materials and Department of Chemistry, The University of Hong Kong, Pokfulam Road, Hong Kong, China

Tsz Lung Lam
The Hong Kong Polytechnic University Shenzhen Re Institute, Shenzhen, PR China
Department of Applied Biology and Chemical Technology, The Hong Kong Polytechnic University, Hung Hom, Hong Kong, China

Sharon Lai-Fung Chan
The Hong Kong Polytechnic University Shenzhen Re Institute, Shenzhen, PR China
Department of Applied Biology and Chemical Technology, The Hong Kong Polytechnic University, Hung Hom, Hong Kong, China

Yuan Wu
State Key Laboratory of Synthetic Chemistry, Institute of Molecular Functional Materials, HKU- Joint Laboratory on New Materials and Department of Chemistry, The University of Hong Kong, Pokfulam Road, Hong Kong, China

Chi-Shun Yeung
State Key Laboratory of Synthetic Chemistry, Institute of Molecular Functional Materials, HKU- Joint Laboratory on New Materials and Department of Chemistry, The University of Hong Kong, Pokfulam Road, Hong Kong, China

Xiangguo Guan
State Key Laboratory of Synthetic Chemistry, Institute of Molecular Functional Materials, HKU- Joint Laboratory on New Materials and Department of Chemistry, The University of Hong Kong, Pokfulam Road, Hong Kong, China

Kai Li
State Key Laboratory of Synthetic Chemistry, Institute of Molecular Functional Materials, HKU- Joint Laboratory on New Materials and Department of Chemistry, The University of Hong Kong, Pokfulam Road, Hong Kong, China
HKU Shenzhen Institute of Re and Innovation, Shenzhen, China

Clive Yik-Sham Chung
State Key Laboratory of Synthetic Chemistry, Institute of Molecular Functional Materials, HKU- Joint Laboratory on New Materials and Department of Chemistry, The University of Hong Kong, Pokfulam Road, Hong Kong, China

Cong-Ying Zhou
State Key Laboratory of Synthetic Chemistry, Institute of Molecular Functional Materials, HKU- Joint Laboratory on New Materials and Department of Chemistry, The University of Hong Kong, Pokfulam Road, Hong Kong, China
HKU Shenzhen Institute of Re and Innovation, Shenzhen, China

Taotao Zou
State Key Laboratory of Synthetic Chemistry, Institute of Molecular Functional Materials, HKU- Joint Laboratory on New Materials and Department of Chemistry, The University of Hong Kong, Pokfulam Road, Hong Kong, China

Chi-Ming Che
State Key Laboratory of Synthetic Chemistry, Institute of Molecular Functional Materials, HKU- Joint Laboratory on New Materials and Department of Chemistry, The University of Hong Kong, Pokfulam Road, Hong Kong, China
HKU Shenzhen Institute of Re and Innovation, Shenzhen, China

Fazlur Rahaman
Department of Chemistry, C.M.R. Institute of Technology, Bengaluru 560037, Karnataka, India

B.H.M. Mruthyunjayaswamy
Department of Studies and Research in Chemistry, Gulbarga University, Gulbarga 585106, Karnataka, India

Yeng Ying Lee
School of Chemistry, The University of New South Wales, Sydney, NSW 2052, Australia

D. Barney Walker
School of Chemistry, The University of New South Wales, Sydney, NSW 2052, Australia

J. Justin Gooding
School of Chemistry, The University of New South Wales, Sydney, NSW 2052, Australia

Barbara A. Messerle
School of Chemistry, The University of New South Wales, Sydney, NSW 2052, Australia

Bagdaulet Kenzhalyevich Kenzhaliyev
Kazakh-British Technical University, Almaty, Republic of Kazakhstan

Zamzagul Duisenovna Dosymbaeva
Kazakh-British Technical University, Almaty, Republic of Kazakhstan

Renata Ramilevna Iskhakova
Kazakh-British Technical University, Almaty, Republic of Kazakhstan

Essen Nurgalievich Suleimenov
Kazakh-British Technical University, Almaty, Republic of Kazakhstan

Haitao Wang
Key Laboratory of Nuclear Materials and Safety Assessment, Institute of Metal Research, Chinese Academy of Sciences, Shenyang, 110016, China

Leijiang Yao
Science and Technology on UVA Laboratory, Northwester Polytechnical University, Xi'an, 710065, China

Xiaofeng Li
Key Laboratory of Nuclear Materials and Safety Assessment, Institute of Metal Research, Chinese Academy of Sciences, Shenyang, 110016, China

En-Hou Han
Key Laboratory of Nuclear Materials and Safety Assessment, Institute of Metal Research, Chinese Academy of Sciences, Shenyang, 110016, China

Xiaoyan Tong
Science and Technology on UVA Laboratory, Northwester Polytechnical University, Xi'an, 710065, China

Preface

Electrochemistry has been undergoing noteworthy transformations in the last few decades. Such evolution is due to a number of possibility and applications of its methods in the industry, medicine and in the field of analysis. The book focus on the recent trends and achievements connected to electrochemical processes involving metal complexes.

The first part of the book deals with important aspects of electroplating, including a discussion of co-deposition of metals and formation of alloys. All this processes are widely used in many industrial applications due to their benefits in improving the corrosion fatigue life, conductivity of supercritical fluid electrolytes, in increasing the magnetocrystalline anisotropy and for designing of new nano-catalysts for electro-oxidation reactions.

Other particular emphasis in the book is given to the using of electrochemical techniques to the field of modern analysis. Here are presented papers that report various electroanalytical techniques for determination of anti-HIV drugs, bond energy and heavy metals in solution.

The last part of the book contains researches of the new metal–organic complexes and their applications. In particular, some of the promising examples for optoelectronic applications are discussed. Other interesting applications of the metal complexes are their potential uses as bioimaging and anti-cancer agents, biological sensors and antibacterial agents. The new applications of the electrochemistry for solving of environmental issues as the corrosion of materials and metal losses are also discusses in this part.

The articles collected in this book will be useful, effective, and beneficial to physicists, chemists, engineers and especially electrochemists.

Editor
Maria Emilova Velinova, Ph.D.

CHAPTER 1

STUDY THE EFFECT OF NICKEL COATING ON FATIGUE LIFE OF LOW CARBON STEEL EXPOSED TO CORROSIVE ENVIRONMENTS

Sarmed Abdalrasoul Salih, Abdalrasoul Salihh Mehdi and Ali Safa Nouri Alsaegh

Electrochemical Eng. Department, University of Babylon, Hillah, Iraq

ABSTRACT

Nickel electroplating on low carbon steel are widely used in many industrial application due to its benefit in improving corrosion fatigue life, increasing coating thickness provide good protection against corrosion but in other hand reduce fatigue life in addition to its high cost so applying a relatively thin layer of the electroplated nickel may give an acceptable corrosion protection in low cost. In this paper the effect of relatively thin layer (6 µ m) electroplated nickel deposited by Watts type bath on the corrosion fatigue life of low carbon steel specimens exposed to lab air and 3% NaCl solution as a corrosive environments was studied. Five groups of standard low carbon steel specimens were prepared for fatigue tests, Mechanical grinding was done for the all45 test specimens and after that Nickel electroplating coating was carried out to 18 specimens and mechanical polishing performed on the remaining 27specimens. Fatigue tests were carried out in air for nine specimens immediately after polishing and the remaining specimens were tested after exposing to lab air and 3% NaCl solution as corrosive environments for a period of 30 day using a reverse bending fatigue machine. Results indicate that fatigue life of relatively thin layer Nickel electroplated specimens was shorter by 4% than that of the uncoated mechanical polished specimens immediately tested in air and not exposed to corrosive environments, but it's corrosion fatigue life was higher by 11.46%than that of polished specimens when both exposed to lab air for 30 day and by 21.68% higher when both exposed to 3% NaCl solution for 30 day , this study indicate that relatively thin layer (6 µ m) Nickel electroplating decrease fatigue live of low carbon steel specimens exposed to un corrosive environment, but it improved corrosion fatigue life of low carbon steel specimens exposed to cyclic load and corrosive environment.

INTRODUCTION

In many industrial applications there are a large number of equipments, like pumps, fans, blowers, compressors, mixers, steam and gas turbines work in dry or wet corrosive environments, some components of these equipments such as blades subjected to repeated load failed faster than when it work in a non corrosive environments. Combination of corrosion and fatigue may cause the rapid failure of these components. When corrosion protection break down the pitting corrosion occurs, pitting corrosion reduce fatigue life. Corrosion fatigue due to corrosive environment is very important and more severe and cause an increase in maintenance costs in addition to equipments failure, so good resistance to fatigue with good corrosion protection is required [1, 2, 3, 4, 12] .

Citation: Sarmed Abdalrasoul Salih, Abdalrasoul Salihh Mehdi and Ali Safa NouriAlsaegh, Study the Effect of Nickel Coating on Fatigue Life of Low Carbon Steel Exposed to Corrosive Environments, ISSN 2224-3216.

Many methods are used to protect steel parts from corrosion fatigue such as changing stress of metal surface by shot penning, surface rolling, or adding inhibitors to the wet environment or coating by organic or inorganic materials [5]. Electroplating is used as a coating method to protect steel parts from corrosion. Electroplating generally reduce fatigue life of steel parts due to the initiation of micro cracks in the coating which then penetrate through the substrate[6,7] . Nickel electroplated on low carbon steel are widely used in many industrial application due to its benefit in improving corrosion fatigue life, increasing coating thickness provide good protection against corrosion but in other hand reduce fatigue life in addition to its high cost[8], so applying a relatively thin layer of the electroplated nickel may give an acceptable corrosion protection in low cost[6], influence of fatigue lives of low carbon steel by the coating have been explained by many research studies ,however studies on corrosion fatigue behavior of electroplated Nickel and its alloys on carbon steel are limited [1,13] . Some authors like Sunada H., (1966)[9], studied the effect of thin(100 μ m) Nickel plating deposited by Watts type bath on fatigue strength of carbon steel in air environment he found that the increase in plating thickness lower the fatigue limit.

S. J. Lennon, et al (1988)[8], studied the effect of thick (300 μ m) nickel coating deposited by sulphamate bath on the mechanical and fatigue properties of a turbine-disc steel, they concluded that thick coating decrease fatigue life of carbon steel .

T. A. Khalifa (1989)[6], studied the effect relatively thin electroplated Nickel (5.9 μm) coating with relatively thin layer (3.1) of copper undercoat on the fatigue behavior of carbon steel in dry air environment, he reported that the electroplating did not cause any change in the fatigue limit. Cheng-Hsun Hsu et al, (2004)[2], studied the effect of thickness of electroless Ni-Plated deposit on corrosion fatigue of 7075-T6 Al-alloy, after salt spray corrosion 100,500, and 1000 hr, they reported that the optimum coating thickness was greater than 5 μm and less than 25μm. Cabrera et al (2006)[10], studied the effect of electroless nickel plated on 7075-T6 aluminum alloy on the corrosion fatigue of these components in 3% NaCl solution as corrosive environment, they reported that electroless nickel increased corrosion fatigue life by 60-70% in low cycle fatigue region of 104-105 cycle. Sotomi Ishihara, et al[11], performed fatigue tests in laboratory air and3% NaCl solution on Mg alloy AZ31 plated with electroplated Ni, they concluded that The corrosion fatigue life of Ni-plated alloy in 3%NaCl solution was the same as those in laboratory air, which is mean the effectiveness of Ni-plating to protect Mg alloy from corrosion fatigue.

In this study fatigue tests were carried out on low carbon steel specimens polished and electro-plated by a thin layer (6μm)of Nickel using watts type bath. A group of polished specimens were tested in air directly after polishing to be the base point, the tests for both type were done in air after 700 hr (30 day) exposure to laboratory air and 3% NaCl solution as a corrosion environment. All the results were compared with the mechanical polished specimens which tested in air directly after polishing. Test results are reproduced in the figures and tables.

MATERIALS

The low carbon steels used in this study has a chemical compositions as supplied by the manufacturer are shown in Table 1.

Table (1): chemical composition (wt%)

wt%	C %	Si %	Mn %	P %	S %	Cr %	Ni %	Mo %
Alloy steel	< 0.2	0.26	<0.9	0.01	0.03	<1.3	0.11	0.01

SPECIMENS PREPARATION

Forty five specimenswere prepared from these materials to perform bending fatigue tests, cutting and grinding machine were used to do this job, specimens dimensionsare shown in Figure(1). These samples tempered at 400oC for one hour.

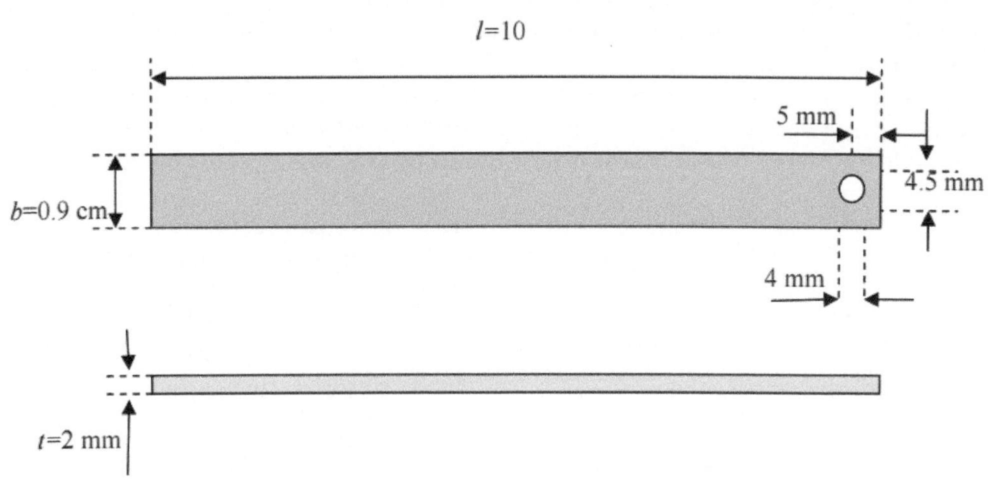

Figure (1): Specimen dimensions.

All specimens were mechanical grinded by grinding machine , smoothed using silicon carbide papers (60, 80 , 180 , 400 , 800) and then cleaned by distilled water followed by alcohol Eighteen from the above specimens were nickel electroplated by Watts type bath containing : 200 g/LNiso4, 50 g/L Mgso4, 70 g/L Na2so4, 30 g/L H3BO3, and 5 g/L NaCl at temperature 25 °C , current density 1.5 A /dm2, PH = 5 , and 20 minute electroplating time to get 6 μm of nickel coating .

Before electroplating the specimens were degreased in alcohol and in alkaline solution and then pickled in HCl solution. After electroplating the samples washed with distilled water, dried at room temperature for 24 hours to avoid hydrogen embrittlement. The remaining 27 specimens mechanical polished using two types of alumina (0.5 micron and 0.3 micron) Then cleaned by distilled water and by alcohol.

FATIGUE TEST

Five groups of standard low carbon steel specimens were prepared for fatigue tests. Fatigue tests were carried out in air for nine specimens immediately after polishing and the remaining specimens were tested after exposing to lab air at 3% NaCl solution as corrosive environment for a period of 30 day using a reverse bending fatigue machine, with a maximum capacity of 800 N/mm2, a maximum speed of 6000rpm and a stress ratio of R= -1.All the tests were performed at room temperature. The specimens were subjected to an applied load from the right side perpendicular to the axis of specimen and hence a bending moment was developed. Therefore, the surface of the specimen was under succession tension and compression stress. The value of the bending stress (σ_b), measured in (N/mm^2), for a known value of deflection (δ), measured in millimeter (mm) was calculated from the relation[4] .

The applied load (P) was calculated by using eq. (4 – 3) from the deflection which set from the test machine.

$$p = \frac{\delta \times E \times b \times t^3}{4 \times l^3} \qquad (4\text{-}3)$$

By substitute the value of P in eq. (4 – 4) then σ_b can be determined from the formula

$$\sigma_b = \frac{6 \times l \times p}{b \times t^3} \qquad (4\text{-}4)$$

Where:

σb: applied stress (MPa) , l= length of the specimen(mm)

b= width of the specimen (mm)

t= thickness of the specimen (mm)

p= applied load (N), δ= applied deflection (mm)

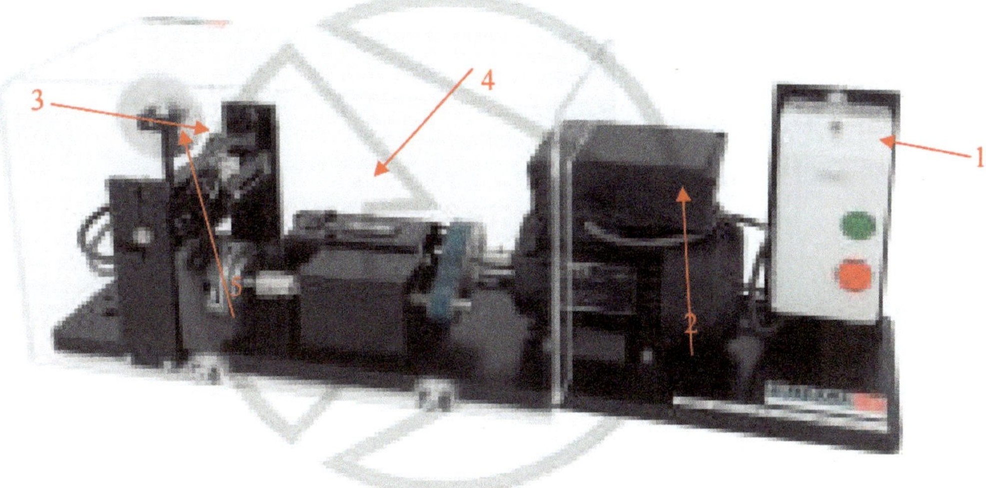

RESULTS AND DISCUSSION

Fig.(1) shows the stress- cycle (S-N) curves of polished specimens measured directly after polishing and after exposing to laboratory air environment for 30 day (700hr) period .The black line is the curve of the specimens tested directly after polishing and blue line is the curve of the specimens tested after 30 day. As can be seen from the figure the corrosion fatigue life of that exposed to 30 day to lab air decreased compared with that measured directly after polishing due to the effect of the corrosion that occurred on the surface of the specimen and then a micro cracks initiated from the corroded area and then propagate through the specimens substrate which lead to failure. Also the figure show that the fatigues limit is 200 MPa for the directly tested specimens and 170 MPa for the specimens which tested after 30 days in air. Fig.(2) shows the comparison of fatigue carves of polished specimens in laboratory air and in 3% NaCl solution environment after 700 hr exposure, these curves indicate that the fatigue life for specimen exposed to 3% NaCl solution for 700 hr was lower (145 MPa) than that in air (170 MPa) because in air fatigue cracks initiated at the whole surface and in3% NaCl solution fatigue cracks initiated at corrosion pits which were the cause of the large decrease in fatigue life relative to air[11] .

Comparison of fatigue curves shown in Figure (3) indicated that fatigue life of nickel coated specimen (192 MPa) was lower by 4% than that of the directly tested polished specimen (200 MPa), Tensile residual stresses, micro sharp cracks in coating acts as a stress raiser for the substrate resulting the initiation and propagation of fatigue cracks leading to the decreasing in fatigue life and steel failure [8] . Comparison of fatigue curves between polished specimens and nickel electroplated specimens shown in fig.(4) indicated the effect of Ni-electroplating on corrosion fatigue life of low carbon steel exposed 30 day to laboratory air environment ,fatigue life of nickel electroplated specimens (192 MPa) were higher by 11.46% than that of polished one (170 MPa). In corrosive air environment nickel electroplated layer protect low carbon steel surface from corrosion. Fig.(5) shows the S-N curves of Ni-electroplated carbon steel after 30 day exposure to lab air (192 MPa) and 3% NaCl solution (190 MPa), these curves indicated that there was no effect of the corrosive environment was found since the fatigue life of them approximately was the same due to the protection effect of the coating . Fig.(6) shows the S-N curves of Ni-electroplated carbon steel samples and polished samples both exposed 30 day to 3% NaCl solution as corrosive environment, it indicated that the corrosion fatigue life of Ni-electroplated steel (190 MPa) was higher than polished one (145 MPa) by 21.68% , this clearly certify the effectiveness of Ni-electroplating to protect low carbon steel from corrosion fatigue.

Table 1: Fatigue test data of uncoated mechanical polished specimens immediately tested

Stress amplitude (MPa)	Number of cycles to failure
400	134824
300	436712
290	1742555
280	2746746
250	3246746
210	4114551
200	9534567
187	unfailed
156	unfailed

Table 2: Fatigue test of uncoated polished specimens tested after 30 day in lab air

Stress amplitude (MPa)	Number of cycles to failure
400	134824
300	436712
270	2485103
240	2993176
200	4791764
180	6019572
170	10534877
185	unfailed
156	unfailed

Table 3: Fatigue test of uncoated mechanical polished specimens tested after 30 day in (3% NaCl solution)

Stress amplitude (MPa)	Number of cycles to failure
400	59086
300	273981
270	1836520
240	2295164
200	3758320
170	5325956
145	10000000
140	unfailed
135	unfailed

Table 4: Fatigue test data of nickel plated steel specimens tested after 30 day in lab air

Stress amplitude (MPa)	Number of cycles to failure
400	115674
300	428987
280	2581621
250	3173682
230	3564329
210	3954325
192	9743105
165	unfailed
180	unfailed

Table 5: shows the data of fatigue tests of nickel plated steel specimens testedafter 30 day in(3% NaCl solution)

Stress amplitude (MPa)	**Number of cycles to failure**
400	**108643**
300	**411876**
280	**2480966**
250	**3007764**
230	**3534278**
210	**3912356**
190	**10000000**
185	**unfailed**
180	**unfailed**

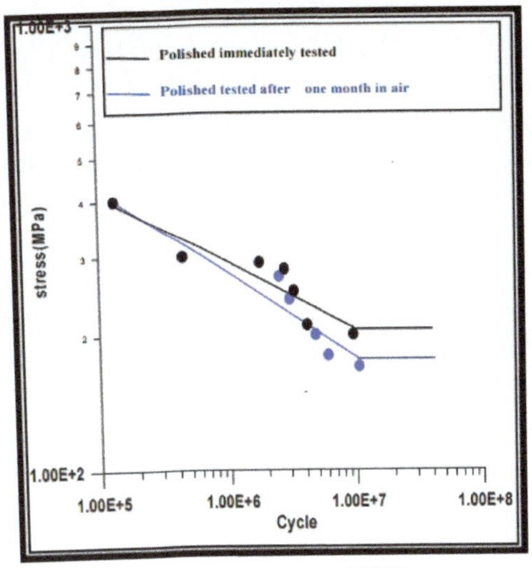

$$\sigma_b = 2332.463 * Nf^{-(0.15072)}$$

$$\sigma_b = 3724.843 * Nf^{-(0.188565)}$$

Figure 1: S.N curve of polished specimens tested directly after polishing and of that exposed 30 day to laboratory air environment.

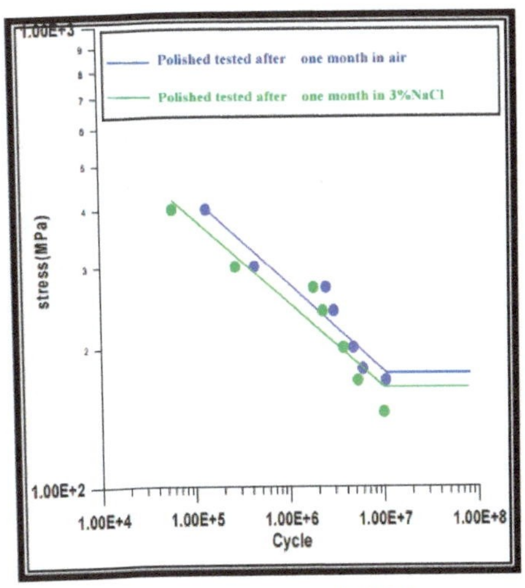

$$\sigma_b = 3724.843 * Nf^{-(0.188565)}$$

$$\sigma_b = 3103.76 * Nf^{-(0.182208)}$$

Figure 2: S.N carves of polished specimens in laboratory air and in 3% NaCl solution environment after 30 day exposure.

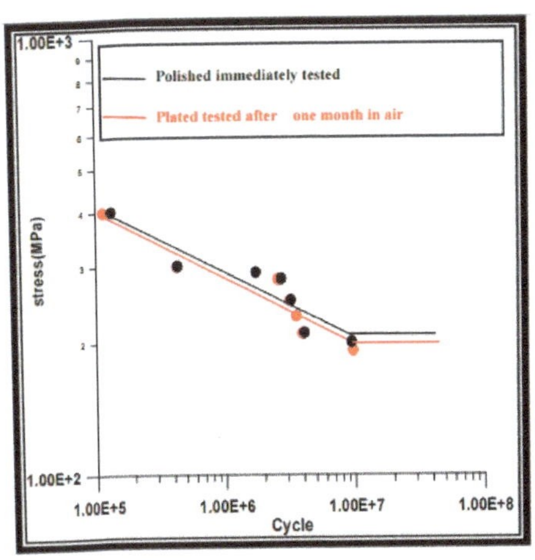

$$\sigma_b = 2332.463 * Nf^{-(0.15072)}$$

$$\sigma_b = 2361.069 * Nf^{-(0.153616)}$$

Figure 3: S.N carves of plated specimens in laboratory air and polished specimens directly tested.

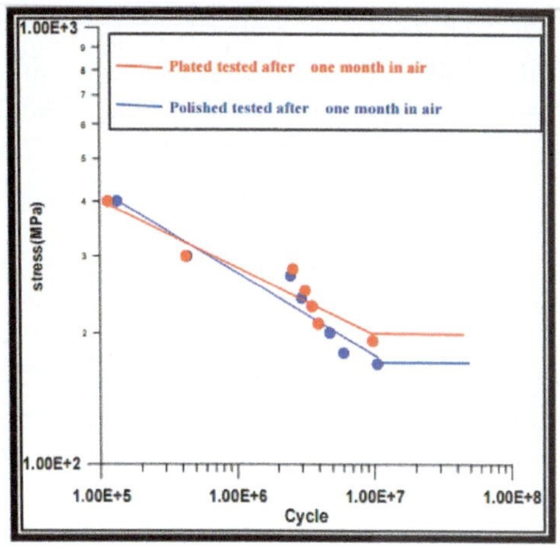

$$\sigma_b = 2361.069 * Nf^{-(0.153616)}$$

$$\sigma_b = 3724.843 * Nf^{-(0.188565)}$$

Figure 4: S.N carves of polished specimens and Nickel plated specimens in air environment after 30 day exposure.

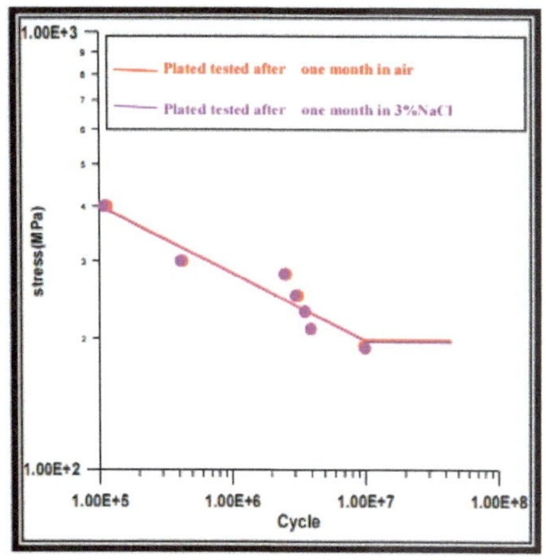

$$\sigma_b = 2361.069 * Nf^{-(0.153616)}$$

$$\sigma_b = 2281.024 * Nf^{-(0.153002)}$$

Figure 5: S.N carves of plated specimens in laboratory air and in 3% NaCl solution environment after 30 day exposure.

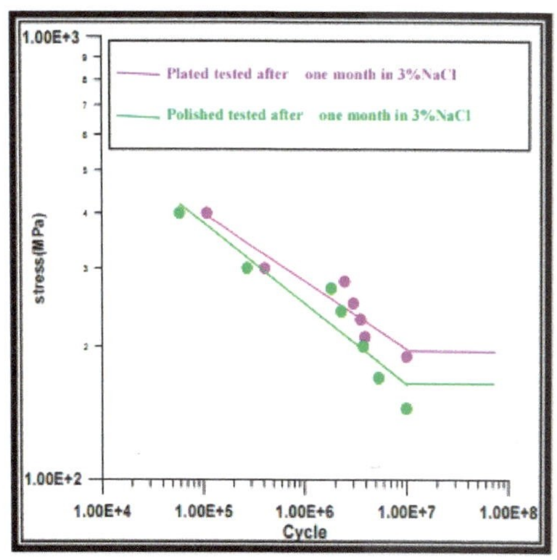

$$\sigma_b = 2281.024 * Nf^{-(0.153002)}$$

$$\sigma_b = 3103.76 * Nf^{-(0.182208)}$$

Figure 6: S.N carves of polished specimens and plated specimens in 3% NaCl solution environment after 30 day exposure.

CONCLUSIONS

Mechanical polished low carbon steel specimens gave good fatigue resistance due to the compressive stress gained by polishing process. Mechanical polished low carbon steel behavior showed low corrosion fatigue life in laboratory air and in 3% NaCl solution. Nickel electroplating in non corrosive environment reduced the fatigue life of low carbon steel specimens by 4% due to the tensile residual stress and micro cracks contained in the coating. In this work 6 μ m layer of electroplated Nickel improved fatigue life of low carbon steel specimen in corrosive environment and exposed to cyclic load in laboratory air by 11.56 % and in 3% NaCl solution by 21.68%.

REFERENCES

1. ChinnaiahMadduri and Raghu V. Prakash (2010) "Corrosion Fatigue Crack Growth Studies in Ni-Cr-Mn Steel", International Journal of Mechanical and Materials Engineering
2. Cheng – Hsun Hsu, et al, (2004) "effect of thickness of electroless Ni-P deposit on corrosion fatigue damage of 7075-T6 Al-alloy under salt spray atmosphere", material transaction, Japan institute of metal, vol.45, No11, pp. 3201-3208,.
3. H.J . C Voorwald et al , (2007) " Influence of electroless nickel interlayer thickness on fatigue strength of chromium plated AISI 4340 steel " "International journal of fatigue " , vol. 29, pp. 895-704
4. ASM Handbook volume 19, (1997) "Fatigue and Fracture" second printing.
5. D. J. Duquettee, (1968) "The Effect on Environment and Potential on Corrosion Fatigue of Low Carbon Steel", PhD theses, Massachusetts Institute of Technology.
6. T. A. Kalifa, (1989) "Effect of a Thin Copper Coating Plus Nickel Coating on the Development of fatigue cracks in a Carbon Steel", Journal of Material Science, pp.1110-1112.
7. P. A. Mikhallov, (1970) " Effect of galvanic coating of fatigue strength of structural steel ", Fisko-khimicheskaya mekanika materialov, vol. 6, No. 3, pp.111-113 .
8. S. J. Lemon, et al, (1988) "The Effect of Thick Nickel Coating on the Mechanical and Fatigue Properties of Turbine Disc Steel", J. S. Afr. Min., metal, Vol. 88, No7. PP. 213-218.
9. Sunada H. , (1967) "the effect of Nickel Plating on Fatigue Strength of Carbon Steel", J. SOC. Material Science, Japan,Vol. 16, pp. 230.
10. E. S. Puchi – Cabrera, et al, (2006) "Fatigue Behavior of a 7075 – T6 Aluminum Alloy Coated with an Electroless Ni – P Deposit ",Int. J. Fatigue, PP.1854-1866.
11. Sotomi Ishihara, et al, "Improvement in Corrosion fatigue resistance of Mg Alloy Due to Plating", Website. [Pdf]Intechweb.org.
12. SergoBaragetti et al, (2011) " A review of the fatigue behavior of component coated with thin hard corrosion resistance coating " , the open corrosion journal 9-17
13. H.J. C. Voorwald, (2005) " effect of electroplated Zn – nickel alloy coating on the fatigue strength of AISI high strength steel ", Journal of material engineering and performance , vol. 14 , pp 249- 259.

CHAPTER 2 ELECTRODEPOSITION FROM SUPERCRITICAL FLUIDS

P. N. Bartlett[a], D. A. Cook [a], M. W. George [b], A. L. Hector [a], J. Ke [b], W. Levason [a], G. Reid [a], D. C. Smith [c] and W. Zhang [a]

[a]Chemistry, The University of Southampton, Southampton SO17 1BJ, UK

[b]School of Chemistry, The University of Nottingham, Nottingham NG7 2RD, UK

[c]Physics & Astronomy, The University of Southampton, Southampton SO17 1BJ, UK

ABSTRACT

Recent studies have shown that it is possible to electrodeposit a range of materials, such as Cu, Ag and Ge, from various supercritical fluids, including hydrofluorocarbons and mixtures of CO_2 with suitable co-solvents. In this perspective we discuss the relatively new field of electrodeposition from supercritical fluids. The perspective focuses on some of the underlying physical chemistry and covers both practical and scientific aspects of electrodeposition from supercritical fluids. We also discuss possible applications for supercritical fluid electrodeposition and suggest some key developments that are required to take the field to the next stage.

INTRODUCTION

The field of supercritical fluid electrodeposition continues a long history of development of materials production and deposition methods which underpin nearly all modern technology. The importance and sophistication of materials deposition techniques can be seen in the cost of silicon fabrication facilities used to produce integrated circuits which are measured in billions of dollars. There now exists a wide range of deposition technologies. Vacuum techniques such as thermal evaporation, sputtering, chemical vapour deposition and molecular beam epitaxy are particularly suited to the deposition of high quality thin films. Such techniques have enabled the engineering of the fundamental quantum mechanical states of electrons in semiconductor structures leading to quantum well lasers in DVD players, quantum cascade lasers for chemical sensing, and entirely new fundamental physics, e.g. the fractional quantum Hall effect. Spin-coating and printing are two deposition techniques which have been used for a very long time, but which have been exploited in new ways recently. Much of this renewed interest has come about because of the discovery and application of organic semiconductors for applications such as flexible electronics. Whilst materials deposition is ubiquitous in technology, specific techniques can be found in particular applications in which their unique characteristics can out perform other techniques. For instance, as supercritical fluids have no surface tension and improved mass transport over liquids they have advantages for thermal deposition into high aspect ratio trenches and pores.[1] Supercritical fluids are also particularly good at swelling polymers in a controllable manner to allow the deposition of nanoparticles into existing polymers structures.[2]

In this perspective we will focus on electrodeposition from supercritical fluids and in particular on the underlying

Citation: P. N. Bartlett, D. A. Cook, M. W. George, A. L. Hector, J. Ke, W. Levason, G. Reid, D. C. Smith and W. Zhang, Electrodeposition from supercritical fluids, DOI: 10.1039/C3CP54955K.

physical chemistry of this new field. Since its introduction in the early 1800s, electrodeposition has become widely exploited because it has a combination of capabilities not found together in other deposition techniques.[3,4] These include: conformal deposition onto and into large complex objects; spatially defined growth allowing deposition onto complex structures only where required and allowing contiguous axial deposition through pores without blocking; direct control over the driving force for deposition through the applied potential; in operando measurement of electrical properties of the deposit during growth allowing active feedback; highly efficient use of reagents.

This perspective focuses on a relatively new development in the field of electrodeposition, supercritical fluid electrodeposition (SCFED). Supercritical fluids have a number of key advantages for electrodeposition. These include (i) a lack of surface tension which enables electrodeposition onto fragile substrates, (ii) good pore penetration and mass transport allowing deposition into high aspect ratio nanopores[1] and small diameter (<4 nm) nanopores,[5] (iii) a range of relatively easily accessible supercritical fluids with high chemical stability and thus large deposition windows, which allow the deposition of reactive materials such as germanium,[6] and (iv) the possibility to increase the deposition temperature with a specific fluid to above the critical temperature. Whilst the technique is still in its infancy, it is possible to envisage technological areas in which SCFED may find unique applications. A likely example is in the production of few nanometre diameter nanowires and nanowire based devices, e.g.Fig. 1.

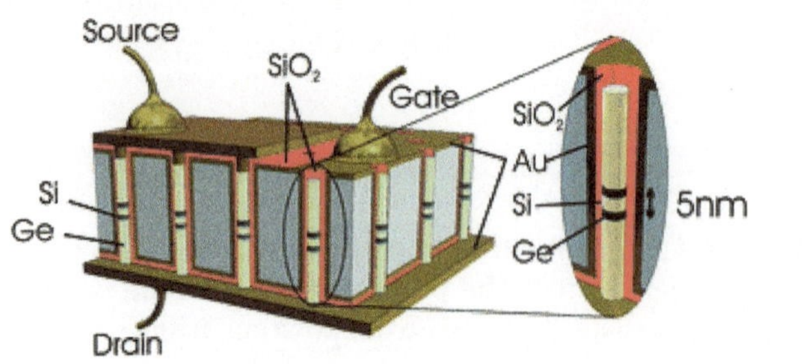

Figure 1: Concept of a device which supercritical fluid electrodeposition may enable in the future. The concept consists of a multiple resonant tunnel diode memory with element separation of 5 nm. Electrodeposition is particularly suited to the production of axial heterostructures like the Si/Ge quantum dot structure shown here. Using a supercritical fluid as the solvent would have advantages for pore penetration, deposition of reactive materials and high temperature deposition for materials quality.

There are many, varied reasons for the interest in nanowires for applications. These include at the simplest level the possibility of further miniaturisation of existing electronic[7] and data storage[8]components. The proximity of the surface to the whole of the bulk of the nanowire makes nanowire chemical and biochemical sensors capable of single molecule sensitivity.[9] In the field of thermoelectric materials nanowires show increased figure of merit over bulk materials because of decreased phononic thermal conductivity.[10] Other routes by which nanowires could contribute to energy efficiency and generation include photovoltaic nanowire devices[11] and piezoelectric nanowires.[12] The fact that nanowires are less susceptible to defects means that it may be possible to produce ultra-strength materials from them.[13] Most of the near to market applications use nanowires with diameters in excess of 10 nm, however initial investigations of ultrathin nanowires[14] suggest that these may be even more exciting in the medium term.

Supercritical fluids may bring a number of advantages for electrodeposition and for the electrodeposition of nanowires and nanostructures. However, there are significant challenges associated with trying to utilise SCFs for electrochemistry and electrodeposition. These include dissolution of a sufficient concentration of ionic species to produce a suitably conductive fluid, identifying sufficiently soluble and stable electrochemical precursors to produce useful deposition rates, ensuring the formation of a homogeneous solution and the development of safe and practical high pressure apparatus. Below we discuss these issues.

PROPERTIES OF SCFS AND MEASUREMENT OF SOLUBILITY AND CONDUCTIVITY

A supercritical fluid is defined as a substance above its critical temperature (T_c) and critical pressure (p_c), and as such they are unique solvents that combine the properties of liquids and gases, Table 1. SCFs have recently been exploited as green alternatives to conventional organic solvents in a variety of applications including: (i) separation processes such as extraction, sorption processes, chromatography, and drying; (ii) mechanical processes such as extrusion, homogenisation, emulsification, micronisation, crystallisation, impregnation and encapsulation, and (iii) chemical and biochemical reactions.[15–21] Many processes utilising sub- and supercritical fluids have been investigated for a wide range of industries including agriculture and food, cosmetics, pharmaceuticals, medicine, coatings, textiles, electronics and semiconductors, and waste treatment.[22–27] The properties of SCFs have received considerable attention, and have been extensively reviewed.[28–33]

Table 1: Density, viscosity, and diffusion coefficient of SCFs, gases and liquids[28,34]

State	Gas	Supercritical fluid	Liquid
Temperature/K	298	T_c	298
Pressure/MPa	0.1	p_c	0.1
Density/g cm^{-3}	0.6–2 × 10^{-3}	0.2–0.5	0.6–2
Diffusion coefficient/cm^2 s^{-1}	0.1–0.4	0.5–4 × 10^{-3}	0.2–2 × 10^{-5}
Viscosity/MPa s	0.01–0.03	0.01–0.03	0.2–20

Unlike conventional solvents, the physical properties of SCFs can be easily tuned by changing either temperature or pressure, or both. This advantage has been widely exploited in supercritical fluid extraction for separating products and in supercritical fluid reactions for achieving high yield and selectivity.[28–33] The tuneable properties of SCFs also have applications in processing nano-materials,e.g. supercritical drying of highly porous β-chitin structures,[35] and preparation of free-standing arrays of CdS nanowires.[36]

In order to successfully electrodeposit from a supercritical fluid we must consider their phase behaviour. SCFs are often illustrated with a p–T phase diagram of a single-component system. One such diagram is presented in Fig. 2 using trifluoromethane (CHF_3) as an example, for which T_c and p_c are 299.1 K and 4.82 MPa, respectively. The supercritical state is represented by the shaded area in Fig. 2. Density is one of the most important physical properties of SCFs; density can be shown as isochores in the p–T phase diagram, e.g. 7 isochores of CHF_3 are depicted in Fig. 2, with the density between 0.25 and 1.45 g cm^{-3}. It is worth pointing out that only a certain combination of temperature and pressure within the shaded area can provide the desired properties for use of SCFs as solvents. For example, consider CHF_3 under the conditions of 400 K and 15 MPa, which is ~100 K above its T_c, and ~10 MPa above its p_c. At 400 K and 15 MPa CHF_3 is in the supercritical state; however its density is only 6.4 mol dm^{-3} and compared with liquid H_2O, having a density of 55.6 mol dm^{-3} at 298 K, or liquid CHF_3, having a density of 20 mol dm^{-3} at its boiling point, the density of CHF_3 at 400 K and 15 MPa is too low for it to be a good solvent for ionic species, and therefore too low for it to be a good solvent for electrodeposition.

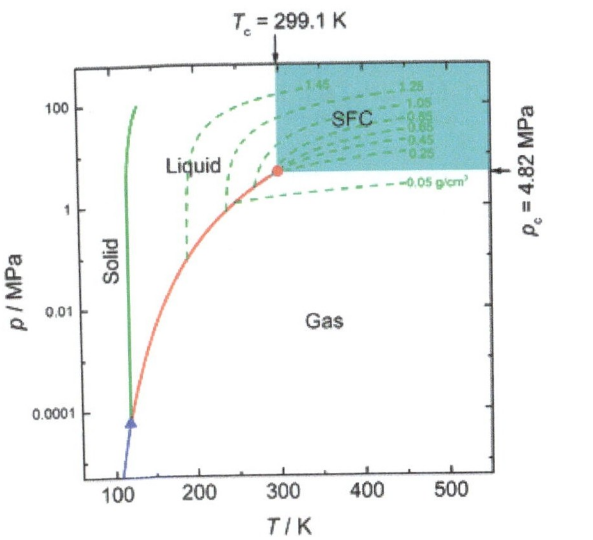

Figure 2: Phase diagram of pure CHF_3. ●, the critical point of CHF_3; ▲, the triple point of CHF_3; SCF, supercritical fluid, shaded in cyan. The red, green and blue curves are the gas–liquid, liquid–solid, and gas–solid phase boundary, respectively. The dashed curves represent the isochores corresponding to the density between 0.05 and 1.45 g cm^{-3}. The critical temperature (T_c) and critical pressure (p_c) are 299.1 K and 4.82 MPa, respectively, indicated by the arrows in the figure.

The dissolution of ionic species in SCFs is one of the key challenges to carry out supercritical fluid electrodeposition because most of the SCFs with easily accessible critical temperatures are non-polar or low polarity molecules and therefore are not good media for ion solvation. Electrostatic interactions contribute more than 80% of the total ion–solvent interactions.[37] Based on the Born equation, the electrostatic solvation energy (ΔG_{ES}), defined as the difference between the electrostatic free energy of an ion in vacuo and that in a solution, can be calculated from the ionic charge, ionic radius, and dielectric constant (ε_r). Fig. 3a shows the calculated ΔG_{ES} for a univalent ion with a radius of 0.2 nm in a medium with a dielectric constant ranging from 1 to 40. It can be seen that ΔG_{ES}decreases very rapidly with the increase of ε_r in the low dielectric constant region ($\varepsilon_r < 7$), but decreases rather slowly in the high dielectric constant region ($\varepsilon_r > 20$). In order for the electrolyte to dissolve and dissociate in the solution ε_r needs to be large enough so that the Gibbs free energy for solvation, ΔG_{ES}, is sufficiently negative to overcome the lattice energy. On the other hand, for those solvents with dielectric constant greater than 20, the difference in ΔG_{ES} is rather small so that the effect of dielectric constant on the solubility is not significant when only the electrostatic solvation energy is taken into account. The ε_r region between 7 and 20 is highlighted in Fig. 3. This is why most non-aqueous electrochemistry is carried out in solvents with a dielectric constant above 10 such as acetonitrile (ε_r = 36.6), N,N′-dimethylformamide (ε_r = 38.3), dimethylsulfoxide (ε_r = 47.2) or propylene carbonate (ε_r = 66.1).

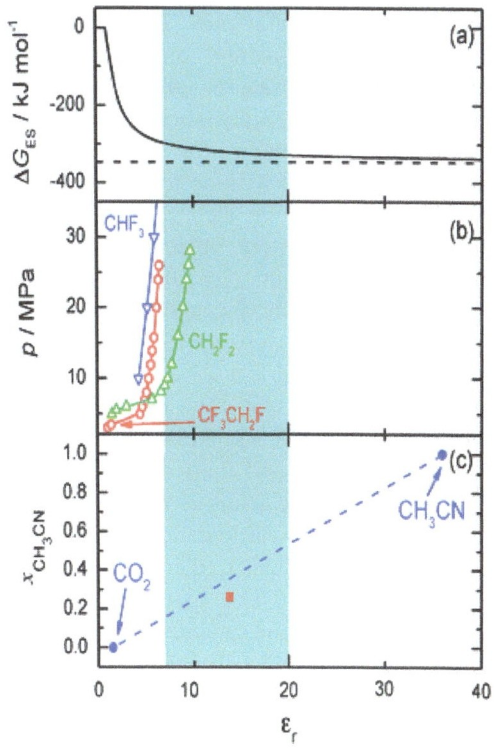

Figure 3: (a) Plot of the electrostatic solvation energy (ΔG_{ES}) of a univalent ion against dielectric constant (ε_r), calculated by the Born equation with the ionic radius of 0.2 nm.[37] (b) The effect of pressure (p) on ε_r for three fluorinated hydrocarbons at $T > T_c$ of each solvent: ▽, CHF_3 at 335 K; ○, CF_3CH_2F at 383 K; Δ, CH_2F_2 at 363 K.[38–40] (c) ε_r of the binary system of CO_2 + CH_3CN. The dashed line represents the linear interpolation of ε_r as a function of the mole fraction of CH_3CN (xCH_3CN).[41,42] ■, ε_r of a mixture of CO_2 + CH_3CN measured at 25 °C and 5.0 MPa.[43] The composition of the mixture is interpolated from the vapour–liquid phase equilibrium data reported by Reighard et al.[44]

Fig. 3b plots the dielectric constant of three fluorinated hydrocarbons (HFCs) as a function of pressure at temperatures above their corresponding T_c. It is clear that the dielectric constant increases with increasing pressure for all three HFCs, suggesting that the solubility of ionic species is higher at high pressures. In addition, the solubility of ionic species can be substantially higher in CH_2F_2 than in the other two HFCs (i.e. CHF_3 and CF_3CH_2F) because the dielectric constant of CH_2F_2 reaches 7 when the pressure is over 8 MPa. Hence, the selection of SCFs with high dielectric constant is the key to increasing the solubility of ionic species.

An alternative method to increase the dielectric constant is the use of mixed solvents, e.g. adding polar co-solvents into non-polar SCFs. This has been a widely accepted concept in all sorts of applications of SCFs.[45–49] In particular, the enhancement of the solubility of polar substances in CO_2 has been extensively studied by using acetone, methanol, ethanol, acetonitrile, etc. as co-solvents.Fig. 3c shows the effect of the mole fraction of CH_3CN ($\varepsilon_r = 36$) on the dielectric constant of the binary system of $CO_2 + CH_3CN$. One can see that it is crucial to add 10–18% (mole fraction) of CH_3CN into CO_2 to achieve a dielectric constant of 7. A much higher concentration of CH_3CN (e.g. $x_{CH3CN} > 0.5$) may not be necessary because it will not significantly decrease ΔG_{ES}, but makes the solution more 'liquid-like', and hence slows down the mass transport.

High-pressure electrochemistry cells are the key equipment for electrodeposition in SCFs. The general design considerations include maximum operating temperature and pressure, electrode insulation and installation, material comparability, and high-pressure safety. (Safety note: the illustrations shown here are purely schematic and should not be regarded as engineering drawings for the construction of safe high-pressure cells. Anyone wishing to build such equipment for themselves must carry out their own safety evaluation.) There are also a number of more subjective requirements, such as stirring the fluid in the cell, dismantling and reassembling the cell frequently and quickly, and convenience of transferring reagents. One cell design is given in Fig. 4. Briefly, the cell consists of a 10 cm^3 cylindrical vessel and a cap at its top, constructed from stainless steel. The cap has a fitted needle valve and 6 ports for SSI (Scientific Systems Inc.) type electrodes, thermocouples and fluid inlet/outlet. The vessel and the cap are assembled using a failsafe clamp system, incorporating a disposable elastomeric O-ring.[50] Recently, a pot made from PEEK (polyether ether ketone) was added to the vessel, together with a PTFE spacer fixed on the cap to reduce contamination from the stainless-steel inner surface.[6]

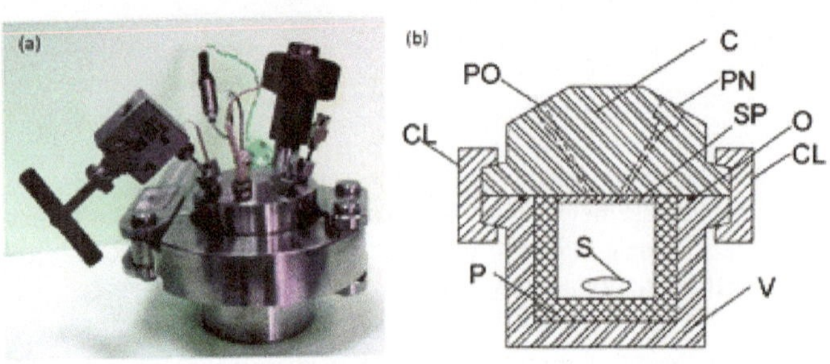

Figure 4: High-pressure electrochemistry cell: (a) the electrochemistry cell with the electrodes installed; (b) schematic diagram, C, cap; CL, clamp; O; O-ring; P, PEEK pot; PO; SSI type port; PN; port for the needle valve; S, magnetic stirrer; SP, PTFE spacer; V, stainless steel vessel. More details can be found from ref. 6 and 34.

Electrodeposition requires working solutions with low ohmic resistance and high concentrations of supporting electrolytes and reagents. Since such data on working solutions are not readily available in the literature for supercritical fluid systems, it is essential to acquire a fundamental understanding on the electrical conductivity and phase behaviour of working solutions at high temperature and high pressure. For this purpose, specialised high pressure equipment has been utilised for studying the conductivity and phase behaviour of multi-component systems; two examples of such equipment are shown in Fig. 5. The detailed description of the equipment can be found from the respective publications.[34,51,52]

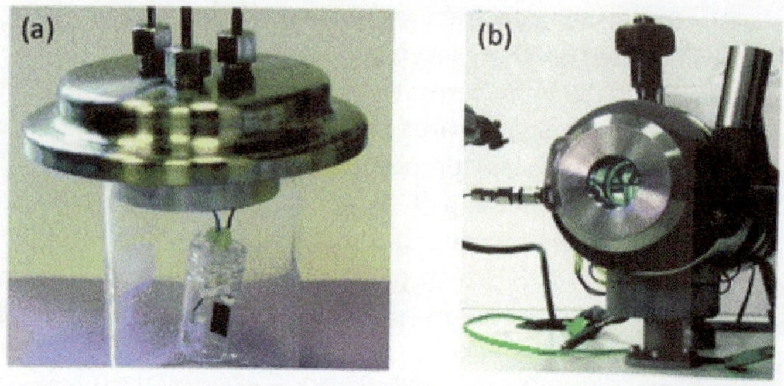

Figure 5: (a) Conductivity cell with two platinum electrodes. The electrode is soldered to a platinum wire, which is insulated with PEEK tubing and sealed into a stainless steel holder for installing in a high-pressure autoclave. (b) Variable-volume view cell for studying the phase behaviour of multicomponent systems. The fluid mixture with a given concentration of supporting electrolyte and reagents is introduced into the cell and sealed. The conditions (temperature and pressure) corresponding to a homogeneous, single-phase state can be determined visually by changing the volume of the cell at a given temperature.

ELECTROCHEMISTRY IN SUPERCRITICAL FLUIDS

It is appropriate to begin this section with a quotation from Darr and Poliakoff which is useful to keep in the forefront of one's mind "Working with supercritical fluids involves high pressures and sometimes high temperatures. Nearly always, it will always be easier to carry out an experiment under conventional conditions than supercritically".[53] This, in our experience, is definitely true in the case of electrochemistry where, when working at elevated temperatures and pressures it is no longer a trivial matter to polish the electrode between measurements, to find well-behaved reference electrodes or to take for granted the function of the counter electrode. On the other hand supercritical fluids have unique properties and sometimes the additional complexity allows results which cannot be achieved another way.

The earliest studies in supercritical fluids other than water appear to have been by Williams and Naiditch who reported the electrodeposition of Ag from "dense gaseous solutions" of $AgNO_3$ in ammonia.[54] These experiments were carried out in sealed glass vessels using two Pt electrodes. Dendritic growth of Ag was observed on the cathode. The deposition was carried out slightly (7 K) above the critical temperature for NH_3 so it is not certain that the fluid was in the supercritical state.

Subsequently in 1981 Silvestri et al. reported results for a preliminary study of supercritical CO_2, bromotrifluoromethane, hydrogen chloride and ammonia as solvents for electrochemistry.[55] Their focus was on the possibility of using these supercritical fluids as solvent for electrosynthesis. They report that "a solution of tetrabutylammonium iodide in CO_2 was a poor conductor in the liquid, as well as in the supercritical state. Bromotrifluoromethane, in which the electrolyte was practically insoluble, also proved to be a very poor conductor". They had greater success with the more polar, and therefore higher dielectric, solvents scHCl and $scNH_3$ and they report long-term electrolyses (4000 C charge passed) for Ag and Fe electrodes in $scNH_3$. scHCl was found to be difficult to work with because it is highly corrosive. Nevertheless they were able to produce elemental iodine by electrolysis of KI in scHCl.

Between 1984 and 1997 Bard's group published a series of studies of electrochemistry in supercritical fluids, NH_3,[56–58] water,[59–62] acetonitrile[63,64] and SO_2.[65] In these studies they developed the techniques for working with supercritical fluids, and particularly corrosive supercritical fluids such as water, and studied the electrochemistry of both inorganic and organic redox systems, including Cu deposition,[62] halide oxidation,[60] oxygen reduction,[60] and hydroquinone oxidation[60] all from scH_2O, and solvated electrons and organic redox couples in $scNH_3$.[56–58]

The systems investigated by Bard's group were all quite polar fluids. These, in general, have the disadvantage of high critical temperatures and pressures. Although less polar supercritical fluids such as $scCO_2$ are more benign to work with, they present the electrochemist with a significant problem in that they have very low dielectric constants and it is therefore difficult to achieve sufficient dissolution and dissociation of the electrolyte to achieve reasonable solution conductivity. Thus, for example, Abbott and Harper[66] briefly studied the electrochemistry of $[Ndodecyl_4]_2[Ni(mnt)_2]$ (mnt = maleonitrile) in $scCO_2$ using hydrophobic electrolytes (tetradecylammonium tetraphenylborate), they found some conductivity (~10^{-6} S cm^{-1}) and obtained some, poorly resolved, voltammetry.

One approach to overcome this problem is to use co-solvents such as methanol or acetonitrile. These mixtures still form single phase supercritical systems, although it is of course essential to characterise the phase behaviour to establish the appropriate conditions in terms of T and p, but the presence of the co-solvents increases the solubility of ionic species and dissociation of the electrolyte, and hence the conductivity of the fluid. To increase the conductivity of $scCO_2$ ($\varepsilon_r = 1.5$ at 308 K and 10 MPa[41]), CH_3OH and CH_3CN have been previously explored as co-solvents.[67] A review of electrochemistry in $scCO_2$ and $scCO_2$ with various co-solvent, covering the literature up to 1998, has been provided by Grinberg and Mazin.[68]

An alternative approach is to use hydrofluorocarbons (HFCs) as supercritical solvents as these are polar and give higher dielectric constant fluids whilst retaining reasonable critical temperatures and pressures. Early work in this area was carried out by Olsen and Tallman[69] who studied the electrochemistry of ferrocene and the cobalticenium cation at microdisc electrodes in $scCHClF_2$ with millimolar concentrations of $[N^nBu_4][BF_4]$ electrolyte. In addition they also investigated the use of $scCHF_3$ and found good voltammetry for ferrocene and the cobalticenium cation with microdisc electrodes (Fig. 6). Subsequently, in a series of papers, Goldfarb and Corti[70–72] studied electrochemistry of decamethylferrocene in $scCHF_3$ containing $[N^nBu_4][PF_6]$ electrolyte.

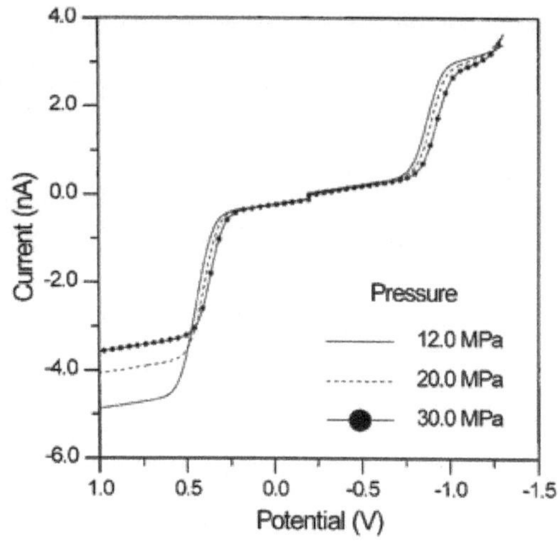

Figure 6: Comparison of voltammetry of ferrocene and the cobalticenium cation as a function of pressure in supercritical chlorodifluoromethane. Conditions: 90 µM Fc, 90 µM Cc^+, 10.0 mM $[N^nBu_4][BF_4]$; 388 K; 25 µm-diameter Pt disk electrode; scan rate, 20 mV s^{-1}. For clarity, only the forward sweep of each voltammogram is displayed, with the ferrocene wave recorded first and the cobalticenium cation wave immediately aft er, each wave originating from −0.20 V. Reprinted with permission from S. A. Olsen and D. E. Tallman, Anal. Chem., 1996, 68, 2054–2061. Copyright 1996 American Chemical Society.

At around the same time Abbott's group also published a series of papers on electrochemistry in supercritical hydrofluorocarbons. Thus they investigated the use of supercritical 1,1,1,2-tetrafluoroethane (CF_3CH_2F) and $scCH_2F_2$ and demonstrated a wide potential window for these solvents with $[N^nBu_4][BF_4]$ or $[N^nBu_4][ClO_4]$ electrolytes allowing oxidation of $[Cs(18\text{-crown-}6)]^+$ and Xe in liquid CF_3CH_2F.[73] They also used supercritical 1,1,1,2-tetrafluoroethane with $[N^nBu_4][BF_4]$ electrolyte as a solvent for the electrochemical reduction of CO_2 at Pt and Pb electrodes.[74]

In addition to these studies of electrode reactions there have been a limited number of studies of more fundamental issues such as diffusion, effects of solvation and the structure of the double layer. For example Goldfarb and Corti[72] have published an in depth study of the diffusion coefficient for decamethylferrocene in $scCHF_3$ with $[N^nBu_4][PF_6]$ electrolyte as a function of the temperature and pressure, taking account of the effects of ion pairing and correlating the results on the basis of the Stokes–Einstein model. Abbott et al.[75] investigated the effects of solute on the viscosity of $scCH_2F_2$ using quartz crystal microbalance measurements and showed that the viscosity could be surprisingly high (8–32 fold increase) at pressures close to the critical value. They also investigated the effects of ion-pairing on the redox potential of ferrocene carboxylic acid in $scCH_2F_2$.[76] Abbott's group are also the only ones to investigate the structure of the double layer in supercritical fluid electrolytes; $scCH_2F_2$ containing $[NR_4][BF_4]$[77] and $scCO_2$ containing long chain quaternary ammonium electrolytes.[78]

NEW ELECTROLYTES

One of the enduring challenges of electrochemistry in supercritical fluids is the choice of electrolyte in order to achieve high solution conductivity and thus reduce iR drop due to the uncompensated solution resistance. This is particularly important for applications in electrosynthesis or electrodeposition where it is necessary to pass larger currents. A number of studies have been reported in the literature on the conductivity of supercritical fluid electrolytes.

Olsen and Tallman[79] reported values for the equivalent conductivity for $[N^nBu_4][BF_4]$ in $scCHClF_2$ at 388 K between 10 and 24 MPa for concentrations between 6 and 13 mM, Fig. 7. They found a non-linear increase in equivalent conductivity with the square root of electrolyte concentration; behaviour indicative of the formation of triple ions (+)(−)(+) or (−)(+)(−) as predicted by the Fuoss–Kraus equation.[80] Abbott and Eardley[81] investigated the conductivity of $[N^nBu_4][BF_4]$ in $scCH_2F_2$ and found similar behaviour with evidence for a contribution from triple ions and Goldfarb and Corti,[71] in an extensive study, found similar effects for $[N^nBu_4][ClO_4]$ in $scCHF_3$. Jun and Fedkiw reported conductivities in the range 10^{-5} to 10^{-4} S cm^{-1} for alkali-metal salts of trifluoroacetates in $scCO_2$–CH_3OH, with the highest values in the single phase region obtained for the lithium salt.[82] In all cases these commercially available electrolytes give solutions with reasonable conductivity but, for electrosynthesis and electrodeposition applications, it would be desirable to find electrolytes with higher conductivity.

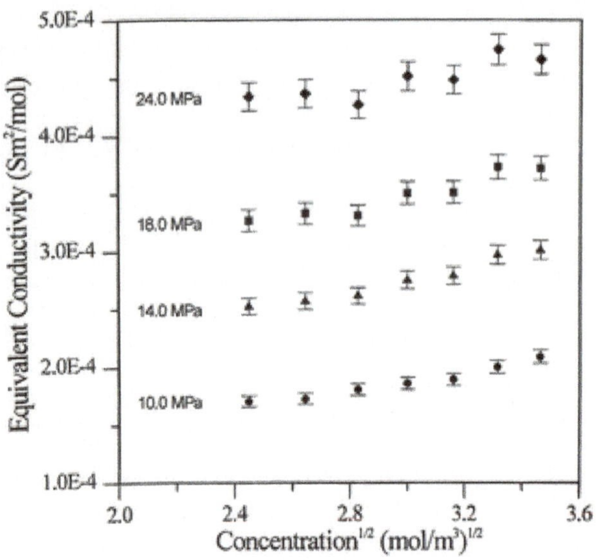

Figure 7: Equivalent conductivity of supercritical $CHClF_2$ at 388 K as a function of the square root of $[N^nBu_4][BF_4]$ concentration. The points (error bars) are the averages (standard deviations) of 5–6 measurements. Data at four densities are shown corresponding to fluid pressures of 10.0 (●), 14.0 (▲), 18.0 (■), and 24.0 MPa (◆). Reprinted with permission from S. A. Olsen and D. E. Tallman, Anal. Chem., 1996, 68, 2054–2061. Copyright 1996 American Chemical Society.

The successful development of electrodeposition from supercritical fluids requires careful choice of both electrolytes and reagents. The electrolytes must (i) have good solubility in the SCF, (ii) be dissociated into the constituent ions in the solution, (iii) have a wide electrochemical window (i.e. redox inactive in the potential range to be used), and (iv) be chemically inert with respect to the reagents added. The low polarity of the SCFs used in this work makes (i) and (ii) in particular, key issues to be addressed. Weakly coordinating anions have long been used in coordination and organometallic chemistry with aims including stabilisation of reactive cations, allowing the coordination of weakly bound ligands to metal centres, and enhancing catalysis rates and activities. Typical examples include $[BF_4]^-$, $[ClO_4]^-$, $[PF_6]^-$, $[SO_3F]^-$, $[BPh_4]^-$ usually in association with alkali metal or large tetra-alkylammonium cations.[83] The search for even more weakly coordinating anions has resulted in development of the chemistries of fluorinated analogues of $[BPh_4]^-$, $[M\{OTeF_5\}_n]^-$ (M = B, As etc.), $[Al\{OC(CF_3)_3\}_4]^-$, and various carboranes.[84] It is important to remember that weakly coordinating anions are not necessarily chemically inert, many examples of fluoride abstraction from $[BF_4]^-$ or $[PF_6]^-$, and B–C bond cleavage in $[BPh_4]^-$ are known.[83] For SCFED, electrolytes based upon tetraalkylammonium cations and weakly coordinating anions seem to be a promising approach. Conductivity studies of $scCO_2$, CH_2F_2 and CHF_2CHF_2 by Abbott et al.[73,81] used $[N^nBu_4][BF_4]$ or $[N^nBu_4][ClO_4]$ as electrolytes. We have used fluorinated tetra-aryl borates ($[B(Ar^F)_4]^-$) which were initially developed by organometallic chemists to address both the tendency of $[BPh_4]^-$ to π-aryl coordination to metal centres and B–C bond fission; they are also much more resistant to fluoride abstraction by electrophilic metals than $[BF_4]^-$.[83,84]

The $[B(Ar^F)_4]^-$ anions with Ar^F = C_6H_4F, C_6F_5, 4-$CF_3C_6H_4$ and 3,5-$(CF_3)_2C_6H_3$, (Fig. 8) are readily made as sodium salts by reaction of $Na[BF_4]$ with the appropriate Grignard reagent $BrMg(Ar^F)$,[85–87] and can be converted to $[NR_4]^+$ salts by metathesis with $[NR_4]I$ in CH_2Cl_2.[88] Various $[B(Ar^F)_4]^-$ salts are commercially available, although very expensive, but we have found that, depending on the metal reagents used, it is often possible to recover the electrolyte after deposition, re-purify it and hence recycle it.

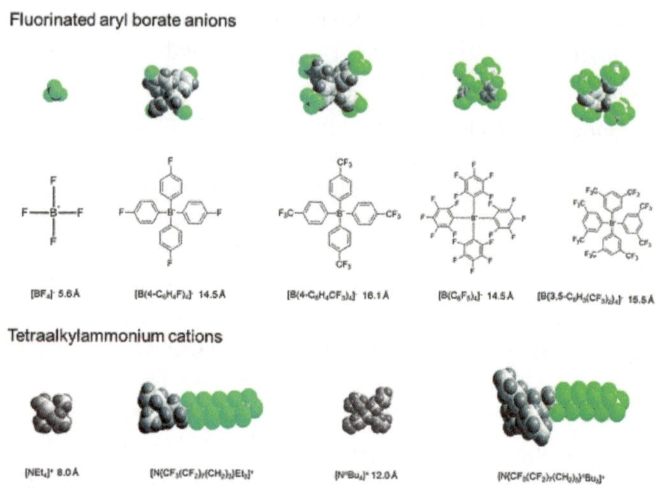

Figure 8: Space filling representations of the cations and anions used as supporting electrolytes (green = F), illustrating their approximate sizes. The dimensions correspond to the estimated diameter across the respective ions based upon van der Waals radii.

We also explored the effects of using alkylammonium cations incorporating fluorous "ponytails" to increase solubility. $[CF_3(CF_2)_7(CH_2)_3N^nBu_3]I$ was made by quaternisation of $^nBu_3N$ with $CF_3(CF_2)_7(CH_2)_3I$, and $[CF_3(CF_2)_7(CH_2)_3NMe_3]I$ from $CF_3(CF_2)_7(CH_2)_3NH_2$ and CH_3I, and metathesis with $Na[B(Ar^F)_4]$ in CH_2Cl_2 gave the appropriate $[CF_3(CF_2)_7(CH_2)_3NR_3][B(Ar^F)_4]$.[88] Thermogravimetric analysis studies showed that the cations were thermally stable up to ~180 °C, although subsequent conductivity measurements showed the cations were not significantly better than $[N^nBu_4]^+$.[88]

As stressed above, it is important to fully characterise the phase behaviour of the system since addition of significant concentrations (>10 mM) of electrolyte is likely to alter the phase behaviour. This can be done using a view cell of the type described above. These studies should then be complemented by measurements of the conductivity to characterise the behaviour of the different electrolytes. In our work[88] we started by comparing the behaviour for mixtures of CO_2 with $[N^nBu_4][BF_4]$ and either methanol (CH_3OH) or acetonitrile (CH_3CN) as the co-solvent. These experiments showed that at similar temperatures and pressures, the solubility of $[N^nBu_4][BF_4]$ was ~5 times higher in $CH_3CN + CO_2$ than in $CH_3OH + CO_2$.

Fig. 9 shows the phase behaviour of a ternary CO_2–$[N^nBu_4][BF_4]$–CH_3CN system and illustrates the potential complexity. For the mixture shown in Fig. 9, the three-phase region is a surface with 2 degrees of freedom as required by the phase rule for a ternary system with three phases ($F = 3 - 3 + 2 = 2$). At high T and p there is a single supercritical phase region which gives way to two and three phase regions as the pressure is decreased. The phase transition is a normal bubble-point transition ($L_1 \rightarrow L_1 + V$) when the temperature is below T*. However, when T is above T*, the sequence of the phase transition on decreasing the pressure isothermally is: L1 → L1 + L2 → L1 + L2 + V → L1 + V.

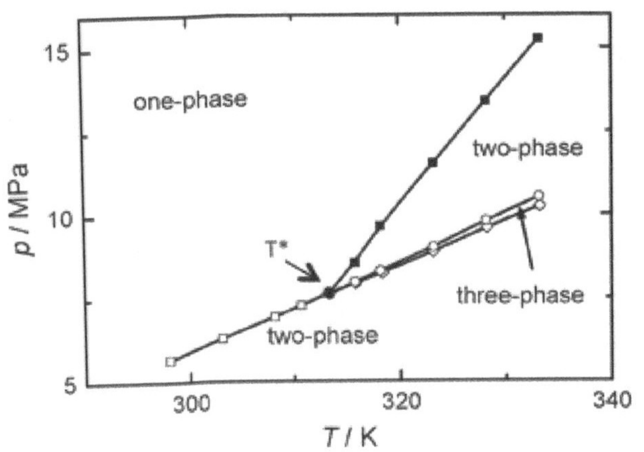

Figure 9: p–T phase diagram of the mixture of CO_2 (1) + CH_3CN (2) + $[N^nBu_4][BF_4]$(3) with $x_1 = 0.871$, $x_2 = 0.129$, and $x_3 = 8.3 \times 10^{-4}$. (■) and (□), the phase boundary between one-, and two-phase region; (○) and (◇) mark the region where three phases are in co-existence (L_1L_2V). The pressure difference between the two three-phase boundaries is less than 0.3 MPa at 333 K.[34]

From these phase behaviour studies for different compositions of the ternary mixture we find that increasing the concentration of $[N^nBu_4][BF_4]$ increases the pressure required to form a homogeneous solution at a given temperature, and furthermore that the more CO_2 there is in the system, the higher the pressure that is required because CO_2 is a poor solvent for the ionic species (in this case $[N^nBu_4][BF_4]$). Again this illustrates the potential complexity of the phase behaviour of these systems and why it is essential to experimentally establish the conditions required to form a single supercritical phase.

In principle, the addition of the electrochemical reagents used in the electrodeposition will also alter the phase behaviour. In practice for this system, as shown in Fig. 10, the effects are much smaller, not least because the concentrations of the dissolved reagents are much smaller – the addition of ~3.5×10^{-4} mole fraction of [Ag(hfac)(cod)] or $[Cu(hfac)_2]$ (cod = cycloocta-1,5-diene, hfacH = $CF_3C(O)CH_2C(O)CF_3$) produces only slight shifts in the supercritical phase boundary (<2 MPa at fixed temperature).

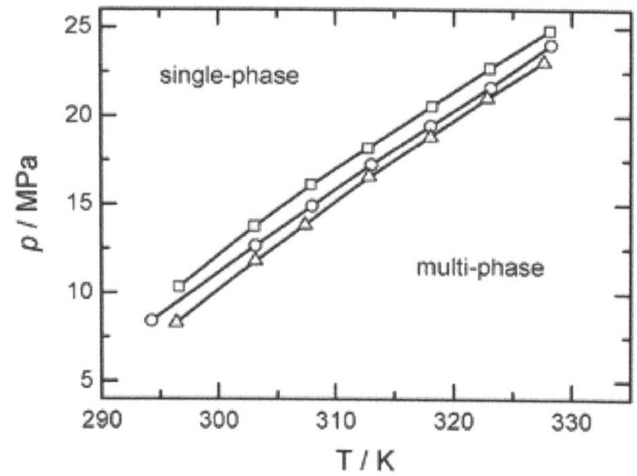

Figure 10: p–T phase diagram of the mixture CO_2 (1) + CH_3CN (2) + $[N^nBu_4][BF_4]$(3) with and without metal salts (4). $x_1 = 0.89$, $x_2 = 0.11$ and $x_3 = 1.34 \times 10^{-3}$. (○) no metal salt; (△) with [Ag(hfac)(cod)], $x_4 = 3.4 \times 10^{-4}$, and (□) with $[Cu(hfac)_2]$, $x_4 = 3.6 \times 10^{-4}$.[34]

In order to improve upon the commercial electrolytes we have synthesised a range of anions and cations selected to achieve greater solubility in $scCO_2$–CH_3CN by increasing the ion size and by increasing the degree of fluorination. In our conductivity studies we found that the effects of adding a fluorous ponytail (replacing $[N^nBu_4]^+$ by $[NR^fMe_3]^+$ or $[NR^{fn}Bu_4]^+$) to the cation were small and did not justify the synthetic effort. This is presumably because the fluorous ponytail does not alter the separation in the ion pair because it is flexible. A similar effect was found by Abbott and Schiffrin[89] for tetrafluoroborate salts of long chain alkylammonium cations in low dielectric solvents, where the interionic distance in the ion pairs is largely independent of the alkyl chain length due to interpenetration.

Table 2 summarises the data, the structures of the ions and their relative sizes are shown in Fig. 8. In terms of the effect of the anions on the molar conductivity, the general trend is $[BF_4]^- \sim [B(4\text{-}C_6H_4F)_4]^- < [B(4\text{-}C_6H_4CF_3)_4]^- \sim [B(C_6F_5)_4]^- < [B\{3,5\text{-}C_6H_3(CF_3)_2\}_4]^-$ with the highest molar conductivity, 22–26 S cm^2 mol^{-1}, achieved using the $[B\{3,5\text{-}C_6H_3(CF_3)_2\}_4]^-$. This is about an order of magnitude improvement over $[N^nBu_4][BF_4]$. For a solution of 0.07 M $[N^nBu_4][B\{3,5\text{-}C_6H_3(CF_3)_2\}_4]$ dissolved in $scCO_2$–CH_3CN (molar ratio $CH_3CN : CO_2$ ~ 0.12) at 20 MPa and 328.15 K we obtain a maximum measured conductivity of ~3 mS cm^{-1}.

Table 2: Molar conductivity of the fluorinated supporting electrolytes in a mixed fluid of CO_2 (1) + CH_3CN (2) with a molar ration of x_2/x_1 = ~0.12 at 328.15 K and 20 MPa[34]

	$[N^nBu_4]^+$	$[NR^fMe_3]^+$	$[NR^{fn}Bu_3]^+$
^{a} The supporting electrolyte represented here consists of a cation and an anion which can be found from the column name and row name respectively. b Molar conductivity in units of S cm^2 mol^{-1}. c The numbers in brackets represent the concentration of electrolytes in mmol dm^{-3}.			
$[BF_4]^-$	2.3^{a,b} (2.4)^c	—	—
$[B(4\text{-}C_6H_4F)_4]^-$	—	3.1 (2.2)	3.6 (2.4)
$[B(4\text{-}C_6H_4CF_3)_4]^-$	—	8.9 (0.9)	10.6 (1.3)
$[B(C_6F_5)_4]^-$	10.6 (2.4)	8.8 (1.0)	6.8 (0.7)
$[B\{3,5\text{-}C_6H_3(CF_3)_2\}_4]^-$	22.7 (2.4)	21.5 (2.4)	26.2 (2.5)

Studies of the conductivity as a function of concentration, Fig. 12, show an increase in molar conductivity with concentration for both $[N^nBu_4][B\{3,5\text{-}C_6H_3(CF_3)_2\}_4]$ and $[NR^{fn}Bu_3][B\{3,5\text{-}C_6H_3(CF_3)_2\}_4]$. At low electrolyte concentrations, Kohlrausch's law predicts that the molar conductivity should decrease with increasing electrolyte concentration.

$$\Lambda = \Lambda_0 - A\sqrt{c} \quad (1)$$

where Λ is the molar conductivity, Λ_0 is the sum of the limiting molar conductivities of the single ions (A^+ and B^-), A is a constant and c the concentration of the electrolyte.

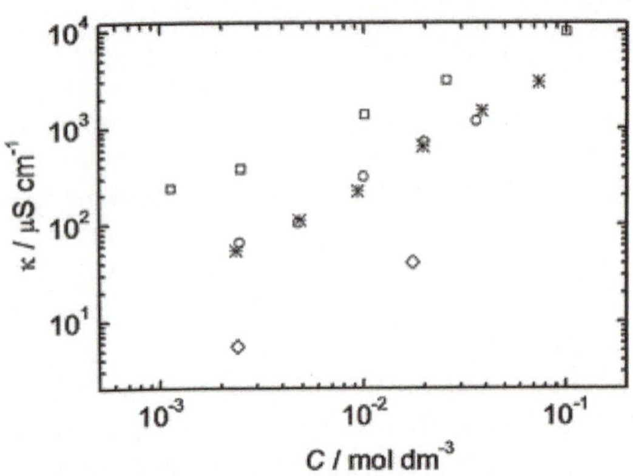

Figure 11: Conductivity of different supercritical fluid electrolyte solutions. Electrical conductivity of supporting electrolytes in a supercritical fluid mixture of CO_2 + CH_3CN (x_{CO2} : x_{CH3CN} = 0.89 : 0.11) at 328.15 K and 20.0 MPa. ◇, $[N^nBu_4][BF_4]$; *, $[N^nBu_4][B\{3,5\text{-}C_6H_3(CF_3)_2\}_4]$; ○, $[NR^{fn}Bu_3][B\{3,5\text{-}C_6H_3(CF_3)_2\}_4]$; □, $[N^nBu_4][BF_4]$ in pure CH_3CN at ambient condition.[5]

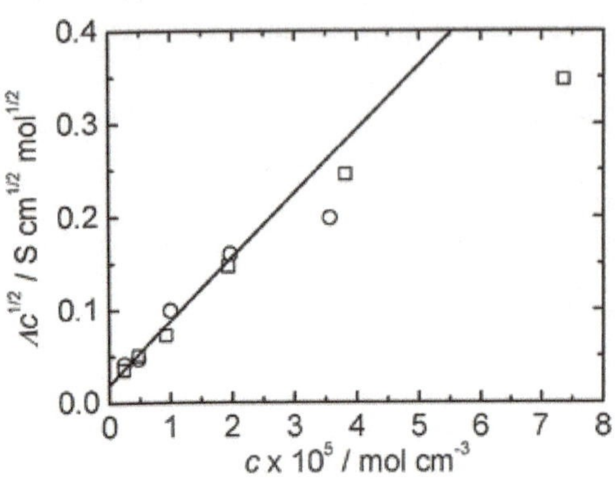

Figure 12: Plot of $\Lambda\sqrt{c}$ against c for the data shown in Fig. 11: (□) $[N^nBu_4][B\{3,5\text{-}C_6H_3(CF_3)_2\}_4]$; and (○) $[NR^{fn}Bu_3][B\{3,5\text{-}C_6H_3(CF_3)_2\}_4]$. The line is the best fit to the low concentration (below 3×10^{-5} mol cm^{-3}) data.[34]

However, for low dielectric constant solutions, such as supercritical fluids, the association of ions to form ion pairs and ion triples becomes significant. According to the Fuoss–Kraus model[80] the conductivity under these conditions is given by

$$\Lambda = \frac{\Lambda_0\sqrt{K}}{\sqrt{c}} + \frac{\lambda_0\sqrt{Kc}}{k} \quad (2)$$

where K is the equilibrium constant describing formation of the AB ion pairs, λ_0 is the sum of the limiting molar conductances of the two kinds of triple ions (ABA^+ and BAB^-) and k is the equilibrium constant describing the formation of the triple ions (assumed to be the same for both types of triple ion for simplicity). As we can see from Fig. 12 there is reasonable agreement between the data and the Fuoss–Kraus model at the lower electrolyte concentrations. The simple Fuoss–Kraus model of eqn (2) neglects the effects of the ion interactions on the conductivity (the electrophoretic and relaxation effects) and assumes that the activity coefficient is unity. More sophisticated treatments have been described in the literature that take these effects into account (see for example the work of Goldfarb and Corti[71] or Abbott and Schiffrin[89] for applications of these) but these require values for the dielectric constant, viscosity or limiting molar conductivity which we do not have for this supercritical system. The results in Fig. 11 and 12 also show that the fluorous ponytail does not have a significant influence on the extent of ion pairing. From these experiments it is clear that triple ions make a significant contribution to conduction in $scCO_2$–CH_3CN.

We have carried out similar phase behaviour and conductivity studies for supercritical CHF_3, CH_2F_2, and CH_2FCF_3 containing $[N^nBu_4][BF_4]$, $[N^nBu_4][B\{3,5\text{-}C_6H_3(CF_3)_2\}_4]$ and $Na[B\{3,5\text{-}C_6H_3(CF_3)_2\}_4]$.[52] As we can see from Table 3 these three hydrofluorocarbons have reasonably accessible critical temperatures and pressures and all have dipole moments greater than 1.6 D. Compared to the CO_2 + CH_3CN system (x_{CO2}/x_{CH3CN} = 0.14) discussed above, the pressures required to form a homogeneous solution for $[N^nBu_4][BF_4]$ electrolyte are much lower (Fig. 13).

Table 3: Physical properties of some hydrofluorocarbons

Physical property	**CHF_3**	**CH_2F_2**	**CH_2FCF_3**
T_c/K	299.29	351.26	374.21
p_c/MPa	4.832	5.782	4.049
ρ_c/kg m^{-3}	526.5	424.00	511.90
μ/D	1.65	1.98 ± 0.02	1.80 ± 0.22

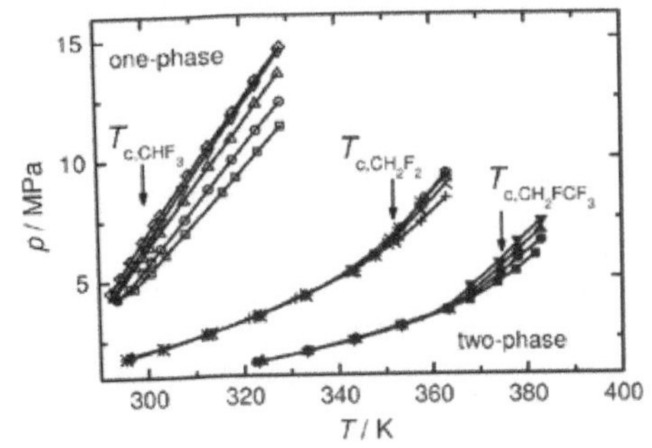

Figure 13: p–T phase diagram of the mixture of [N^nBu_4][BF_4] (1) + HFC (2). HFC = CHF_3: □, $x_1 = 0.82 \times 10^{-3}$; $x_1 = 1.59 \times 10^{-3}$; Δ, $x_1 = 2.44 \times 10^{-3}$; ▽, $x_1 = 4.83 \times 10^{-3}$; ◇, $x_1 = 6.90 \times 10^{-3}$. HFC = CH_2F_2: +, $x_1 = 1.67 \times 10^{-3}$; ×, $x_1 = 2.55 \times 10^{-3}$; *, $x_1 = 3.29 \times 10^{-3}$; ⊕, $x_1 = 4.17 \times 10^{-3}$; HFC = CHF_2CF_3: ■, $x_1 = 1.33 \times 10^{-3}$; ●, $x_1 = 2.62 \times 10^{-3}$; ▲, $x_1 = 3.91 \times 10^{-3}$; ▼, $x_1 = 5.14 \times 10^{-3}$. The arrows indicate the critical temperature of the HFCs.[52]

For all three HFCs we find that when the temperature is below the corresponding T_c (marked by the arrows in Fig. 13) the p–T boundaries are overlapped for the mixtures with different composition of [N^nBu_4][BF_4] but at higher temperature the solubility of [N^nBu_4][BF_4] in the HFC fluid phase increases with increasing pressure. Although all three HFCs are good solvents for [N^nBu_4][BF_4], CH_2F_2 gives the lowest reduced pressure, p_r (where $p_r = p/p_{c,HFC}$ the ratio of the pressure on the phase to the critical pressure of the corresponding HFC), for dissolving the same amount of [N^nBu_4][BF_4]. This may be explained by the fact that CH_2F_2 has the highest dipole moment of the three (Table 3).

Fig. 14 shows p–T phase diagrams for three binary mixtures of CH_2F_2 and three different electrolytes, [N^nBu_4][BF_4], [N^nBu_4][B{3,5-$C_6H_3(CF_3)_2$}$_4$], and Na[B{3,5-$C_6H_3(CF_3)_2$}$_4$], at similar mole fraction ($x_{electrolyte} = \sim 1.65 \times 10^{-3}$). The p–T phase boundaries for all of the mixtures are almost superimposed with only a slight difference in the phase separation pressure observed when T > 353 K. It is not surprising that two binary mixtures with [N^nBu_4][BF_4] and [N^nBu_4][B{3,5-$C_6H_3(CF_3)_2$}$_4$] show similar behaviour because CH_2F_2 is such a good solvent for tetrabutylammonium salts and, consequently, the effect of replacing [BF_4]$^-$ by the [B{3,5-$C_6H_3(CF_3)_2$}$_4$]$^-$ anion is not very significant. However, it is surprising that Na[B{3,5-$C_6H_3(CF_3)_2$}$_4$] shows such similar behaviour, given that sodium salts often have low solubility in organic solvents. Here it would appear that the [B{3,5-$C_6H_3(CF_3)_2$}$_4$]$^-$ anion is a key component to enhance the solubility of Na[B{3,5-$C_6H_3(CF_3)_2$}$_4$] in CH_2F_2, and the large size difference between Na^+ and [B{3,5-$C_6H_3(CF_3)_2$}$_4$]$^-$ is probably important. Similar results are found for the same three supporting electrolytes in CH_2FCF_3 at a fixed mole fraction of $\sim 2.60 \times 10^{-3}$, also shown in Fig. 14.

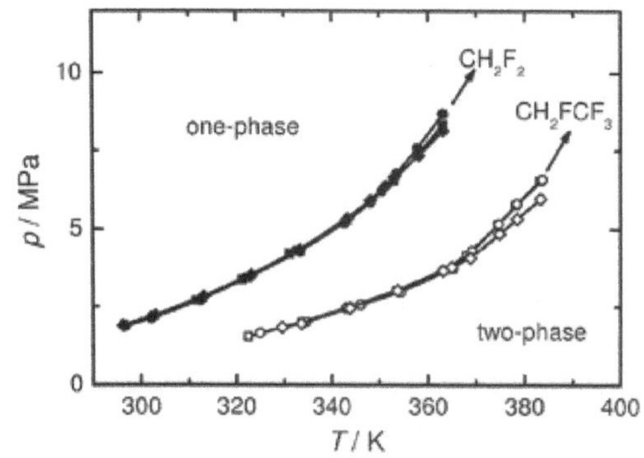

Figure 14: p–T phase diagram of the mixtures of electrolyte (1) + CH_2F_2 (2): ■, [NBu^n_4][BF_4], $x_1 = 1.67 \times 10^{-3}$; ●, [$NBu^n_4$][B{3,5-$C_6H_3(CF_3)_2$}$_4$], $x_1 = 1.65 \times 10^{-3}$; ◆, Na[B{3,5-$C_6H_3(CF_3)_2$}$_4$], $x_1 = 1.62 \times 10^{-3}$; and of electrolyte (1) + CHF_2CF_3 (2): □, [NBu^n_4][BF_4], $x_1 = 2.62 \times 10^{-3}$; ○, [$NBu^n_4$][B{3,5-$C_6H_3(CF_3)_2$}$_4$], $x_1 = 2.58 \times 10^{-3}$; ◇, Na[B{3,5-$C_6H_3(CF_3)_2$}$_4$], $x_1 = 2.59 \times 10^{-3}$.[52]

Studies of the conductivity of [N^nBu_4][B(C_6F_5)$_4$], [N^nBu_4][B{3,5-$C_6H_3(CF_3)_2$}$_4$], [$NR^{fn}Bu_3$][B{3,5-$C_6H_3(CF_3)_2$}$_4$] (R^f = $(CH_2)_3C_7F_{15}$), and Na[B{3,5-$C_6H_3(CF_3)_2$}$_4$] in scCH_2F_2, Fig. 15 and Table 4, show that these salts are more conducting than [N^nBu_4][BF_4] under the same conditions although the increase is much less significant than that for supercritical CO_2 + CH_3CN. Consequently, either [N^nBu_4][BF_4] or the corresponding [B(Ar^F)$_4$]$^-$ salts are good choices for supporting electrolytes for electrodeposition from scCH_2F_2.

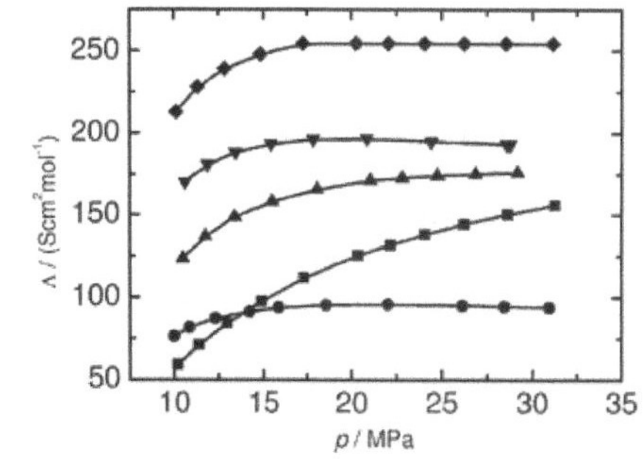

Figure 15: Molar conductivity of the supporting electrolytes in CH_2F_2 at 363.15 K. The molar concentration of the electrolytes is ~9.0 mol m^{-3}. ■, [N^nBu_4][BF_4]; ●, [$NR^{fn}Bu_3$][B{3,5-$C_6H_3(CF_3)_2$}$_4$]; [N^nBu_4][B(C_6F_5)$_4$]; ▲, [N^nBu_4][B{3,5-$C_6H_3(CF_3)_2$}$_4$]; ▼, Na[B{3,5-$C_6H_3(CF_3)_2$}$_4$]; ◆, Na[B{3,5-$C_6H_3(CF_3)_2$}$_4$].[52]

Table 4: Molar conductivity of the supporting electrolytes in CH_2F_2 at 363 K and 20 MPa. The concentration of all the electrolytes is ~9.0 mol m^{-3}. Piecewise cubic Hermite polynomials were used to interpolate the data shown in Fig. 15 to the fixed pressure (20 MPa)

Electrolyte	Λ/S cm^2 mol^{-1}
a $R^f = CF_3(CF_2)_7(CH_2)_3$.[52]	
$[N^nBu_4][BF_4]$	124
$[NR^{fn}Bu_3][B\{3,5\text{-}C_6H_3(CF_3)_2\}_4]^a$	95
$[N^nBu_4][B(C_6F_5)_4]$	170
$[N^nBu_4][B\{3,5\text{-}C_6H_3(CF_3)_2\}_4]$	197
$Na[B\{3,5\text{-}C_6H_3(CF_3)_2\}_4]$	255

Typical conductivities for 1 M electrolyte solutions in acetonitrile[90] are in the range 20–55 S cm^2 mol^{-1} so the values we obtain for the supercritical fluids are quite respectable for a non-aqueous electrolyte and sufficient to carry out electrochemical studies.

REAGENTS

The reagents used in SCFED also require a number of key properties. In addition to solubility in the SCF, they need to be both stable to the electrolyte and at the working temperature, and ideally readily synthesised in high purity. Unwanted electrochemical side reactions which could result in fouling of the electrode or counter-electrode by the reagents and reactive liberated ligands must also be avoided. For example, $[Ag(cod)(hfac)]BF_4$ (cod = cyclooctadiene; hfac = hexafluoroacetylacetonyl) is unsuitable for Ag deposition, probably because the cod fouls the counter electrode, whilst if $[Cu(hfac)_2]$ is used for copper deposition in $scCO_2$–CH_3CN the deposited copper can redissolve due to the reaction[91]

$$[Cu(hfac)_2] + Cu + 8CH_3CN \rightarrow 2[Cu(CH_3CN)_4]^+ + 2hfac^- \quad (3)$$

For deposition of transition metals from $scCO_2$–CH_3CN the nitrile complexes $[M(CH_3CN)_x][BF_4]_n$ (M = Fe, Co, Cu, Ag, etc.) are attractive reagents since the anion can be present in the electrolyte and the deposition liberates CH_3CN which is present as co-solvent, hence no chemical incompatibilities are presented.[5,91] However in $scCH_2F_2$, the complexes are unstable in the absence of added CH_3CN, which limits their usefulness, whilst $[Cu(hfac)_2]$ is stable in $scCH_2F_2$.[52] The nitrile complexes are readily made from the metal powders and $NO[BF_4]$ in CH_3CN.[92,93] In the case of Cu(I), $[Cu(CH_3CN)_4][BF_4]$ is obtainable from Cu_2O and aqueous HBF_4 in CH_3CN, or by refluxing $Cu[BF_4]_2$ with copper in CH_3CN.[94] The nitrile complexes can also be converted to $[B(Ar^F)_4]^-$ salts to increase solubility in the SCF if required.[92,93] For silver electrodeposition several reagents were considered for use in $scCO_2$–CH_3CN including $[Ag(hfac)(cod)]$, $[Ag(CH_3CN)_4][BF_4]$, and $[CF_3(CF_2)_6CO_2Ag(PPh_3)_2]$ and $[Ag(PPh_3)_4][BF_4]$, in which the arylphosphine groups and the perfluorocarboxylate ligand were introduced to increase solubility in hydrofluorocarbons. Of these, $[Ag(CH_3CN)_4][BF_4]$ was found to be the most suitable for silver deposition from $scCO_2$–CH_3CN (see below).[95]

The choice of reagents to electrodeposit germanium is less obvious than for copper or silver. Germanium forms compounds in two oxidation states, Ge(II) and Ge(IV), and the chemical stability of its compounds varies widely with the ligands present.[6,96] Many Ge(IV) complexes are moisture sensitive, whilst the thermal stability of Ge(II) complexes is often poor. Among the Ge(II) complexes explored were the commercially available $GeBr_2$, a complex of the functionalised diimine $[GeCl_2(4,4'\text{-didecyl-}2,2'\text{-bipyridyl})]$,[6] the trichlorogermanate(II) salt $[N^nBu_4][GeCl_3]$, made by reduction of $GeCl_4$ in aqueous hydrochloric acid with H_3PO_2, followed by addition of $[N^nBu_4]Cl$,[6] and $[Ge(12\text{-crown-}4)_2][CF_3SO_3]_2$, made from $[GeCl_2(dioxane)]$ and 12-crown-4. The first two proved to be too poorly soluble in $scCH_2F_2$, and whilst it was soluble, $[Ge(12\text{-crown-}4)_2][CF_3SO_3]_2$ degraded in the same solvent.[6] $[N^nBu_4][GeCl_3]$ was both soluble and stable in $scCH_2F_2$ and its electrochemistry is described below. Among the Ge(IV) reagents assessed were $[GeF_4(CH_3CN)_2]$, made by passing GeF_4 gas into dry CH_3CN, and $[GeF_4(Me_2NCH_2CH_2NMe_2)]$, prepared from $[GeF_4(CH_3CN)_2]$ and $Me_2NCH_2CH_2NMe_2$ in CH_2Cl_2.[97] These two reagents were chosen as they are thermally relatively stable, and since they are coordinatively saturated, they are moisture stable solids and relatively resistant to hydrolysis in solution. Unfortunately, they proved to be insufficiently soluble in the SCFs. As a non-polar liquid, $GeCl_4$ (B.P. 365 K) was expected to be readily miscible with SCFs, but it is moisture sensitive and readily hydrolysed. It is commercially available and easily dried and purified by distillation from a mixture of Na_2CO_3 and CaH_2.

ELECTRODEPOSITION

There is very little published work on electrodeposition from supercritical fluids. As described above, the earliest work is probably that of Williams and Naiditch[54] in 1976 and the work of Silvestri et al.[55] five years later. Finally, MacDonald et al.[62] in 1986 reported deposition of Cu from $CuCl_2$ under supercritical, or close to supercritical, H_2O containing KCl at 573 K and 8.2 MPa. Apart from the three papers on electrodeposition of metals there are also a few papers which describe the electrodeposition of conducting polymers from supercritical fluids. This is slightly different because in this case the neutral solution species (the heterocyclic monomer) is oxidised to produce a charged product (the oxidised polymer) which deposits with the counter ion on the electrode surface. The first paper claiming to electrodeposit poly(pyrrole) and poly(aniline) from a supercritical fluid appeared in 2002.[98] This group used CO_2 containing 13.1% acetonitrile, 0.16 M pyrrole and 0.16 M $[N^nBu_4][PF_6]$ at 323 K and 9.65 MPa. Subsequent work published in 2005 by Yan et al.[99] using a view cell showed that under these conditions the system was biphasic and not in the supercritical state. Yan et al. were, however, able to deposit poly(pyrrole) films from supercritical fluid using CO_2 containing 13.1% acetonitrile, 0.16 M pyrrole and 0.04 M $[N^nBu_4][PF_6]$ at 323 K and 10 MPa. This illustrates one of the potential pitfalls of working with, or trying to work with, supercritical fluids. It is not sufficient to rely on the critical temperature and critical pressure of the pure fluid, but rather it is essential to properly characterise the phase behaviour of the whole system. The honour of reporting the first successful electropolymerisation of a conducting polymer from a supercritical fluid consequently falls to Atobe et al.[100] who, in 2004, reported the electropolymerisation of poly(pyrrole) and poly(thiophene) from $scCHF_3$ containing 10 mM pyrrole or thiophene with 40 mM $[N^nBu_4][PF_6]$ at 323 K and 15 MPa.

Before moving on to discuss our recent work, it is appropriate to mention the work of Sone's group on electrodeposition from dense emulsions of CO_2. The group has published extensively, starting in 2002,[101] on the electrodeposition of Ni from a dense emulsion of an aqueous Ni plating (Watts) bath formed in $scCO_2$ using surfactants (typical conditions 323 K and 15 MPa and 60 vol% of the aqueous plating bath stabilised with 1 vol% surfactant).[102] This is not a single phase supercritical fluid. The advantages of the system are that it produces high quality Ni films with higher uniformity, smaller (sub 100 nm) grain size and significantly higher Vickers hardness than conventional electroplating from aqueous solution.[103] This is attributed, at least in part, to the enhanced desorption of hydrogen gas bubbles during deposition.[104] The same group has also applied this approach to the deposition of poly(pyrrole)[105] and copper.[106]

In our work we have studied the electrodeposition of copper from supercritical fluids.[5,91] Fig. 16 shows cyclic voltammetry for $[Cu(CH_3CN)_4][BF_4]$ in $scCO_2$–CH_3CN with $[N^nBu_4][BF_4]$ electrolyte. The voltammetry shows the characteristic features expected for electrodeposition with mass transport limited reduction of Cu(I) to Cu(0) at negative potentials, accompanied by a stripping peak for Cu on the return scan. The stripping peak is sharp and undistorted, showing that iR drop is insignificant in these experiments. The inset in Fig. 16 shows a plot of the mass transport limited currents recorded at a microdisc electrode for different concentrations of $[Cu(CH_3CN)_4][BF_4]$ added to the solution. This is a very sensitive way to determine the solubility of the redox active species since the mass transport limiting current at the microdisc is given by

$$I_L = 4nFaDc \qquad (4)$$

where n is the number of electrons transferred and a is the radius of the microdisc electrode. Thus, assuming the diffusion coefficient, D, does not change with concentration, the limiting current should be proportional to the concentration of species dissolved in the solution. From the inset in Fig. 16 we can see that at low concentrations the current increases linearly with the concentration, as expected, but then abruptly reaches a plateau value, indicating that the solution is saturated. From the intersection of the two lines we obtain an estimate of the solubility of $[Cu(CH_3CN)_4][BF_4]$ under these conditions of 0.49 mM. From the slope of the initial part of the curve, using the microdisc equation (eqn (4)), we obtain a diffusion coefficient of 3.5×10^{-5} cm^2 s^{-1}. This is approximately 1.6 times larger than the value measured for the same complex in acetonitrile at 311 K (2.2×10^{-5} cm^2 s^{-1}).

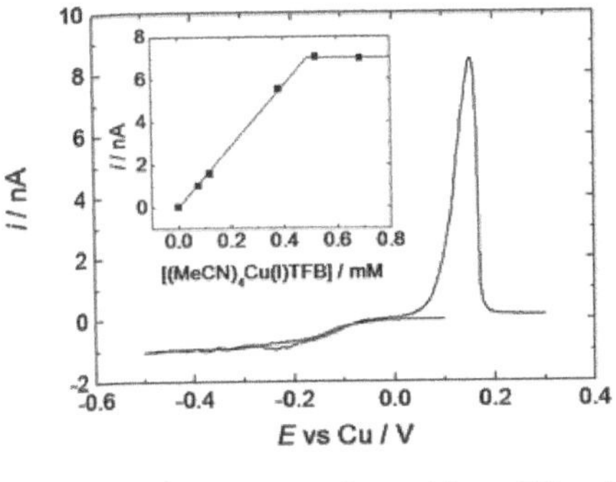

Figure 16: Copper voltammetry performed in $scCO_2$ with 12.1 wt% CH_3CN and $[N^nBu]_4[BF_4]$ (20 mM); at 310 K and 17.24 MPa. Electrodes were: 25 μm diameter platinum disc working electrode, 0.5 mm diameter platinum disc pseudo reference, and 0.5 mm diameter platinum wire counter electrode. The sweep rate was 20 mV s^{-1}. The insert shows the limiting current as a function of the concentration of the copper complex.[5]

By changing the counter anion from $[BF_4]^-$ to $[B\{3,5\text{-}C_6H_3(CF_3)_2\}_4]^-$ it is possible to significantly increase (>30 times) the solubility of $[Cu(CH_3CN)_4]^+$ in $scCO_2$–CH_3CN. This illustrates a significant advantage of careful choice and design of the electrolyte to enhance electrodeposition from supercritical fluids, the higher solubility of the reagent allows a greater rate of metal plating. For $[Cu(CH_3CN)_4]^+$ in supercritical CO_2 with 12.1 wt% CH_3CN and 20 mM $[N^nBu_4][B\{3,5\text{-}C_6H_3(CF_3)_2\}_4]$ electrolyte at 310–311 K the diffusion coefficient is 3.3×10^{-5} cm^2 s^{-1}. This is ~1.5 times larger than the corresponding value in $scCO_2$–CH_3CN with 20 mM $[N^nBu_4][BF_4]$, suggesting that there may be less ion pairing between the copper and the large $[B\{3,5\text{-}C_6H_3(CF_3)_2\}_4]^-$ ion, leading to higher ion mobility. It is worth noting that these diffusion coefficients are relatively low compared to values for complexes dissolved in supercritical CO_2.[107] This is probably due to the addition of 12–13 wt% acetonitrile as a co-solvent which significantly increases the viscosity of the solution.[108,109]

Using $[Cu(CH_3CN)_4]^+$ in $scCO_2$–CH_3CN it is possible to electroplate smooth, reflective copper films onto 1 cm^2 electrodes. Fig. 17 shows SEM images of films deposited at different potentials. The morphology of the deposits varies with the applied potential as expected. At high overpotentials, when the deposition is mass transport limited, the films are rough and dendritic, whereas at low overpotentials the films are smooth, shiny and adherent. EDX, SIMS, and Auger studies all show that the films are high purity copper. Four-point probe measurements gave film resistivities for the best films of 4×10^{-6} Ω cm, which is comparable to values reported for copper electrodeposition from aqueous solution (1.75 to 2×10^{-6} Ω cm depending on plating bath additives[110]) or by chemical deposition in supercritical CO_2 as reported by Watkins et al.[111] (2×10^{-6} Ω cm). Thus the quality of the electrodeposited films is high and already close to that required for device applications.

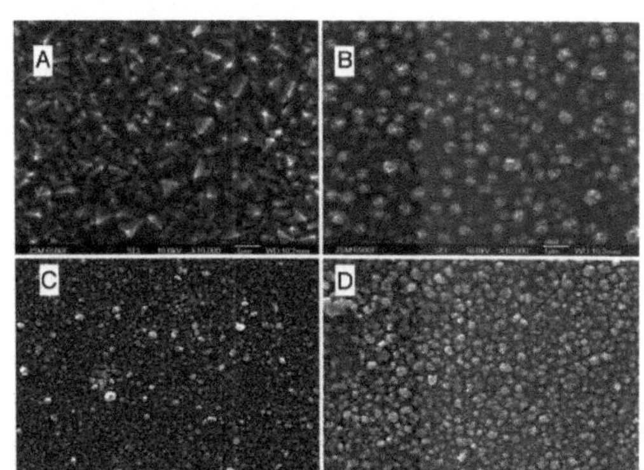

Figure 17: Electroplated copper SEM images of copper films deposited from a saturated solution of $[Cu(CH_3CN)_4]BF_4$ in 87.4 wt% CO_2, 12 wt% acetonitrile, 0.6 wt% $[N^nBu_4][BF_4]$ 309–311 K and 13.8 MPa at (A) −0.6 V, (B) −0.9 V, (C) −1.5 V, and (D) −1.7 V vs. Pt.[5]

An important feature of the copper plating bath described above is that it uses the Cu(I) complex. The relative stability of the Cu(I) and Cu(II) redox states strongly depends on the choice of ligands. For $[Cu(CH_3CN)_4]^+$ the redox potential for oxidation to $[Cu(CH_3CN)_4]^{2+}$ is high (>1.25 V vs. Cu in CH_3CN). It is possible to electroplate copper from the $[Cu(hfac)_2]$ from $scCO_2$–CH_3CN, Fig. 18. Now however the comproportionation reaction, eqn (3), complicates the process so that this is not a good choice of reagent in this case. EDX analyses of these films shows contamination with fluorine and carbon.

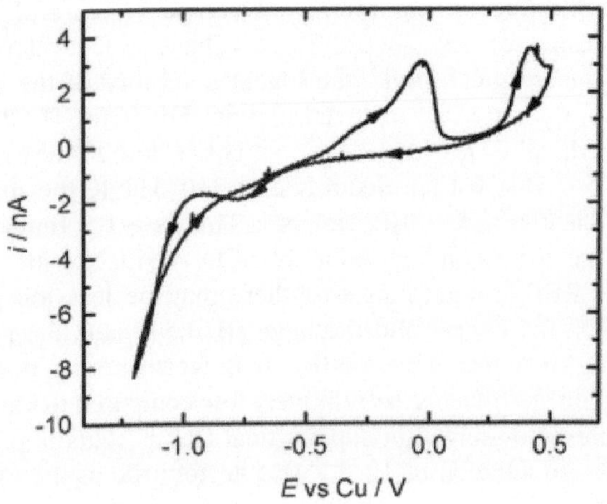

Figure 18: Cyclic voltammetry for 1.2 mM $[Cu(hfac)_2]$ in supercritical CO_2 with 12.1 wt% CH_3CN containing 20 mM $[N^nBu_4][BF_4]$ electrolyte. T = 310 K. p = 17.2 MPa. The working electrode was a 25 μm diameter platinum disc, the counter electrode was a platinum grid and the reference a 0.5 mm polished copper disc. Sweep rate 20 mV s^{-1}.[50]

We have also electrodeposited copper from supercritical hydrofluorocarbons. Studies of the phase behaviour of CH_2F_2 containing $[N^nBu_4][BF_4]$ and $[Cu(CH_3CN)_4][BF_4]$ showed that the copper complex was unstable in the absence of added CH_3CN.[52] Consequently, a small amount of CH_3CN was added to the CH_2F_2 not only to increase the solubility of $[Cu(CH_3CN)_4][BF_4]$, but also to stabilise it. Phase behaviour studies show that addition of $[N^nBu_4][BF_4]$ to the $[Cu(CH_3CN)_4][BF_4]$ system in $scCH_2F_2$ drastically shifts the phase boundary to high pressures at $T > T_{c,CH2F2}$.[52] We believe that this is caused by selective solvation of the $[N^nBu_4]^+$ and $[BF_4]^-$ ions by CH_3CN, depleting the CH_3CN molecules in the bulk phase, so that the density, and hence the pressure of the system, needs to be increased considerably for the quaternary mixture to form a homogeneous phase. Again this illustrates the importance of carefully characterising the phase behaviour of the electrolyte system.

$[Cu(hfac)_2]$ is more soluble and more stable than $[Cu(CH_3CN)_4][BF_4]$ in $scCHF_3$ and it only increases the phase-separation pressure by a moderate amount when added to the CHF_3 solution with $[N^nBu_4][BF_4]$ as the supporting electrolyte.[91] Fig. 19 shows a cyclic voltammogram of $[Cu(hfac)_2]$ in $scCHF_3$ with free hfacH ligand added to the solution. The voltammetry is similar to that in CH_3CN and in $scCO_2$–CH_3CN, with the $[Cu(hfac)_2]$ reduced to metallic copper in one step. In this case there is a significant cathodic background current beyond −0.7 V. Deposition onto large area (~1 cm^2) electrodes was found to give good copper films with resistivities as low as 5.8×10^{-6} Ω cm. The best films were obtained with stirring and in the presence of the free ligand (hfacH).

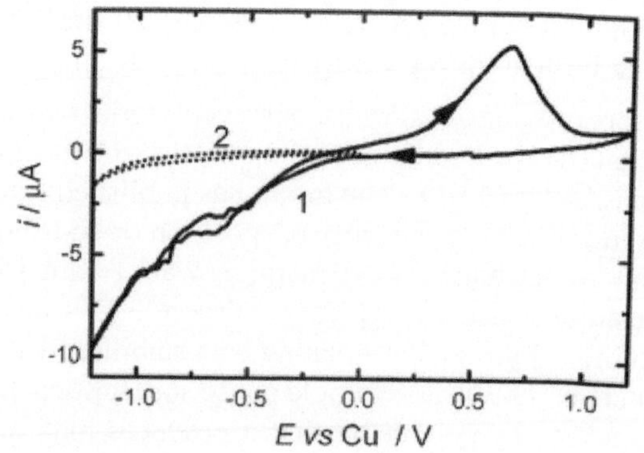

Figure 19: Scan 1, cyclic voltammetry for 5.3 mM $[Cu(hfac)_2]$ in $scCHF_3$ with 20 mM $[N^nBu_4][BF_4]$ and 0.1 M hfacH at 311 K and 17 MPa. Scan 2 shows the background current in the absence of the Cu complex. The working electrode was a 0.37 mm diameter platinum disc, and the counter and reference electrodes were 0.5 mm diameter copper wires. Sweep rate 100 mV s^{-1}.[50]

We have also studied the electrodeposition of silver from $scCO_2$–CH_3CN.[95] Five silver reagents were investigated. Of these the most suitable was found to be $[Ag(CH_3CN)_4][BF_4]$ which gives good quality metallic deposits without any complicating electrochemical reactions. Of the other complexes [Ag(hfac)(cod)] was found to give silver deposits, but polymerisation of the cod ligand on the counter electrode caused problems, and the silver deposition from $[CF_3(CF_2)_6CO_2Ag(PPh_3)_2]$, $[Ag(hfac)(PPh_3)]$ and $[Ag(PPh_3)_4][BF_4]$ were all accompanied by electrochemical reduction of the PPh_3 ligand leading to a complex set of reactions with the electrolyte.

To exploit supercritical fluid electrodeposition to make devices it will be necessary to electrodeposit not only simple metals but also functional materials such as semiconductors. Germanium is an extremely important material in a wide range of technologies and single crystalline SiGe layers are widely used in integrated "silicon" electronics.[112] Electrodeposition of nanostructured germanium would offer a number of advantages over other deposition techniques if sufficient material quality could be achieved. In addition, electrodeposition of germanium is a stepping stone to the electrodeposition of silicon–germanium and silicon layers.[113] We have investigated several reagents for Ge deposition from $scCO_2$–CH_3CN and $scCH_2F_2$.[6] For Ge(II) reagents the most promising results were obtained using $[N^nBu_4][GeCl_3]$. Better results were obtained using $GeCl_4$. Fig. 20 shows the phase behaviour for binary mixtures of CH_2F_2 with $GeCl_4$ and $[N^nBu_4]Cl$. At 357 K and 15 MPa the system forms a single phase, supercritical system. Fig. 21 shows typical cyclic voltammograms of $GeCl_4$ in $scCH_2F_2$ under these conditions. At the 250 μm Au disc electrode reduction starts at ~0.7 V vs. Pt followed by a less well resolved increase in current cathodic of −1.5 V, presumably corresponding to the two-step reduction of $GeCl_4$. In the limiting current region there are large, random fluctuations in the reduction current. These fluctuations arise from unsteady convective flow in the high-pressure electrochemistry cell because of the low viscosity of the fluid when the operating conditions are close to the critical point of CH_2F_2. We can also see that there is significant iR drop in this case. This is because the conductivity of the $[N^nBu_4]Cl$ supporting electrolyte is low in $scCH_2F_2$. These conditions can be used for bulk deposition of Ge, the 'as deposited' material was amorphous, but crystallised under high intensity laser illumination.[6]

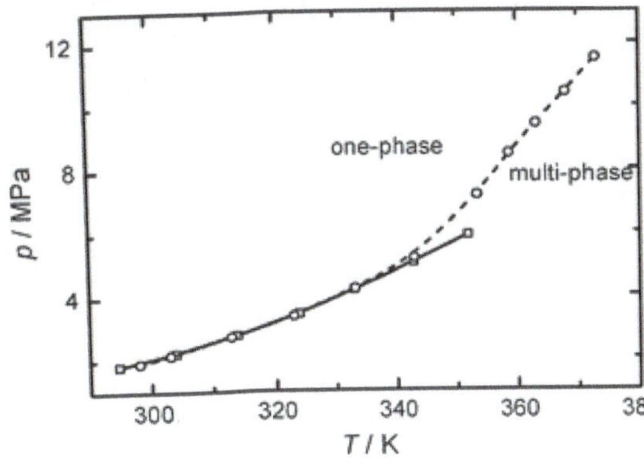

Figure 20: p–T phase diagrams of the binary mixtures of CH_2F_2 + A: A = $GeCl_4$ (□), $x_A = 4.4 \times 10^{-3}$; and A = $[N^nBu_4]Cl$ (○), $x_A = 2.95 \times 10^{-3}$.[6]

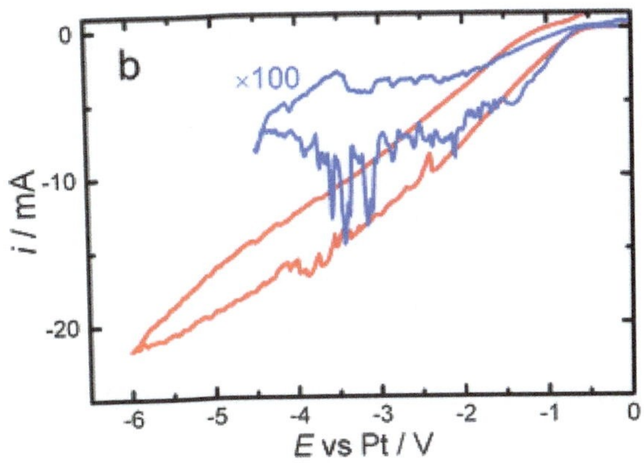

Figure 21: Cyclic voltammogram of 0.06 mol dm^{-3} $GeCl_4$ in $scCH_2F_2$ with 0.06 mol dm^{-3} $[N^nBu_4]Cl$ as the supporting electrolyte. The working electrodes were a 250 µm Au microdisc (blue line, scan rate 0.1 V s^{-1}), and a 0.36 cm^2 gold-coated glass slide (red line, scan rate 0.1 V s^{-1}). The measurements were c arried out at 357 K and 15 MPa.[6]

TEMPLATES

Electrodeposition into small and high aspect ratio pore structures is a specific advantage of SCFED, and provides opportunities to grow high aspect ratio nanowires or to construct multi-component devices in electrically directed locations. Aligned pore structures will need to be used in future to deliver the promise of linking properties to designed nanostructures.

Electrochemical reaction of aluminium in an acidic environment is well known to produce porous anodic aluminium oxide (AAO) films and well-ordered hexagonal AAO films (Fig. 22a) are achievable if the anodisation conditions are well optimised.[114–116] These materials can be produced with pore sizes from around 5 nm[117] to 500 nm,[118] and the resulting templates are electrically and thermally insulating, optically transparent and robust. Typically they are grown on the surface of a solid aluminium foil electrodes, but can also be produced in sputtered aluminium films on other surfaces such as ITO-glass,[119] suggesting opportunities to make porous structures on various substrates. Several authors have published electrodeposition of metals and other materials into these structures from more conventional electrolytes.[120,121] Many of these porous films are now commercially produced, where they are cleaved from the aluminium surface to produce self-supporting membranes. The chemistry to produce similar structures in other oxides, especially titanium dioxide, is also now well developed.[122]

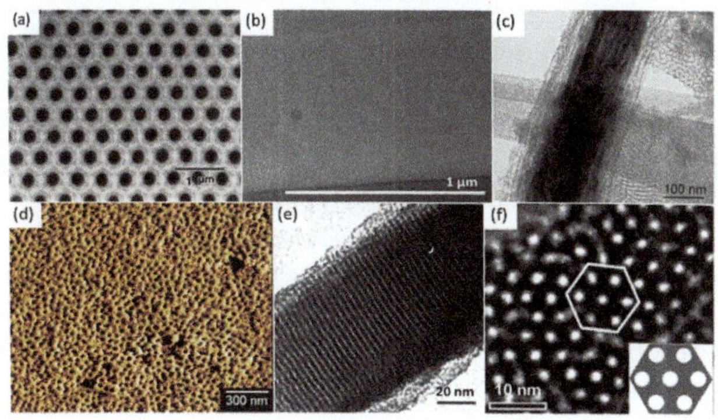

Figure 22: Images of ordered porous structures. (a) SEM of anodically oxidised alumina with a hexagonal array of large ordered pores[118] (Copyright 1998 The Japan Society of Applied Physics). (b) Cross-sectional TEM of long range ordered pores parallel to substrate surface produced from a block copolymer–homopolymer blend[123] (1 µm scale bar; reprinted (adapted) with permission from V. R. Tirumala et al., Chem. Mater., 2007, **19**, 5868. Copyright 2007, American Chemical Society). (c) Cu nanowires in columnar mesoporous silica originally grown in AAO[124] (reprinted (adapted) with permission from A. Keilbach et al., Chem. Mater., 2010, **22**, 5430. Copyright 2007, American Chemical Society). (d) Phase scanning force micrograph of a block copolymer film with cylindrical domains perpendicular to the surface infused with silica[125] (© WILEY-VCH Verlag GmbH & Co. KGaA, Weinheim). (e) A cross-sectional TEM of an EASA-derived film with well-aligned perpendicular pores[126] (reprinted by permission from Macmillan Publishers Ltd.: A. Walcarius et al., Nat. Mater., **6**, 602, copyright 2007). (f) TEM of a flake of Stöber silica film showing the hexagonal pore structure[127] (20 nm scale bar; © 2012 Wiley-VCH Verlag GmbH & Co. KGaA, Weinheim).

Surfactant- and block copolymer-templated porous films are the commonest porous materials with small pore sizes, with pores ranging from below 3 nm to above 10 nm. It is possible to produce a variety of 1-, 2- or 3-dimensional pore structures controlled by a combination of surfactant type, concentration and temperature. Evaporation induced self-assembly (EISA) is a process in which the concentration increase that results from solvent evaporation during coating drives both the formation of micelles from the surfactant and condensation of the silica species in the sol. However, the EISA process is difficult to control. The linear pores that are present in hexagonal phase silica are particularly attractive hosts for electrodeposition and are the most heavily studied morphology. The alignment of the pores is controlled by the relative strength of the interaction between the micelle surfaces and the substrate compared with micelle–micelle interactions. In most cases the pores are either randomly oriented (referred to as "worm-like") or are parallel to the substrate.[128] Parallel pores can be brought into long range alignment (Fig. 22b) by the addition of a low molar mass homopolymer to a block copolymer template film and then infiltrating a tetraethylorthosilicate (TEOS) precursor as a supercritical CO_2 solution.[123] Filling pores in this orientation is very interesting from an applications viewpoint due to the anisotropy that is present in the plane of the film, but pore access is more limited than in films with pores perpendicular to the film surface.

The tendency for pores to align with the substrate surface has been used to good effect in formation of mesoporous silica inside AAO membranes.[129] These have well-formed larger pores that are perpendicular to the face of the membrane. When a silica sol containing a surfactant at an appropriate concentration is infiltrated into one of these membranes the mesopores can be oriented along the length of the host pore (columnar growth) or can form toroidal or helical channels depending on deposition conditions. High humidity is frequently the key parameter in obtaining columnar growth, which is the most accessible morphology for electrodeposition (although electrodeposition into the other structure types has also been demonstrated). Copper, silver and tellurium nanowires have already been deposited in larger (P123-templated) columnar pores synthesised in this way (Fig. 22c).[124]

Thin hexagonal mesoporous films with pores perpendicular to the plane of the film have been grown at an oil–water interface using CTAB (cetyltrimethylammonium bromide) surfactant.[130] These films were very thin and not useful as electrodeposition hosts. However, Nagarajan et al. have shown that supercritical CO_2 infiltration of TEOS into spin-coated films of the amphiphilic block copolymer, poly(α-methyl styrene-β-hydroxy styrene), which exhibits orientation of the cylindrical microdomains perpendicular to the film surface, results in perpendicular pore structures with pore sizes of 20 nm or larger (Fig. 22d).[125] These could be grown on conductive surfaces and hence be useful electrodeposition hosts. Recently three approaches have used more conventional surfactant–silica sol mixtures to make well-ordered hexagonal pore structures. Tolbert used a (111)-oriented cubic porous titania film as the substrate for growth of hexagonal porous silica, with the surfactant chosen to give a lattice match such that the perpendicular pore structure is inherently favoured from the base of the pores.[131] Walcarius has used an "Electro-assisted self-assembly" (EASA) method to control pore orientation.[126,132] The application of an electric field has two roles in this method, it aids the formation of micelles from the cationic surfactant (CTAB) on the electrode at the same time as generating hydroxyl ions via water oxidation which aids local condensation of the sol particles at the electrode surface. This method produces well-aligned pores with ~3 nm diameter (Fig. 22e) in films with thickness up to ~100 nm. A key advantage is that a conductive substrate, typically indium tin oxide (ITO), is used in growth and so is inherently available as the electrode for electrodeposition. Thicker films with similar pore size can be grown by a related route developed by Zhao and carried out with a dilute Stöber silica solution containing CTAB.[127] The method relies on the formation of CTAB micelles at the substrate surface under weakly basic conditions in which TEOS undergoes slow hydrolysis and again produces a hexagonal array of pores perpendicular to the substrate surface (Fig. 22f). Whilst these films do not need to be grown on conductive surfaces the authors report the best ordered films on ITO.

Fig. 23 shows results for the deposition of copper into nanoscale (<10 nm) pores. The templates were prepared by dip coating ITO slides in a solution of Brij56, TMOS, 0.5 M HCl and MeOH (1 : 1.8 : 1 : 3.8 by weight). After calcining, this produces an adherent mesoporous silica film roughly 200 nm thick containing a regular hexagonal array of approximately 3 nm diameter cylindrical pores approximately 6 nm apart but randomly aligned with respect to the substrate. Electrochemical deposition of copper was carried out from $[Cu(CH_3CN)_4][B\{3,5\text{-}C_6H_3(CF_3)_2\}_4]$ in $scCO_2$–CH_3CN containing 20 mM $[N^nBu_4][B\{3,5\text{-}C_6H_3(CF_3)_2\}_4]$. The TEM image in Fig. 23A shows the regular array of cylindrical pores clearly visible in cross section on the left hand side of the image. Evidence that the pores are filled with copper is provided by the EDX analysis, which shows strong copper as well as the expected silicon and oxygen signals, Fig. 23B, and by the selected area electron diffraction results, Fig. 23C, which show a combination of diffuse rings attributed to silica together with rings that are consistent with the expected (111), (200), (220), and (311) diffraction peaks for copper. These results demonstrate the proof-of-principle, but much remains to be done to improve alignment and produce nanowire devices.

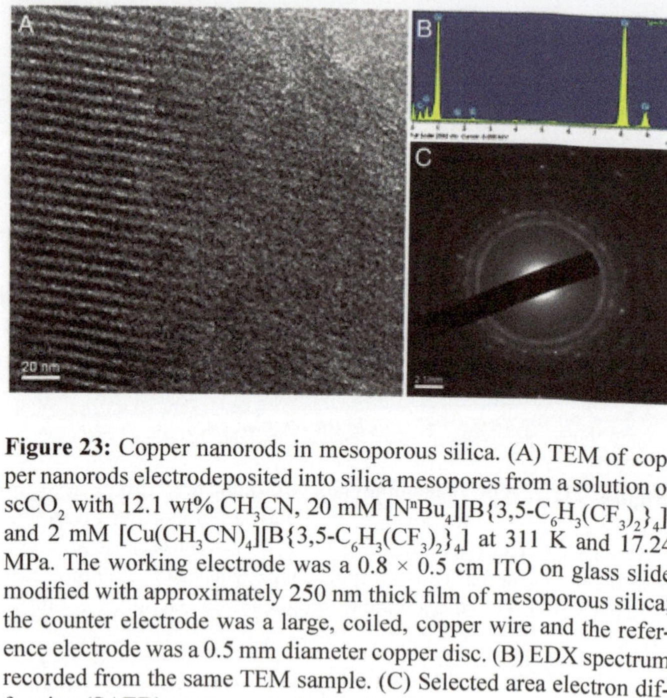

Figure 23: Copper nanorods in mesoporous silica. (A) TEM of copper nanorods electrodeposited into silica mesopores from a solution of $scCO_2$ with 12.1 wt% CH_3CN, 20 mM $[N^nBu_4][B\{3,5\text{-}C_6H_3(CF_3)_2\}_4]$, and 2 mM $[Cu(CH_3CN)_4][B\{3,5\text{-}C_6H_3(CF_3)_2\}_4]$ at 311 K and 17.24 MPa. The working electrode was a 0.8 × 0.5 cm ITO on glass slide modified with approximately 250 nm thick film of mesoporous silica; the counter electrode was a large, coiled, copper wire and the reference electrode was a 0.5 mm diameter copper disc. (B) EDX spectrum recorded from the same TEM sample. (C) Selected area electron diffraction (SAED) pattern recorded on 1 of the copp er rods.[5]

CONCLUSIONS AND FUTURE PROSPECTS

The barriers to successful electrodeposition from a supercritical fluid are significant, but the field has now progressed beyond the stage of preliminary studies to the point where generally applicable methods for supercritical fluid electrodeposition have been developed. An important contribution has been to establish a range of supporting electrolytes capable of achieving high ionic conductivities in the low dielectric constant fluids. Supercritical fluid electrodeposition has been used to deposit a range of different materials, e.g. Ag, Cu, and Ge, from complex ions with different overall charges, different ligand types and with different oxidation states. It has been demonstrated that a range of supercritical fluids, including both those with moderate dielectric constants and those with low dielectric constant when combined with a suitable co-solvent, can be used to electrodeposit films. Whilst not every combination of supercritical fluid and target material will be possible, the existing work clearly indicates that the range of materials that can be deposited could be expanded significantly, for example to a wider range of metals, semiconductors, binary and ternary compounds and alloys. The work so far has demonstrated that, by suitable design and choice of the combination of anion and cation, it is possible to achieve supercritical fluid electrolytes with surprisingly large conductivities that will allow deposition at sufficiently high rates to be useful. The field of supercritical fluid electrodeposition is still at an early stage but progress has been good and the prospects are very encouraging. In taking this work forward our increasing knowledge of the underpinning physical chemistry, in terms of the phase behaviour of the supercritical fluid mixtures, the solubility, stability and conductivity of electrolytes and complex reagents in supercritical fluids, the phase behaviour of supercritical fluids in confined spaces, the structure of the double layer at the supercritical fluid/electrode interface, the kinetics of ligand exchange and electron transfer reactions in supercritical fluids, etc., will be very important.

Considerable advances have been achieved in the field of supercritical fluid electrodeposition, but it is important not to lose sight of the fact that electrodeposition from a supercritical fluid will always involve significant practical challenges due to the high pressures involved. This means that supercritical fluid electrodeposition will only become technologically important if there are very clear advantages and novel applications which only supercritical fluid electrodeposition can fulfil. The particular properties of supercritical fluids strongly suggest that just such novel applications will exist, most probably in the area of complex and extreme nanostructures. However, as yet definitive proof of this has not been demonstrated and this has to be one of the key priorities for the field. What are the most likely candidates for these novel applications? One way of viewing supercritical fluid electrodeposition is as an extension of non-aqueous electrodeposition to higher temperature. Whilst lower deposition temperatures are often advantageous economically, better material quality, e.g. crystallinity, often requires an elevated deposition temperature. For any specific solvent the critical temperature is a fundamental limit above which electrodeposition can only be performed using the supercritical fluid electrodeposition technique. One area in which supercritical fluids are already showing promise is their ability to penetrate into very small pores. In particular supercritical fluid electrochemistry has been used to electrodeposit into 3 nm pores within mesoporous silicas.

It will take some time, and more work, to discover whether supercritical fluid electrodeposition can open up novel areas in nanomaterials and structures; however there are a number of shorter term goals which will enable the field to develop into new and exciting areas. These include, from a physical chemistry perspective, a better understanding of the structure of the double layer and of electrolyte solvation in supercritical fluids, studies of electron transfer kinetics at the electrode supercritical fluid interface, studies of speciation in supercritical fluids, and a greater understanding of the phase behaviour of supercritical fluids and supercritical fluid electrolytes in confined spaces. In addition there is much scope for practical advances in the design of new cells, in solvent purification and for the application of supercritical fluids in electrosynthesis.

In conclusion, supercritical fluid electrochemistry has come a long way in a relatively short time and has achieved a number of goals which many people thought were not possible. It is an exciting field with significant scientific challenges and we hope that this perspective will stimulate others to consider how electrochemistry in supercritical fluids might be applied not only in materials deposition, but also in other areas.

ACKNOWLEDGEMENTS

This work is part of the Supercritical Fluid Electrodeposition project (www.scfed.net) which is a multidisciplinary collaboration of British universities investigating the fundamental and applied aspects of supercritical fluids funding by a Programme Grant from the EPSRC (EP/I013394/1). MWG gratefully acknowledges receipt of a Wolfson Merit award. We thank K. Thaller for help in preparation of the manuscript.

REFERENCES

1. J. M. Blackburn, D. P. Long, A. Cabanas and J. J. Watkins, Science, 2001, 294, 141–145.
2. V. K. Popov, V. N. Bagratashvili, L. I. Krotova, A. O. Rybaltovskii, D. C. Smith, P. S. Timashev, J. Yang, Y. S. Zavorotnii and S. M. Howdle, Green Chem., 2011, 13, 2696–2700.
3. M. Schlesinger and M. Paunovic, Modern Electroplating, Wiley, New York, 5th edn, 2010.
4. M. Paunovic and M. Schlesinger, Fundamentals of Electrochemical Deposition, Wiley, New York, 2006.
5. J. Ke, W. T. Su, S. M. Howdle, M. W. George, D. Cook, M. Perdjon-Abel, P. N. Bartlett, W. Zhang, F. Cheng, W. Levason, G. Reid, J. Hyde, J. Wilson, D. C. Smith, K. Mallik and P. Sazio, Proc. Natl. Acad. Sci. U. S. A., 2009, 106, 14768–14772.
6. J. Ke, P. N. Bartlett, D. Cook, T. L. Easun, M. W. George, W. Levason, G. Reid, D. Smith, W. Su and W. Zhang, Phys. Chem. Chem. Phys., 2012, 14, 1517–1528.
7. W. Lu and C. M. Lieber, Nat. Mater., 2007, 6, 841–850.
8. R. Skomski, J. Phys.: Condens. Matter, 2003, 15, R841–R896.
9. J. Wang, Biosens. Bioelectron., 2006, 21, 1887–1892.
10. C. J. Vineis, A. Shakouri, A. Majumdar and M. G. Kanatzidis, Adv. Mater., 2010, 22, 3970–3980.
11. B. Tian, T. J. Kempa and C. M. Lieber, Chem. Soc. Rev., 2009, 38, 16–24.
12. Z. L. Wang, Mater. Today, 2007, 10, 20–28.
13. T. Zhu and J. Li, Prog. Mater. Sci., 2010, 55, 710–757.
14. L. Cademartiri and G. A. Ozin, Adv. Mater., 2009, 21, 1013–1020.
15. R. D. Smith, B. W. Wright and C. R. Yonker, Anal. Chem., 1988, 60, 1323A.
16. M. Poliakoff, S. M. Howdle and S. G. Kazarian, Angew. Chem., Int. Ed. Engl., 1995, 34, 1275–1295.
17. R. S. Oakes, A. A. Clifford and C. M. Rayner, J. Chem. Soc., Perkin Trans. 1, 2001, 917–941.
18. D. J. Cole-Hamilton, Adv. Synth. Catal., 2006, 348, 1341–1351.
19. P. Licence and M. Poliakoff, NATO Sci. Ser., II, 2008, 246, 171–191.
20. K. P. Johnston and R. P. da Rocha Sandro, J. Supercrit. Fluids, 2009, 47, 523–530.

21. J. Yang, T. Hasell, D. C. Smith and S. M. Howdle, J. Mater. Chem., 2009, 19, 8560–8570.
22. M. Perrut, Ind. Eng. Chem. Res., 2000, 39, 4531–4535.
23. M. Herrero, A. Cifuentes and E. Ibanez, Food Chem., 2006, 98, 136–148.
24. J. J. A. Barry, M. M. C. G. Silva, V. K. Popov, K. M. Shakesheff and S. M. Howdle, Philos. Trans. R. Soc. London, Ser. A, 2006, 364, 249–261.
25. M. Banchero, S. Sicardi, A. Ferri and L. Manna, Text. Res. J., 2008, 78, 217–223.
26. J. O'Regan, S. Preston and A. Dunne, 13th European Biosolids & Organic Resources Conference and Workshop, 2008.
27. C. J. Tighe, R. Q. Cabrera, R. I. Gruar and J. A. Darr, Ind. Eng. Chem. Res., 2013, 52, 5522–5528.
28. M. A. McHugh and V. J. Krukonis, Supercritical Fluid Extraction: Principles and Practice, Butterworth, Oxford, 1986.
29. J. F. Brennecke and C. A. Eckert, AIChE J., 1989, 35, 1409–1427.
30. E. Kiran and J. F. Brennecke, Supercritical Fluid Engineering Science: Fundamentals and Applications, ACS Symp. Ser. 514, ACS, 1993.
31. A. A. Clifford, Chromatogr.: Princ. Pract., 1999, 2, 1–14.
32. N. R. Foster, F. P. Lucien and R. Mammucari, Handbook of Green Chemistry, Wiley-VCH, 2010, vol. 4, pp. 77–100.
33. M. Skerget, Z. Knez and M. Knez-Hrncic, J. Chem. Eng. Data, 2011, 56, 694–719.
34. P. N. Bartlett, D. C. Cook, M. W. George, J. Ke, W. Levason, G. Reid, W. Su and W. Zhang, Phys. Chem. Chem. Phys., 2010, 12, 492–501.
35. P. Ratanajiajaroen and M. Ohshima, J. Supercrit. Fluids, 2012, 68, 31–38.
36. Y. Liang, C. Zhen, D. Zou and D. Xu, J. Am. Chem. Soc., 2004, 126, 16338–16339.
37. K. Izutsu, Electrochemistry in nonaqueous solutions, John Wiley & Sons, 2009.
38. A. P. Abbott, C. A. Eardley and R. Tooth, J. Chem. Eng. Data, 1999, 44, 112–115.
39. D. Bertolini, M. Cassettari and G. Salvetti, J. Mol. Liq., 1984, 28, 175–189.
40. K. Reuter, S. Rosenzweig and E. U. Franck, Physica A, 1989, 156, 294–302.
41. F. G. Keyes and J. G. Kirkwood, Phys. Rev., 1930, 36, 754–761.
42. G. Ritzoulis, N. Papadopoulos and D. Jannakoudakis, J. Chem. Eng. Data, 1986, 31, 146–148.
43. A. P. Abbott, E. G. Hope, R. Mistry and A. M. Stuart, Green Chem., 2009, 11, 1530–1535.
44. T. S. Reighard, S. T. Lee and S. V. Olesik, Fluid Phase Equilib., 1996, 123, 215–230.
45. A. K. Dillow, J. S. Brown, C. L. Liotta and C. A. Eckert, J. Phys. Chem. A, 1998, 102, 7609–7617.
46. H.-K. Bae, J.-H. Jeon and H. Lee, Fluid Phase Equilib., 2004, 222–223, 119–125.
47. M. Caravati, J.-D. Grunwaldt and A. Baiker, Appl. Catal., A, 2006, 298, 50–56.
48. J. R. Forero-Mendieta, H. I. Castro-Vargas, F. ParadaAlfonso and J. A. Guerrero-Dallos, J. Supercrit. Fluids, 2012, 68, 64–70.
49. J. Liu, E. L. Regalado, I. Mergelsberg and C. J. Welch, Org. Biomol. Chem., 2013, 11, 4925–4929.
50. D. Cook, P. N. Bartlett, W. Zhang, W. Levason, G. Reid, J. Ke, W. Su, M. W. George, J. Wilson, D. Smith, K. Mallik, E. Barrett and P. Sazio, Phys. Chem. Chem. Phys., 2010, 12, 11744–11752.

CHAPTER 3

DYSPROSIUM ELECTRODEPOSITION FROM A HEXAALKYLGUANIDINIUM-BASED IONIC LIQUID

Claudia A. Berger[a], Maria Arkhipova[b], Gerhard Maas[b] and Timo Jacob[ac]

[a]Institute of Electrochemistry, Ulm University, Albert-Einstein-Allee 47, 89081 Ulm, Germany

[b]Institute of Organic Chemistry I, Ulm University, Ulm, Germany

[c]Helmholtz-Institute-Ulm (HIU), Ulm, Germany

ABSTRACT

The rare-earth element dysprosium (Dy) is an important additive that increases the magnetocrystalline anisotropy of neodymium magnets and additionally prevents from demagnetizing at high temperatures. Therefore, it is one of the most important elements for high-tech industries and is mainly used in permanent magnetic applications, for example in electric vehicles, industrial motors and direct-drive wind turbines. In an effort to develop a more efficient electrochemical technique for depositing Dy on Nd-magnets in contrast to commonly used costly physical vapor deposition, we investigated the electrochemical behavior of dysprosium(III) trifluoromethanesulfonate in a custom-made guanidinium-based room-temperature ionic liquid (RTIL). We first examined the electrodeposition of Dy on an Au(111) model electrode. The investigation was carried out by means of cyclic voltammetry (CV) and X-ray photoelectron spectroscopy (XPS). The initial stages of metal deposition were followed by *in situ* scanning tunneling microscopy (STM). CV measurements revealed a large cathodic reduction peak, which corresponds to the growth of monoatomic high islands, based on STM images taken during the initial stages of deposition. XPS identified these deposited islands as dysprosium. A similar reduction peak was also observed on an Nd–Fe–B substrate, and positively identified as deposited Dy using XPS. Finally, we varied the concentration of the Dy precursor, electrolyte flow and temperature during Dy deposition and demonstrated that each of these parameters could be used to increase the thickness of the Dy deposit, suggesting that these parameters could be tuned simultaneously in a temperature-controlled flow cell to enhance the thickness of the Dy layer.

INTRODUCTION

The rare-earth element dysprosium (Dy) is rather unique amongst the magnetic materials, exhibiting both the highest magnetic moment per atom ($10\mu_B$) and the highest saturation magnetization of any element.[1] Therefore, it is no surprise that in its most prominent industrial application it is alloyed into neodymium–iron–boron magnets[2] in order to increase the magnetocrystalline anisotropy.[3]Neodymium–iron–boron magnets were invented by Sagawa *et al.* in 1984 [4] and are now widely employed as magnetic field sources in hybrid electric vehicles and wind turbines,[5] for instance. Indeed, as early as 2007, 30 000 tons per year of neodymium magnets were being produced worldwide.[6] Current industrial methods rely on a combination of physical vapor deposition (PVD)[7] and

Citation: Claudia A. Berger, Maria Arkhipova, Gerhard Maas and Timo Jacob, Dysprosium electrodeposition from a hexaalkylguanidinium-based ionic liquid, DOI: 10.1039/C6NR01351A.

sputtering techniques to produce Dy layers. The high energy cost of these methods, due to the high-vacuum conditions they require, makes them ecologically unsustainable, particularly at large scales. However, the electrochemical deposition of dysprosium from non-aqueous solvents has been put forward as a promising alternative to present industrial methods.[8] Room-temperature ionic liquids (RTILs) are organic salts, which are in the molten state at room-temperature.[9] Seemingly almost innumerable combinations of different anions and cations are possible, resulting in a vast variety of potential RTILs with tunable physicochemical properties that can be tailored for specific applications.[10] Besides well-investigated ionic liquids (ILs) with common imidazolium, pyridinium or piperidinium cations, there are new classes of ILs which are still being developed. One example here are RTILs, which already contain lanthanides in their anions and cations.[11–15] Another promising new class of ILs are hexaalkyl-substituted guanidinium salts, because of the flexibility that can be achieved by tuning each of the six alkyl groups independently.[16] In addition, they exhibit good thermal stability and are unreactive toward bases and most nucleophilic reagents due to the excellent charge stabilization in the peralkylated cation. Applications in electrochemistry have driven recent interest in RTILs, making their large electrochemical windows and high ionic conductivities two of their most important features along with their low vapor pressures and high thermal stabilities.[17–22] Indeed, the fact that RTILs enable access to the electrodeposition of non-noble metals, which cannot be deposited from aqueous solutions, more than compensates for their high viscosities and costs, when considering their usefulness in a wide range of applications. One of these non-noble metals is dysprosium, the electrodeposition of which several groups have demonstrated from commercially available ionic liquids,*e.g.* phosphonium[23] or pyrrolidinium-based[8] RTILs. However, the deposition rates achieved in these studies were not yet sufficient, especially with a view to industrial applications. To address this need we investigated the electrodeposition of Dy from its triflate in a custom-made guanidinium-based ionic liquid. In the present work we characterize the electrodeposition of dysprosium first on Au(111) single crystal electrodes and then on Nd–Fe–B-magnets. Based on these results we discuss strategies for increasing the amount of deposited dysprosium on Nd–Fe–B magnets to achieve better permanent magnetic properties by variation of the concentration, flow of the electrolyte and temperature.

EXPERIMENTAL

Various guanidinium-based ionic liquids with the general composition $[C(NR^1_2)(NR^2_2)(NR^3_2)]^+A^-$ (R = alkyl, cycloalkyl; A^- = anion) were synthesized and examined with respect to their suitability for dysprosium deposition (see Table 1). Here, the two most significant factors were the width of the electrochemical potential window, especially in the cathodic potential region, and the amount of dissolved $Dy(OTf)_3$ without forming a precipitate in solution. Details of the synthesis and characterization of the applied guanidinium-based ILs can be found in the ESI† and in ref. 16, 24 and 25.

Table 1: Overview of features of synthesized guanidinium-based ionic liquids

Ionic liquid[a]	Cathodic limit *vs.* Pt/mV	Anodic limit *vs.* Pt/mV	Solubility $Dy(OTf)_3$/mol L^{-1}
a See the ESI for structural formulae.			
$[N_{11}N_{22}N_{44}Gua]TFSI$	−1700	1000	>0.007
$[N_{11}N_{11}N_{44}Gua]OTf$	−800	1400	<0.026
$[N_{22}N_{44}N_{66}Gua]BF_4$	−1900	1300	>0.025
$[N_{11}N_{11}N_{pip}Gua]TFSI$	−2500	1500	<0.017
$[N_{11}N_{pip}N_{pip}Gua]TFSI$	−1800	1700	<0.200

Based on the broad electrochemical potential window, a reasonable solubility of $Dy(OTf)_3$ and a relatively low viscosity of 108 mPa s (25 °C),[26] $N_{11}N_{11}N_{pip}GuaTFSI$ (pip = piperidin-1-yl) (see Fig. 1) was chosen as an appropriate electrolyte for dysprosium deposition. Additionally, a commercially available RTIL was used in the STM study, namely 1-butyl-3-methylimidazolium bis(trifluoromethylsulfonyl)imide (BMITFSI) (Merck KGaA, purity ≥ 99.5%, water ≤ 100 ppm, halides ≤ 100 ppm). The ionic liquids were vacuum-dried for 24 h at elevated temperatures (80 °C, 2×10^{-4} mbar) before adding the precursor dysprosium triflate (98% $Dy(OTf)_3$, Sigma-Aldrich) to reach a concentration between 0.02 and 0.20 mol L^{-1} for the metal deposition.

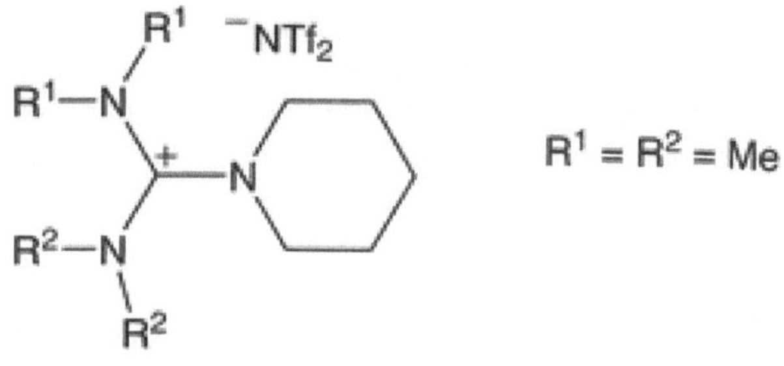

Figure 1: Structure of the applied guanidinium-based ionic liquid, $[N_{11}N_{11}N_{pip}Gua]TFSI$.

CV and STM experiments, including handling and storage of the used substances, were performed inside a glove box under an argon atmosphere. During the transfer of the sample to the XPS chamber, air contact was inevitable due to the absence of an appropriate transfer vessel. Electrochemical investigations were carried out in self-designed cells made of KelF, each with a volume of 150 µl, and an Au(111) single crystal (MaTeck GmbH, Jülich, FRG) with a 12 mm diameter and Nd–Fe–B sintered square magnets with 10 mm edge lengths and 3 mm thickness (Vacuumschmelze GmbH & Co. KG, Hanau, Germany) acting as the working electrodes. Nd–Fe–B substrates were produced *via* sintering processes, resulting in rather rough surfaces. In this study, the calculated theoretical surface area was used for all charge determinations and current density evaluations, although the real surface area is supposedly larger. As counter and quasi-reference electrodes Pt wires (MaTeck GmbH, Jülich, FRG) were used; thus, all potentials are reported with respect to the Pt reference electrode. Prior to the measurements all electrodes (except Nd–Fe–B substrates) were annealed in a hydrogen flame and cooled down slowly in an argon stream.

All CVs were recorded with a Zahner IM6 potentiostat from Zahner Elektrik controlled by the Thales Z 1.20 USB software. Electrochemical *in situ* STM studies were performed with a Topometrix Discoverer TMX 2010. For the preparation of the STM tips, Pt/Ir wires (80 : 20) were electrochemically etched in 3.5 M NaCN and coated with BASF electrophoretic paint (ZQ84-3225) to reduce the faradaic current. All images were recorded in the constant-current mode with a tip current between 1–3 nA.

The elemental composition of the sample surfaces was determined by X-ray photoelectron spectroscopy (XPS) measurements using monochromatized Al K_α (1486.6 eV) radiation (PHI 5800 MultiTechnique ESCA System, Physical Electronics). A surface spot of 0.8 × 0.8 mm^2 was used for all analyses. The measurements were performed with a detection angle of 45°, using pass energies at the analyzer of 93.9 and 29.35 eV for survey and detailed spectra, respectively. To calibrate the binding energies the C(1s) peak was set to 284.8 eV.

RESULTS AND DISCUSSION

$Dy(OTf)_3/N_{11}N_{11}N_{pip}GuaTFSI$ on Au(111)

Cyclic voltammetry

Fig. 2 shows the cyclic voltammograms of pure $N_{11}N_{11}N_{pip}GuaTFSI$ (dashed line) and with addition of 0.02 mol L^{-1} $Dy(OTf)_3$ (solid line) on Au(111) recorded with a scan rate of 50 mV s^{-1}. The CV of the pure ionic liquid $N_{11}N_{11}N_{pip}GuaTFSI$ shows a broad double-layer-type behavior over a potential range of about 3 V with a few small humps due to surface-related processes or minor impurities. The potential where the irreversible decomposition of the electrolyte takes place lies outside this potential range and remarkably occurs at low current densities around the reversal potentials in the CV. After adding 0.02 mol L^{-1} of $Dy(OTf)_3$, one can observe a single broad cathodic hump C_1 at −1 V *vs.* Pt, which indicates a possible reduction of dysprosium. We assume that Dy reduction proceeds in a single multi-electron transfer step from Dy(III) to Dy(0), as firstly no other cathodic processes could be observed, and secondly this behavior is reported in the literature for Dy deposition from other types of RTILs.[1,8,23] The reduction peak corresponds to a charge of 400 µC cm^{-2} as derived from the CV (Fig. 2). The charge complies with approximately 8 × 10^{14} Dy atoms, indicating slightly less than one monolayer of deposited Dy at this scan rate.

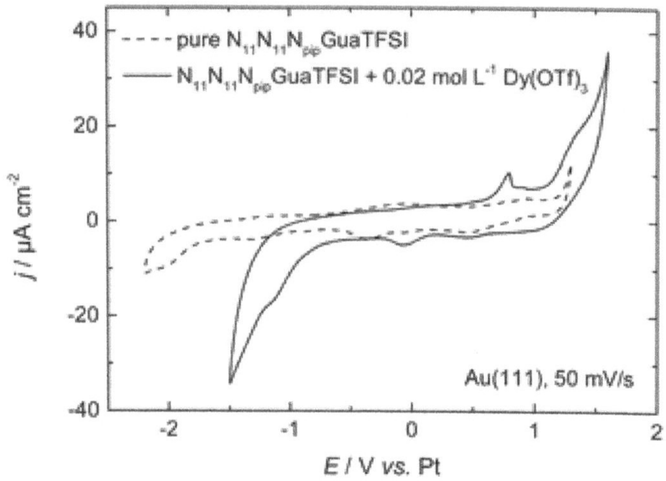

Figure 2: CV of $N_{11}N_{11}N_{pip}GuaTFSI$ (dashed line) and $N_{11}N_{11}N_{pip}$-GuaTFSI + 0.02 mol L^{-1} $Dy(OTf)_3$(solid line) on Au(111) at a scan rate of 50 mV s^{-1}.

In the positive potential region we observe a large anodic peak, whose related charge is comparable to the cathodic peak. Based on the CV measurements alone, one cannot clearly distinguish whether this peak is caused by oxidation of Dy or decomposition of the ionic liquid. However, previous studies in other types of RTILs[1,8] claim the deposition of dysprosium to be an irreversible process, so we assume that the anodic process indicates more likely the oxidation of the ionic liquid. The comparison with the CV of the pure RTIL confirms this assumption, due to the increase of the current density at the positive reversal potential, which indicates the oxidation of the electrolyte.

In situ scanning tunneling microscopy

For a more detailed characterization of the Dy deposition, *in situ* scanning tunneling microscopy (STM) measurements were performed. The obtained STM images are shown in Fig. 3. Fig. 3(a) shows the bare Au surface with several step edges at positive potentials, *i.e.* 300 mV *vs.* Pt. Fig. 3(b) is recorded at very negative potentials (−1.6 V*vs.* Pt), which means potentials in the region of the reduction peak in the CV (see Fig. 2). Here, the beginning of the growth of monoatomic high islands on the Au terraces can be observed, which could indicate Dy deposition. Unfortunately, we were not able to record further STM images of the deposition process in this system. We suppose that the guanidinium cations form a dense and disordered adsorbate structure at very negative potentials, which compromises the STM measurement by blurring of the STM images. Therefore, the STM investigation of this system is a formidable task. However, investigations in an alternative RTIL, namely 1-butyl-3-methylimidazolium bis(trifluoromethylsulfonyl)imide (BMITFSI), showed more details of the dysprosium deposition (Fig. 4).

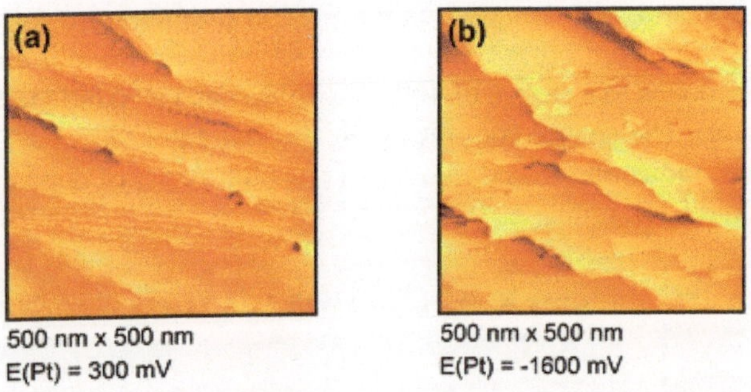

Figure 3: *In situ* STM images of $Dy(OTf)_3/N_{11}N_{11}N_{pip}$GuaTFSI on Au(111) at (a) 300 mV and (b) −1600 mV.

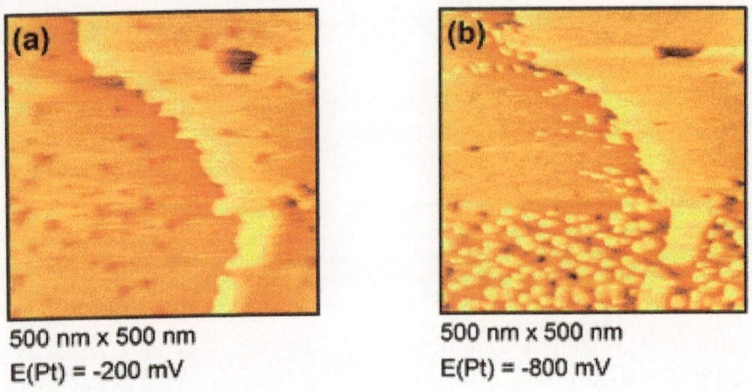

Figure 4: *In situ* STM images of $Dy(OTf)_3$/BMITFSI on Au(111) at (a) −200 mV and (b) −800 mV.

Fig. 4(a) shows the free Au surface at relatively positive potentials (−200 mV *vs.* Pt), comparable to $Dy(OTf)_3$ in $N_{11}N_{11}N_{pip}$GuaTFSI. After decreasing the potential the formation of islands can be observed (see Fig. 4(b)). The formation of the islands seems to start at step edges and then to spread over the Au terraces. The derived height of the islands from this STM image is approximately 300 nm.

X-ray photoelectron spectroscopy

Further insight into the chemical composition of the sample was gained by XPS measurements. Fig. 5 shows the detailed XP survey spectra of Dy(4d) and O(1s) signals obtained directly following the electrodeposition of Dy from $Dy(OTf)_3/N_{11}N_{11}N_{pip}$GuaTFSI on Au(111). As mentioned before, we expect to detect dysprosium in the form of Dy_2O_3due to the inevitable air contact during transportation to the XPS device. In the detailed spectrum, the Dy(4d) peak shows a more complex shape than expected from sole spin–orbit splitting, which has been reported in the literature and is attributed to final state multiplet coupling between the 4d and 4f states.[27] Here, the peak with the highest intensity is centered at a binding energy of 157.0 eV and it possesses an additional shoulder of lower intensity around 154.9 eV. The binding energy of the absolute maximum is in relatively good agreement with binding energies for Dy_2O_3 reported in the literature.[28,29] Further, the detailed spectrum for O(1s) is shown in Fig. 5. Here, the O(1s) peak is observable at a binding energy of 532.0 eV. Comparable values have been reported in the literature[28,29]and were assigned to the adsorbed OH-groups and (bi)carbonates on the Dy_2O_3 deposit due to the reactivity of lanthanides toward air and moisture.

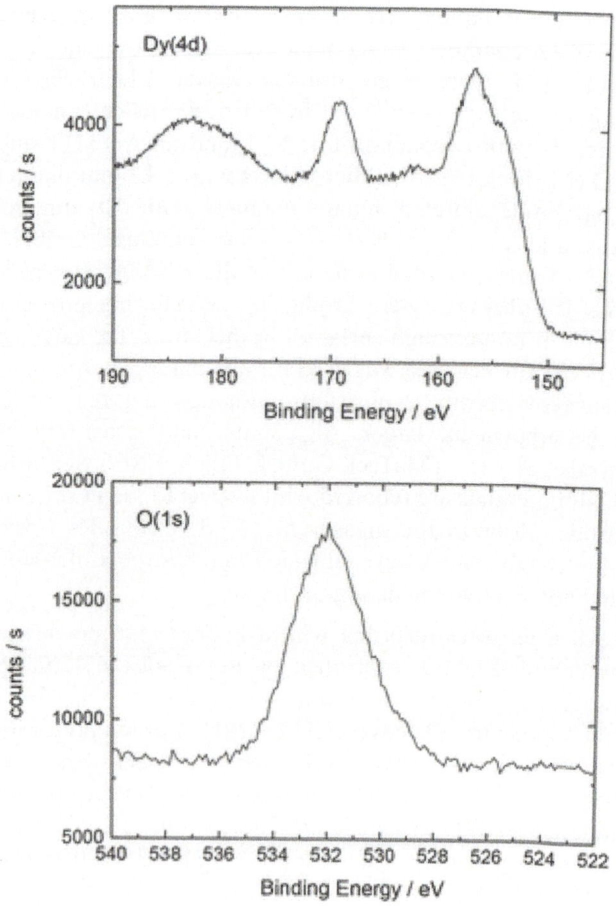

Figure 5: Detailed XP spectrum of the Dy deposit on Au(111).

Additionally, Table 2 shows the determined atomic concentrations on the sample surface after electrochemical Dy deposition on Au(111). Here, the ratio of O/Dy is around 5, which is much higher than the expected value for stoichiometric Dy_2O_3. This observation has also been reported in the literature[29] and could be due to the presence of excess oxygen on the surface of the deposited films, as mentioned before.

Table 2: Atomic concentration of Dy on Au(111)

O(1s)	C(1s)	F(1s)	Dy(4d)	S(2p)	Au(4f)
51.0	16.3	15.7	10.6	5.0	1.4

$Dy(OTf)_3/N_{11}N_{11}N_{pip}$GuaTFSI on NdFeB

Cyclic voltammetry

Fig. 6 shows the first cycles of the cyclic voltammograms of $N_{11}N_{11}N_{pip}$GuaTFSI containing 0.063 mol L^{-1} of $Dy(OTf)_3$ (black curve) and 0.200 mol L^{-1} of $Dy(OTf)_3$(red curve) on a Nd–Fe–B magnet recorded with a scan rate of 10 mV s^{-1}. After adding 0.063 mol L^{-1} of $Dy(OTf)_3$ to the RTIL, one can observe a single broad cathodic hump, C_1, at around −1.2 V *vs.*Pt, which also indicates a possible deposition of dysprosium. According to literature, the electrodeposition is supposed to proceed from Dy(III) to Dy(0) in a single one-step electron transfer.[1,8,23] This reduction peak corresponds to a charge of 60 mC cm^{-2} during the deposition process as derived from the CV (Fig. 6), which complies with 10^{17} Dy atoms per cm^2 (geometrically calculated area). Here, no clear oxidation peak was detected, but a slight increase in the current density at potentials >0.2 V can be observed, which could be an indication of a partial dissolution of Dy.

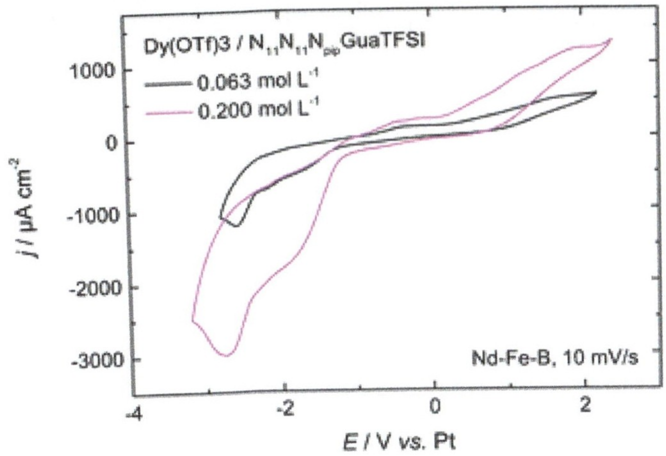

Figure 6: CV of 0.063 mol L^{-1} and 0.200 mol L^{-1} of $Dy(OTf)_3$ + $N_{11}N_{11}N_{pip}$GuaTFSI on Nd–Fe–B, recorded with a scan rate of 10 mV s^{-1}.

As the first attempt to increase the amount of deposited dysprosium, and thereby improve the permanent magnetic properties of the material, the concentration of the dissolved Dy salt in the RTIL was increased by a factor of 3. The red curve in Fig. 6 shows the CV of 0.200 mol L^{-1} of $Dy(OTf)_3$/ $N_{11}N_{11}N_{pip}$GuaTFSI on Nd–Fe–B. Here, one can clearly notice an increase of the charge flow during the deposition process and dissolution, respectively. This also confirms the assumption of Dy oxidation at positive potentials of 0.2 V. The charge associated with the cathodic processes was enhanced from 60 mC cm^{-2} (Fig. 6, black curve) to 150 mC cm^{-2} (Fig. 6, red curve) over the same potential range, *i.e.* the charge and thus the amount of deposited Dy atoms (3×10^{17} per cm^2) are almost tripled.

X-ray photoelectron spectroscopy

Here, also XPS measurements were performed in order to identify the obtained deposit. Fig. 7 shows the detailed XP survey spectra of Dy(4d) and O(1s) obtained directly after electrodeposition of Dy from $Dy(OTf)_3$/$N_{11}N_{11}N_{pip}$GuaTFSI on Nd–Fe–B. Analogous to section 3.1.3, we expect to detect dysprosium in the form of Dy_2O_3 due to the inevitable air contact during transportation to the XPS chamber. In the detailed spectrum, the Dy(4d) peak shows also the complex shape, which was observed in Fig. 5. The main peak appears at a binding energy of 157.0 eV, which very likely indicates Dy in the form of Dy_2O_3, according to the literature.[28,29] Further, the detailed spectrum for O(1s) is shown in Fig. 7. Here, the O(1s) peak is observable at a binding energy of 531.5 eV, which is also in good agreement with the literature and is an indication for (bi)carbonates and OH-groups adsorbed on the surface.[28,29]

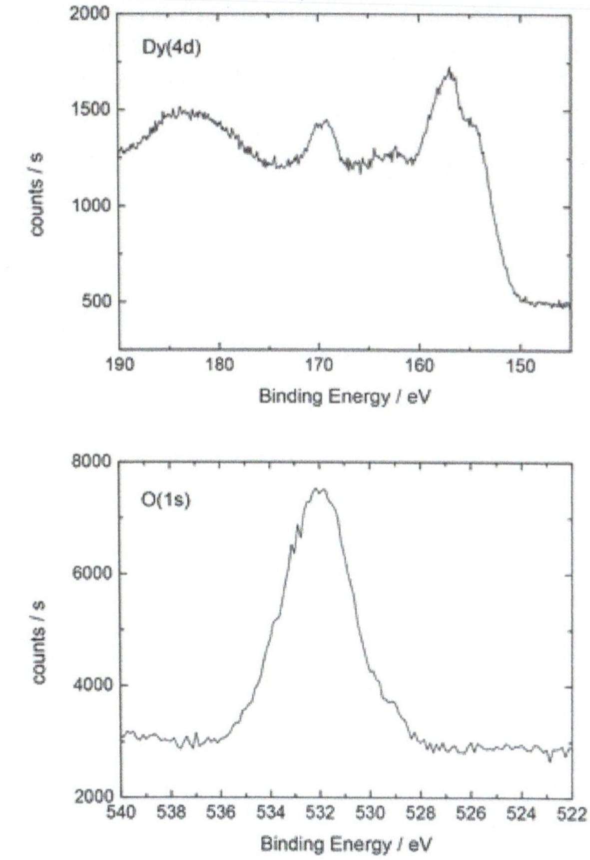

Figure 7: Detailed XP spectrum of the Dy deposit on the Nd–Fe–B magnet.

Additionally, Table 3 shows the determined atomic concentrations on the sample surface after electrodeposition of Dy on Nd–Fe–B. Here, the ratio of O/Dy is around 2.5, which is again higher than the expected value for stoichiometric Dy_2O_3. We assume that this observation could be due to the presence of excess oxygen in the form of adsorbed OH-groups or carbonates on the surface of the deposited films or at least partial oxidation of the various components of the substrate.

Table 3: Relative atomic concentration of Dy on Nd–Fe–B

C(1s)	O(1s)	Dy(4d)	F(1s)	Fe(2p)	Nd(4d)
47.2	32.3	12.6	6.3	1.1	0.5

Increasing Dy layer thickness on Nd–Fe–B

Flow cell

After the basic electrochemical characterization, the influence of a flow cell on dysprosium deposition from $Dy(OTf)_3$/ $N_{11}N_{11}N_{pip}$GuaTFSI on Nd–Fe–B was investigated. For this purpose, a potential of 2 V *vs.* Pt was applied to the system, *i.e.* the potential region of Dy deposition, and the resulting current was recorded over time. Here, all single measurements were performed at room temperature. Fig. 8 shows *I*–*t* transients for flow rates of the electrolyte of 0 ml h^{-1} (grey curve), 3 ml h^{-1} (black curve) and 4 ml h^{-1} (red curve). It has to be mentioned that the aberration in the red curve is caused by a brief disruption of the electrolyte flow. The charges related to Dy reduction were determined by integration of the particular transient over the first 800 seconds. For no electrolyte flow (0 ml h^{-1}), we determined a charge of 2.3 mC cm^{-2} for Dy deposition. For a flow rate of 3 ml h^{-1}, the derived charge amounts to 38.5 mC cm^{-2}, which signifies a massive increase. Finally, a 4 ml h^{-1} flow rate leads to a maximum charge of 122.5 mC cm^{-2}, which means about 2.5 × 10^{17} Dy atoms per cm^2 of the Nd–Fe–B substrate. According to our experience implementation of even higher flow rates of the electrolyte is not manageable due to the generally high viscosity of ILs.

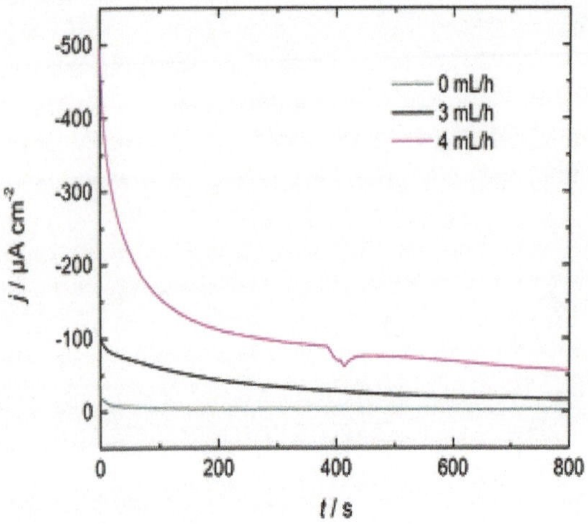

Figure 8 Flow rate dependence of *I–t*-transients during Dy deposition from $Dy(OTf)_3/N_{11}N_{11}N_{pip}$GuaTFSI on Nd–Fe–B.

Temperature-controlled cell

Fig. 9 shows *I–t* transients recorded over 4 hours for the system $Dy(OTf)_3/N_{11}N_{11}N_{pip}$GuaTFSI on Nd–Fe–B. The *I vs. t* measurements were performed analogous to section 3.3.1. The sole difference is the absence of electrolyte flow, which means constant electrolyte volume, and the differing temperatures of the single experiments. The overall charge flow due to dysprosium reduction was determined for the first 30 minutes by integration of the *I–t* transients for 25 °C (grey curve), 28 °C (black curve) and 35 °C (red curve). While we obtain 32.0 mC cm^{-2} for 25 °C, increasing the temperature by only three degrees to 28 °C results in an increased charge of 69.8 mC cm^{-2}. A further increase of the temperature up to 35 °C results in a maximum charge of 192.4 mC cm^{-2}, translating to an overall deposition coverage of 145 × 10^{16} dysprosium atoms per cm^2 of the Nd–Fe–B substrate.

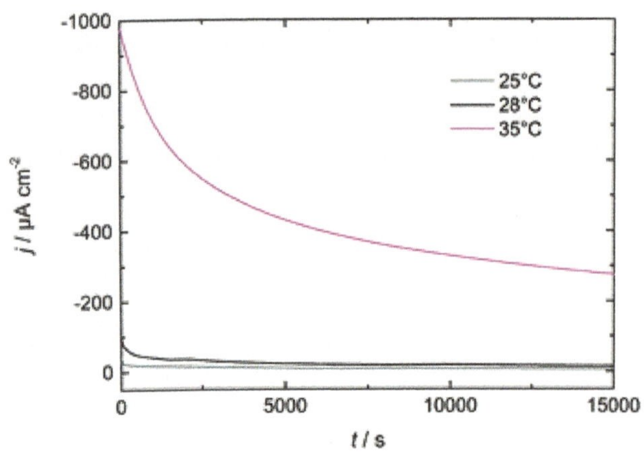

Figure 9: Temperature dependence of *I–t*-transients during Dy deposition from $Dy(OTf)_3/N_{11}N_{11}N_{pip}$GuaTFSI on Nd–Fe–B.

CONCLUSIONS

In this paper we have investigated Dy electrodeposition from a tailored guanidinium-based RTIL on Au(111) as well as Nd–Fe–B substrates. In the presence of Au(111) we found that the reduction of Dy(III) to Dy(0) occurs in a single step at <–1 V *vs.* Pt. The reduction of Dy coincides with the formation of monoatomic high islands on the Au terraces which were observed in *in situ* STM and positively identified as Dy deposits on the Au(111) electrode surface in XPS measurements.

A single reduction step at <–1 V *vs.* Pt was also observed in the presence of the Nd–Fe–B substrate, which is comparable to the behavior obtained for Au(111). The XP spectra confirmed the presence of Dy on the magnet›s surface. We were able to enhance the thickness of the deposited Dy layer by either increasing the electrolyte flow or increasing the temperature. Thus, we expect that a combination of both approaches could be used to further optimize the thickness of the deposited Dy for industrial applications. This optimization could be realized *via* a temperature-controlled flow cell, resulting in improvements in the magnetic properties of the Nd–Fe–B substrate at a reduced energetic cost.

ACKNOWLEDGEMENTS

The authors thank Dr M. Rührig and Dr A. Kanitz for the supporting assistance and the Siemens AG for the financial support. The authors also thank the DFG (Deutsche Forschungsgemeinschaft) for their support through the Project KO 576/28-1 as well as the European Research Council through the ERC – Starting Grant THEOFUN (Grant Agreement No. 259608). Finally, we thank Dr T. Diemant (Institute of Surface Chemistry and Catalysis, Ulm University) for the XPS service measurements.

REFERENCES

1. J. Lodermeyer, M. Multerer, M. Zistler, S. Jordan, H. J. Gores, W. Kipferl, E. Diaconu, M. Sperl and G. Bayreuther, *J. Electrochem. Soc.*, 2006, 153, C242–C248 .
2. N. Oono, M. Sagawa, R. Kasada, H. Matsui and A. Kimura, *J. Magn. Magn. Mater.*, 2011, 323, 297–300 .
3. M. Sagawa, S. Fujimura, H. Yamamoto, Y. Matsuura and K. Hiraga, *IEEE Trans. Magn.*, 1984, MAG-20, 1584 .
4. M. Sagawa, S. Fujimura, N. Togawa, H. Yamamoto and Y. Matsuura, *J. Appl. Phys.*, 1984, 55, 2083–2087 .
5. S. Hoenderdaal, L. T. Espinoza, F. Marscheider-Weidemann and W. Graus, *Energy*, 2013, 49, 344–355 .
6. K. Miura, M. Itoh and K.-I. Machida, *J. Alloys Compd.*, 2008, 466, 228–232 .
7. H. Sepehri-Amin, T. Ohkubo and K. Hono, *J. Appl. Phys.*, 2010, 107, 09A745 .
8. G. Suppan, M. Ruehrig, A. Kanitz and H. J. Gores, *J. Electrochem. Soc.*, 2015, 162, D382–D388 .
9. K. R. Seddon, *J. Chem. Technol. Biotechnol.*, 1997, 68, 351–356 .
10. H. Tokuda, K. Hayamizu, K. Ishii, M. A. Bin Hasan Susan and M. Watanabe, *J. Phys. Chem. B*, 2005, 109, 6103–6110 .

11. P. Nockemann, B. Thijs, N. Postelmans, K. V. Hecke, L. V. Meervelt and K. Binnemans, *J. Am. Chem. Soc.*, 2006, 128, 13658–13659 .

12. P. Nockemann, B. Thijs, S. Pittois, J. Thoen, C. Glorieux, K. V. Hecke, L. V. Meervelt, B. Kirchner and K. Binnemans, *J. Phys. Chem. B*, 2006, 110, 20978–20992 .

13. B. Mallick, B. Balke, C. Felser and A.-V. Mudring, *Angew. Chem., Int. Ed.*, 2008, 47, 7635–7638 .

14. D. Prodius, F. Macaev, Y. Lan, G. Novitchi, S. Pogrebnoi, E. Stingaci, V. Mereacre, C. E. Anson and A. K. Powell, *Chem. Commun.*, 2013, 49, 9215–9217 .

15. C. Lu, S. Das, N. Siraj, P. K. S. Magut, M. Li and I. M. Warner, *J. Phys. Chem. A*, 2015, 119, 4780–4786 .

16. H. Kunkel and G. Maas, *Eur. J. Org. Chem.*, 2007, 3746–3757 .

17. F. Endres, *ChemPhysChem*, 2002, 3, 144–154 .

18. R. D. Rogers and K. R. Seddon, *Science*, 2003, 302, 792–793 .

19. M. J. Earle, J. M. S. S. Esperanca, M. A. Gilea, J. N. Canongia Lopes, L. P. N. Rebelo, J. W. Magee, K. R. Seddon and J. A. Widegren, *Nature*, 2006, 439, 831–834 .

20. F. Endres and S. Zein El Abedin, *Phys. Chem. Chem. Phys.*, 2006, 8, 2101–2116 .

21. F. Endres, D. MacFarlane and A. Abbott, *Electrodeposition from Ionic Liquids*, Wiley-VCH, Weinheim, 2008 Search .

22. M. Freemantle, *An Introduction to Ionic Liquids*, Royal Society of Chemistry, Cambridge, 2009 Search .

23. A. Kurachi, M. Matsumiya, K. Tsunashima and S. Kodama, *J. Appl. Electrochem.*, 2012, 42, 961–968 .

24. W. Kantlehner, E. Haug, W. W. Mergen, P. Speh, T. Maier, J. J. Kapassakalidis, H.-J. Bräuner and H. Hagen, *Liebigs Ann. Chem.*, 1984, 1984, 108–126 .

25. M. Gnahm, C. Berger, M. Arkhipova, H. Kunkel, T. Pajkossy, G. Maas and D. M. Kolb, *Phys. Chem. Chem. Phys.*, 2012, 14, 10647–10652 .

26. N. Bucher, S. Hartung, M. Arkhipova, D. Yu, P. Kratzer, G. Maas, M. Srinivasan and H. E. Hoster, *RSC Adv.*, 2014, 4, 1996–2003 .

27. H. Ogasawara, A. Kotani and B. T. Thole, *Phys. Rev. B: Condens. Matter Mater. Phys.*, 1994, 50, 12332 .

28. D. Barreca, A. Gasparotto, A. Milanov, E. Tondello, A. Devi and R. A. Fischer, *Surf. Sci. Spectra*, 2007, 14, 52–59 .

29. A. P. Milanov, T. Toader, H. Parala, D. Barreca, A. Gasparotto, C. Bock, H.-W. Becker, D. K. Ngwashi, R. Cross and S. Paul, *et al.*, *Chem. Mater.*, 2009, 21, 5443–5455 .

CHAPTER 4

AU–PT ALLOY NANOCATALYSTS FOR ELECTRO-OXIDATION OF METHANOL AND THEIR APPLICATION FOR FAST-RESPONSE NON-ENZYMATIC ALCOHOL SENSING

Liyan Zhao, Joseph P. Thomas, Nina F. Heinig, Marwa Abd-Ellah, Xiongyao Wang and K. T. Leung

WATLab and Department of Chemistry, University of Waterloo, Waterloo, Ontario, Canada N2L 3G1

ABSTRACT

Electrodeposited Au–Pt alloy nanoparticles supported on ITO-glass substrates have been used as a model catalyst to study the electro-oxidation of methanol (0.5 M) in a KOH (0.5 M) electrolyte. Using cyclic voltammetry, we show that the Au–Pt alloy nanoparticles exhibit a significantly higher catalytic activity than the Pt nanoparticles, while the Au nanoparticles show no catalytic affinity toward electro-oxidation of methanol. The Au–Pt alloy nanoparticles are also found to be catalytically more reactive than bimetallic Au-core Pt-shell and Pt-core Au-shell nanoparticles. We further show that these Au–Pt alloy nanoparticles working in alkaline solution could act as an amperometric sensor for the determination of alcohol concentration. The exceptional performance of this nanoalloy sensor, including its short response time (2 s), relatively large linear range (1–11 mM), low detection limit (0.1 mM), high sensitivity (0.043 mA cm^{-2} mM^{-1}) and long-term stability (over 90 days), are also characterized. The present nanoalloy sensor also has the lowest working potential (0.25 V), making it less prone to an interference effect.

INTRODUCTION

The study of methanol electro-oxidation has attracted a lot of attention due to its broad applications in fuel cells.[1,2] Platinum nanoparticles (NPs) have been traditionally used as an effective catalyst,[3,4]and it is well established that electro-oxidation of methanol on a Pt surface follows a dual-path mechanism.[5] In particular, methanol can be oxidized directly to CO_2 through a reactive intermediate, or indirectly first to CO at a relatively low potential and then to CO_2 only at a higher potential. The earlier work has shown that in addition to CO, formic acid and formaldehyde could act as intermediates during electro-oxidation of methanol.[6] One important challenge is the poisoning of Pt catalysts by adsorbed CO-like species.[7] As Au NPs have been reported to be extraordinary catalysts for CO oxidation,[8,9] which is part of the indirect reaction pathway, this affinity for CO oxidation has created a strong interest in Au–M bimetallic or alloy NPs (where M is, *e.g.*, Pt,[10–12] Pd,[13] Rh,[14] and Cu[15]) for methanol electro-oxidation. In fuel cell application, carbon black is often used as a support for the noble metal electrode. However, the chemisorption of surface oxygen complexes on the high surface area of this type of support could complicate the study of these catalysts, particularly in probing the mechanism of electro-oxidation of methanol at the

Citation: Liyan Zhao , Joseph P. Thomas , Nina F. Heinig , Marwa Abd-Ellah , Xiongyao Wang and K. T. Leung, Au-Pt alloy nanocatalysts for electro-oxidation of methanol and their application for fast-response non-enzymatic alcohol sensing, DOI: 10.1039/C3TC32317J.

active sites of the noble metal. In this work, we investigate the methanol electro-oxidation in alkaline medium using Au, Pt and Au–Pt NPs as model catalysts electrodeposited onto ITO-glass substrates (as the support). Such systems offer relatively simple model systems to investigate mechanistically the surface chemistry involved because the ITO-glass support has been found to have no effect on methanol electro-oxidation. We also demonstrate the superior catalytic performance of Au–Pt alloy NPs compared to those of bimetallic Au on Pt and Pt on Au NPs. It is well known that methanol electro-oxidation could occur in acidic (H_2SO_4) medium. We focus our present study on alkaline medium (KOH) because fuel cell performance has been found to be significantly better in alkaline medium.[16–18]

Monitoring alcohol concentration in a solution with a high sensitivity and a short response time is important in many industries. Electrochemical alcohol sensing is one of the most promising methods due to its simplicity and accuracy. Two types of sensors based on their electrode materials have been widely studied. The enzyme-modified electrode employs either alcohol oxidase[19,20] or alcohol dehydrogenase,[21,22] and it is known to be not only expensive but also bio-inactive at room temperature. A non-enzymatic sensor is based on a metal or metal oxide electrode, and there are a number of such electrochemical sensors, all with their advantages and challenges. For example, an Au electrode modified by co-electrodeposited Pt and WO_3 (ref. 23) has a long response time and a short lifetime due to the long-term dissolution of WO_3. Such a sensor, however, has a very high sensitivity, and its reproducibility and accuracy of electrochemical sensing are comparable to traditional methods such as the near-infrared methods.[24] A RuO_2-modified Ni electrode[25] and a Ni/Pt/Ti nanostructured thin film sputtered on an Al_2O_3 substrate[26] also exhibit a relatively long response time. A CuO-modified copper disc electrode has shown a very poor detection limit, despite its good reproducibility.[27] A Ni–Pt alloy film electrodeposited on an Au/Al_2O_3 electrode can work as a sensor with a long stability of over 63 days, but it has a poor detection limit.[28] For alcohol sensors based on an electrodeposited Ni–B thin film[29] and on PtRuNi ternary alloy NPs electrodeposited on a glassy carbon electrode pre-coated with a multi-walled carbon nanotubes–ionic liquid gel film,[30] their common disadvantage is the requirement to operate at a high over-potential, as is also needed for all of the aforementioned sensors (0.4–0.65 V). A high over-potential has been found to lead to interference effect and is therefore less desirable in sensor application. While a Pt electrode electrodeposited with a mixed oxide film of Co–Ni could work at a lower potential (0.27 V), it suffers from an extremely short lifetime of just 3 days.[31]

It is well known that Pt is a widely used electro-catalyst for electro-oxidation of alcohol. Here, by introducing Au into Pt in a simple, one-step co-electrodeposition method, we produce Au–Pt alloy NPs on an ITO-glass substrate and demonstrate its superior performance as an alcohol sensor. Linear sweep voltammetry and chronoamperometry are used to characterize the electrochemical behavior of Au–Pt alloy NPs and their catalytic activity in the electro-oxidation of alcohol.

EXPERIMENTAL DETAILS

Electrodeposition of Pt NPs, Au NPs, and Au–Pt alloy NPs were performed in a three-electrode cell with a potentio/galvanostat electrochemical workstation (CH Instruments 660A). An ITO-glass chip (5 × 15 mm^2, 0.7 mm thick, with a sheet resistivity of 4–8 Ω per square) was used as the working electrode, along with a standard Ag/AgCl reference electrode and a Pt-wire counter electrode. Three deoxygenated aqueous solutions of 1 mM $PtCl_4$, 1 mM $AuCl_3$, and 1 mM $AuCl_3$ plus 1 mM $PtCl_4$, each along with 0.1 M $NaClO_4$ used as the supporting electrolyte, were used to prepare Pt NPs, Au NPs, and Au–Pt alloy NPs, respectively. The respective NPs were deposited onto ITO-glass substrates by amperometry at a constant potential of −0.3 V (*vs.* Ag/AgCl, 3 M KCl) for a selected amount of time. After the deposition, the sample was thoroughly rinsed with filtered deionized water (with a resistivity of 18.2 MΩ cm) and dried under N_2 before further analysis. The surface morphology and crystal structures of the resulting electrodeposits were characterized by field-emission scanning electron microscopy (SEM) using a LEO FESEM 1530 microscope and glancing-incidence X-ray diffraction (GIXRD) using a PANalytical X›pert Pro MRD diffractometer, respectively. The corresponding chemical-state composition was analyzed by X-ray photoelectron spectroscopy (XPS) using a Thermo-VG Scientific ESCALab 250 Microprobe equipped with a monochromatic Al Kα X-ray source (1486.6 eV). High resolution transmission electron microscopy (HRTEM), using a Zeiss Libra 200 MC microscope, has been used to investigate the structure of the AuPt alloy NPs, which were removed from the ITO-glass substrate by carefully scratching the surface. UV-Visible spectra of NPs on ITO-glass substrates were collected using a Lambda 1050 UV/VIS/NIR spectrometer equipped with an integrating sphere accessory.

The as-electrodeposited Pt NPs, Au NPs, and Au–Pt alloy NPs on ITO-glass substrates were used as the working electrodes in the subsequent electro-oxidation studies conducted with the same electrochemical workstation. We employed cyclic voltammetry to detect the oxidation features of these samples in a deoxygenated solution of 0.5 M KOH with and without 0.5 M methanol. The potential was cycled between −0.8 V and 1.0 V (*vs.* Ag/AgCl) with a scan rate of 10 mV s^{-1}. For the alcohol sensing experiments, we employed the Au–Pt alloy NP working electrode and also conducted linear sweep voltammetry between 0 and 0.6 V at the same scan rate (10 mV s^{-1}) in a deoxygenated solution of 0.5 M KOH and 6 mM ethanol. After selecting the appropriate applied potential, amperometric measurement was performed in a 0.5 M KOH electrolyte at 0.25 V (*vs* Ag/AgCl), with constant stirring in order to ensure transport of the analyte to the surface of the working electrode. Once the background current became stabilized, a desired amount of alcohol was injected into the solution and the amperometric response current was recorded.

RESULTS AND DISCUSSION

Fig. 1a shows that spherical Pt NPs were obtained by a 1 s electrodeposition. These NPs are well dispersed on the ITO-substrate and they exhibit a narrow size distribution with an average size of 4 nm (Fig. 1b). A few sparsely distributed clusters of Pt NPs with an average cluster size of 25 nm are also observed. The XPS spectrum of the Pt 4f region (Fig. 1c) reveals the characteristic Pt $4f_{7/2}$ (Pt $4f_{5/2}$) peak at 71.3 eV (74.7 eV) for metallic Pt.[32] The weak feature near 78.5 eV corresponds to In 4p of the ITO-glass substrate. For the 1 s

Au electrodeposition, very uniformly dispersed spherical NPs (Fig. 1d) with a broader size distribution and a larger average size of 18 nm are obtained (Fig. 1e). The corresponding Au 4f spectrum (Fig. 1f) shows the characteristic Au $4f_{7/2}$ (Au $4f_{5/2}$) peak at 84.0 eV (87.7 eV) for metallic Au.[32] For the 1 s electrodeposition carried out in the electrolyte containing both $PtCl_4$ and $AuCl_3$, the morphology of the electrodeposited NPs and their size distribution shown in Fig. 1g and h, respectively, reveal two groups of spherical NPs, one with an average size of 6 nm and the other with an average size of 40 nm. When the electrolyte contains $PtCl_4$ (Fig. 1a and g), electrodeposition appears to follow a progressive growth mode. The Pt $4f_{7/2}$ (Pt $4f_{5/2}$) feature at 70.8 eV (74.1 eV) for these NPs (Fig. 1i) is found to be 0.5 eV lower than that for metallic Pt and can be assigned to the Au–Pt alloy. The formation of alloy is also consistent with the Au $4f_{7/2}$ (Au $4f_{5/2}$) feature at 83.9 eV (87.6 eV), which is 0.1 eV lower than that for metallic Au. The lower binding energies found for both Au $4f_{7/2}$ and Pt $4f_{7/2}$ features of the Au–Pt alloy NPs, relative to the respective metallic Au and Pt features, are in good agreement with the earlier reports, [33–36] which is also supported by theoretical calculation using the local density functional method within the initial state approximation.[33] From the atomic ratio of Au to Pt (2.59) estimated from the XPS peak intensities corrected by the respective relative atomic sensitivity factors, we deduce the alloy composition to be $Au_{72}Pt_{28}$. By increasing the deposition time to 5 s, a nanocrystalline Au–Pt alloy film is obtained and XPS analysis reveals the alloy composition to be $Au_{71}Pt_{29}$ (ESI, Fig. S1†). The longer deposition has resulted in an increase in the metal loading, which is shown to be a better sensor as discussed below.

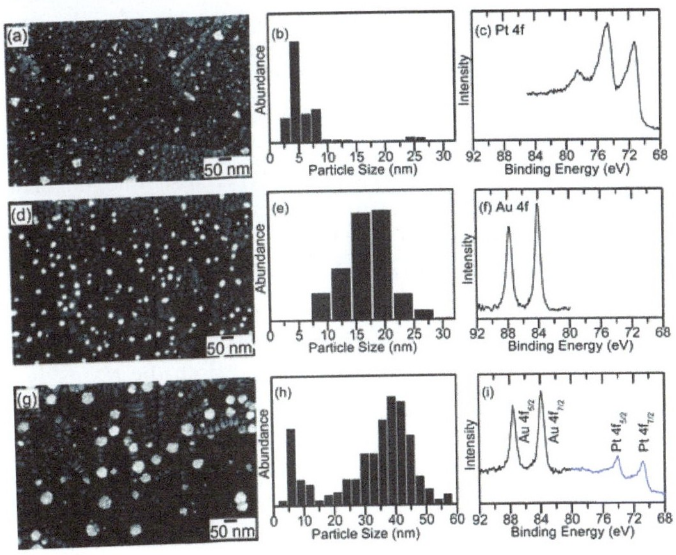

Figure 1: SEM images (left column), particle size distributions (center column), and XPS spectra (right column) of the Au 4f and/or Pt 4f regions for (a–c) Pt NPs, (d–f) Au NPs, and (g–i) Au–Pt alloy NPs, all supported on ITO-glass substrates.

GIXRD patterns of the as-prepared Pt NPs, Au NPs and Au–Pt alloy NPs are shown in Fig. 2. Except for the diffraction features due to the ITO-glass substrates, all the other diffraction peaks for Pt NPs and for Au NPs can be assigned to the fcc phases of metallic Pt (JCPDS 04-0802) and of metallic Au (JCPDS 04-0784), respectively. For the Au–Pt alloy NPs, the more prominent (111) peak at 38.92° is located in between the Au(111) feature at 38.19° and the Pt(111) feature at 39.81°, which is consistent with the alloy formation. The corresponding interplanar spacing for the (111) reflection of the Au–Pt alloy NPs can be calculated to be 2.312 Å according to Bragg›s law. The HRTEM image of a typical Au–Pt alloy NP (shown in Fig. S2, ESI†) indicates that the NP is single-crystalline, with a (111) interplanar spacing of 2.32 Å, consistent with our GIXRD result.

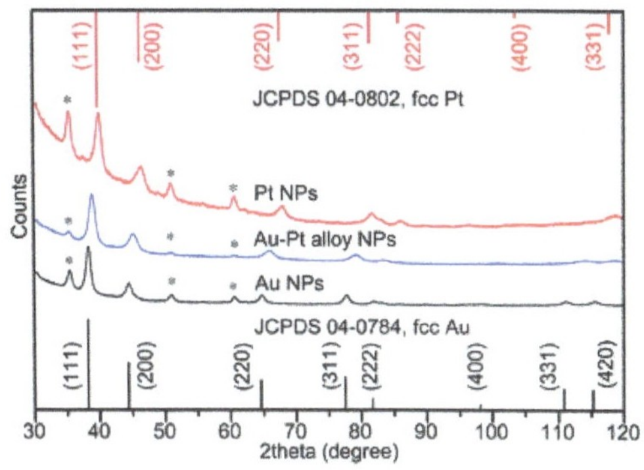

Figure 2: GIXRD patterns of Pt NPs, Au NPs and Au–Pt alloy NPs on ITO-glass substrates and the reference patterns of fcc Au (JCPDS 04-0784) and fcc Pt (JCPDS 04-0802). The peaks corresponding to the ITO-glass substrate are marked with asterisks (*).

The NP-modified ITO-glass electrodes need to be activated by positive potential under either potentiostatic or potentiodynamic conditions. We have activated the electrodes (potentiodynamically) by performing cyclic voltammetric scans of Pt NPs, Au NPs, and Au–Pt alloy NPs on ITO-glass electrodes in a solution of 0.5 M KOH and 0.5 M methanol (Fig. 3). For Pt NPs on the ITO-glass electrode in the methanol-free electrolyte, oxygen evolution occurs at *ca.* 0.6 V, and the reduction peak at *ca.* −0.4 V in the backward scan is due to oxygen reduction reaction on the surface of the Pt NPs (Fig. 3b). In the presence of methanol (Fig. 3a), the additional sharp, intense oxidation peak at −0.19 V is attributed to methanol oxidation. The methanol oxidation current is found to be small for the first scan. Upon subsequent scans, the methanol oxidation potential is shifted to −0.22 V and the current increases till maximum catalytic activity is reached. In a separate experiment, we verify that no methanol electro-oxidation feature can be found for the pristine ITO-glass substrate, *i.e.* without any deposited NPs (ESI, Fig. S3†). This clearly illustrates the catalytic activity of the as-prepared Pt NPs for electro-oxidation of methanol. For Au NPs (Fig. 3d), the voltammogram for the methanol-free electrolyte shows a small, broad oxidation peak at 0.8 V and a reduction peak at *ca.* 0.1 V, which are attributed to, respectively, the formation and reduction of surface gold oxide on the NPs. The more intense reduction peak located near −0.2 V is due to the oxygen reduction reaction. Upon activation, the intensity of the oxidation peak at 0.8 V is increased. In the presence of methanol, the voltammograms are found to be similar to that for the absence of methanol, except for the notably weaker intensities. This similarity indicates that the deposited Au NPs even after activation is not catalytic for the methanol electro-oxidation. As the most inert noble metal, Au possesses very weak chemisorption properties due to the absence of vacancies in its d-bands. Methanol is therefore not chemisorbed onto the Au NPs, making methanol oxidation unlikely. Tang *et al.* also reported the lack of any activity of Au NPs (with an average size greater than 50 nm) toward methanol oxidation in a 0.5 M methanol/0.5 M KOH solution.[37] However, Liu *et al.* found that smaller Au NPs (with an average diameter of 1.7 nm) showed a high catalytic activity while larger Au NPs (with a diameter larger than 100 nm) did not.[38] Other earlier studies also reported that Au NPs with diameters of 2 nm,[39] 2.5 nm,[40] and 4 nm (ref. 41) were catalytically active in methanol electro-oxidation in an alkaline medium. Au NPs with different sizes therefore have different catalytic properties towards methanol electro-oxidation. As Au NPs prepared in the present work are found to be 18 nm in size, the observed lack of activity toward methanol electro-oxidation is therefore consistent with these earlier studies.

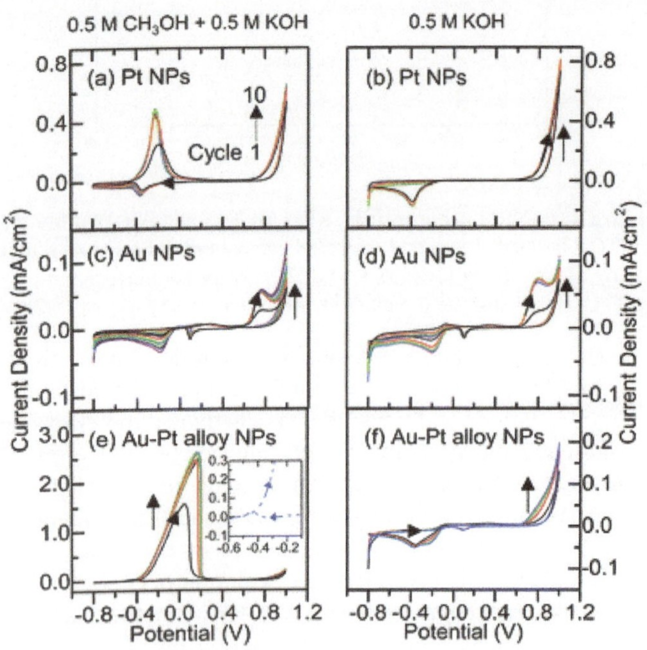

Figure 3: Cyclic voltammograms initiated in the forward (positive-going) scan direction (marked by arrows) at a scan rate of 10 mV s^{-1} for (a) and (b) Pt NPs; (c) and (d) Au NPs and (e) and (f) Au–Pt alloy NPs on ITO-glass substrates in 0.5 M KOH solution with and without 0.5 M methanol. Magnified view of the −0.6 to −0.1 V region for activated Au–Pt alloy NPs in methanol solution is shown in the inset of (e).

In contrast to the Au NPs, no Au oxidation peak is observed in the same solution without methanol for the Au–Pt alloy NPs (Fig. 3f), which suggests a different form of Au in its alloy state. In a solution containing 0.5 M methanol and 0.5 M KOH, a very intense anodic peak at 0.05 V appears during the forward scan in the first cycle, which can be assigned to the methanol oxidation (Fig. 3e). Upon subsequent scans the methanol oxidation potential is shifted to 0.17 V and the current increases until the maximum catalytic activity is reached. A smaller methanol oxidation peak at −0.45 V (shown in the inset of Fig. 3e) was also observed during the backward scan and this phenomenon is not found for metallic Pt NPs, which illustrates the synergistic effect of Au–Pt alloy NPs. The occurrence of methanol electro-oxidation during the backward scan indicates the resistance of the catalysts to CO adsorption,[42] thus validating that the formation of Au–Pt alloy could reduce CO poisoning on the surface of Pt. The intensity of this strong oxidation peak during the forward scan is dependent on the amount of metal NPs present (metal loading) on the ITO-glass substrates. For the supported Pt NPs, quantitative analysis of the Pt content, based on the XPS peak area intensity appropriately normalized by the corresponding relative sensitivity factor, reveals a 6.2% Pt surface concentration. The corresponding surface concentration of the metals (Au + Pt) for the supported Au–Pt alloy NPs is 7.6%. This indicates similar metal loading for the Pt NPs and Au–Pt alloy NPs on the surface of the ITO-glass substrate. Since the maximum current density of methanol electro-oxidation from the Au–Pt alloy NPs (2.63 mA cm^{-2}) is over five times that from Pt NPs (0.47 mA cm^{-2}), despite their similar average sizes, this therefore shows the superior methanol oxidation capacity of Au–Pt alloy NPs ($Au_{72}Pt_{28}$) compared to Pt NPs. This is an important result because achieving a higher activity at a lower Pt composition is of particular interest in fuel cell research due to the reduction in the cost of fuel cell catalysts. Tang *et al.* also investigated the electro-catalytic behavior of methanol in alkaline electrolytes for a series of Au–Pt alloy NPs with different compositions (prepared by electrodeposition onto fluorinated tin oxide substrates).[37] No information about metal loading is, however, provided to allow proper comparison of their catalytic behavior. Mott *et al.* reported that Au–Pt alloy NPs with a substantial Au composition (65–85%) have similar mass activities to metallic Pt NPs for methanol electro-oxidation in the alkaline electrolyte,[43,44] in contrast to the result in the present work. This difference could be due to post-annealing (at 400–600 °C) employed after their NP preparation, which is generally expected to have a significant effect on the core and the surface crystallinity of the nanocrystals, and thus their catalytic behavior.

To investigate the general catalytic properties of a bimetallic Au–Pt system, we also prepared, in separate experiments, bimetallic NPs by sequential electrodeposition, *i.e.*, Au NPs were first electrodeposited at −0.3 V for 1 s in 1 mM $AuCl_3$ and 0.1 M $NaClO_4$ solutions (step A), which was then followed by deposition of Pt at −0.3 V for 1 s in a 1 mM $PtCl_4$ and 0.1 M $NaClO_4$ solution (step B). The as-prepared sample, denoted as Pt/Au NPs, corresponds to distorted NPs (20 nm in size) uniformly distributed on the ITO-glass substrate (Fig. 4a). Alternatively, we obtained Pt NPs first (step B) followed by deposition of Au (step A). Labelled as Au/Pt NPs, this sample corresponds to spherical NPs with average sizes of 4 nm and 15 nm in the size distribution (Fig. 4b). Both the respective XPS spectra (Fig. 4d) for Pt/Au NPs and Au/Pt NPs show a Pt $4f_{7/2}$ ($4f_{5/2}$) feature at 71.2 eV (74.6 eV) and a Au $4f_{7/2}$ ($4f_{5/2}$) feature at 84.0 eV (87.7 eV), corresponding to metallic Pt and Au, respectively. The absence of any binding energy shift, in contrast to those found for the Au–Pt alloy NPs with Pt $4f_{7/2}$ ($4f_{5/2}$) at 70.8 eV (74.1 eV) and Au $4f_{7/2}$ ($4f_{5/2}$) at 83.9 eV (87.6 eV) (Fig. 1i), therefore indicates that alloying does not occur for the two sequential electrodepositions. GIXRD patterns of Pt/Au NPs and Au/Pt NPs also show two distinct sets of diffraction patterns corresponding to Au and Pt, which confirms the bimetallic nature and the absence of alloy formation (ESI, Fig. S4†). The electro-oxidation of methanol for Pt/Au NPs and Au/Pt NPs are also studied and compared with the Au–Pt alloy NPs (and shown in Fig. 4e). For Pt/Au NPs, the cyclic voltammogram shows a methanol electro-oxidation peak at −0.18 V with a current density of 0.79 mA cm^{-2}. The absence of an Au oxidation peak confirms that the subsequent Pt deposition occurred on top of the individual

Au NPs, *i.e.* the formation of Au-core Pt-shell NPs. Yang *et al.* reported a much better electro-catalytic activity for Au–Pt core–shell NPs than Pt NPs in methanol electro-oxidation reaction.[45] Their observation is consistent with our results for Pt/Au NPs (*i.e.* with an Au core and a Pt shell), which also show a higher electro-catalytic activity (0.79 mA cm^{-2}) than the Pt NPs. For Au/Pt NPs, a methanol oxidation peak at −0.18 V with a current density of 0.66 mA cm^{-2} and an Au oxidation peak at 0.7 V are observed. This indicates that the subsequent deposition of Au did not completely cover the Pt core. Both bimetallic Pt/Au and Au/Pt systems exhibit much lower current densities than the Au–Pt alloy NPs, indicating lower electro-catalytic activities than the Au–Pt alloy NPs. The corresponding surface concentrations of the metals (Au + Pt) for the bimetallic Pt/Au and Au/Pt systems estimated from their corresponding XPS intensities are 11.7% and 25.6%, respectively, which are larger than those for the supported Au–Pt alloy NPs (7.6%). The activity of Au–Pt alloy NPs toward methanol electro-oxidation should therefore be higher, even given due consideration to the metal loading effect.

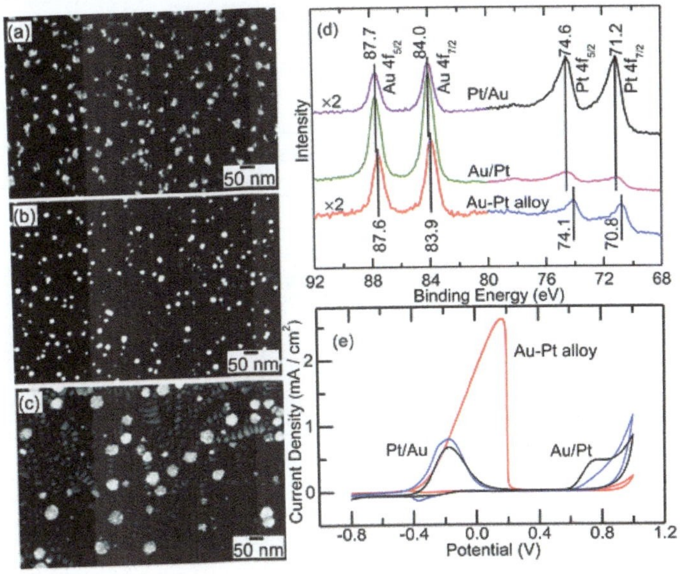

Figure 4: SEM images of (a) Pt/Au NPs, (b) Au/Pt NPs, and (c) Au–Pt alloy NPs supported on ITO-glass substrates, and (d) their corresponding XPS spectra of the Au 4f and Pt 4f regions and (e) cyclic voltammograms in a solution of 0.5 M methanol and 0.5 M KOH. The scan directions are marked by arrows.

We have also measured the UV-Visible spectra of the Au–Pt alloy NPs, Pt/Au NPs, and Au/Pt NPs, along with those for Pt NPs and Au NPs, all on ITO substrates (ESI, Fig. S5a†). Other than the spectrum for Au NPs on the ITO substrate with an additional weak broad peak at 549 nm (due to the Au surface plasmon), the spectra for all the NPs appear to be very similar to one another and to the blank ITO substrate.

Due to the formation of the Au–Pt alloy, the d-band center of Pt is modified and the resulting changes in the electronic structures could cause the observed high catalytic activity for electro-oxidation of methanol. As Au has been shown to work as a catalyst for CO oxidation,[8,9] the presence of the Au component reduces the buildup of CO and surface oxides on Pt, therefore diminishing the effect of catalyst poisoning on the Au–Pt surface. This synergistic effect is not present on the surface of just Pt NPs but is clearly operative for Au–Pt alloy NPs, which exhibit a very high catalytic activity towards electro-oxidation of methanol.

The aforementioned studies on methanol can be extended to other alcohols. Xie *et al.* have studied electro-oxidation of various alcohols on Pt electrodes and found that ethanol could be more easily electro-oxidized than methanol.[46] Indeed, ethanol electro-oxidation is important to such promising applications as direct ethanol fuel cells[47] and electrochemical sensing. For the ethanol electro-oxidation on Au–Pt alloy NPs, our results (shown in Fig. S6, ESI†) indicate that a higher activity can also be obtained for Au–Pt alloy NPs than Au NPs, Pt NPs, and Au–Pt bimetallic NPs prepared by sequential electrodeposition.

To further demonstrate the superior catalytic properties of these Au–Pt alloy NPs, we characterize the efficacy of these Au–Pt alloy NPs as the catalyst material in an alcohol sensor for monitoring ethanol concentration in a solution. The inset in Fig. 5a shows a linear sweep voltammetric curve obtained for Au–Pt alloy NPs in a solution of 6 mM ethanol and 0.5 M KOH. The intense oxidation peak that appeared at 0.25 V corresponds to ethanol oxidation and can therefore be used as a measure of the catalytic activity of Au–Pt alloy NPs to ethanol electro-oxidation. In particular, the oxidation current obtained at 0.25 V can be used for determination of the ethanol concentration. The sensor sensitivity is evaluated by performing amperometric experiments at 0.25 V using a Au–Pt alloy NP-modified ITO-glass electrode. The Au–Pt alloy NPs are electrodeposited at −0.3 V for 1 s on the ITO-glass substrate. Eleven replicate injections (100 µL each) of a pre-selected concentration of aqueous ethanol solution are administered to the 0.5 M KOH supporting electrolyte (20 mL) sequentially. Five ethanol concentrations of 0.5, 1.0, 2.0, 4.0 and 6.0 mM are used. The successive addition of a 100 µL ethanol aliquot induces a small rise in the oxidation current, and the resulting current–time profiles are plotted in Fig. 5a. When the ethanol concentration is increased, a higher jump in the response current is observed. A clear increase in the response current is found even with the addition of alcohol with a concentration as small as 0.5 mM. For successive addition of alcohol with concentrations of 0.5, 1 and 2 mM to the supporting electrolyte, a steady-state response current can be reached after just 2 s, which indicates an excellent electro-catalytic response time. For successive addition of 4 mM and 6 mM ethanol, however, a steady-state current cannot be reached. This is because stirring of the solution used in our experiment is not sufficiently powerful to quickly remove the concentration gradient and to expose uniform concentration of alcohol evenly to the electrode surface. The relationship between the response current density and ethanol concentration (Fig. 5b) is found to be linear for the ethanol concentration range from 1 mM to 11 mM, and the corresponding slope gives a detection sensitivity of 0.043 mA cm^{-2} mM^{-1}. The detection limit is found to be 0.1 mM by performing the cyclic voltammetric scan in the alcohol solution including 0.5 M KOH, which gives a signal three times the background noise. We also repeat the same sensor evaluation over a period of several months. The Au–Pt alloy NPs electrode retains 95% of its initial activity even after 3 months, which indicates excellent long-term stability of the sensor at room temperature. For the Au–Pt nanoalloy film deposited on ITO-glass at −0.3 V for 5 s, the linear sweep voltammetric scan in a solution of 5 mM ethanol and 0.5 M KOH shows that the maximum oxidation current density is three times that obtained with a 1 s deposition time (ESI, Fig. S7†). This means increasing the metal loading should produce a better sensor. In comparison with other alcohol sensors based on metal or metal oxide electrodes, the present alcohol sensor therefore exhibits an excellent sensitivity, a shorter response time and a considerably more extended long-term stability (Table 1). Furthermore, this Au–Pt nanoalloy sensor can be operated at a low potential of 0.25 V, which will reduce the interference effect arising from oxidizing other species in the analyte solution.

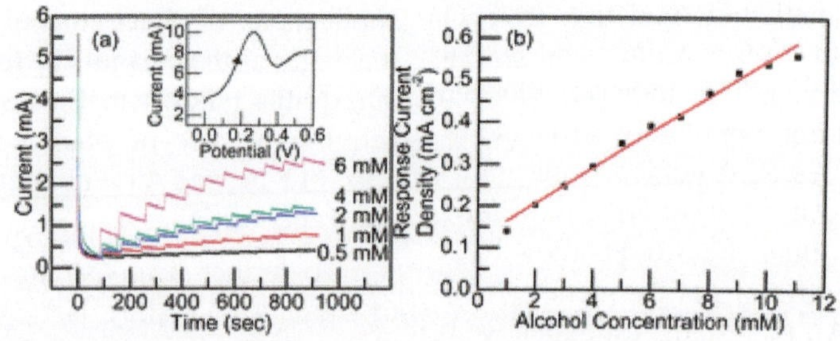

Figure 5: (a) Typical amperometric response of the Au–Pt alloy NP-modified ITO-glass electrode to successive additions of a 100 μL aliquot of aqueous ethanol solution with preselected concentration in an actively stirred electrolyte solution (20 mL 0.5 M KOH). Five ethanol concentrations (0.5, 1.0, 2.0, 4.0 and 6.0 mM) are used. The current is measured at an applied potential of 0.25 V. The inset shows a linear sweep voltammetric scan of the Au–Pt alloy NP-modified ITO-glass electrode in a solution of 6 mM ethanol and 0.5 M KOH. The scan direction is marked by the arrow. (b) The corresponding response current density as a function of ethanol concentration. The Au–Pt alloy NPs are prepared by electrodeposition at −0.3 V for 1 s in a solution of 1 mM $PtCl_4$, 1 mM $AuCl_3$and 0.1 M $NaClO_4$ on an ITO-glass substrate.

Table 1: Performance comparison of reported alcohol sensors based on metal or metal oxide electrodes

Working electrode	Applied potential (V)	Concentration range (mM)	Detection limit (mM)	Sensitivity (mA $cm^{-2}mM^{-1}$)	Response time (s)	Lifetime (days)	Ref.
a **PtRuNi ternary alloy nanoparticles electrodeposited on a glassy carbon electrode pre-coated with multi-walled carbon nanotubes (MWNTs) and an ionic liquid (1-octyl-3-methylimidazolium hexafluorophosphate) gel film.**							
Pt electrodeposited with mixed Co–Ni oxide film	0.27	1.5–79	0.032		<25	3	28
Au modified by co-electrodeposition of Pt and WO_3	0.4	34–170		0.28	60	37	20
RuO_2-modified Ni	0.5	2.17–21.74		0.23	13		22
Ni–Pt alloy film electrodeposited on Au/Al_2O_3	0.52	1.09–6.52	1.09	0.060	13	63	25
Electrodeposited Ni–B thin film	0.56	2.17–13.04		0.28	9		26
CuO-modified Cu	0.6	1.7–850	0.85				24
Ni/Pt/Ti sputtered on Al_2O_3	0.61	0.083–0.50		3.08	27		23
PtRuNi NPs electrodeposited on glassy carbon coated with MWNT–ionic liquid gel[a]	0.65	0.05–38	0.05	0.021	3		27
Au–Pt alloy NPs on ITO-glass	0.25	1–11	0.1	0.043	2	90	This work

In summary, we have prepared Pt NPs, Au NPs, and Au–Pt alloy NPs by simple electrodeposition onto ITO-glass substrates and we used them as model catalysts to study the electro-oxidation of methanol in a 0.5 M KOH supporting electrolyte. While Au NPs exhibit no catalytic property for the electro-oxidation of methanol, the Au–Pt alloy NPs show a significantly higher catalytic activity than pristine Pt NPs, and bimetallic Au/Pt and Pt/Au NP systems. We further demonstrate that these Au–Pt alloy NPs can be used as a high-performance alcohol sensor for analyzing alcohol concentration in a solution. This nanoalloy sensor shows excellent linearity in the alcohol concentration range of 1–11 mM, a low detection limit of 0.1 mM, and a good detection sensitivity of 0.043 mA cm^{-2} mM^{-1}. This sensor can be operated at a low potential of 0.25 V, which is desirable for reducing the interference effect arising from other oxidizing species in the analyte solution. The present sensor also has a very fast response time of 2 s and an extremely long lifetime of up to 3 months at room temperature. Improved control of the size and composition of the Au–Pt alloy NPs are necessary to further optimize the sensor performance.

ACKNOWLEDGEMENTS

This work was supported by the Natural Sciences and Engineering Research Council of Canada.

REFERENCES

1. T. Iwasita, Methanol and CO electrooxidation, *Handbook of Fuel Cells*, ed. W. Vielstich*et al.*, Wiley, Chichester, UK, 2003, vol. 2, p. 603 .
2. J. Müller, G. Frank, K. Colbow and D. Wilkinson, Transport/Kinetic Limitations and Efficiency Losses, *Handbook of Fuel Cells*, ed. W. Vielstich*et al.*, Wiley, Chichester, UK, 2003, vol. 4, p. 847 .
3. S. Wasmus and A. Küver, Methanol Oxidation and Direct Methanol Fuel Cells: a Selective Review, *J. Electroanal. Chem.*, 1999, 461, 14–31.
4. *Electrocatalysis, Frontiers in Electrochemistry*, ed. J. Lipkowski and P. N. Ross, Jr, Wiley-VCH, New York, 1998 .
5. E. Herrero, W. Chrzanowski and A. Wieckowski, Dual Path Mechanism in Methanol Electrooxidation on a Platinum Electrode, *J. Phys. Chem.*, 1995, 99, 10423–10424.
6. A. Hamnett, Mechanism and Electrocatalysis in the Direct Methanol Fuel Cell, *Catal. Today*, 1997, 38, 445–457.
7. E. Antolini, Formation of Carbon-supported PtM Alloys for Low Temperature Fuel Cells: a Review, *Mater. Chem. Phys.*, 2003, 78, 563–573.
8. M. Haruta, When Gold Is Not Noble: Catalysis by Nanoparticles, *Chem. Rec.*, 2003, 3, 75–87 .
9. M. Haruta, Size- and Support-dependency in the Catalysis of Gold, *Catal. Today*, 1997, 36, 153–166.
10. W. C. Ye, H. H. Kou, Q. Z. Liu, J. F. Yan, F. Zhou and C. M. Wang, Electrochemical Deposition of Au–Pt Alloy Particles with Cauliflower-like Microstructures for Electrocatalytic Methanol Oxidation, *Int. J. Hydrogen Energy*, 2012, 37, 4088–4097 .
11. F. Terzi, C. Zanardi, S. Daolio, M. Fabrizio and R. Seeber, Au–Pt Nanoparticle Systems in Methanol and Carbon Monoxide Electroxidation, *Electrochim. Acta*, 2011, 56, 3673–3678 .
12. M. Khalid, N. Wasio, T. Chase and K. Bandyopadhyay, In Situ Generation of Two-Dimensional Au–Pt Core–Shell Nanoparticle Assemblies, *Nanoscale Res. Lett.*, 2010, 5, 61–67 .
13. X. Y. Lang, H. Guo, L. Y. Chen, A. Kudo, J. S. Yu, W. Zhang, A. Inoue and M. W. Chen, Novel Nanoporous Au – Pd Alloy with High Catalytic Activity and Excellent Electrochemical Stability,*J. Phys. Chem. C*, 2010, 114, 2600–2603 .
14. R. Imbeault, D. Reyter, S. Garbarino, L. Roué and D. Guay, Metastable Au_xRh_{100-x} Thin Films Prepared by Pulsed Laser Deposition for the Electrooxidation of Methanol, *J. Phys. Chem. C*, 2012, 116, 5262–5269 .
15. X. F. Xing, D. Q. Han, Y. F. Wu, Y. Guan, N. Bao and X. H. Xu, Fabrication and Electrochemical Property of Hierarchically Porous Au–Cu Films, *Mater. Lett.*, 2012, 71, 108–110 .
16. A. D. Modestov, M. R. Tarasevich, A. Y. Leykin and V. Y. Filimonov, MEA for Alkaline Direct Ethanol Fuel Cell with Alkali Doped PBI Membrane and Non-platinum Electrodes, *J. Power Sources*, 2009, 188, 502–506 .
17. N. Fujiwara, Z. Siroma, S. Yamazaki, T. Ioroi, H. Senoh and K. Yasuda, Direct Ethanol Fuel Cells Using an Anion Exchange Membrane, *J. Power Sources*, 2008, 185, 621–626 .
18. J. R. Varcoe and R. C. T. Slade, Prospects for Alkaline Anion-Exchange Membranes in Low Temperature Fuel Cells, *Fuel Cells*, 2005, 5, 187–200.
19. A. R. Vijayakumar, E. Csöregi, A. Heller and L. Gorton, Alcohol Biosensors Based on Coupled Oxidase-peroxidase Systems, *Anal. Chim. Acta*, 1996, 327, 223–234.
20. M. Boujtita, J. P. Hart and R. Pittson, Development of a Disposable Ethanol Biosensor Based on a Chemically Modified Screen-printed Electrode Coated with Alcohol Oxidase for the Analysis of Beer, *Biosens. Bioelectron.*, 2000, 15, 257–263.
21. M. J. L. tanon, A. J. M. Ordieres and P. T. Blanco, Amperometric Detection of Ethanol with Poly(*o*-phenylenediamine)-modified Enzyme Electrodes, *Biosens. Bioelectron.*, 1997, 12, 511–520.
22. J. Katrlík, J. Švorc, M. Streďanský and S. Miertuš, Composite Alcohol Biosensors Based on Solid Binding Matrix, *Biosens. Bioelectron.*, 1998, 13, 181–191 .
23. Y. Chen, K. Y. Chen and A. C. C. Tseung, An Electrochemical Alcohol Sensor Based on a Co-electrodeposited Pt|WO_3 Electrode, *J. Electroanal. Chem.*, 1999, 471, 151–155.
24. L. Ciaccheri, E. E. Samano-Baca, M. T. Russo, H. Ottevaere, H. Thienpont and A. G. Mignani, Diffuse-light absorption spectroscopy for beer classification and prediction of alcoholic content,*Optical Sensing and Detection II*, ed. F. Berghmans, A. G. Mignani and P. De Moor, Proc. of SPIE, 2012, vol. 8439, p. 84392E .
25. Y. C. Weng and T. C. Chou, Ethanol Sensors by Using RuO_2-modified Ni Electrode, *Sens. Actuators, B*, 2002, 85, 246–255.
26. Y. C. Weng, J. F. Rick and T. C. Chou, A Sputtered Thin Film of Nanostructured Ni/Pt/Ti on Al_2O_3 Substrate for Ethanol Sensing, *Biosens. Bioelectron.*, 2004, 20, 41–51 .
27. T. R. L. C. Paixão and M. Bertotti, Development of a Breath Alcohol Sensor Using a Copper Electrode in an Alkaline Medium, *J. Electroanal. Chem.*, 2004, 571, 101–109.
28. J. J. Huang, W. S. Hwang, Y. C. Weng and T. C. Chou, Determination of Alcohols Using a Ni–Pt Alloy Amperometric Sensor, *Thin Solid Films*, 2008, 516, 5210–5216 .
29. Y. C. Weng and T. C. Chou, Electrodeposited Nickel–Boron Thin-Film Ethanol Sensor Sensors and Displays: Principles, Materials, and Processing, *J. Electrochem. Soc.*, 2006, 153, H127–H132 .

30. F. Xiao, F. Q. Zhao, J. J. Zeng and B. Z. Zeng, Novel Alcohol Sensor Based on PtRuNi Ternary Alloy Nanoparticles–multi-walled Carbon Nanotube–ionic Liquid Composite Coated Electrode,*Electrochem. Commun.*, 2009, 11, 1550–1553 .

31. E. T. Hayes, B. K. Bellingham, H. B. Mark and A. Galal, An Amperometric Aqueous Ethanol Sensor Based on the Electrocatalytic Oxidation at a Cobalt–nickel Oxide Electrode, *Electrochim. Acta*, 1996, 41, 337–344 .

32. J. F. Moulder, W. F. Stickle, P. E. Sobol and K. D. Bomben*Handbook of X-ray Photoelectron Spectroscopy*, ed. J. Chastain, Perkin-Elmer Corp., 1992, 2nd edn .

33. L. Y. Zhao, N. F. Heinig and K. T. Leung, Formation of Au–Pt alloy Nanoparticles on a Si Substrate by Simple Dip-coating at Room Temperature, *Langmuir*, 2013, 29, 927–931 .

34. Z. Bastl and S. Pick, Angle resolved X-ray Photoelectron Spectroscopy Study of Au deposited on Pt and Re Surfaces, *Surf. Sci.*, 2004, 566–568, 832–836 .

35. X. B. Ge, X. L. Yan, R. Y. Wang, F. Tian and Y. Ding, Tailoring the Structure and Property of Pt-Decorated Nanoporous Gold by Thermal Annealing, *J. Phys. Chem. C*, 2009, 113, 7379–7384 .

36. C. X. Xu, R. Y. Wang, M. W. Chen, Y. Zhang and Y. Ding, Dealloying to Nanoporous Au–Pt Alloys and Their Structure Sensitive Electrocatalytic Properties, *Phys. Chem. Chem. Phys.*, 2010, 12, 239–246 .

37. W. Tang, S. Jayaraman, T. F. Jaramillo, G. D. Stucky and E. W. McFarland, Electrocatalytic Activity of Gold – Platinum Clusters for Low Temperature Fuel Cell Applications, *J. Phys. Chem. C*, 2009, 113, 5014–5024 .

38. Z. H. Wei and C. J. Liu, Synthesis of Monodisperse Gold Nanoparticles in Ionic Liquid by Applying Room Temperature Plasma, *Mater. Lett.*, 2011, 65, 353–355 .

39. J. Luo, M. M. Maye, N. N. Kariuki, L. Y. Wang, P. Njoki, Y. Lin, M. Schadt, H. R. Naslund and C. J. Zhong, Electrocatalytic Oxidation of Methanol: Carbon-supported Gold–platinum Nanoparticle Catalysts Prepared by Two-phase Protocol, *Catal. Today*, 2005, 99, 291–297 .

40. F. Terzi, C. Zanardi, S. Daolio, M. Fabrizio and R. Seeber, Au–Pt Nanoparticle Systems in Methanol and Carbon Monoxide Electroxidation, *Electrochim. Acta*, 2011, 56, 3673–3678 .

41. J. Hernández, J. Solla-Gullón, E. Herrero, A. Aldaz and J. M. Feliu, Methanol Oxidation on Gold Nanoparticles in Alkaline Media: Unusual Electrocatalytic Activity, *Electrochim. Acta*, 2006, 52, 1662–1669.

42. M. Yin, Y. J. Huang, L. Liang, J. H. Liao, C. P. Liu and W. Xing, Inhibiting CO Formation by Adjusting Surface Composition in PtAu Alloys for Methanol Electrooxidation, *Chem. Commun.*, 2011, 47, 8172–8174 .

43. D. Mott, J. Luo, P. N. Njoki, Y. Lin, L. Y. Wang and C. J. Zhong, Synergistic Activity of Gold–platinum Alloy Nanoparticle Catalysts, *Catal. Today*, 2007, 122, 378–385 .

44. J. Luo, P. N. Njoki, Y. Lin, D. Mott, L. Y. Wang and C. J. Zhong, Characterization of Carbon-supported AuPt Nanoparticles for Electrocatalytic Methanol Oxidation Reaction, *Langmuir*, 2006,22, 2892–2898 .

45. L. Yang, J. H. Chen, X. X. Zhong, K. Z. Cui, Y. Xu and Y. F. Kuang, Au@Pt Nanoparticles Prepared by One-phase Protocol and Their Electrocatalytic Properties for Methanol Oxidation,*Colloids Surf., A*, 2007, 295, 21–26 .

46. S. W. Xie, S. Chen, Z. Q. Liu and C. W. Xu, Comparison of Alcohol Electrooxidation on Pt and Pd Electrodes in Alkaline Medium, *Int. J. Electrochem. Sci.*, 2011, 6, 882–888 .

47. D. Y. Wang, J. P. Liu, Z. Y. Wu, J. H. Zhang, Y. Z. Su, Z. L. Liu and C. W. Xu, Electrooxidation of Methanol, Ethanol and 1-Propanol on Pd Electrode in Alkaline Medium, *Int. J. Electrochem. Sci.*, 2009, 4, 1672–1678 .

CHAPTER 5

CORROSION PERFORMANCE OF FE-CR-NI ALLOYS IN ARTIFICIAL SALIVA AND MOUTHWASH SOLUTION

J. Porcayo-Calderon[1,2] M. Casales-Diaz[2] V. M. Salinas-Bravo[3] and L. Martinez-Gomez[2,4]

[1]CIICAp, Universidad Autónoma del Estado de Morelos, Avenida Universidad 1001, 62209 Cuernavaca, MOR, Mexico

[2]Instituto de Ciencias Físicas, Universidad Nacional Autónoma de México, Avenida Universidad s/n, 62210 Cuernavaca, MOR, Mexico

[3]Instituto de Investigaciones Eléctricas, Avenida Reforma 113, Colonia Palmira, 62490 Cuernavaca, MOR, Mexico

[4]Corrosion y Protección (CyP), Buffon 46, 11590 México City, DF, Mexico

ABSTRACT

Several austenitic stainless steels suitable for high temperature applications because of their high corrosion resistance and excellent mechanical properties were investigated as biomaterials for dental use. The steels were evaluated by electrochemical techniques such as potentiodynamic polarization curves, cyclic polarization curves, measurements of open circuit potential, and linear polarization resistance. The performance of steels was evaluated in two types of environments: artificial saliva and mouthwash solution at 37°C for 48 hours. In order to compare the behavior of steels, titanium a material commonly used in dental applications was also tested in the same conditions. Results show that tested steels have characteristics that may make them attractive as biomaterials for dental applications. Contents of Cr, Ni, and other minor alloying elements (Mo, Ti, and Nb) determine the performance of stainless steels. In artificial saliva steels show a corrosion rate of the same order of magnitude as titanium and in mouthwash have greater corrosion resistance than titanium.

INTRODUCTION

Biomaterials are materials used in the manufacture of devices that interact with biological systems and coexist for a long period of service with minimal failure. They are widely used for repairing or replacement of components of the tissue-skeletal system such as bones, joints, and teeth. Suitable materials for implants are those endured by the body and they should be able to withstand cyclic loading in the aggressive environment of the body. Metals and alloys have been widely used as biomaterials providing the mechanical strength and corrosion resistance required. Biomaterials are generally made of one of these three types of materials: austenitic stainless steel, chromium-cobalt alloys, and titanium and its alloys.

Oral environment is suitable for the formation of corrosion products. The mouth is wet and it is continually exposed to temperature fluctuations. Foods and beverages cause important changes in environmental chemistry. The food and fluid ingested have wide ranges of pH. Acids are released during the decomposition of foodstuffs. Food remains often adhere tenaciously to metal restorations providing a localized condition favoring an accelerated reaction between oral media and metal or alloy [1]. One of the fundamental conditions of any metal used in the mouth is that they should not produce corrosion products that are harmful to the person. Corrosion is a chemical or electrochemical process by which metal

Citation: J. Porcayo-Calderon, M. Casales-Diaz, V. M. Salinas-Bravo, and L. Martinez-Gomez, "Corrosion Performance of Fe-Cr-Ni Alloys in Artificial Saliva and Mouthwash Solution," Bioinorganic Chemistry and Applications, vol. 2015, Article ID 930802, 14 pages, 2015. doi:10.1155/2015/930802.

is attacked by ordinary substances, resulting in partial or complete dissolution, deterioration, and weakness. Corrosion causes catastrophic disintegration of metals. Some metallic elements are completely safe but some can form dangerous ions or even toxic compounds. Additionally, degradation of an alloy should be limited in order to guarantee its life [2]. Various sulfides such as hydrogen sulfide or ammonium compounds corrode silver, mercury, and similar metals found in an amalgam. Water, oxygen, and chloride ions are in saliva and contribute to corrosion attack. The presence of various acids such as phosphoric, acetic, and lactic acid can lead to corrosion having the appropriate concentration and pH in the environment. Specific ions can play an important role in the corrosion of certain alloys; for example, oxygen and chlorine ions are involved in the corrosion of amalgams in the tooth interface and within the alloy. Therefore, it is important to consider the use of materials with high corrosion resistance in order to withstand the environments of the oral cavity.

Titanium and titanium alloys are widely used for many biomedical applications due to their low density, excellent biocompatibility, corrosion resistance, and mechanical properties. In the passive state, these alloys are not completely stable and under certain circumstances the passive film breaks down producing localized corrosion. This fault calls to modify the material surface to increase the corrosion and wear resistance without affecting their mechanical properties [3]. Austenitic stainless steel, specially type AISI 316-L, is the most widely used steel for implants, including orthodontic treatments because of its low cost, ease to fabricate, and welding if compared to Co-Cr and Ti alloys. Corrosion resistance is the most important characteristic of austenitic stainless steel due to formation of a passive film of Cr_2O_3 on its surface [4, 5]. 316-L type stainless steel has an acceptable corrosion resistance, biocompatibility, strength, and fatigue resistance that makes it a desirable material as biomaterial [6].

In recent years, there has been a rapid increase in the utilization of fluorinated prophylactic gels and mouthwash in the odontology field. The negative influence of fluoride on the corrosion of titanium and its alloys has been reported earlier [7–9]. Similarly, fluoride ions are aggressive ions that degrade the protective oxide layer formed on a stainless steels surface [10]. On the other hand, many mouthwashes contain both fluorine ions and hydrogen peroxide, and therefore the corrosion performance of biomaterials can be affected by the combined presence of both compounds. Antiseptic rinses are commonly used with antiplaque, anticaries, antibacterial, and whitening properties. Biomaterials and various surrounding tissues may get exposed to antiseptic rinses for several minutes every day, depending upon usage habits.

The objective of the present study was to evaluate the corrosion behavior of different austenitic stainless steels widely utilized in corrosive environments at high temperatures for dental applications in artificial saliva and mouthwash solution containing NaF and H_2O_2. The study was conducted using conventional electrochemical techniques.

EXPERIMENTAL PROCEDURE

Materials

Stainless steels are divided into five families, each with its own microstructure, alloying elements, and specific mechanical properties. These stainless steels are classified as ferritic, austenitic, duplex, martensitic, and precipitation hardened. In particular, austenitic steels are the largest family of stainless steels; they cannot be hardened by heat treatment, are nonmagnetic, and have a face-centered cubic structure; in addition they have excellent ductility and good weldability. Six different austenitic stainless steels widely utilized in corrosive environments at high temperatures were evaluated. Table 1 shows the materials evaluated and the limits of chemical composition as established by ASTM A213/A213M [11]. From Table 1 it can be seen that the contents of C, Mn, P, S, and Si are the same, and then it is possible that the stainless steels performance is based on the content of either Cr and Ni or other minor elements (Mo, Nb, and Ti). For comparison purposes commercial pure titanium (CP-Ti Grade 1) was also evaluated.

Table 1: Chemical composition of austenitic stainless steels evaluated as reported by ASTM A213/A213M [11]

Grade	UNS designation	Chemical composition									
		C	Mn	P	S	Si	Cr	Ni	Mo	Nb	Ti
TP304H	S30409	0.04–0.10	2.00	0.045	0.030	1.00	18.0–20.0	8.0–11.0			
TP309H	S30909	0.04–0.10	2.00	0.045	0.030	1.00	22.0–24.0	12.0–15.0			
TP310H	S31009	0.04–0.10	2.00	0.045	0.030	1.00	24.0–26.0	19.0–22.0			
TP316H	S31609	0.04–0.10	2.00	0.045	0.030	1.00	16.0–18.0	11.0–14.0	2.00–3.00		
TP321H	S32109	0.04–0.10	2.00	0.045	0.030	1.00	17.0–19.0	9.0–12.0			4(C + N) – 0.7
TP347H	S34709	0.04–0.10	2.00	0.045	0.030	1.00	17.0–19.0	9.0–12.0		8C – 1.10	

Sample Preparation

Test specimens were cut in squares of 5.0 × 5.0 × 3.0 mm using a diamond tipped blade. For electrical connection, specimens were spot-welded to a Ni20Cr wire and then mounted in thermosetting resin. Sample surfaces were metallographically polished; the grinding process began with 120-grit sandpaper down to 1200-grit sandpaper. After polishing, the materials were washed with distilled water, and after with ethanol in an ultrasonic bath for 10 minutes. In this condition the materials were used as working electrode (WE).

Corrosive Electrolytes

The most important fluid in the oral environment is the natural saliva. However, the unstable nature of natural saliva makes it unsuitable for in vitro studies and instead artificial saliva is used. In this work the corrosion performance of materials was evaluated in artificial saliva proposed by Duffó and Quezada-Castillo [12]. Table2 reports the chemical composition of artificial saliva. Similarly, materials were evaluated in mouthwash widely used for hygiene purposes. Chemical composition of mouthwash reported by the manufacturer indicates it contains water, alcohol 8.8%, sorbitol, hydrogen peroxide 2%, sodium fluoride 0.022% (100 ppm de fluorine), PEG hydrogenated castor oil, sodium saccharin 0.119%, sucralose, phosphoric acid, disodium phosphate, poloxamer 407, flavor, and menthol. Test conditions were pH 6.5 and 37°C temperature, which is equivalent to human body temperature.

Table 2: Chemical composition of artificial saliva (pH = 6.5) [12]

Compound	Content [g/L]
NaCl	0.600
KCl	0.720
$CaCl_2 \cdot 2H_2O$	0.220
KH_2PO_4	0.680
$Na_2HPO_4 \cdot 12H_2O$	0.856
KSCN	0.060
$NaHCO_3$	1.500
Citric acid	0.030

Electrochemical Tests

Electrochemical tests were carried out using an ACM instruments zero-resistance ammeter (ZRA) coupled to a personal computer. A typical three-electrode arrangement was used where the reference electrode (RE) was a saturated calomel electrode (SCE, 0.242 V versus SHE) and the counter electrode (CE) was a high-purity graphite rod (Pt cannot be used because it decomposes H_2O_2). All potentials described in the text are relative to the SCE, unless stated differently. The volume of solution in the cell was 100 mL, following the recommendation V/A specified in ASTM G31-90 [13]. To determine the corrosion resistance of materials, corrosion current density (I_{corr}) was calculated by potentiodynamic polarization test from −400 mV to 1500 mV with respect to open circuit potential (E_{corr}) according to standard ASTM G5-94 (reapproved 2004) [14] and ASTM G-3-89 (reapproved 2004) [15]. These tests allow determining the corrosion potential and the corrosion rate by extrapolation of the Tafel slopes from the curves obtained. For all materials, potentiodynamic polarization tests were performed at a sweep rate of 1 mV/s. Before starting the test, samples were left to stabilize for 20 minutes. The current density values (I_{corr}) and anodic and cathodic slopes and corrosion potential (E_{corr}) were calculated using the extrapolation Tafel method considering an extrapolation potential of ±250 mV around the value of the corrosion potential (E_{corr}).

A cyclic polarization technique was used to evaluate pitting corrosion resistance of all materials tested according to the ASTM F2129-01 [16] and ASTM G3-89 [15]. The working electrode was stabilized at the free corrosion potential for an hour, and then a potential sweep (0.166 mV/s) was made in the anodic direction until the potential reaches a predetermined value. At this point, the scan direction is reversed until the hysteresis loop closes. To assess the ability of the different materials to form a protective oxide scale on their surfaces upon immersion in the simulated body fluid environment, free corrosion potential as a function of time of the working electrodes E_{corr} was measured versus a SCE for 50 hours. Linear polarization curves were obtained by polarizing the specimens from −30 to 30 mV with respect to the free corrosion potential value, E_{corr}, at a scanning rate of 10 mV/min; measurements were made for 50 hours. After testing, corroded specimens were analyzed in a DSM 960 Carl Zeiss scanning electronic microscope (SEM).

RESULTS AND DISCUSSION

Potentiodynamic Polarization Curves

Figure 1 shows potentiodynamic polarization curves of materials evaluated in artificial saliva at 37°C. It can be observed that titanium has the nobler E_{corr} and stainless steels the most active E_{corr}, but, at potentials above E_{corr}, stainless steels reach a pseudopassivation zone at lower potentials. Above its corrosion potential, titanium shows a passive region from 250 mV to 1200 mV; this is due to the growth of a protective TiO_2 layer on its surface. Stainless steels show virtually the same comportment. Small variations observed in stainless steels may correspond to the variations in their chemical composition. Generally it can be seen that above its corrosion potential all steels show a tendency to slow down corrosion by increasing the applied potential. This trend seems to correspond to an attempt to develop a passive region because of the formation of a Cr_2O_3 protective film on its surface. Extending this apparent passive region behavior is different for each one of steels. At end of the passive region an abrupt increase is observed in the current density that may correspond to the onset of pitting on the surface of steels. The active region of polarization curves is due to the electrochemical reaction of the sample in the electrolyte medium and formation of a porous or defective oxide layer, whereas the passive region is associated with the formation of one or more protective oxide films.

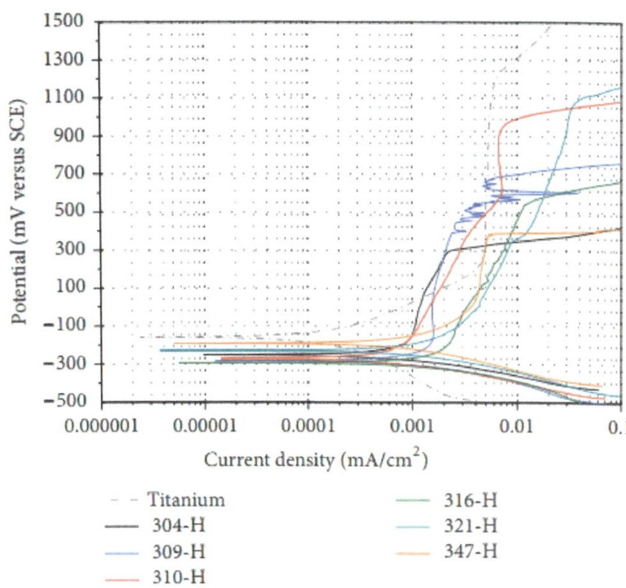

Figure 1: Polarization curves of materials in artificial saliva at 37°C, dE/dt = 1.0 mV/s.

Particularly, steel 309-H clearly shows the breakdown and repassivation of the protective layer before reaching its pitting potential. Other steels do not show this tendency. This could be because a potential sweep of 1 mV/s was applied; possibly slower sweep rates (0.167 mV/s) could define the passive region and more precise values of pitting potential and other parameters of interest to biomaterials. It is to be noted that steels 316-H, 321-H, and 347-H, with addition of Mo, Ti, and Nd, are materials whose apparent passivation areas are located at higher current densities compared to those containing only Cr and Ni (304-H, 309-H, and 310-H). In general, stainless steels showed the highest corrosion rates, and the material with greater corrosion resistance was titanium. Table 3 shows electrochemical parameters obtained from polarization curves shown in Figure 1.

Table 3: Electrochemical parameters of materials in artificial saliva at 37°C

Material	E_{corr} (mV)	B_a (mV/Dec)	B_c (mV/Dec)	I_{corr} (mA/cm^2)
304-H	−252	2248	104	0.00090
309-H	−280	8572	138	0.0015
310-H	−268	965	111	0.00075
316-H	−292	1000	127	0.00176
321-H	−227	493	137	0.00148
347-H	−189	245	129	0.00083
Titanium	−159	260	286	0.00019

Figure 2 shows potentiodynamic polarization curves of materials evaluated in mouthwash at 37°C. Titanium shows the most active E_{corr} but quickly develops a passive zone at more anodic potentials. Stainless steels having similar E_{corr} values show a more homogeneous behavior than that observed in the presence of artificial saliva (Figure 1). Moreover, their anodic branches are almost the same with a passive region between 850 and 1050 mV. Small variations observed correspond to differences in chemical composition of each material. Stainless steels show lower corrosion rates and titanium the highest. Table 4 shows the electrochemical parameters obtained from the polarization curves of Figure 2.

Table 4: Electrochemical parameters of the materials evaluated in mouthwash at 37°C

Material	E_{corr} (mV)	B_a (mV/Dec)	B_c (mV/Dec)	I_{corr} (mA/cm^2)
304-H	317	581	133	0.001713
309-H	330	356	129	0.00111
310-H	333	306	121	0.00127
316-H	302	642	140	0.0023
321-H	304	333	138	0.0027
347-H	310	371	126	0.0021
Titanium	−96	356	176	0.0116

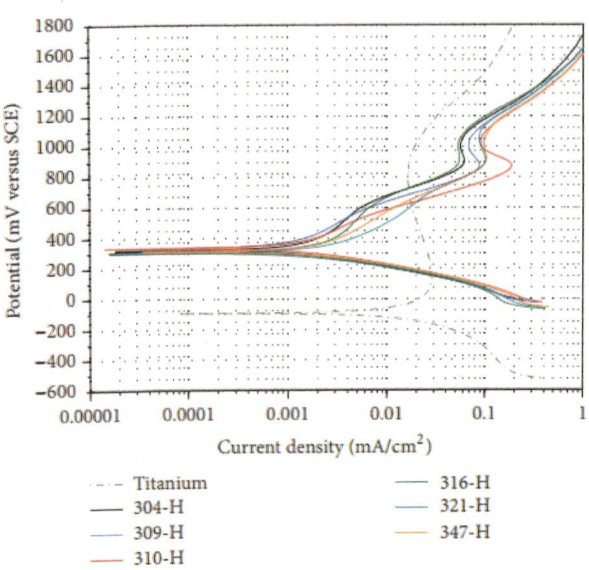

Figure 2: Polarization curves of materials in mouthwash at 37°C, dE/dt = 1.0 mV/s.

Although stainless steels have the same tendency, its particular position on the graph and the particular values of their electrochemical parameters depend on its chemical composition. Thus in Figure 3 the effect of Cr and Ni content on E_{corr} values is shown and in Figure 4 the effect of Cr and Ni content on $_{corr}$ values is also shown. The Cr and Ni values plotted in both figures correspond to the minimum content of these elements in stainless steels according to ASTM A213/A213M-04 (Table 1).

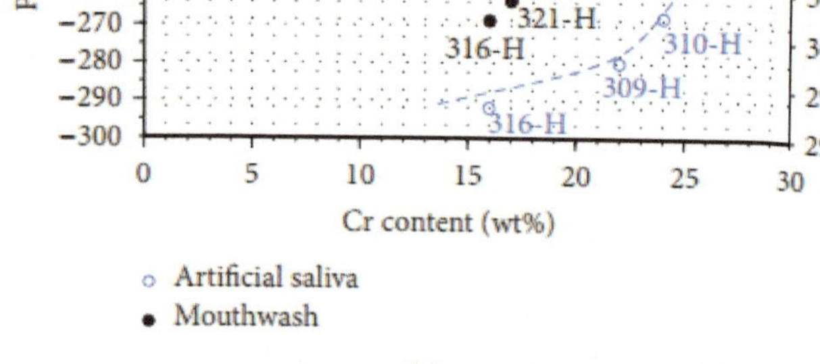

(a)

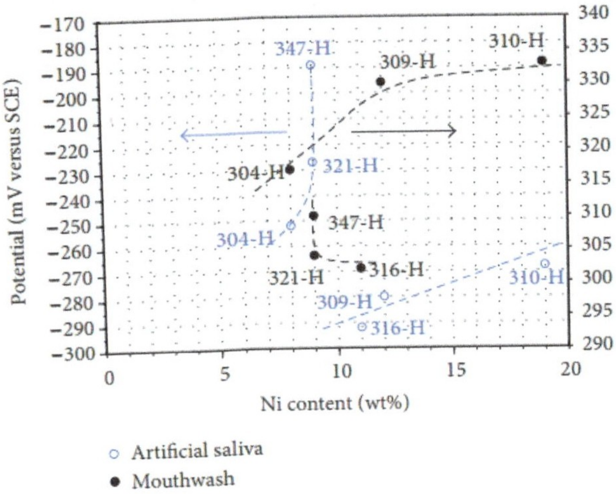

(b)

Figure 3: Effect of Cr and Ni content of stainless steels on E_{corr} values.

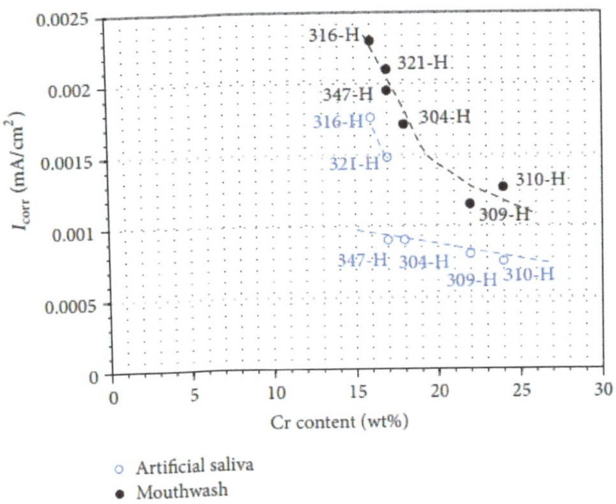

(a)

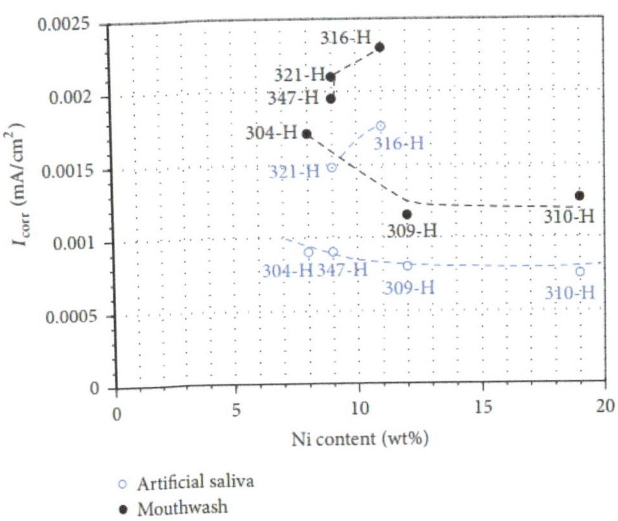

(b)

Figure 4: Effect of the Cr and Ni content of stainless steels on I_{corr} values.

In artificial saliva (Figure 3(a)) steels 316-H, 309-H, and 310-H show that Cr contents above 16%, E_{corr} of steels, become nobler. However, steels 304-H, 321-H, and 347-H are out of this trend, showing a nobler E_{corr} having a Cr content between 17 and 18%. This discrepancy is associated with the Ni content of these steels (Figure 3(b)). It is observed that those steels with less Ni content exhibit a nobler E_{corr}. Nevertheless, the difference in Ni content between steels 304-H, 321-H, and 347-H does not justify such a significant difference among their E_{corr} values. This can be explained because steels 321-H and 347-H contain Ti and Nb, respectively, which are added as stabilizing elements. Nb seems to have a favorable effect on E_{corr} of steel 347-H, followed by Ti in steel 321-H. This is consistent with the behavior of Ti according to Figure 1 where it was noted that Ti has a nobler E_{corr} than that of stainless steels.

Figure 4(a) shows a direct relationship between Cr content of stainless steels and corrosion rate. Steels having higher Cr content have small values of I_{corr}. It is also seen that, in the particular case of 316-H and 321-H steels, they have higher I_{corr} values than steels having similar content of Cr and Ni like 347-H and 304-H steels. In these cases, these steels contain additional elements such as Mo for the case of 316-H and Ti in case of 321-H as shown in Table 1. This indicates that the incorporation of these elements brings a detriment in the corrosion resistance of these steels in artificial saliva environments. Furthermore, it is noted that even though 347-H contains Nb, it does not significantly affect the steel corrosion resistance. Regarding the Ni content a similar trend is observed as that described for the Cr content. Therefore it can be established that the corrosion potential (E_{corr}) of stainless steels is affected by the Cr and Ni contents. Cr has a beneficial influence on the passive behavior of steels with lower Ni content. The addition of stabilizing elements such as Nb and Ti has a greater effect than the increase in the concentration of Cr. On the other hand, the corrosion rate of steels (I_{corr}) is directly related to the content of Cr and Ni in the alloy. Higher content of both elements increases the corrosion resistance; however, addition of Mo and Ti negatively affects the corrosion resistance.

In mouthwash (Figure 3(a)) a clear effect of Cr content of stainless steels on E_{corr} values is observed. Higher content of Cr in steels results in nobler corrosion potentials. However, observing the Ni content (Figure 3(b)), only in steels 304-H, 309-H, and 310-H, the tendency of greater E_{corr} with higher Ni content is maintained. In case of steels 347-H, 321-H, and 316-H, this trend is not followed. Examining the chemical composition for the late steels it can be seen that they contain Nb, Ti, and Mo, respectively. Then the addition of Mo creates a greater impact in decreasing E_{corr} of 316-H steel (high Cr and Ni content compared to 321-H and 347-H), followed by Ti in 321-H (same content of Cr and Ni as in 347-H). This is consistent with the results of the polarization curves where it was observed that Ti showed the more active behavior (Figure 2).

Regarding their corrosion rate, Figure 4(a) shows a direct relationship between the Cr content of the stainless steels and corrosion rate. Steels with higher content of Cr have lower I_{corr} values. Concerning the effect of the Ni, Figure 4(b) shows that in case of steels 304-H, 309-H, and 310-H there is a link (not so marked as in the case of Cr) between higher Ni content and lower I_{corr}. In case of steels

316-H, 321-H, and 347-H the trend is not followed. This can be due to the content of Mo, Ti, and Nb affecting adversely the corrosion resistance. Therefore an increase in the Cr and Ni content favors a nobler behavior (>E_{corr}), the effect being greater with increasing Cr; addition of Mo, Nb, and Ti causes more active behavior. The worst performance observed by titanium is because the presence of fluorine ions and H_2O_2 has a negative influence on the corrosion rate. It is known that the presence of fluoride ions causes surface films to become less stable and eventually a breakdown of the film may occur [17]. On the other hand, the presence of H_2O_2 causes the dissolution of titanium, formation of a titanium oxide on the metal surface, and Ti catalyses decomposition of peroxide [18].

Cyclic Polarization Curves

Because any material or alloy proposed for use as a biomaterial must exhibit excellent resistance to localized corrosion, cyclic polarization measurements were performed in order to determine the tendency of the alloys to undergo pitting or crevice corrosion when submerged in the electrolyte solutions. A plot of potential versus log of the current density reveals the tendency of the material to undergo localized attack. Differences between the forward and reverse scans result in hysteresis loops. If the current density during the reverse scan is higher than that for the forward scan at any given potential (positive hysteresis loop), the area under the loop indicates the amount of localized corrosion incurred by the material. Conversely, if the current density during the reverse scan is less than that for the forward scan at any given potential (negative hysteresis loop) it indicates high resistance to localized corrosion. Two important potentials are also used to characterize the hysteresis loop, the pitting nucleation, or breakdown potential (E_{np}) and the protection potential (E_{pp}). E_{np} is defined as the potential at which the pitting or crevice corrosion or both will initiate and propagate and in the polarization curve appears as an abrupt increase in the anodic current density at the point where the passive zone ends. Pitting is characterized by a rapid increase in current with a very small change in potential. Above this potential, pits initiate and propagate. An increase in the resistance to pitting corrosion is associated with a shift of E_{np} to nobler values. E_{pp} is defined as the potential at which the forward and the reverse scans intersect. At this potential, localized attack stops and the current decreases significantly. This value is always lower than the E_{np} value [19]. Figures 5 and 6 show the cyclic polarization curves for materials tested in artificial saliva and mouthwash solution, respectively. In general by increasing anodic polarization, the formation and repair of an anodic film are observed. Nevertheless, the irreversible breakdown of the passive film is clearly identified by the abrupt increase of the current density. The differences in composition of the alloys affected the polarization and the passivation behaviors of tested materials.

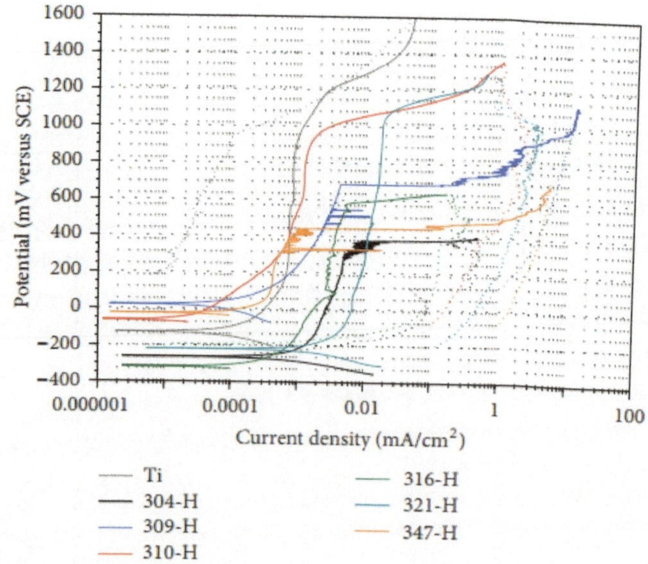

Figure 5: Cyclic polarization curves recorded for materials in artificial saliva at 37°C, dE/dt= 0.167 mV/s.

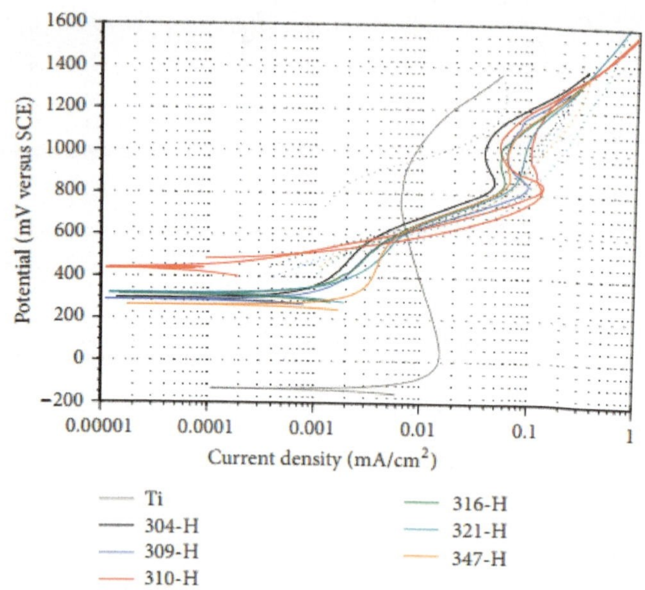

Figure 6: Cyclic polarization curves recorded for materials in mouthwash solution at 37°C,dE/dt = 0.167 mV/s.

Figure 5 shows that in artificial saliva the anodic branch of the cyclic polarization curves of all materials exhibits an active-passive transition. It is known that by increasing anodic polarization several processes may take place on the materials' surface, namely, the formation and repair of an anodic film, water oxidation leading to electrogenerated oxygen bubbles, gas evolution from the oxidation reaction of some entities (e.g., Cl_2production through $2Cl^- \rightarrow Cl_2 + 2e^-$), and the onset of the breakdown at critical potential [20]. Ti, 310-H, and 321-H steels showed the largest passive zone compared to that shown by the rest of the stainless steels. 304-H, 309-H, 316-H, and 347-H steels show a passive zone with minor disturbances. As reported by other studies, these disturbances are created because of the breakdown and regeneration of the passive layer formed [21–23]. In the passive-pitting zone (E_{np}), Ti, 310-H, and 321-H steels showed a slow increase in current with a very small change in potential. However, the rest of the stainless steels showed an abrupt increase; this indicates both a greater growth and propagation of pitting. Ti has higher resistance to localized corrosion because the current density during the reverse scan is lower than that for the forward scan. However, stainless steels showed a positive hysteresis loop and the reverse scan continues at potentials below the corrosion potential. Only 304-H and 316-H steels showed an intersection with the forward scan but at a potential near to its corrosion potential. These features indicate a high susceptibility to pitting of stainless steels in artificial saliva.

Figure 6 shows that Ti has more active behavior and 321-H the noblest in mouthwash solution. Ti showed the largest passive zone, with E_{pp} close to its E_{np} (around 930 mV). On the other hand, stainless steels showed a similar behavior; a small passive zone was observed at potentials away from its corrosion potential. The $_{np}$ values showed slight variations with values around 1000–1070 mV. Similar behavior had the E_{pp} values around 850–900 mV. In all cases, a slow increase in current with a small change in potential was observed after E_{np} zone. This indicates that, in mouthwash solution, both the growth and propagation of pitting are less than those observed in artificial saliva.

In order to determine the effect of chemical composition of stainless steels with respect to the pitting potential, Figure 7 shows the relation of the chromium and nickel content on the E_{np} value. The lowest Cr and Ni content as reported by ASTM A213/A213M were plotted. It is evident that the effect is more noticeable in artificial saliva than that observed in mouthwash solution.

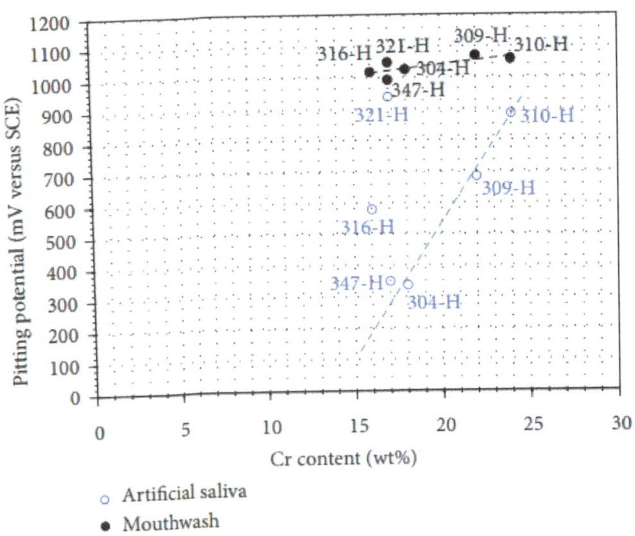

(a)

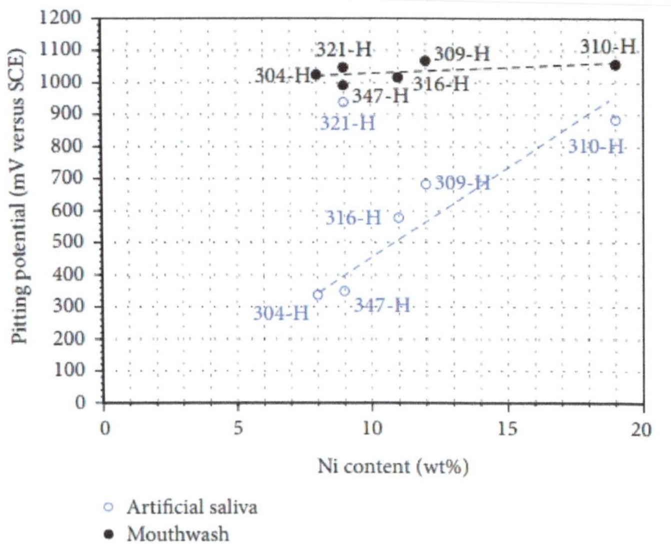

(b)

Figure 7: Effect of the Cr and Ni content of stainless steels on pitting potential.

In artificial saliva it is observed that the pitting potential increases by increasing the chromium content. However, 347-H, 316-H, and 321-H have a lower chromium content and do not show the same behavior. This is because Nb, Mo, and Ti content affect the pitting potential. The effect is more marked with the addition of Ti; that is, the addition of Ti to steel 321-H gives it greater resistance to pitting corrosion than that provided by the addition of Mo to 316-H. On the other hand, it is observed that the nickel content also affects the pitting potential. In this case, the Nb and Mo content have not a significant effect; however, the titanium content shows a marked effect. In mouthwash solution, it is observed that by increasing chromium and nickel content the pitting potential shows a slight increase, and in this case, Nb, Mo, and Ti content have not a significant influence.

Free Corrosion Potential Curves

Establishing the chemical interaction of a biomaterial with the body fluid environment is essential in order to understand its stability in the human body. E_{corr} gives a relative thermodynamic rating of a metal or alloy in a given environment. A rise of E_{corr} in the positive direction indicates the formation of a passive film, and a steady E_{corr} indicates that the film remains intact and protective, but a drop of E_{corr} in the negative direction indicates breaks in the film, dissolution of the film, or no film formation [24]. Figures 8 and 9 show the variation of E_{corr} values as a function of time for materials tested in artificial saliva and mouthwash solution, respectively.

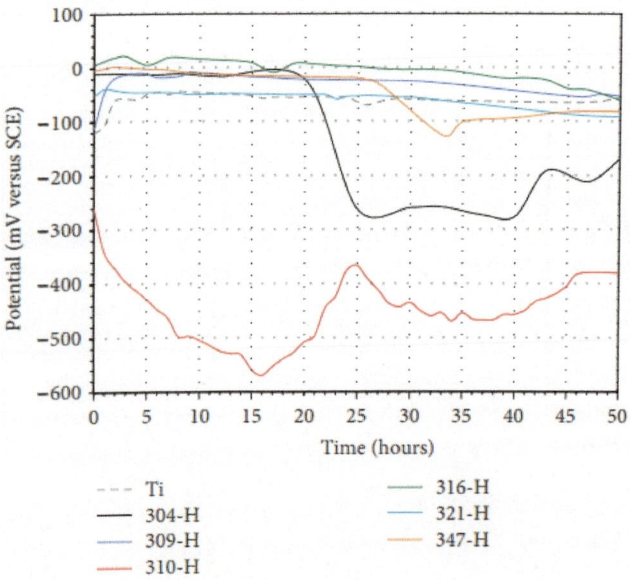

Figure 8: E_{corr} values after testing time for materials in artificial saliva at 37°C.

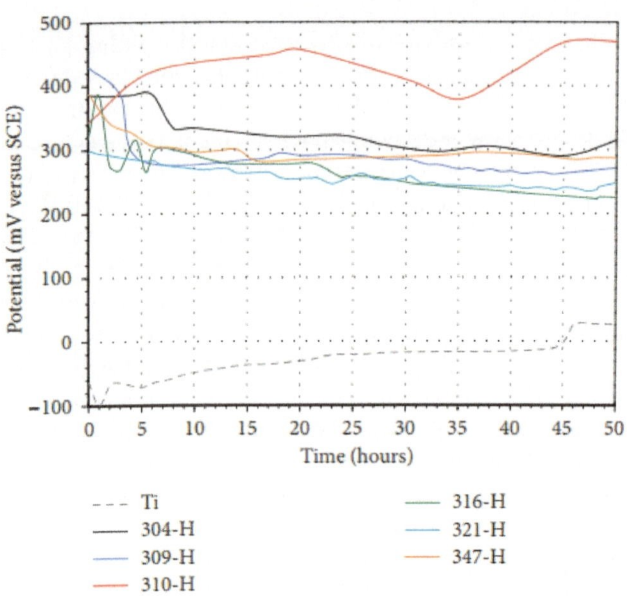

Figure 9: E_{corr} values after testing time for materials in mouthwash solution at 37°C.

According to the concepts stated above, it can be seen that Ti shows a passive behavior in artificial saliva. In the first 2 hours, Ti shows a significant increase in its E_{corr} values and then a slow increase until the end of the test. This passive tendency is due to the formation of a protective layer on its surface. Stainless steels showed an active behavior; in particular 309-H, 316-H, and 321-H showed a more stable behavior compared to 304-H, 310-H, and 347-H. 304-H and 347-H steels showed an abrupt decrease in its E_{corr} values at 20 and 27 hours of test, respectively, and then an increase in its E_{corr} values. 310-H steel showed continuous increments and decrements in E_{corr} values throughout the test and only after 45 hours showed a stable trend. The abrupt drop in E_{corr} values may be due to the susceptibility of these materials to pitting corrosion in chloride-rich environments, but it is evident that, in all cases, materials exhibit an immediate tendency to self-healing of its passive layer. The passivity of the alloys depend on the rates of several processes; such as formation of oxide at the metal/oxide interface, ionic transport across oxide, and dissolution of oxide at the oxide/electrolyte interface [25]. Steel 321-H showed E_{corr} values very similar to Ti. 316-H steel presented nobler E_{corr} values; this is expected because the presence of Mo in the steel provides greater resistance to corrosion in chloride-rich environments.

In mouthwash, Ti showed a more active behavior than that observed in artificial saliva. After the first hour Ti shows a slight decrease in its E_{corr} values and then a steady increase until the end of the test. Titanium is prone to become covered with an oxide film in noncomplex solutions of pH ranging from 3.5 to 7.5. However, the initial behavior could be explained because the fluoride ions caused a localized corrosion and partial dissolution of the protective film. The subsequent passive trend observed could be explained because of the combined effect of the fluoride ion with the organic active ingredients of the mouthwash solution as it has been suggested [25]. Stainless steels showed a passive-active behavior in mouthwash solution. An active trend was observed in the first 45 hours and subsequently a passive trend. 310-H showed the noblest behavior, with a continuous increment in its E_{corr} values until 20 hours of immersion and then a decrease until 35 hours, and finally a continuous increase. This material contains the highest content of chromium and nickel of the stainless steels evaluated. Regarding stainless steels, it is interesting to note that the trend observed in the E_{corr} values in mouthwash is completely opposite to the trend observed in artificial saliva (Figure 8). Steels that showed nobler behavior in artificial saliva in mouthwash showed the most active behavior and vice versa. The addition of Mo and Ti seems to be not beneficial for obtaining nobler E_{corr} values in stainless steels.

Linear Polarization Curves

Once polarization resistance is determined, calculation of I_{corr} requires knowledge of the Tafel constants, and these constants are determined from experimental polarization curves. However, in the absence of the Tafel constants values, an approximation is sometimes used, and the expected error in the calculated value of I_{corr} should be less than a factor of two [26]. When the polarization resistance values are in the same order of magnitude, it is necessary to use more precise values of these constants in order to perform a reliable analysis of the results [27]. Progress of I_{corr} obtained by linear polarization measurements over time for materials evaluated in artificial saliva and mouthwash solution is showed in Figures 10 and 11. Data was obtained from the polarization resistance measurements using Stern-Geary expression:

$$i_{corr} = \frac{b_a b_c}{2.303 R_p (b_a + b_c)}, \quad (1)$$

where the b_a and b_c values were those reported in Tables 3 and 4.

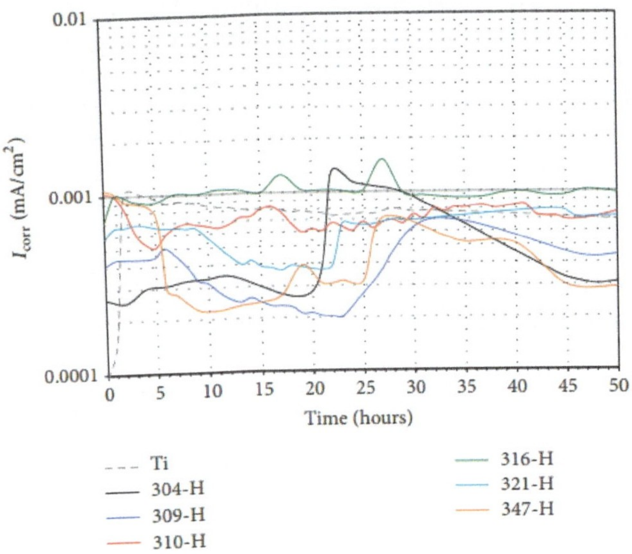

Figure 10: Change of I_{corr} values with time of materials in artificial saliva at 37°C.

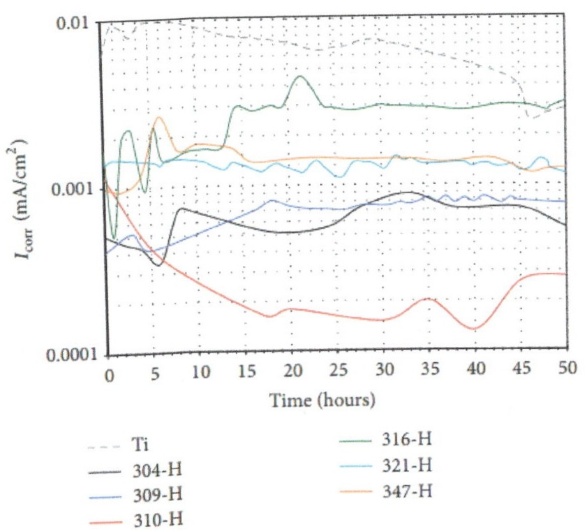

Figure 11: Change of I_{corr} values with time of materials in mouthwash solution at 37°C.

It is observed that, in artificial saliva in the first 35 hours, materials showed constant fluctuations in their I_{corr} values, and then a steady behavior was displayed (Figure 10). This performance is consistent with measurements, where an active-passive condition was observed. Ti showed an abrupt increase in its I_{corr} values in the first 2 hours and then stable decrements until the end of the test. This can be explained because of dissolution of the passive layer and then self-healing of the titanium surface. The excellence of titanium as biomaterial is due to the passive layer that forms on its surface which protects it from electrochemical attack in the human body. This layer is composed of amorphous titanium oxides from Ti_2O to TiO_2, having variable thickness between 0.5 and 10 nm according to the treatment, surface finish, corrosive medium, and so forth [28–30]. It is reported that the passivation layer is formed naturally after a few milliseconds of contact with a medium with oxygen [31]. Stainless steels showed that, in long time immersion test, the lowest I_{corr} values were exhibited by 304-H, 309-H, and 347-H, and the higher I_{corr} values were showed by 316-H, 310-H and 321-H, having similar I_{corr} values to Ti at the end of the test.

Figure 11 shows I_{corr} values for materials in mouthwash solution. All materials showed a more stable behavior than that exhibited in artificial saliva. Ti showed an increase in its I_{corr} values in the first hour and then a steady decline until the end of the test. According to I_{corr} values, the corrosion resistance of the stainless steels is in the following order: 310-H > 304-H > 309-H > 321-H ≈ 347-H > 316-H. This means that steels with Ti, Nd, or Mo had the highest corrosion rates.

Figure 12 shows the average I_{corr} values corresponding to the last 5 hours of testing for each material according to its minimum content of Cr and Ni as reported in ASTM A213/A213M.

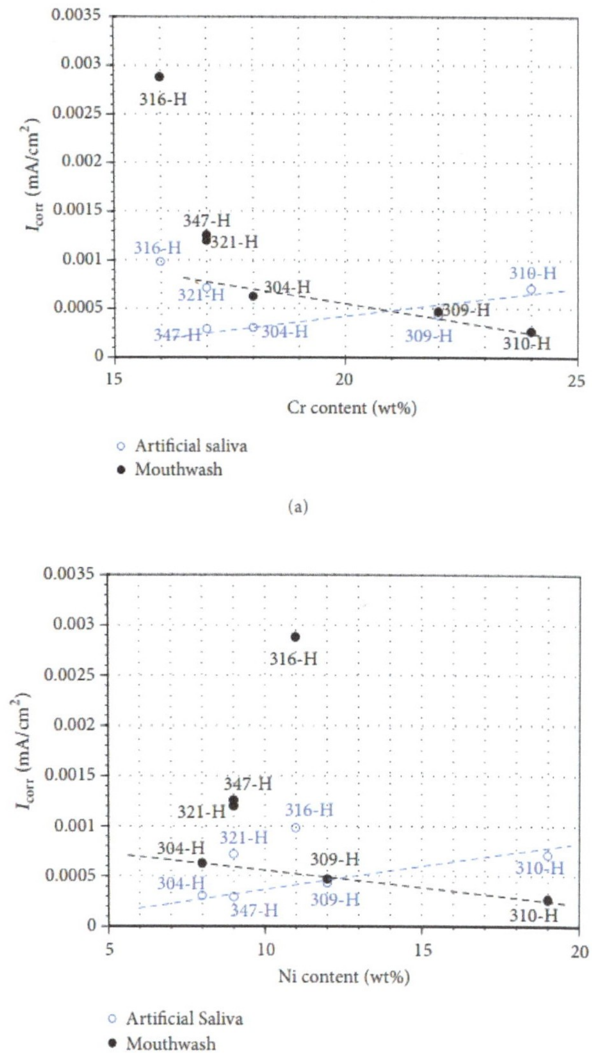

Figure 12: Effect of Cr and Ni content of stainless steels on I_{corr} values for long time tests.

In artificial saliva it is observed that the corrosion resistance of stainless steels increases with the contents of chromium and nickel. Steels 316-H and 321-H do not follow this trend; both exhibit higher corrosion rates than expected according to trends estimated by their Cr and Ni contents. Both steels contain additional elements such as Mo (316-H) and Ti (321-H). This suggests that the presence of these elements causes a detriment to corrosion resistance in environments of artificial saliva. On the other hand, 347-H steel also contains an additional element (Nb) and its presence did not affect the corrosion resistance. Similar results were obtained in potentiodynamic polarization tests. However in this case, the effect of the Cr and Ni content was opposite to that observed in the potentiodynamic polarization tests.

Although electrochemical methods are extremely useful in studying corrosion processes, they alone do not provide enough information to elucidate the mechanism of the system under study. Therefore the use of complementary techniques has been suggested, that is, scanning electron microscopy (SEM) and auger electron spectroscopy (AES), among others in order to clarify both the morphology of the attack and the chemical composition and distribution of the elements present. Combination of these methods provides the information for understanding the reactions occurring on the surface [32]. Consequently, in this study the surface of the working electrodes was studied using scanning electron microscopy (SEM) in order to clarify the morphology of attack. Figure 13 shows the superficial aspect of titanium and 310-H and 321-H stainless steel after the corrosion test in artificial saliva.

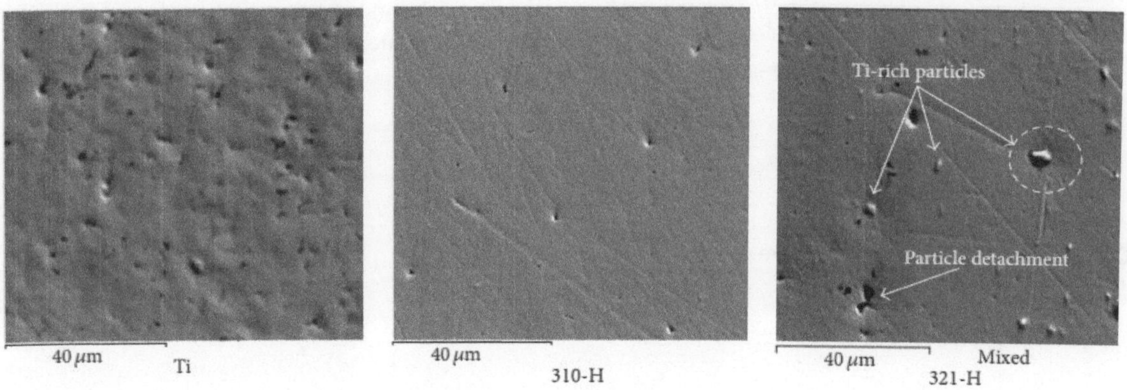

Figure 13: Superficial aspect for Ti, 310-H, and 321-H stainless steel after corrosion test in artificial saliva.

Surface appearance shown by Ti after the corrosion test shows a smooth surface with the presence of slight pitting. Pit location is associated with lines resulting from surface preparation. This is consistent with studies indicating that any imperfections in the material surface can modify its resistance to pitting [33]. In particular a fine surface finish increases the corrosion resistance of the material and moves its pitting potential to more anodic potential values. The presence of both surface roughness and residual stress changes its electrochemical response [34, 35]. Furthermore, even though the roughness of the implant promotes better fixation, this can be detrimental to denture elements as it facilitates biofilm buildup that causes inflammation of the tissues [36].

Surface appearance of 310-H steel shows a striped surface which was made by the surface preparation process with some pits associated with these lines. Similar features were shown by stainless steels 304-H, 309-H, and 347-H. In the latter steel the presence of Nd-rich precipitates was observed but they did not affect its corrosion resistance. Small pits can be due to MnS particles in the alloy. It has been demonstrated that MnS inclusions perform as initiating sites for pitting [37]. The surface appearance shown by the 321-H stainless steel indicates the presence of additional corrosion processes to those observed in 310-H stainless steel. In this case, the presence of Ti-rich precipitates acting as cathode sites supports corrosion of the Fe-Cr-Ni matrix. In the area bounded by the dashed circle shown in Figure 13, the depression around the Ti-rich particles caused by the corrosion process created by the anode-cathode pair can be seen. In some areas this corrosion process caused the detachment of Ti-rich particles. Similar appearance was observed with the Mo-rich precipitates in 316-H stainless steel. The fact that the presence of the Nd-rich precipitates observed in the 347-H stainless steel did not increase its corrosion rate was possibly because the phases that form the anode-cathode pair have similar corrosion potential.

Austenitic stainless steels are essentially alloys of Fe-Cr-Ni, and the chromium addition has been known to improve their corrosion resistance. Nickel is the basic substitutional element used for austenite stabilization at all temperatures. Often, alloying elements, either interstitial such as C or N or substitutional such as Mo, Mn, Ti, and Nb, are used to obtain the required properties. However, the addition of alloying elements often results in the formation of carbides, nitrides, and intermetallic compounds. Molybdenum is a ferrite stabilizer and it also facilitates carbide precipitation. Stabilizing elements such as Nd and Ti greatly improve the creep strength of austenitic stainless steels, mainly by precipitating fine carbides intragranularly. These phases are not always desirable because they can provoke deterioration of mechanical or chemical properties. Nitrogen is a strong austenite stabilizer; it can act like carbon in stabilized stainless steels by precipitating in form of titanium or niobium nitrides [38]. It can be said that passivity of the austenitic stainless steels arises from the high corrosion resistance shown by the Cr(III) oxide-hydroxides present in the passivating layers. Chromium forms insoluble Cr_2O_3, which prevents the dissolution of iron. It has been reported that in solutions containing chloride ions Ni is rarely observed in the passive films of the stainless steels, and NiO is less effective than Cr_2O_3 in anchoring the water molecules and forming hydroxyl ions complex species [39].

However, according to the results shown here (Figures 12 and 13), it is evident that an increase in Ni content in stainless steels leads to an increase in corrosion rate. This may be due to the presence of a NiO layer on the passive (Cr_2O_3). This originates the presence of defects that favored the attack by chloride ions. It has been reported that, in NiO forming alloys, chloride ions can migrate through the passivating layer by replacing water and hydroxyl groups gathering in the metal oxide interface causing localized corrosion [10].

Figure 12 shows that in mouthwash solution the corrosion resistance of stainless steels increases with its chromium and nickel content but steels 316-H and 321-H do not obey this trend. Both steels have higher corrosion rates than expected according to Cr and Ni contents. Both steels contain additional elements such as Mo (316-H) and Ti (321-H). This suggests that the presence of these elements causes a detriment to corrosion resistance in artificial saliva environments. However, 347-H also contains an additional element (Nb) and its presence did not affect its corrosion resistance. As explained in the case of artificial saliva, this was possibly because the phases that form the anode-cathode pair have similar corrosion potentials. Analogous results were obtained in potentiodynamic polarization tests. However in this case the effect of Cr and Ni content was contrary to that observed in potentiodynamic tests. These analyses make clear that the corrosion behavior of materials is defined more consistently in long duration tests. Quick tests as potentiodynamic polarization curves are a tool that allows us to establish a relative corrosion resistance of materials but the initial trend may change in long time exposure because the corrosion process may cause changes in the local chemistry of the electrolyte making the environment more aggressive. Another consideration in long time tests is that changes occur when the residence time is long enough to reach a material-electrolyte thermodynamic equilibrium between the actual dynamic variables (I_{corr}, E_{corr}) and the corrosion process.

Figure 14 shows the superficial aspect of titanium, 310-H, and 321-H stainless steel after the corrosion test in mouthwash solution. Surface appearance shown by Ti shows a homogeneous surface with isolated pits on its surface.

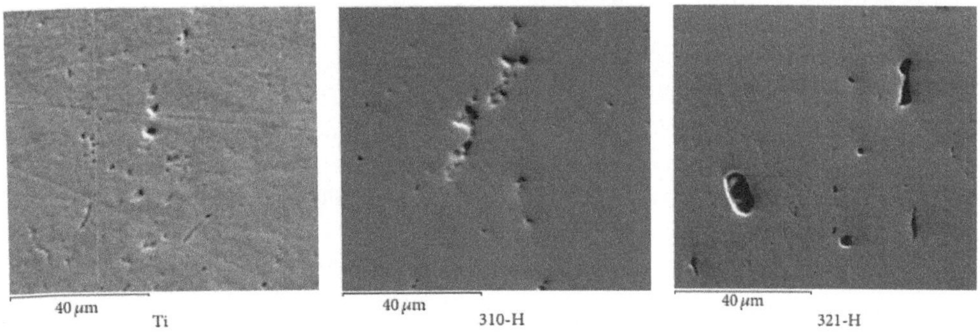

Figure 14: Superficial aspect of Ti, 310-H, and 321-H stainless steel after corrosion test in mouthwash solution.

Surface appearance observed in 310-H steel shows a surface with isolated localized attack associated with surface lines originated by the specimen preparation process. Similar issues were shown by 304-H and 309-H steels. The surface appearance shown by stainless steel 321-H shows the presence of large voids corresponding to the spaces left by the detachment of Ti-rich precipitates. Similar features were observed for 306-H and 347-H steels. This indicates that the presence of precipitates containing Ti, Mo, and Nd acted as cathodic sites promoting corrosion of Fe-Cr-Ni matrix.

It is known that a biological environment can be highly aggressive for metallic materials, since some ions can cause localized corrosion much more favorably than others. The passive layers formed on the materials surface are usually assumed to be composed with an inner oxide layer, being the barrier layer inhibiting cations transfer, and an outer hydroxide being an exchange layer with the electrolyte. On the other hand, passivity of materials will depend on the ions present in the corrosive medium. In this sense it can be said that the aggressive ions may be categorized into two classes depending on the properties of their surrounding hydration shell. Chaotrope ions are ionic species referred to as water-structure breaking ions; they are usually large and generate weak electric fields around them, possessing a loose hydration shell which can be easily removed. The chaotrope ions are capable of initiating pitting (i.e., Cl^-). On the other hand, the cosmotrope ions are ionic species referred to as water-structure-making ions and they exhibit a high surface charge density generating high electric fields at short distances such that they are able to bind water molecules in their vicinity more strongly than water itself. The cosmotrope ions do not break the passive layer and therefore do not initiate pitting (i.e., F^-) [20]. This indicates that the surface features observed in the case of stainless steels in the presence of mouthwash are due primarily to the anode-cathode corrosion arising between the inclusions and Fe-Ni-Cr matrix.

In general, the passivity of the austenitic steels arises from the high corrosion resistance exhibited by the Cr(III) oxide-hydroxides present in the passivating layers. It is considered that chromium forms insoluble Cr_2O_3 which prevents the dissolution of iron. However, the addition of fluoride ions causes the surface films to become less stable. Fluoride ions are aggressive ions that degrade the protective oxide layer formed onto stainless steels surface. The complex formation of metal-fluoride molecules onto alloy surface can reduce the potential range for passivity and a further growth of the oxide film is hindered as the metal dissolution becomes the dominant electrochemical process [10]. In the presence of fluoride ion, the dissolution of the Cr_2O_3 may be due to the following reaction:

$$Cr_2O_3 + 12F^- + 8H^+ \longrightarrow 2CrHF_6^{2-} + 3H_2O \tag{2}$$

The formation and repair of the anodic film could be according to the reaction:

$$2Cr + 3H_2O \longrightarrow Cr_2O_3 + 6H^+ + 6e^- \tag{3}$$

However, the corrosion performance of stainless steels is also influenced by the presence of H_2O_2 in the mouthwash solution. Among many oxidants, hydrogen peroxide (H_2O_2) is quite unique. It has high active oxygen content (47%) and produces water as the only by-product [40, 41]. Reaction between H_2O_2 and Cr of the stainless steels is

$$2Cr + 2H_2O_2 + 2e^- \longrightarrow Cr_2O_3 + H_2O + 2H^+ \quad (4)$$

In relation to titanium, it has been reported that its corrosion in the presence of a solution with fluoride ions is superficial and transitional because the passive layer created on titanium surface reassembles quickly [42]. The negative influence of fluoride on the corrosion of titanium and its alloys has been reported earlier [7–9]. The surface of Ti consists of a thin oxide layer (2–6 nm), composed mainly of TiO_2, and its corrosion is due to the alteration of the passivation oxide film [43]. TiO_2-based passive film is destroyed by the attack of fluoride ions via the formation of a soluble Ti-F complex compound, $[TiF_6]^{2-}$ that form soluble salts [44–46]. Once the passive oxide layer on Ti is destroyed by fluoride attack, the rate of regeneration of passive oxide layer is a function of the dissolved oxygen concentration [17]. In the presence of fluoride ion, the dissolution of the Ti passive layer may be due to the following reaction:

$$TiO_2 + 6F^- + 4H^+ \longrightarrow TiF_6^{2-} + 2H_2O \quad (5)$$

The formation and repair of the anodic film of TiO_2 could be according to the reaction:

$$Ti + 2H_2O \longrightarrow TiO_2 + 4H^+ + 4e^- \quad (6)$$

When Ti is brought into contact with H_2O_2 two phenomena will take place at the metal interface: (i) the Ti-catalyzed decomposition of hydrogen peroxide and (ii) the corrosion of the metal which will involve the metal dissolution into the electrolyte and the formation of a titanium oxide. This oxide is far thicker than the one obtained by simple immersion of the metal in saline solutions for similar periods and is composed of two layers: an outer layer highly porous and hydroxylated and an inner layer, much thinner and with insulating characteristics [18]. Reactions between H_2O_2 and titanium are [47]

$$2TiO_2 + H_2O_2 + 2e^- \longrightarrow Ti_2O_3 + H_2O + O_2 \quad (7)$$

$$Ti_2O_3 + H_2O_2 \longrightarrow 2TiO_2 + H_2O + 2e^- \quad (8)$$

Therefore, the increase in E_{corr} and decrease in I_{corr} values observed in Figures 9 and 11 may be due to the presence of H_2O_2. Because H_2O_2 is a strong oxidizer, its presence in the mouthwash solution could have some effects on the properties of the passive film formed on Ti. The potential shift to the positive direction was observed in a similar study and it was attributed to an enhanced oxide film growth rate [47].

CONCLUSIONS

Potentiodynamic polarization tests in artificial saliva show that stainless steels have corrosion rates higher than titanium. However in mouthwash its behavior was reversed. In general it was observed that in the short-term tests corrosion resistance of stainless steels increases with increasing Cr and Ni and the presence of precipitates (Ti-rich, Nd, or Mo) increases its corrosion rate.

In artificial saliva stainless steels have a high susceptibility to pitting and its pitting potential increases by increasing the content of Cr, Ni, and Ti. In mouthwash the pitting potential of steels is only slightly affected by the contents of Cr and Ni and no significant influence was observed by the presence of Ti, Mo, and Nd.

The long-term tests in artificial saliva indicate that corrosion resistance of steel decreases by increasing its content of Cr and Ni and that the presence of precipitates rich in Ti and Mo increases the corrosion rate. In mouthwash the corrosion resistance of steels increases with increasing contents of Cr and Ni but precipitates of Ti, Mo, and Nd increase its corrosion rate. Precipitated phases function like cathodes favoring corrosion of Fe-Cr-Ni matrix which acts as the anode. The above takes place until the precipitates detach leaving large voids on the surface.

ACKNOWLEDGMENTS

The authors acknowledge the support of C. I. Alvarez-Alvarez (Instituto Tecnologico de Zacatepec) during his Professional Residence at the Instituto de Ciencias Físicas-UNAM. Financial support from Consejo Nacional de Ciencia y Tecnologia (CONACYT, México) (Project 198687) is gratefully acknowledged.

REFERENCES

1. M. Fathi and V. Mortazavi, "A review on dental amalgam corrosion and its consequences," Journal of Research in Medical Sciences, vol. 9, no. 1, pp. 42–51, 2004.
2. G. S. Duffó and S. B. Farina, "Corrosion behaviour of a dental alloy in some beverages and drinks,"Materials Chemistry and Physics, vol. 115, no. 1, pp. 235–238, 2009.
3. H. S. Dobbs, "Fracture of titanium orthopaedic implants," Journal of Materials Science, vol. 17, no. 8, pp. 2398–2404, 1982.
4. G. George and H. Shaikh, Introduction to Austenitic Stainless Steels, Corrosion of Austenitic Stainless Steels, Mechanism, Mitigation and Monitoring, edited by: H. S. Khatak, B. Raj, Woodhead Publishing House, Cambridge, UK, 2002.
5. A. J. Sedriks, Corrosion of Stainless Steels, Wiley, New York, NY, USA, 2nd edition, 1996.
6. U. K. Mudali, T. M. Sridhar, and B. Raj, "Corrosion of bio implants," Sadhana, vol. 28, no. 3-4, pp. 601–637, 2003.
7. M. Nakagawa, S. Matsuya, and K. Udoh, "Corrosion behavior of pure titanium and titanium alloys in fluoride-containing solutions," Dental Materials Journal, vol. 20, no. 4, pp. 305–314, 2001.
8. H.-H. Huang, "Effect of fluoride and albumin concentration on the corrosion behavior of Ti-6Al-4V alloy," Biomaterials, vol. 24, no. 2, pp. 275–282, 2003.
9. G. K. Stookey, "Critical evaluation of the composition and use of topical fluorides," Journal of Dental Research, vol. 69, pp. 805–812, 1990.
10. A. Kocijan, D. K. Merl, and M. Jenko, "The corrosion behaviour of austenitic and duplex stainless steels in

artificial saliva with the addition of fluoride," Corrosion Science, vol. 53, no. 2, pp. 776–783, 2011.

11. ASTM International, ASTM A213/A213M—04. Standard Specification for Seamless Ferritic and Austenitic Alloy-Steel Boiler, Superheater, and Heat-Exchanger Tubes, ASTM International, West Conshohocken, Pa, USA, 2004.

12. G. S. Duffó and E. Quezada-Castillo, "Development of an artificial saliva solution for studying the corrosion behavior of dental alloys," Corrosion, vol. 60, no. 6, pp. 594–602, 2004.

13. ASTM, "Standard practice for laboratory immersion corrosion testing metals," ASTM G31-90, ASTM International, West Conshohocken, Pa, USA, 2004.

14. ASTM, "Standard reference test method for making potentiostatic and potentiodynamic anodic polarization measurements," ASTM G 5-94, ASTM International, West Conshohocken, Pa, USA, 2004.

15. ASTM International, "Standard practice for conventions applicable to electrochemical measurements in corrosion testing," ASTM G-3-89, ASTM International, West Conshohocken, Pa, USA, 2004.

16. ASTM International, ASTM F 2129-01. Standard Test Method for Conducting Cyclic Potentiodynamic Polarization Measurements to Determine the Corrosion Susceptibility of Small Implant Devices, ASTM International, West Conshohocken, Pa, USA, 2008.

17. S. Kumar, T. S. N. Sankara Narayanan, and S. S. Kumar, "Influence of fluoride ion on the electrochemical behaviour of β-Ti alloy for dental implant application," Corrosion Science, vol. 52, no. 5, pp. 1721–1727, 2010.

18. C. Fonseca and M. A. Barbosa, "Corrosion behaviour of titanium in biofluids containing H_2O_2 studied by electrochemical impedance spectroscopy," Corrosion Science, vol. 43, no. 3, pp. 547–559, 2001.

19. S. Frangini and N. De Cristofaro, "Analysis of the galvanostatic polarization method for determining reliable pitting potentials on stainless steels in crevice-free conditions," Corrosion Science, vol. 45, no. 12, pp. 2769–2786, 2003.

20. J. L. Trompette, L. Massot, L. Arurault, and S. Fontorbes, "Influence of the anion specificity on the anodic polarization of titanium," Corrosion Science, vol. 53, no. 4, pp. 1262–1268, 2011.

21. X. Cheng and S. G. Roscoe, "Corrosion behavior of titanium in the presence of calcium phosphate and serum proteins," Biomaterials, vol. 26, no. 35, pp. 7350–7356, 2005.

22. S. Tamilselvi, V. Raman, and N. Rajendran, "Corrosion behaviour of Ti-6Al-7Nb and Ti-6Al-4V ELI alloys in the simulated body fluid solution by electrochemical impedance spectroscopy," Electrochimica Acta, vol. 52, no. 3, pp. 839–846, 2006.

23. S. L. de Assis, S. Wolynec, and I. Costa, "Corrosion characterization of titanium alloys by electrochemical techniques," Electrochimica Acta, vol. 51, no. 8-9, pp. 1815–1819, 2006.

24. I. Gurappa, "Characterization of different materials for corrosion resistance under simulated body fluid conditions," Materials Characterization, vol. 49, no. 1, pp. 73–79, 2002.

25. M. C. R. A. Rezende, A. P. R. Alves, E. N. Codaro, and C. A. M. Dutra, "Effect of commercial mouthwashes on the corrosion resistance of Ti-10Mo experimental alloy," Journal of Materials Science: Materials in Medicine, vol. 18, no. 1, pp. 149–154, 2007.

26. E. E. Stansbury and R. A. Buchanan, Fundamentals of Electrochemical Corrosion, ASM International, Materials Park, Ohio, USA, 2000.

27. A. Pardo, S. Feliu Jr., M. C. Merino, R. Arrabal, and E. Matykina, "Electrochemical estimation of the corrosion rate of magnesium/aluminium alloys," International Journal of Corrosion, vol. 2010, Article ID 953850, 8 pages, 2010.

28. R. J. Solar, "Corrosion resistance of titanium surgical implant alloys: a review," ASTM STP 684, ASTM, Baltimore, Md, USA, 1979.

29. C. da Fonseca, A. Traverse, A. Tadjeddine, and M. D. C. Belo, "A characterization of titanium anodic oxides by X-ray absorption spectroscopy and grazing X-ray diffraction," Journal of Electroanalytical Chemistry, vol. 388, no. 1-2, pp. 115–122, 1995.

30. J. Pan, D. Thierry, and C. Leygraf, "Electrochemical impedance spectroscopy study of the passive oxide film on titanium for implant application," Electrochimica Acta, vol. 41, no. 7-8, pp. 1143–1153, 1996.

31. F. J. Gil, M. P. Ginebra, A. Padrós, and J. A. Planell, "Obtención y caracterización de capas pasivadas en implantes dentales de titanio," Revista Española Odontoestomatológica de Implantes, vol. 2, pp. 113–119, 1993.

32. Y. Van Ingelgem, A. Hubin, and J. Vereecken, "Investigation of the first stages of the localized corrosion of pure copper combining EIS, FE-SEM and FE-AES," Electrochimica Acta, vol. 52, no. 27, pp. 7642–7650, 2007.

33. K. M. Speck and A. C. Fraker, "Anodic polarization behavior of Ti-Ni and Ti-6A1-4V in simulated physiological solutions," Journal of Dental Research, vol. 59, no. 10, pp. 1590–1595, 1980.

34. R. W. Revie and N. D. Greene, "Corrosion behaviour of surgical implant materials: II. Effects of surface preparation," Corrosion Science, vol. 9, no. 10, pp. 763–770, 1969.

35. C. Aparicio, F. J. Gil, C. Fonseca, M. Barbosa, and J. A. Planell, "Corrosion behaviour of commercially pure titanium shot blasted with different materials and sizes of shot particles for dental implant applications," Biomaterials, vol. 24, no. 2, pp. 263–273, 2003.

36. J. C. M. de Souza, Biotribocorrosion behavior of titanium in simulated oral environments [Ph.D. thesis], Escola de Engenharia, Universidade do Minho, Braga, Portugal, 2009.

37. N. Hara, K. Hirabayashi, Y. Sugawara, and I. Muto, "Improvement of pitting corrosion resistance of type

316L stainless steel by potentiostatic removal of surface MnS inclusions," International Journal of Corrosion, vol. 2012, Article ID 482730, 6 pages, 2012.

38. T. Sourmail, "Precipitation in creep resistant austenitic stainless steels," Materials Science and Technology, vol. 17, no. 1, pp. 1–14, 2001.

39. C. R. Clayton and Y. C. Lu, "A bipolar model of the passivity of stainless-steel—the role of Mo addition,"Journal of the Electrochemical Society, vol. 133, no. 12, pp. 2465–2473, 1986.

40. M. G. Clerici and P. Ingallina, "Oxidation reactions with in situ generated oxidants," Catalysis Today, vol. 41, no. 4, pp. 351–364, 1998.

41. T. M. Salama, Z. M. El-Bahy, and F. I. Zidan, "Aqueous H_2O_2 as an oxidant for CO over Pt- and Au–NaY catalysts," Journal of Molecular Catalysis A: Chemical, vol. 264, no. 1-2, pp. 128–134, 2007.

42. A. C. L. Faria, A. R. D. V. Bordin, V. Pedrazzi, R. C. S. Rodrigues, and R. F. Ribeiro, "Effect of whitening toothpaste on titanium and titanium alloy surfaces," Brazilian Oral Research, vol. 26, no. 6, pp. 498–504, 2012.

43. G. Mabilleau, S. Bourdon, M. L. Joly-Guillou, R. Filmon, M. F. Baslé, and D. Chappard, "Influence of fluoride, hydrogen peroxide and lactic acid on the corrosion resistance of commercially pure titanium,"Acta Biomaterialia, vol. 2, no. 1, pp. 121–129, 2006.

44. L. Reclaru and J.-M. Meyer, "Effects of fluorides on titanium and other dental alloys in dentistry,"Biomaterials, vol. 19, no. 1–3, pp. 85–92, 1998.

45. H.-H. Huang and T.-H. Lee, "Electrochemical impedance spectroscopy study of Ti–6Al–4V alloy in artificial saliva with fluoride and/or bovine albumin," Dental Materials, vol. 21, no. 8, pp. 749–755, 2005.

46. A. Robin and J. P. Meirelis, "Influence of fluoride concentration and pH on corrosion behavior of titanium in artificial saliva," Journal of Applied Electrochemistry, vol. 37, no. 4, pp. 511–517, 2007.

47. N. A. Al-Mobarak, A. M. Al-Mayouf, and A. A. Al-Swayih, "The effect of hydrogen peroxide on the electrochemical behavior of Ti and some of its alloys for dental applications," Materials Chemistry and Physics, vol. 99, no. 2-3, pp. 333–340, 2006.

CHAPTER 6

A REVIEW OF ELECTROANALYTICAL TECHNIQUES FOR DETERMINATION OF ANTI-HIV DRUGS

Burçin Bozal, Bengi Uslu, and Sibel A. Özkan

Department of Analytical Chemistry, Faculty of Pharmacy, Ankara University, Tandogan, 06100 Ankara, Turkey

ABSTRACT

Until now after the human immunodeficiency virus (HIV) was discovered as the then tentative aetiological agent of acquired immune deficiency syndrome (AIDS), exactly 25 anti-HIV compounds have been formally approved for clinical use in the treatment of AIDS. These compounds fall into six categories: nucleoside reverse transcriptase inhibitors (NRTIs: zidovudine, didanosine, zalcitabine, lamivudine, abacavir, stavudine, and emtricitabine), nucleotide reverse transcriptase inhibitors (NtRTIs: tenofovir), nonnucleoside reverse transcriptase inhibitors (NNRTIs: efavirenz, nevirapine, delavirdine, and etravirine), protease inhibitors (PIs: ritonavir, indinavir, saquinavir, nelfinavir, amprenavir, lopinavir, fosamprenavir, atazanavir, tipranavir and darunavir), fusion inhibitors (FIs: enfuvirtide), coreceptor inhibitors (CRIs: maraviroc), and integrase inhibitors (INIs: raltegravir). The present paper submitted the use of various electroanalytical techniques for the determination of anti-HIV drugs. This paper covers the time period from 1990 to 2010 including voltammetric techniques that were reported. Presented application concerns analysis of anti-HIV drugs from pharmaceutical dosage forms and biological samples.

INTRODUCTION

In these early days of anti-HIV drug research, it could hardly be foreseen that within 27 years of the virus being discovered we would now, in 2010, have at hand 25 anti-HIV compounds licensed (thus formally approved) for the treatment of AIDS according to different categories (Table 1). (a)Nucleoside reverse transcriptase inhibitors (NRTIs): the reverse transcriptase associated with HIV is actually the target for three classes of inhibitors: NRTIs, NtRTIs, and NNRTIs. The NRTIs and NtRTIs interact with the catalytic site of the enzyme, whereas the NNRTIs interact with an allosteric site located at a short distance from catalytic site [1]. There are at present (in 2010) 7 NRTIs that have been formally approved for the treatment of HIV infections: zidovudine, didanosine, zalcitabine, lamivudine, abacavir, stavudine and emtricitabine (Table 1). All the NRTIs could be considered as 2′,3′-dideoxynucleoside analogues and act in a similar fashion. (b)Nucleotide reverse transcriptase inhibitors (NtRTIs): NtRTIs should be clearly distinguished from the NRTIs as they are nucleotide analogues, which means that they only need two phosphorylation steps to be converted to their active form. Also, they contain a phosphonate group that cannot be cleaved by esterases, which would make it more difficult to cleave off these compounds, once incorporated at the 3′-terminal end,

Citation: Burçin Bozal, Bengi Uslu, and Sibel A. Özkan, "A Review of Electroanalytical Techniques for Determination of Anti-HIV Drugs," International Journal of Electrochemistry, vol. 2011, Article ID 343947, 17 pages, 2011. doi:10.4061/2011/343947.

compared with their regular nucleotide counterpart. At present, there is only one NtRTI that has been formally approved for the treatment of HIV infections: tenofovir (Table 1).(c) Nonnucleoside reverse transcriptase inhibitors (NNRTIs): NNRTIs attack the same target enzyme as NRTIs. However, rather than integrating themselves into the transcribed DNA, NNRTIs attach themselves to reverse transcriptase and prevent the enzyme from converting RNA to DNA. Unlike NRTIs, which must be phosphorylated to prevent HIV from infecting the cell, NNRTIs are active in the form administered.

Table 1: Approved anti-HIV drugs and their structure formulae

Categories	Generic name	Structure formulae	Brand name	Manufacturer
	Zidovudine	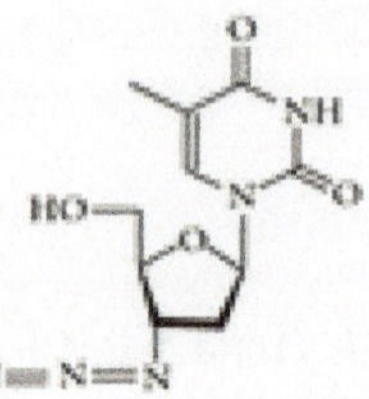	Retrovir	GlaxoSmithKline
	Didanosine	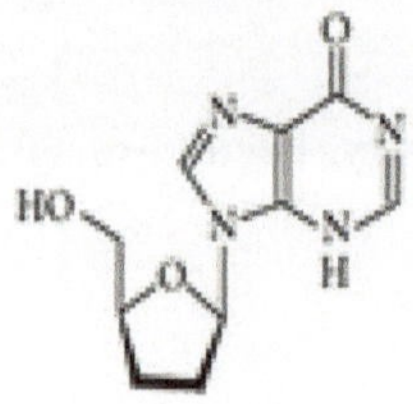	Videx (tablet) Videx EC (capsule)	Bristol-Myers Squibb Bristol-Myers Squibb
	Zalcitabine	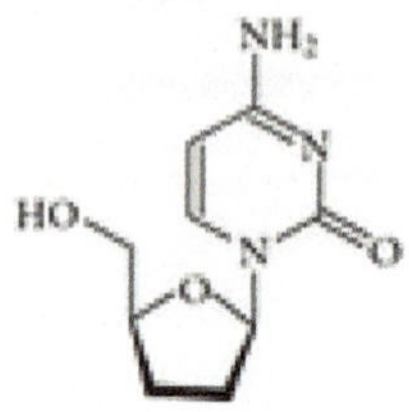	Hivid	Hoffmann-La Roche
Nucleoside reverse transcriptase inhibitors (NRTIs)	Stavudine	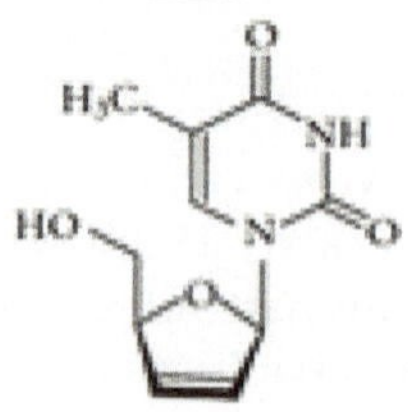	Zerit	Bristol-Myers Squibb
	Lamivudine	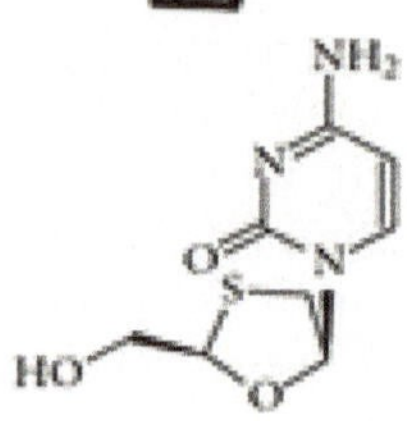	Epivir	GlaxoSmithKline
	Abacavir	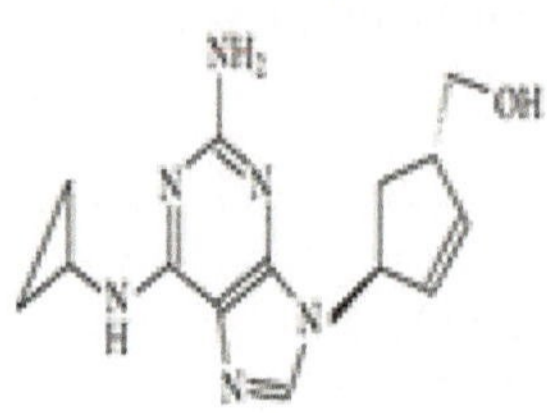	Ziagen	GlaxoSmithKline
	Emtricitabine	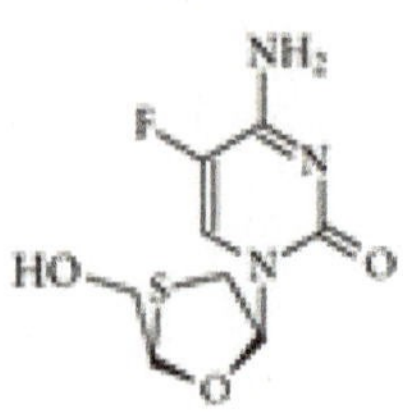	Emtriva	Gilead Sciences

Nucleotide reverse transcriptase inhibitors (NtRTIs)	Tenofovir	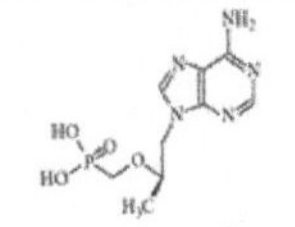	Viread	Gilead Sciences
Nonnucleoside reverse transcriptase inhibitors (NNRTIs)	Nevirapine	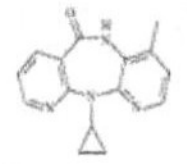	Viramune	Boehringer Ingelheim
	Delavirdine	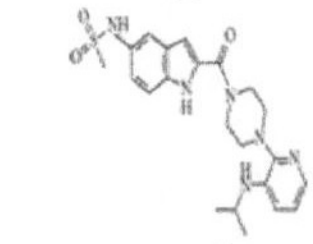	Rescriptor	Pfizer
	Efavirenz	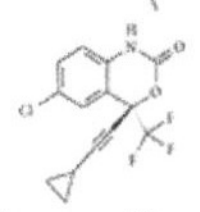	Sustiva Stocrin	Bristol-Myers SquibbMerck
	Etravirine	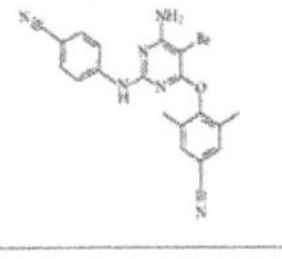	Intelence	Tibotec Therapeutics
Protease inhibitors (PIs)	Saquinavir	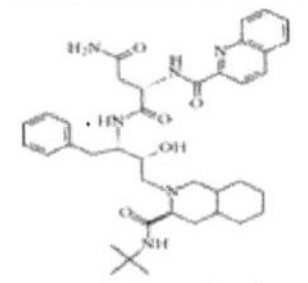	Invirase (hard gel capsule) Fortovase (soft gel capsule)	Hoffmam-La Roche Hoffmann-La Roche
	Ritonavir	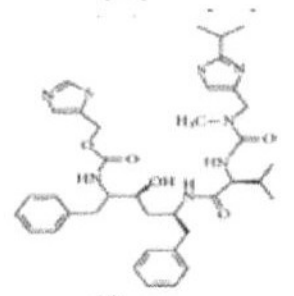	Norvir	Abbott Laboratories
	Indinavir	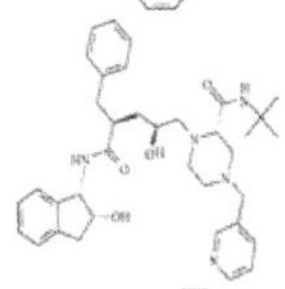	Crixivan	Merck
	Nelfinavir	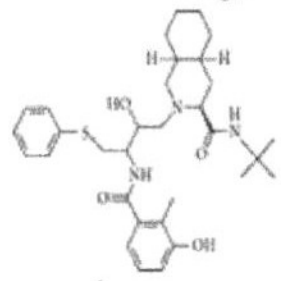	Viracept	Agouron Pharmaceuticals
	Amprenavir	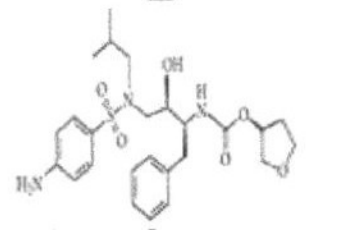	Agenerase	GlaxoSmithKline
	Lopinavir (Tritonavir)	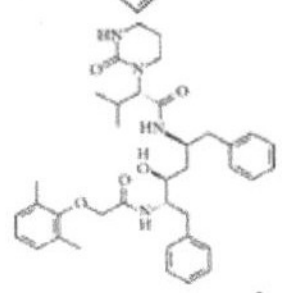	Kaletra	Abbott Laboratories
	Atazanavir	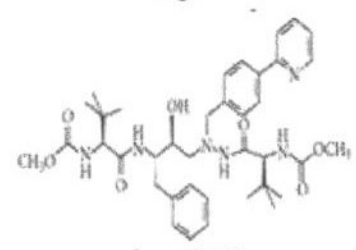	Reyetaz	Bristol-Myers Squibb
	Fosamprenavir	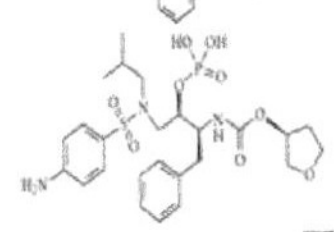	Lexiva Telzir	GlaxoSmithKline GlaxoSmithKline
	Tipranavir	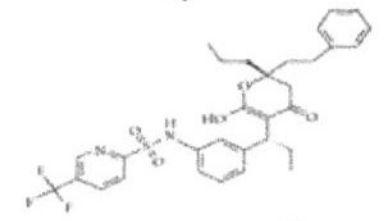	Aptivus	Boehringer Ingelheim
	Darunavir	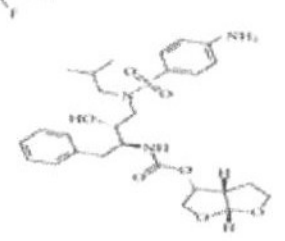	Prezista	Tibotec, Inc.
Integrase inhibitors (INIs)	Raltegravir	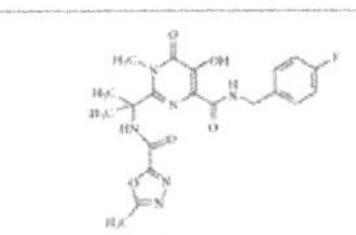	Isentress	Merck & Co., Inc.
	Maraviroc	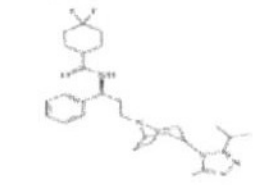	Celsentri Selzentry	Pfizer Pfizer

Cell entry inhibitors and co-receptor inhibitors	Enfuvirtide	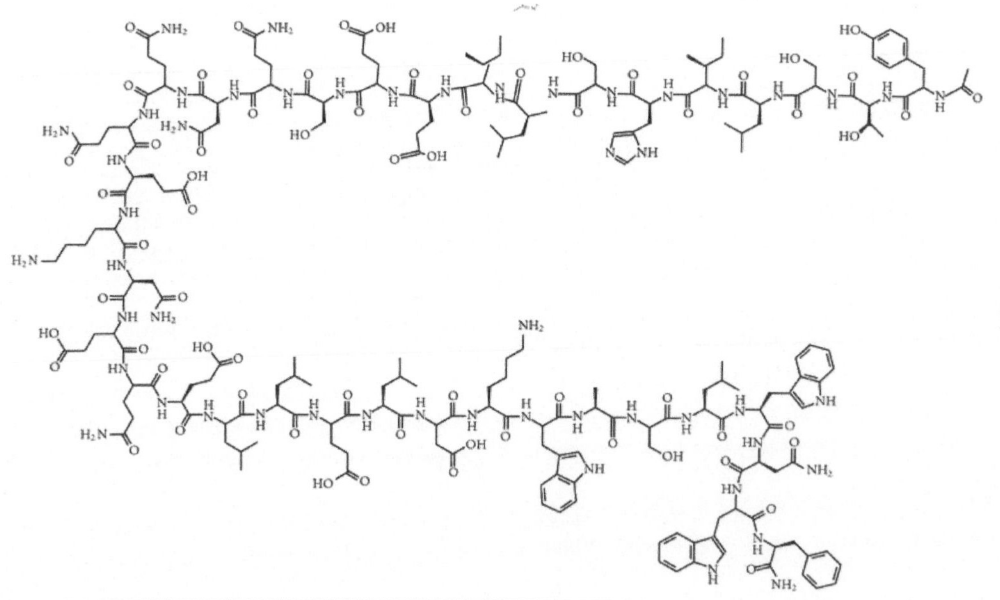	Fuzeon	Hoffmann-La Roche & Trineris

The four NNRTIs presently available for the treatment of HIV infections are efavirenz, nevirapine, delavirdine, and etravirine [1, 2] (Table 1).(d)Protease inhibitors (PIs): during the later stages of the HIV growth cycle, the Gag and Gag-Pol gene products are translated into polyproteins, and these become immature budding particles. Protease is responsible for cleaving these precursor molecules to produce the final structural proteins of the mature virion care. By preventing posttranslational cleavage of the Gag-Pol polyprotein, PIs prevent the processing of viral proteins into functional conformations, resulting in the production of immature, noninfectious viral particles. PIs are active against both HIV-1 and HIV-2; unlike the NRTIs, however, they do not need intracellular activation. The ten PIs currently available for the treatment of HIV infections are ritonavir, indinavir, saquinavir, nelfinavir, amprenavir, lopinavir, fosamprenavir, atazanavir, tipranavir, and darunavir [1–6].(e)Entry inhibitors (fusion inhibitors and coreceptor inhibitors): the process of HIV-1 entry into host cell is complex; each step forms a potential target for inhibition. Viral attachment to the host cell entails binding of the viral envelope glycoprotein complex gp160 to its cellular receptor CD4. At present, there are only two entry inhibitors that have been formally approved for the treatment of HIV infections: enfuvirtide and maraviroc (Table 1) [1–6].(f) Integrase inhibitors (INIs): although integrase has been pursued for many years as a potential target for the development of new anti-HIV compounds, the first integrase inhibitor licensed for clinical use, raltegravir, has only recently been approved. Raltegravir is a pyrimidinone analog that binds integrase, a viral enzyme essential to the replication of both HIV-1 and HIV-2. It is licensed for use in treatment-experienced adult patients infected with strains of HIV-1 resistant to multiple other agents.

In addition, new therapies are continually being sought that exploit new viral targets, have activity against resistant viral strains, have a lower incidence of adverse effects, and offer convenient dosing. New agents of existing classes are currently in advanced stages of clinical development [1–6].

The growing demand for these agents stimulates a search for new even more effective drugs, but also calls for higher level of quality control of these therapeutic substances and preparations, so that they are in the highest possible degree free from any impurities that may come from the production process, as well as from decompositions products of active or auxiliary substances. Therefore, it seems appropriate to develop new analytical methods regarding their qualitative and quantitative analysis. For this aim, different analytical methods are used for determining anti-HIV drugs. One that has been gaining more and more applications is the electroanalytical methods. Electroanalytical methods are widely used in scientific studies and in monitoring of industrial materials, pharmaceutical compounds, biological samples and the environment. The most widespread electroanalytical methods are voltammetry and polarography. These methods permit the screening and determination of a great number of organic compounds especially pharmaceutical substances with a detection limit of $\sim 10^{-7}$ M. The voltammetric methods used today in analytical chemistry laboratories were made possible by recent advances in instrumentation, by computerized processing of analytical data, and in particular, by innovative electrochemists. Electrochemistry involves in its broad sense a chemical phenomenon associated with charge separation at an electrode surface. As voltammetric methods continue to develop, the range of working electrode materials continues to expand, and the analyst must remain aware of what is available. Both the geometry and composition of the working electrode material must be considered since they will influence the performance of the voltammetric method. Also, the physical form of the working electrode may influence the diffusion process and the electron transfer process involved in the detection of the analyte [7–15]. The voltammetric and polarographic analysis of drugs in pharmaceutical preparations by far is the most common use of electrochemistry for analytical-pharmaceutical problems. As a rule, many of active compounds of the formulations can be readily oxidized or reduced. As has been pointed out, electrochemical techniques are well suited for the determination of drugs in various samples, that is, raw material, pharmaceutical dosage forms even those involving a complex matrix such as syrups, tablets, creams, suppositories, or ointments, or else in biological fluids. The principal advantage of the modern electroanalytical methods is that the excipients do not interfere, and generally the separation and extraction is not necessary. Thus, sample preparation usually consists of dissolving out the active substance from the pharmaceutical dosage form with a suitable solvent and performing direct analysis on an aliquot portion of this solution [16]. Electrochemical techniques are most suitable to investigate the redox properties of a new drug (like new anti-HIV agents); this can give insights into its metabolic fate. Electrochemical data are often correlated to the molecular structure and the pharmacological activity. Also, interactions of drugs with metal ions or with proteins as well as degredation processes occurring on photo- or pH-sensitive drugs may be studied using modern electrochemical techniques [9, 16, 17].

Anti-HIV drugs are the recent developments of drugs and there is a great need to review the analytical work reported so far in the literatures. Efforts have been made to collect the literature from 1990 up to the present. For electroanalytical determinations, CV, LSV, DPV, SWV, and polarography have been used.

ELECTROANALYTICAL METHODS FOR ANTI-HIV DRUG DETERMINATION

Efavirenz

EFV, (S)-6-chloro-4-(cyclopropylethynyl)-1, 4-dihydro-4-(trifluoromethyl)-2H-3,1-benzoxazin-2-one, is an HIV-1 spesific, nonnucleoside reverse transcriptase inhibitor. EFV was approved by the FDA on September 21, 1998, making it the 14th approved antiretroviral drug. It is not effective against HIV-2, as the pocket of the HIV-2 reverse transcriptase has a different structure, which confers intrinsic resistance to the NNRTI class. EFV is metabolized in the liver and possesses both inhibitory and including effects on the 3A4 isoform of the cytochrome P450 system. This means EFV may interact with other drugs metabolized in the liver, requiring either increased or decreased dosages [18, 19]. Electrochemical techniques also help for the identification of the redox mechanism of drug compounds and provide important information about DNA-drug interaction [20]. There is only one electroanalytical method for the determination of EFV in dosage forms. Dogan-Topal et al. [20] investigated the voltammetric behavior of EFV at a PGE and presented DNA biosensor. The aim of this study was to establish the experimental conditions such as EFV concentration, its interaction time dsDNA, dsDNA concentration, and the effect of the ionic strength. These results showed that this DNA biosensor could be used for the sensitive, rapid, simple, and cost-effective detection and determination of EFV-dsDNA interaction. The linearity was between 2 and 24 ppm of EFV concentration on guanine signal decreasing curve. This work also aimed to develop a new, selective, simple voltammetric method for the direct determination of EFV in raw material and pharmaceutics without any time-consuming extraction, evaporation, and separation steps prior to drug assay. Also DPV method was used at bare and dsDNA modified disposable PGE. EFV showed an irreversible oxidation behavior at all pH values, and AdsDPV method was developed for determination of EFV. Under these conditions, the current showed a linear dependence with concentration in the range between 0.018 and 2.56 ppm.

Abacavir

ABA, {((1S,cis)-4-[2-amino-6-(cyclopropylamino)-9H-purin-9-yl]cyclopent-2-enyl)methanol}, is an NRTI with antiretroviral activity against HIV. It is a carbocyclic synthetic nucleoside analog. ABA is rapidly absorbed following oral administration with a bioavailability of about 80%. It is about 50% bound to plasma proteins. ABA is metabolized into the pharmacological inactive 5′-glucuronide and the 5′-carboxylate.

ABA has been studied and determined using anodic voltammetry by Uslu and Özkan [21]. The goal of this work was the development of new voltammetric methods (DPV and SWV) for the direct determination of ABA in pharmaceutical dosage forms, raw materials, spiked human serum, and urine samples. These two voltammetric techniques for the determination of ABA in Britton-Robinson buffer at pH 2.0, which allows quantitation over the 8×10^{-7} to 2×10^{-4} M range in supporting electrolyte for both methods were proposed. The linear response was obtained in Britton-Robinson buffer in the ranges of 1×10^{-5} to 1×10^{-4} M for spiked urine sample at pH 2.0 and 2×10^{-5} to 2×10^{-4} M for spiked serum samples at pH 3.0 for both techniques.

ABA is electrochemically reduced at the dropping mercury electrode in a four-electron process, similar to structurally related adenine, and adenosine triphosphate [22]. Procedures of analyses of tablets containing ABA were established and validated, based on peak currents obtained by LSV, DPV, or SWV with an HMDE as indicator electrode. The procedure proved to be more sensitive and more reliable than based on oxidation on a glassy carbon electrode.

Fosamprenavir

FOS, {(2R, 3S)-1-[N-(2-methylpropyl) (4-aminobenzene) sulfonamido]-3-({[(3S)-oxolan-3-yloxy]carbonyl}amino) 4-phenyl butan-2-yloxy] carbonyl}amino)-4-phenylbutan-2-yl]oxy} phosphonic acid is a prodrug of antiretroviral protease inhibitor amprenavir. Amprenavir is metabolized in the liver by CYP3A4 enzyme system. It is an inhibitor of HIV-1 protease. Amprenavir binds to the active site of HIV-1 protease resulting in the formation of immature noninfectious viral particles [18, 23, 24]. The two major metabolites result from oxidation of the tetrahydrofuran and aniline moieties. Several analytical methods have been published for the determination of amprenavir with other HIV protease inhibitors in biological fluids, based on HPLC with MS detection, with UV detection, and with fluorescence detection. Also there is only one published literature related to electroanalytical studies of amprenavir. In this study, the anodic oxidation of fosamprenavir was investigated using CV and LSV at boron-doped diamond and glasy carbon electrodes. In CV, depending on pH values, FOS showed one sharp irreversible oxidation peak or wave depending on the electrode. The aim of this study was to determine FOS levels in pharmaceutical formulations and biological samples by means of electrochemical methods. DPV and SWV techniques were 0,1 M H_2SO_4 and phosphate buffer at pH 2.0 which allow quantitation over a 4×10^{-6} to 8×10^{-5} M range using boron-doped diamond and 1×10^{-5} to 1×10^{-4} M range using glassy carbon electrodes, respectively, in supporting electrolyte [25].

Didanosine

2′-3′-Dideoxyinosine (DDI), 9-[(2R, 5S)-5-(hydroxymethyl) oxolan-2-yl]-6,9-dihydro-3H-purin-6-one, is a dideoxy analogue of the purine nucleoside inosine, which shows high antiretroviral activity against HIV, the agent of AIDS [26]. DDI was approved as a drug in 1991 by the FDA for the treatment of adults and children with advanced HIV infection, who are intolerant or resistant to azidothymidine therapy or whose health has been deteriorated by azidothymidine use [27].

Electroanalytical determination of DDI is possible; however up to now, there is only one published paper. In this study, cobalt (II) phthalocyanine and iron (II) phthalocyanine complexes have been used as an electrocatalysts for the oxidation of DDI [28]. The detection limits were calculated to be 8.80×10^{-7} M for CoPc-CPE with a linear concentration range between $2.0 \times 10^{-6} - -7.0 \times 10^{-4}$ M, whereas for the FePc-CPE the LOD is 3.50×10^{-7} M with a linear concentration range between 1.0×10^{-6} and 8.0×10^{-4} M in pH 7.4 phosphate buffer solution. Also the main advantage of the proposed method over the other known methods described for DDI is a possibility of its determination directly, without prior separation, with high precision, rapidity, stability, and low consumption of sample and buffer.

Indinavir

IND, [1(1S,2R),5(S)]-2,3,5-trideoxy-N-(2,3-di-hydro-2-hydroxy-1H-inden-1-yl)-5-[2-[[(1,1-dimethyl ethyl)-amino] carbonyl]-4-(3-pyridinylmethyl)-1-piperazinyl]-2-phenylmethyl)-D-erythro-pentonamide sulfat is an inhibitor of the HIV protease. IND has a low protein binding ranging from 60 to 70% in species and is extensively metabolized in all species by the cytochrome P450 system. It has been determined by spectrophotometric, capillary electrophoretic, liquid chromatographic methods; however, all the reported methods are laborious, time-consuming and require highly sophisticated instrumentation. Only one paper reported anodic voltammetry using glassy carbon electrode. In this study, pH 6.0 Britton-Robinson buffer solution was selected for analytical medium in which IND exhibited diffusion controlled oxidative peak at +0.89 V. Also the peak current varied linearly with drug concentration in the range between 2.8×10^{-7} and 5.0×10^{-5} M. The proposed DPV and SWV techniques have been applied to the determination of IND in pharmaceutical dosage forms with good recoveries [29]. Besides, there is only one electroanalytical study about the determination and reduction mechanism of IND in bulk form, pharmaceutical dosage forms, and biological fluids. In this study, IND exhibited irreversible cathodic waves over the pH range of 2.0–12.00 in different supporting electrolytes. The current-concentration plot was rectilinear over the range from 8.0×10^{-7} to 8.0×10^{-6} M for DPV and 8.0×10^{-7}–1.0×10^{-5} M for SWV in Britton-Robinson buffer at pH 10.00. The wave was characterized as being irreversible and diffusion controlled also. The proposed methods were fully validated and successfully applied to the determination of IND in capsules and spiked human serum samples with good recoveries [30]. Ignaszak et al. [31] developed a novel therapeutic biosensor for the determination of IND. An amperometric drug metabolism biosensor consisting of CYP3A4 encapsulated in a didodecyldimethyl ammonium bromide vesicular system on a platinum disk electrode was fabricated. CV, SWV, and pulse voltammetric responses of the bioelectrode showed quasireversible electrochemistry of the Fe^{3+}/Fe^{2+} redox species of the heme thiolate CYP3A4 enzyme under aerobic and anaerobic conditions. The new biosensor exhibited excellent response to IND with a detection limit and response time of 6.158×10^{-2} mgL^{-1} and 40 s, respectively. Also detection limit was well below the plasma concentration of IND (8 h after intake) which ranges from 0.13 to 8.6 mgL^{-1} [31]. Fizzano et al. [32] developed an HPLC method with electrochemical detection for the quantification of IND in cell culture. For the electrochemical detector, the working parameters were +400 mV for the first electrode and +750 mV for the second; these setting were found to provide optimal detection conditions. The signals generated by the second electrode were used for the quantitation. Also the proposed HPLC assay was utilized to directly evaluate the capability of p-glycoprotein expressing multidrug resistant cells in meading the transport and efflux of protease inhibitor IND, a basic compound in AIDS care.

Lamivudine

LAM, 4-amino-1-[(2R,5S)-2-(hydroxymethyl)-1,3-oxathiolan-5-yl]-1,2-dihydro pyrimidin-2-one, is a synthetic nucleoside analogue with activity against HIV-1 and hepatitis B virus. It is an NRTI structurally related to cytosine with activity against retroviruses including HIV. It is used usually in combination with other antiretroviral drugs. Following oral administration, LAM is rapidly absorbed and peak plasma concentrations are achieved in about 1 hour. Binding to plasma protein is reported to be less than 36% [18, 33].

Nowadays, there are two published literatures related to electrochemical study of LAM in pharmaceutical dosage forms. One of them was studied by Dogan et al. [34]. The aim of this study was to determine LAM levels in serum and pharmaceutical formulations, by means of electrochemical methods using HMDE. On this electrode, LAM undergoes irreversible reduction at the peak potential near Ep −1.26 V. Reduction LAM signals were measured DPV and SWV techniques. DPV and SWV techniques for the determination of LAM in asetate buffer at pH 4.5, which allows quantitation over the $4.0 \times 10^{-6} - -1.0 \times 10^{-4}$ M range for both methods, were proposed. The linear response was obtained in acetate buffer in the ranges of 2.0×10^{-6} to 2.0×10^{-4} M for spiked serum samples at pH 4.5 for both techniques.

Another study was developed by Jain et al. [35]. The electrochemical reduction and adsorption of LAM were studied in a phosphate buffer at pH 3.4 at an HMDE. The reduction was irreversible and exhibited diffusion-controlled adsorption. The responses were evaluated with respect to preconcentration time, pH effect, accumulation potential, accumulation time, and scan rate. The calibration plot of the peak current versus the concentration was found to be linear over the range of 500 ng/mL to 10 μg/mL for LAM in the SWV method. Also the number of electrons transferred in the reduction process was calculated and the probable reduction mechanism was proposed.

Zidovudine (Azidothymidine)

ZDV, 3′-azido-3′-deoxythymidine, is the cornerstone drug for treatment of patients with HIV-1. ZDV is derived from thymidine where the hydroxyl group at carbon 3 of the sugar moiety is replaced by the azide. The azido group is responsible not only for the in vivo antiviral activity but also for the electrochemical reduction signal of ZDV on a mercury electrode. The electrochemical properties of ZDV were investigated using CV, DPV, and chronocoulometry [36]. These methods were used to determine the reduction pathway of ZDV and the number of protons and electrons involved in the reduction process. Experiments using a narrow range of moderate scan rates indicate that the electrochemical reduction of ZDV proceeds via a four-electron, two-proton mechanism to a reduced species that undergoes a fast chemical reaction to give a species that precipitates on the electrode surface. The diffusion coefficient for ZDV was calculated to be $4.90 \times 10^{-6}\,cm^2s^{-1}$ using the irreversible CV equation and an n value of four electrons. The electrochemical investigation was studied by SWV, LSV, and elimination voltammetry with linear scan using HMDE. This paper explores the possibility of determining ZDV in the presence of native (dsDNA) or denatured calf thymus DNA (ssDNA), and/or same synthetic oligodeoxynucleotides. The detection limit of ZDV in the absence and in the presence of ssDNA (10 µg/mL) is 1 and 250 nM, respectively. It was found that the signal of ZDV is not substantially affected by the presence of DNA. While ZDV provides an irreversible reduction peak at a potential near −1.1 V, the DNA gives the reduction peak due to adenine and cytosine residues at a potential of ca. −1.4 V [37].

Other similar experiment was developed by Vacek et al. [38]. The aim of this study was to determine ZDV levels in natural samples (urine, serum, whole blood, and cell cultures) without their mineralization and/or purification, by means of electrochemical methods using HMDE. Reduction ZDV signals were measured by CV, DPV, SWV, and constant current chronopotentiometric stripping analysis. In phosphate buffer at pH 8.0, the SWV yielded the best ZDV signal with the detection limit of 1 nM. For monitoring the influence of these compounds, ZDV reduction was performed in the presence of 10 µg/mL calf thymus ssDNA and/or 100 µg/mL bovine serum albumin. In these cases, the detection limit increased to 0.25 µM. The work showed that SWV is a suitable electrochemical method for the determination of ZDV in various biological samples, such as cell cultures and body liquids of HIV-positive patients. Also the SWV method can be recommended as a useful tool for pharmacological and metabolic studies of ZDV.

New simple and direct electroanalytical method was developed for the determination of ZDV in commercial pharmaceutical preparations by Peckova et al. [39]. It is based on DPV at p-AgSAE or m-AgSAE. Achieved LOQs are in the 10^{-7} M concentration range for both amalgam electrodes. This is more than one order of magnitude higher than using HMDE, nevertheless, still attractive for determination of ZDV in pharmaceutical preparations and in biological matrices, especially after a preconcentration step using solid phase or liquid-liquid extraction.

Another DPV method has been described for the quantitative determination of ZDV by Leandro et al. [40]. In this study, the reduction of ZDV at an HMDE was at −0.96 V at pH 8.0 phosphate buffer. Under these optimized conditions, the ZDV peak current varied linearly with its concentration from 0.25 to 1.25 mg/L. Detection and determination limits of 0.0025 and 0.025 mg/L, respectively, and recovery of 99.88% with an RSD of 0.95% were obtained.

Mahmoud and Luong developed highly sensitive screening assay based on electrochemical impedance spectroscopy [41]. The assay format was based on the immobilization of the thiol terminated ferrocene (Fc)-pepstatin conjugate on a single-walled carbon nanotube/gold nanoparticle modified gold electrode. This study has demonstrated for the first time the use of nanomaterials in impedance spectroscopy for detecting lopinavir, ritonavir, saquinavir, indinavir, and fosamprenavir at picomolar levels. Thiolated single-walled carbon nanotubes together with gold nanoparticles and a ferrocene conjugate serve as sensitive enhancer tools in this assay format. On the basis of its applicability for evaluation of low-molecular weight compounds, this technique is a definite asset for the expedited development of effective HIV-1 protease inhibitors. This strategy might be extended for the sensing of other interfacial interactions such as enzyme-DNA, antibody-antigen, and cell receptor.

SIMULTANEOUS ELECTROANALYTICAL DETERMINATION OF ANALYZED DRUGS

Besides the analytical methods for single components, some of the simultaneous determinations are also included in this paper. Lamivudine, didanosine, and saquinavir were simultaneously determined by capillary zone electrophoresis [42]. In this study, the anti-HIV drug mixture of lamivudine, didanosine, and saquinavir was separated and quantitated in human serum with capillary zone electrophoresis. The effects of various factors, such as run buffer type, buffer concentration, and pH on the separation were investigated. All analytes were separated within 10 min with a voltage of +20 kV and a current around 30 µA in the pH 2.5 phosphate buffer. The method was validated over the range of 0.4–37.8 µg/mL for lamivudine, 1.4–34.0 µg/mL for didanosine, and 0.5–24.4 µg/mL for saquinavir. Baseline separation of lamivudine, didanosine, and saquinavir was achieved with retention times of 4.4, 10.8, and 6.9 min, respectively. The internal standard diltiazem gave a migration time of 5.6 min.

MEKC was used for determination of stavudine, didanosine, saquinavir (mixture A) and stavudine, didanosine, efavirenz (mixture B) in human serum samples [43]. The optimized resolution of both mixture was achieved with a run buffer containing 18 mM sodium dodecylsulfate in 15 mM phosphate and borate buffer (pH 9.0). All analytes were separated within 15 min with a voltage of +15 kV and a current around 30 µA. The methods were validated over the range of 0.7–35.3 µg/mL for stavudine, 0.8–18.5 µg/mL for didanosine, 0.5–12.2 µg/mL for saquinavir in mixture A and 0.7–35.3 µg/mL for stavudine, 0.8–18.5 µg/mL for didanosine, and 0.6–31.9 µg/mL for efavirenz in mixture B.

Another MEKC method was developed by Fan and Stewart [44]. In this study, the anti-HIV drug mixtures containing zidovudine/didanosine/nevirapine (mixture A) and zidovudine/didanosine/ritonavir (mixture B) were quantitated in human serum samples. All analytes were separated within

14 min with a voltage of +15 kV and a current around 30 μA. The methods were validated over the range of 0.5–25.0 μg/mL for zidovudine, 0.8–18.5 μg/mL for didanosine, 0.5–22.8 μg/mL for nevirapine in mixture A and the range of 0.5–25.0 μg/mL for zidovudine, 0.8–18.5 μg/mL for didanosine, and 1.2–28.8 μg/mL for ritonavir in mixture B.

VALIDATION OF THE ELECTROANALYTICAL METHODS

Validation of a method is the planned and documented procedure to establish its performance characteristics. Typical parameters that characterize each electroanalytical method include selectivity, specificity, range, linearity, accuracy (recovery), LOD, LOQ, precision, robustness, and ruggedness [45–52]. They define what the method can do under optimized conditions of matrix solution, instrumental settings, active ingredient isolation, and other experimental features.

The electroanalytical procedures presented in this paper have been validated in terms of basic parameters, that is, linearity, LOD, LOQ, precision, and recovery. Validation parameters of the electroanalytical methods used for the determination of efavirenz, abacavir, fosamprenavir, didanosine, indinavir, lamivudine, zidovudine, saquinavir, stavudine, nevirapine, and ritonavir are presented in Table 2. Comparing validation parameters of already researched techniques, it can be concluded which one of them is more sensitive (low LOD and LOQ values), accurate (precision and recovery) and allows marking in a broad linearity scope.

Table 2: Validation data for analyzed drugs

Compound	Method	Linear range	LOD	LOQ	Precision (RSD %)		Recovery (%)	Ref.
					Intraday	Interday		
Efavirenz	AdsDPV	2.0–24.00 ppm	0.599 ppm	1.995 ppm	0.96	2.22	101.67	[20]
Abacavir	DPV	8.0×10^{-7}–2.0×10^{-4} M	2.20×10^{-7} M	7.34×10^{-7} M	1.02	1.93	99.37	[21]
	SWV	8.0×10^{-7}–2.0×10^{-4} M	1.18×10^{-7} M	3.93×10^{-7} M	0.46	1.76	99.83	
Abacavir	CV	1.0×10^{-7}–1.0×10^{-5} M	2.11×10^{-8} M	—	0.24	0.87	99.99	[22]
	DPV	1.0×10^{-7}–1.0×10^{-5} M	2.41×10^{-8} M	—	0.43	1.17	99.84	
	SWV	1.0×10^{-7}–1.0×10^{-5} M	2.69×10^{-8} M	—	0.43	1.10	100.21	
Fosamprenavir	DPV (BDDE)	4.0×10^{-6}–1.0×10^{-4} M	2.17×10^{-7} M	6.59×10^{-7} M	0.19	1.40	100.10	[25]
	DPV (GCE)	1.0×10^{-5}–1.0×10^{-4} M	5.05×10^{-7} M	1.53×10^{-6} M	0.58	0.66	99.88	
	SWV (BDDE)	4.0×10^{-6}–1.0×10^{-4} M	2.97×10^{-7} M	9.00×10^{-7} M	1.07	1.85	100.02	
	SWV (GCE)	1.0×10^{-5}–1.0×10^{-4} M	3.78×10^{-7} M	1.16×10^{-6} M	0.47	0.72	100.54	
Didanosine	DPV (CoPc-CPE)	2.0×10^{-6}–7.0×10^{-4} M	8.80×10^{-7} M	—	—	—	99.76	[28]
	DPV (FePc-CPE)	1.0×10^{-6}–8.0×10^{-4} M	3.50×10^{-7} M	—	—	—	99.79	
Indinavir	CV	2.8×10^{-7}–5.0×10^{-5} M	1.2×10^{-8} M	3.8×10^{-7} M	—	0.35	98.7	[29]
	LSV	2.8×10^{-7}–5.0×10^{-5} M	1.2×10^{-8} M	7.0×10^{-7} M	—	0.46	98.7	
	DPV	2.8×10^{-7}–5.0×10^{-5} M	4.4×10^{-8} M	2.7×10^{-7} M	—	0.64	99.0	
	SWV	2.8×10^{-7}–5.0×10^{-5} M	4.4×10^{-8} M	2.5×10^{-7} M	—	0.86	98.2	
Indinavir	DPV	8.0×10^{-7}–8.0×10^{-6} M	1.26×10^{-7} M	4.21×10^{-7} M	0.55	1.45	100.36	[30]
	SWV	8.0×10^{-7}–1.0×10^{-5} M	1.57×10^{-7} M	5.23×10^{-7} M	0.44	0.72	100.80	
Indinavir	Amperometry	0.13–8.60 mg/L	6.16×10^{-2} mg/L	—	—	—	—	[31]
Indinavir	HPLC (ED)	6–1000 ng/mL	2 ng/mL	4 ng/mL	1.04	3.31	99.9	[32]
Lamivudine	DPV	4.0×10^{-6}–1.0×10^{-4} M	6.28×10^{-8} M	2.09×10^{-7} M	0.30	0.14	99.45 (tablet)	[34]
							99.98 (oral solution)	
	SWV	4.0×10^{-6}–1.0×10^{-4} M	2.02×10^{-8} M	6.72×10^{-8} M	0.32	0.10	99.53 (tablet)	
							99.96 (oral solution)	

Lamivudine	CAdSV	500 ng/mL–10 μg/mL	69 ng/mL	500 ng/mL	0.55	0.72	102.0	[35]
	DPV	2.5 μg/mL–40 μg/mL	245 ng/mL	2500 ng/mL	0.98	1.5	101.8	
Zidovudine	SWV	—	1 nM (absence of ssDNA)	—	—	—	—	[37]
		—	250 nM (presence of ssDNA)	—	—	—	—	
Zidovudine	SWV	500 pM–1 μM	1 nM	—	—	—	—	[38] —
Zidovudine	DPV (m-Ag-SAE)	0.4–1500 μmol/L	0.12 μmol/L	0.41 μmol/L	1.7	—	101.4	[39]
	DPV (p-AgSAE)	0.6–1500 μmol/L	0.20 μmol/L	0.67 μmol/L	0.3	—	100.3	
	DPV (HMDE)	0.03–1900 μmol/L	0.007 μmol/L	0.024 μmol/L	0.3	—	102.0	
Zidovudine	DPV	0.25–1.25 mg/L	0.0025 mg/L	0.025 mg/L	—	—	99.88	[40]
Lamivudine		0.4–37.8 μg/mL	0.2 μg/mL	0.4 μg/mL	< 13.3	< 13.3	>79.0	[42]
Didanosine	CZE	1.4–34.0 μg/mL	0.9 μg/mL	1.4 μg/mL				
Saquinavir		0.5–24.4 μg/mL	0.2 μg/mL	0.5 μg/mL				
Stavudine	MEKC	0.7–35.3 μg/mL	0.3 μg/mL	0.7 μg/mL	< 14.5	< 14.5	>77.3	[43]
Didanosine		0.8–18.5 μg/mL	0.4 μg/mL	0.8 μg/mL				
Saquinavir		0.5–12.2 μg/mL	0.3 μg/mL	0.5 μg/mL				
Efavirenz		0.6–31.9 μg/mL	0.3 μg/mL	0.6 μg/mL				
Zidovudine	MEKC	0.5–25.0 μg/mL	0.2 μg/mL	0.5 μg/mL	< 13.9	< 13.9	>75.9	[44]
Didanosine		0.8–18.5 μg/mL	0.4 μg/mL	0.8 μg/mL				
Nevirapine		0.2–22.8 μg/mL	0.1 μg/mL	0.2 μg/mL				
Ritonavir		1.2–28.8 μg/mL	0.6 μg/mL	1.2 μg/mL				

CONCLUSIONS

This paper presents electroanalytical methods applied for the determination of anti-HIV drugs between 1990 and 2010. A great number of studies on zidovudine, indinavir, and lamivudine can be noted, whereas for efavirenz, abacavir, fosamprenavir, and didanosine are only a few. Among all of the published methods, DPV and SWV techniques are the most popular techniques which are used both for the analysis of pharmaceutical dosage forms and biological samples. Other techniques, that is, CV, LSV, and amperometry are not so popular. They are used only for understanding redox mechanism. Another technique used for mixture components separation is zone electrophoresis, in which after separation, the individual components can be analyzed in the buffer solution using a spectrophotometric or electrochemical dedector. MEKC is a mode of capillary electrophoresis that is capable of separating uncharged compounds. It is viewed as a chromatographic technique in which migrating charged micelles act as a pseudostationary phase. MEKC offers analyte partitioning between micelles formed by a surfactant in the run buffer, such that mixtures of charged and uncharged anti-HIV drugs can be satisfactorily separated. Up to date a lot of new modified electrodes and biosensors were developed for determination of anti-HIV drugs. The ultimate goal is to obtain results with more and more precision and accuracy and at increasingly lower concentration levels of the substances being determined.

REFERENCES

1. E. De Clercq, "Anti-HIV drugs: 25 compounds approved within 25 years after the discovery of HIV,"International Journal of Antimicrobial Agents, vol. 33, no. 4, pp. 307–320, 2009.
2. J. E. Gallant, "Initial therapy of HIV infection," Journal of Clinical Virology, vol. 25, no. 3, pp. 317–333, 2002.
3. B. G. Kalzung, S. B. Masters, and A. J. Trevor, Eds., Basic & Clinical Pharmacology, McGraw Hill Medical, New York, NY, USA, 11th edition, 2010.
4. M. R. Sadaie, R. Mayner, and J. Doniger, "A novel approach to develop anti-HIV drugs: adapting non-nucleoside anticancer chemotherapeutics," Antiviral Research, vol. 61, no. 1, pp. 1–18, 2004.
5. D. J. Back, S. H. Khoo, S. E. Gibbons, H. Reynolds, J. F. Tjia, and C. Merry, "Therapeutic drug monitoring of anti-HIV drugs," International Congress Series, vol. 1220, pp. 145–160, 2001.
6. X. Tan, C. K. Chu, and F. D. Boudinot, "Development and optimization of anti-HIV nucleoside analogs and prodrugs: a review of their cellular pharmacology, structure-activity relationships and pharmacokinetics," Advanced Drug Delivery Reviews, vol. 39, no. 1–3, pp. 117–151, 1999.
7. P. T. Kissinger and W. R. Heineman, Eds., Laboratory Techniques in Electroanalytical Chemistry, Marcel Dekker, New York, NY, USA, 2nd edition, 1996.
8. J. Wang, Analytical Electrochemistry, Wiley-VCH, New York, NY, USA, 2000.

9. M. R. Smyth and J. G. Vos, Analytical Voltammetry, vol. XXVII, Elsevier Science, Amsterdam, The Netherlands, 1992.
10. J. P. Hart, Ed., Electroanalysis of Biologically Important Compounds, Ellis Harwood, New York, NY, USA, 1990.
11. R. Kellner, J. M. Mermet, M. Otto, M. Valcarcel, and H. M. Widmer, Eds., Analytical Chemistry: A Modern Approach to Analytical Science, Wiley-VCH, Weinheim, Germany, 2nd edition, 2004.
12. J. Wang, Ed., Electroanalytical Techniques in Clinical Chemistry and Laboratory Medicine, Wiley-VCH, New York, NY, USA, 1988.
13. D. K. Gosser Jr., Ed., Cyclic Voltammetry: Simulation and Analysis of Reaction Mechanism, Wiley-VCH, New York, NY, USA, 1993.
14. G. Christian, Ed., Analytical Chemistry, John Wiley & Sons, Hoboken, NJ, USA, 6th edition, 2004.
15. B. Uslu and S. A. Ozkan, "Solid electrodes in electroanalytical chemistry: present applications and prospects for high throughput screening of drug compounds," Combinatorial Chemistry and High Throughput Screening, vol. 10, no. 7, pp. 495–513, 2007.
16. S. A. Özkan, B. Uslu, and H. Y. Aboul-Enein, "Analysis of pharmaceuticals and biological fluids using modern electroanalytical techniques," Critical Reviews in Analytical Chemistry, vol. 33, no. 3, pp. 155–181, 2003.
17. B. Uslu and S. A. Ozkan, "Electroanalytical application of carbon based electrodes to the pharmaceuticals," Analytical Letters, vol. 40, no. 5, pp. 817–853, 2007.
18. S. C. Sweetman, Ed., Martindale the Complete Drug Reference, Pharmaceutical Press, 35th edition, 2007.
19. K. D. Sanbar, Ed., Physicans' Desk Reference (PDR), Thomson PDR, Montvale, NJ, USA, 61st edition, 2007.
20. B. Dogan-Topal, B. Uslu, and S. A. Ozkan, "Voltammetric studies on the HIV-1 inhibitory drug Efavirenz: the interaction between dsDNA and drug using electrochemical DNA biosensor and adsorptive stripping voltammetric determination on disposable pencil graphite electrode," Biosensors and Bioelectronics, vol. 24, no. 8, pp. 2358–2364, 2009.
21. B. Uslu and S. A. Özkan, "Anodic voltammetry of abacavir and its determination in pharmaceuticals and biological fluids," Electrochimica Acta, vol. 49, no. 25, pp. 4321–4329, 2004.
22. B. Dogan, B. Uslu, S. A. Ozkan, and P. Zuman, "Electrochemical determination of HIV drug abacavir based on its reduction," Analytical Chemistry, vol. 80, no. 1, pp. 209–216, 2008.
23. J. Weiss, J. Rose, C. H. Storch et al., "Modulation of human BCRP (ABCG2) activity by anti-HIV drugs,"Journal of Antimicrobial Chemotherapy, vol. 59, no. 2, pp. 238–245, 2007.
24. Lexiva® US prescribing information, GlaxoSmithKline, 2008.
25. M. Gumustas and S. A. Ozkan, "Electrochemical evaluation and determination of antiretroviral drug fosamprenavir using boron-doped diamond and glassy carbon electrodes," Analytical and Bioanalytical Chemistry, vol. 397, no. 1, pp. 189–203, 2010.
26. M. Nasr, C. Litterst, and J. McGowan, "Computer-assisted structure-activity correlations of dideoxynucleoside analogs as potential anti-HIV drugs," Antiviral Research, vol. 14, no. 3, pp. 125–148, 1990.
27. G. D. Morse, M. J. Shelton, and A. M. O›Donnell, "Comparative pharmacokinetics of antiviral nucleoside analogues," Clinical Pharmacokinetics, vol. 24, no. 2, pp. 101–123, 1993.
28. K. I. Ozoemena, R. I. Stefan-von Stoden, and T. Nyokong, "Metallophthalocyanina based carbon paste electrodes for the determination of 2›-3›-dideoxyinosine," Electroanalysis, vol. 21, pp. 1651–1654, 2009.
29. N. Erk, "Voltammetric behavior of indinavir and determination in pharmaceutical dosage forms,"Analytical Letters, vol. 37, no. 1, pp. 47–63, 2004.
30. B. Dogan, D. Canbaz, S. A. Ozkan, and B. Uslu, "Electrochemical methods for determination of the protease inhibitor indinavir sulfate in pharmaceuticals and human serum," Pharmazie, vol. 61, no. 5, pp. 409–413, 2006.
31. A. Ignaszak, N. Hendricks, T. Waryo et al., "Novel therapeutic biosensor for indinavir—a protease inhibitor antiretroviral drug," Journal of Pharmaceutical and Biomedical Analysis, vol. 49, no. 2, pp. 498–501, 2009.
32. M. R. Fizzano, L. Valvo, M. L. Dupuis, V. Mennella, and M. Cianfriglia, "LC determination of Indinavir in biological matrices with electrochemical detection," Journal of Pharmaceutical and Biomedical Analysis, vol. 22, no. 2, pp. 307–314, 2000.
33. PDR®, Electronic Library, Thomson Medical Economics, Mantuale, NJ, USA, 2003.
34. B. Dogan, B. Uslu, S. Suzen, and S. A. Ozkan, "Electrochemical evaluation of nucleoside analogue lamivudine in pharmaceutical dosage forms and human serum," Electroanalysis, vol. 17, no. 20, pp. 1886–1894, 2005.
35. R. Jain, N. Jadon, and K. Radhapyari, "Cathodic adsorptive stripping voltammetric studies on lamivudine: an antiretroviral drug," Journal of Colloid and Interface Science, vol. 313, no. 1, pp. 254–260, 2007.
36. G. C. Barone, H. B. Halsall, and W. R. Heineman, "Electrochemistry of azidothymidine," Analytica Chimica Acta, vol. 248, no. 2, pp. 399–407, 1991.
37. L. Trnková, R. Kizek, and J. Vacek, "Square wave and elimination voltammetric analysis of azidothymidine in the presence of oligonucleotides and chromosomal DNA," Bioelectrochemistry, vol. 63, no. 1-2, pp. 31–36, 2004.
38. J. Vacek, Z. Andrysík, L. Trnková, and R. Kizek, "Determination of azidothymidine—an antiproliferative and virostatic drug by square-wave voltammetry," Electroanalysis, vol. 16, no. 3, pp. 224–230, 2004.
39. K. Pecková, T. Navrátil, B. Yosypchuk, J. C. Moreira,

K. C. Leandro, and J. Bareka Jiri, "Voltammetric determination of azidothymidine using silver solid amalgam electrodes," Electroanalysis, vol. 21, no. 15, pp. 1750–1757, 2009.

40. K. C. Leandro, J. C. Moreira, and P. A. M. Farias, "Determination of zidovudine in pharmaceuticals by differential pulse voltammetry," Analytical Letters, vol. 43, no. 12, pp. 1951–1957, 2010.

41. K. A. Mahmoud and J. H. T. Luong, "Impedance method for detecting HIV-1 protease and screening for its inhibitors using ferrocene-peptide conjugate/Au nanoparticle/single-walled carbon nanotube modified electrode," Analytical Chemistry, vol. 80, no. 18, pp. 7056–7062, 2008.

42. B. Fan and J. T. Stewart, "Determination of lamivudine/didanosine/saquinavir in human serum using capillary zone electrophoresis," Journal of Liquid Chromatography and Related Technologies, vol. 25, no. 2, pp. 241–249, 2002.

43. B. Fan and J. T. Stewart, "Determination of stavudine/didanosine/saquinavir and stavudine/didanosine/efavirenz in human serum by micellar electrokinetic chromatography (MEKC),"Journal of Liquid Chromatography and Related Technologies, vol. 25, no. 6, pp. 937–947, 2002.

44. B. Fan and J. T. Stewart, "Determinations of zidovudine/didanosine/nevirapine and zidovudine/didanosine/ritonavir in human serum by micellar electrokinetic chromatography," Journal of Pharmaceutical and Biomedical Analysis, vol. 30, no. 4, pp. 955–960, 2002.

45. ICH Harmonised Tripartite Guideline, ICH Steering Committee, November 1996, İnc. November 2005.

46. B. Uslu and S. A. Ozkan, "Solid electrodes in electroanalytical chemistry: present applications and prospects for high throughput screening of drug compounds," Combinatorial Chemistry and High Throughput Screening, vol. 10, no. 7, pp. 495–513, 2007.

47. B. Bozal, B. Doĝan-Topal, B. Uslu, S. A. Özkan, and H. Y. Aboul-Enein, "Quantitative analysis of irbesartan in pharmaceuticals and human biological fluids by voltammetry," Analytical Letters, vol. 42, no. 14, pp. 2322–2338, 2009.

48. B. Dogan-Topal, B. Bozal, B. Tolga Demircigil, B. Uslu, and S. A. Ozkan, "Electroanalytical studies and simultaneous determination of amlodipine besylate and atorvastatine calcium in binary mixtures using first derivative of the ratio-voltammetric methods," Electroanalysis, vol. 21, no. 22, pp. 2427–2439, 2009.

49. B. Uslu and D. Canbaz, "Anodic voltammetry of zolmitriptan at boron-doped diamond electrode and its analytical applications," Pharmazie, vol. 65, no. 4, pp. 245–250, 2010.

50. D. Kul, M. Gumustas, B. Uslu, and S. A. Ozkan, "Electroanalytical characteristics of antipsychotic drug ziprasidone and its determination in pharmaceuticals and serum samples on solid electrodes," Talanta, vol. 82, pp. 286–295, 2010.

51. Y. Altun, B. Uslu, and S. A. Ozkan, "Electroanalytical characteristics of lercanidipine and its voltammetric determination in pharmaceuticals and human serum on boron-doped diamond electrode,"Analytical Letters, vol. 43, no. 12, pp. 1958–1975, 2010.

52. B. Bozal and B. Uslu, "Applications of carbon based electrodes for voltammetric determination of lornoxicam in pharmaceutical dosage form and human serum," Combinatorial Chemistry and High Throughput Screening, vol. 13, no. 7, pp. 599–609, 2010.

CHAPTER 7 ELECTROCHEMISTRY OF AUII AND AUIII PINCER COMPLEXES: DETERMINATION OF THE AUII–AUIIBOND ENERGY

Thomas Dann, Dragoş-Adrian Roşca, Joseph A. Wright, Gregory G. Wildgoose and Manfred Bochmann

Wolfson Materials and Catalysis Centre, School of Chemistry, University of East Anglia, Norwich, NR4 7TJ, UK

ABSTRACT

The bond energy of the unsupported Au–Au bond in the Au(II) dimer [(C^N^C)Au]$_2$ and the difference between Au^{III}–OH and Au^{III}–H bond enthalpies have been determined experimentally by electrochemical methods, with Au–OH and Au–H complexes showing unexpected differences in their reduction pathways, supported by DFT modelling.

INTRODUCTION

The chemistry of gold complexes is dominated by compounds in oxidation states I and III.[1] There are comparatively few Au(II) complexes, and most of these possess a binuclear Au^{II}–Au^{II} core supported by bridging ligands. Without such ligands Au(II) has a tendency to disproportionate into Au(I) and Au(III).[2,3] There are only few examples of compounds which contain an unsupported Au^{II}–Au^{II} bond.[4–7] Zopes *et al.* recently reported the formation of [Au^{II}(CF$_3$)$_2$(py)]$_2$,[6] and Xiong and Pyykkö suggested a 6s5d$_{xy}$6p$_z$-type hybridization for the Au–Au bond and calculated an Au–Au bond energy of about 200 kJ mol^{-1}, depending on the ligands.[8] [Au^{II}(CF$_3$)$_2$(py)]$_2$ spontaneously disproportionates in solution into Au(I) and Au(III) products.[6] By contrast, we have recently shown that the Au(III) pincer complexes (C^N^C)AuOH (**1**)[9] and (C^N^C)AuH (**2**) undergo reductive condensation to give the temperature- and air-stable Au(II) complex **4** (eqn (1)).[7] This unusual stability caused us to investigate the bonding in this compound in more detail and to determine the Au–Au bond energy by electrochemical and computational methods. Unsupported Au–Au bonds show significant variations in metal–metal distances (and therefore Au–Au bond energies), ranging from 2.6405(8) Å in [{Au(dppn)I}$_2$](PF$_6$)$_2$[4b] to 2.5062(9) Å in [Au(CF$_3$)$_2$(py)]$_2$ (py = pyridine; dppn = 1,8-bis(diphenylphosphino)naphthalene).[6]

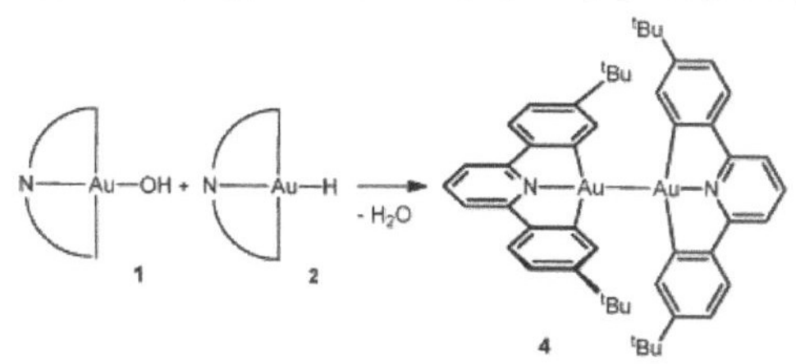

(1)

We report here the voltammetric characterisation of the Au(III) pincer complexes (C^N^C)AuH **1**, (C^N^C)AuOH **2**, and (C^N^C)AuCl, **3**[10] and the Au(II) complex **4**, together with digital simulations of the recorded voltammetry to

Citation: Thomas Dann, Dragoş-Adrian Roşca, Joseph A. Wright, Gregory G. Wildgoose and Manfred Bochmann, Electrochemistry of AuII and AuIII pincer complexes: determination of the AuII–AuIIbond energy, DOI: 10.1039/C3CC45984E.

yield globally optimised parameters for the electrochemistry describing the kinetic and thermodynamic redox properties of these complexes.[11,12] The thermodynamic data gleaned from the digital simulation of the voltammetric data alongside DFT calculations are used to estimate the Au^{II}–Au^{II} bond strength and to rationalise the unusual reactivity and chemistry of (C^N^C)Au^{II}X complexes (X = H or OH).

The direct cyclic voltammetric characterisation of the gold pincer complexes **1–4** was carried out in CH_2Cl_2 using 0.1 M $[NBu_4][B(C_6F_5)_4]$ as supporting electrolyte. Initially survey scans were obtained by first scanning from potentials where no Faradaic current flows in a reductive direction to the edge of the solvent window before reversing the potential up to the oxidative edge of the solvent window at a scan rate of 100 mV s^{-1} (Fig. 1).

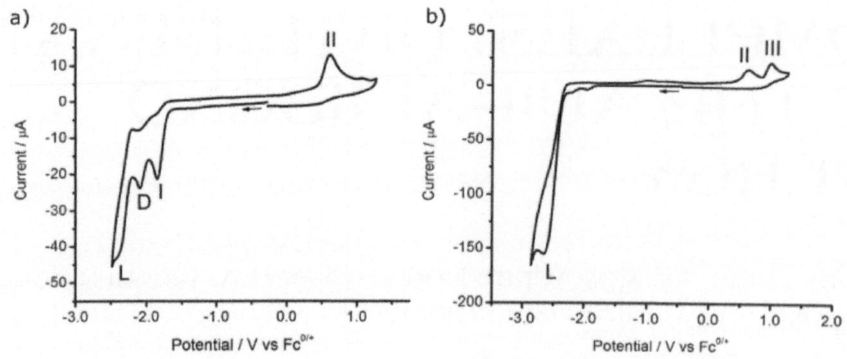

Figure 1: Full range cyclic voltammograms of **3** (a) and **4** (b) in CH_2Cl_2 (1.5 mM, 0.05 M $[nBu_4N][B(C_6F_5)_4]$) at a scan rate of 100 mV s^{-1}.

In the case of **1** two reduction waves are observed with peak potentials at −2.39 and −2.66 V *vs.* $Fc^{0/+}$ respectively.[13] In these survey scans no corresponding oxidation waves are observed upon reversing the scan direction either at the reductive limit of solvent breakdown or sequentially after passing each reduction peak potential. This suggests that the reduction waves correspond to either electrochemically or chemically irreversible processes. An irreversible oxidation wave is observed at +0.59 V *vs.* $Fc^{0/+}$, but only when the potential was first scanned to more negative potentials than that of the first reduction wave at −2.39 V. By comparison with the voltammetry of the dimeric $[(C\text{^}N\text{^}C)Au^{II}]_2$ (**4**) we can assign the oxidation wave at +0.59 V to be the oxidation of **4**, which we postulate is formed as a product of the initial reduction of **1** at −2.39 V. Hence we assign the first reduction wave to the reduction of the Au^{III} metal centre. By comparison with the voltammetry of structurally related alkynyl Au^{III} pincer complexes reported by Yam *et al.*[10,11] the large, irreversible reduction wave at −2.66 V *vs.* $Fc^{0/+}$ (labelled L) may be attributed to a complex multi-electron ligand-based reduction process and will not be considered further.

In the case of LAu^{III}H (**2**) three reduction waves are observed with peak potentials at −2.00 V (corresponding to a metal centred Au(III → II) reduction, labelled I), −2.25 V and −2.60 V *vs.* $Fc^{0/+}$, respectively. We assign the −2.25 V wave (labelled D in Scheme 1 and Fig. 1) tentatively to the reduction of a bridged complex of the type [(C^N^C)Au^{II}–H–Au^{III}(C^N^C)] formed from the reaction of**2** with a (C^N^C) Au^{II} radical. Such an intermediate is only observed for the hydride **2** and chloride **3**, but not for the hydroxide **1**. The third reduction at −2.60 V (labelled L in Scheme 1 and Fig. 1) again corresponds to ligand based multi-electron processes. As was the case with **1**, no corresponding oxidation waves are seen upon reversing the scan direction from the solvent window or sequentially after each reduction potential, indicating that each reduction exhibits electrochemically or chemically irreversible behaviour. The same irreversible oxidation wave at +0.59 V *vs.* $Fc^{0/+}$ is observed with **2** as it was with **1**, and again this oxidation wave is only observed after first scanning the potential beyond the reduction wave at −2.25 V, supporting our assignment of a metal-centred Au(III → II) reduction with subsequent dimerisation of the (C^N^C)Au^{II} radical intermediates to form **4** (Scheme 1).

(C^N^C)Au^{III}–X + e^- ⟶ (C^N^C)$Au^{II\bullet}$ + X^- (*I*)

(C^N^C)$Au^{II\bullet}$ + X–Au^{III}(C^N^C) ⟶ (C^N^C)Au^{II}–X–Au^{III}(C^N^C)

(C^N^C)Au^{II}–X–Au^{III}(C^N^C) + e^- ⟶ [(C^N^C)Au^{II}–X–Au^{II}(C^N^C)]$^-$ (*D*)

[(C^N^C)Au^{II}–X–Au^{II}(C^N^C)]$^-$ ⟶ (C^N^C)Au^{II}–Au^{II}(C^N^C) + X^-

Scheme 1: Electron transfer steps for the reduction steps I and D; X = H, Cl.

(C^N^C)Au^{III}Cl **3** was characterised by voltammetry as a control to account for the possibility of halogen abstraction from the CH_2Cl_2 solvent. We observed similar voltammetry to **2**, with three reductions at −1.85, −2.10 and −2.45 V *vs.* $Fc^{0/+}$, respectively (Fig. 1a). The first is assigned to reduction step I, the second to D, the third to the ligand based reduction L.

The dimer **4** shows the ligand reduction at −2.63 V *vs.* $Fc^{0/+}$, as well as an irreversible oxidation wave at +0.59 V *vs.* $Fc^{0/+}$, due to the direct oxidation of **4** at the electrode (Fig. 1b). A second oxidation wave (labeled III) is also observed at +0.97 V *vs.* $Fc^{0/+}$, that is not seen in the case of **1** or **2**, which only appears after the ligand-based reduction (L) has occurred. We therefore assign this to the oxidation product of ligand decomposition.

To further investigate the electrochemical behaviour of **1–4**, cyclic voltammetry was performed at varying scan rates from 50 to 1000 mV s^{-1}. In the case of **1**, at scan rates >100 mV s^{-1} a corresponding re-oxidation wave for the metal-centred reduction at −2.39 V *vs.* $Fc^{0/+}$ becomes apparent. This suggests that in the electrochemical reduction of **1** the initial heterogeneous electron transfer is quasi-reversible but is followed by an irreversible, homogeneous chemical step involving the cleavage of the Au–OH bond and subsequent dimerisation of the (C^N^C)Au^{II} radical (EC mechanism using Testa–Reinmuth notation[14]). As the voltage scan rate is increased, the kinetics of the follow-up chemical step(s) begin to be outrun on the voltammetric timescale, such that some – but not all – of the Au(II) intermediate is subsequently re-oxidized to the parent LAu^{III}OH complex before the Au–OH bond is cleaved and dimerisation can occur. The fact that the dimer oxidation wave at +0.59 V is still observed at all scan rates indicates that some of the LAu^{II} intermediates are still able to dimerise even at the higher scan rate of 1 V s^{-1}.

The behaviour of **1** is in contrast to that of the hydride **2**, in which no back-peak corresponding to the initial metal-centred reduction is observed at any scan rate up to 1 V s^{-1}. This suggests that the Au^{II}–H bond is broken more rapidly than the Au^{II}–OH bond, such that any follow-up homogenous chemical step(s) are not being outrun on the voltammetric timescale (up to scan rates of 1 V s^{-1}). The faster cleavage of the Au^{II}–H compared to Au^{II}–OH contrasts with the bond energies of the parent Au^{III}–X compounds, where the Au^{III}–OH bond was calculated to be weaker than the Au^{III}–H bond (279 *vs.* 317 kJ mol^{-1}).[15]

This may be explained by different reduction pathways for the two species. DFT calculations show that for X = H (Scheme 1) the reaction of [(C^N^C)Au^{III}–X] with a (C^N^C)Au^{II} fragment to give a [(C^N^C)Au–X–Au(C^N^C)] intermediate is exothermic by 49 kJ mol^{-1}. This species has a roughly symmetrical Au–H–Au core which is retained on reduction. In contrast, when X = OH no OH-bridged molecular intermediate could be identified: the second gold fragment associates only weakly with the oxygen atom (Au···O distance 2.487 Å).

In all cases the oxidation of **4** (+0.59 V *vs.* Fc$^{0/+}$) is irreversible at all voltage scan rates. This would suggest that the product of the oxidation of the Au(II) dimer is unstable and likely undergoes rapid follow-up chemistry, probably involving cleavage of the Au^{II}–Au^{II} bond to form redox inactive products. There was no evidence for detectable amounts of a mixed-valence radical cation intermediate [(C^N^C)Au^{II}–Au^{III}(C^N^C)]$^+$ in the oxidation process.

Digital simulations of the metal centred reductions of **1** and **2** and the oxidation of **4** provided kinetic and thermodynamic parameters and support for the mechanisms of the observed electrochemical processes. The best fit between experiment and theory (Fig. 2) is consistent with the mechanisms discussed above, where the Au(III) centres each undergo a one-electron reduction concomitant with Au–X bond cleavage (X = H$^-$ or OH$^-$) at the electrode (E-step), followed by rapid dimerisation between two (C^N^C)Au^{II} radicals (C-step). This dimer **4** can then undergo a one-electron oxidation (E-step) at more positive potentials. The identity of the product of this oxidation is currently unknown but, given the irreversible nature of the oxidation, we postulate that it likely involves cleavage of the Au^{II}–Au^{II} bond. As we found earlier,[7] (C^N^C)Au^{II} pincer complexes are resistant to disproportionation, and any mechanism involving a disproportionation step gave poor fits between experiment and simulation and were thus discounted. The overall mechanism can therefore be described as ECEC, in which the chemical step is a bimolecular dimerisation. Table 1 summarises the globally optimised parameters for the metal-centred reduction of **1** and **2** and oxidation of **4** (formal potential, E^0_f; charge transfer coefficient, ; standard electron transfer rate constant, k^0; diffusion coefficient, D_0).[16]

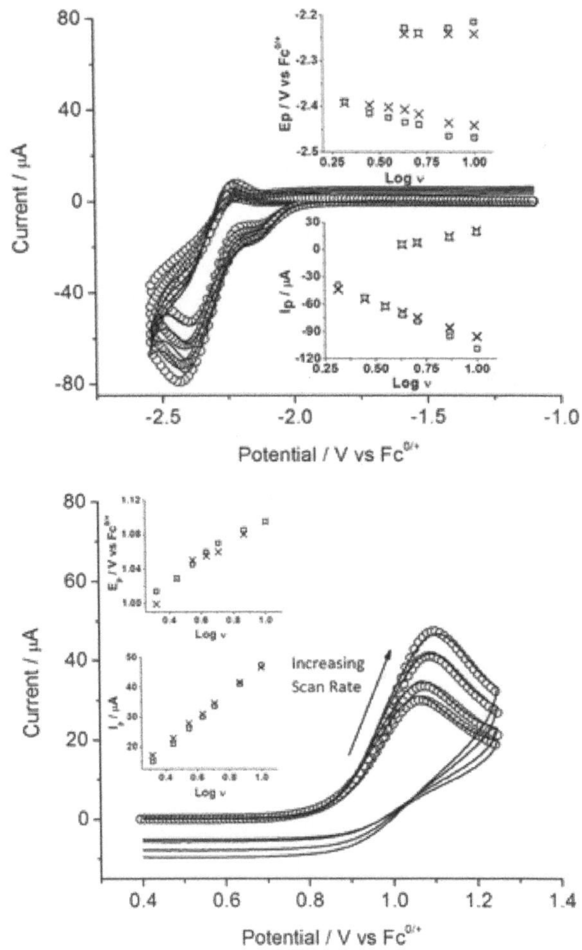

Figure 2: Experimental (solid line) and simulated (open circles) overlaid cyclic voltammograms of **1**(top), **4** (bottom), (all redox active species initially present at 1.5 mM; 0.05 M [nBu$_4$N][B(C$_6$F$_5$)$_4$] supporting electrolyte) at scan rates of 200–500 mV s^{-1}. Inset: plots comparing simulated (open squares) and experimental (crosses) peak potentials (E_p) and peak currents (I_p).

Table 1: Simulated electrochemical parameters for the reduction of **1** and **2** and oxidation of **4**

Parameter	LAuOH (1)	LAuH (2)	LAu–AuL (4)
E^0/V *vs.* Fc$^{0/+}$	−2.30	−1.91	0.60
α	0.3	0.35	0.6
k^0/10^{-3} cm s^{-1}	5.4 ± 0.5	2.0 ± 0.1	68 ± 0.2
D_0/10^{-5} cm^2 s^{-1}	1.51 ± 0.1	1.60 ± 0.1	0.70 ± 0.1

The Au^{II}–Au^{II} bond energy of **4**, along with Au–H/OH bond energy difference between **1** and **2** were estimated using the formal potentials obtained above (corrected to the standard hydrogen electrode, SHE; the minor difference between formal and standard potential, E^0, which accounts for non-ideality in activities, is neglected and is likely within the error range reported). Neglecting entropic contributions, the Au^{II}–Au^{II} bond enthalpy is estimated to be 198 ± 1 kJ mol^{-1}. This value is in good agreement with the bond energy of **4** calculated by DFT methods, 225 kJ mol^{-1}, and the value of approximately 200 kJ mol^{-1} recently calculated by Xiong and Pyykkö[8] for structurally similar $[Au^{II}X_2(py)]_2$ complexes. If we neglect differences in solvation between hydride and hydroxide ions, then a similar Hess cycle can be used to estimate the difference between the Au–OH and Au–H bond enthalpy, which shows that the Au^{III}–H bond is more stable by *ca.* 19 kJ mol^{-1}. This value is reassuringly similar to that predicted by our DFT calculations.[15]

The Au(II) dimer **4** is air stable and does not insert O_2 into the Au–Au bond, even though the reaction **4** + O_2 → (C^N^C)Au–OO–Au(C^N^C) is exothermic by 136 kJ mol^{-1} and the resulting peroxide product is accessible by other routes.[15] It is stable to disproportionation under thermal conditions.

In summary, we have reported the first experimental determination of the bond energy of an unsupported Au^{II}–Au^{II} bond. The value is in agreement with DFT calculations, as are the relative Au^{III}–H and Au^{III}–OH bond energies. The electrochemical reduction pathway of (C^N^C)Au^{III}–X pincer complexes shows unexpected differences, with X = H proceeding *via* a bridged intermediate, whereas X = OH does not.

NOTES AND REFERENCES

1. M. C. Gimeno and A. Laguna, in *Comprehensive Coordination Chemistry II*, ed. J. A. McCleverty and T. J. Meyer, Elsevier, Amsterdam, 2003, vol. 6, pp. 911–1145 .
2. (*a*) A. Laguna and M. Laguna, *Coord. Chem. Rev.*, 1999, 193–195, 837 CAS; (*b*) A. A. Mohamed, H. E. Abou and J. P. Fackler Jr, *Coord. Chem. Rev.*, 2010, 254, 1253 .
3. E. Tkatchouk, N. P. Mankad, D. Benitez, W. A. Goddard III and F. D. Toste, *J. Am. Chem. Soc.*, 2011, 133, 14293 .
4. (*a*) V. W.-W. Yam, S. W.-K. Choi and K.-K. Cheung, *J. Chem. Soc., Chem. Commun.*, 1996, 1173 RSC; (*b*) V. W.-W. Yam, C.-K. Li, C.-L. Chan and K.-K. Cheung, *Inorg. Chem.*, 2001, 40, 7054 CAS .
5. J. Coetzee, W. F. Gabrielli, K. Coetzee, O. Schuster, S. D. Nogai, S. Cronje and H. G. Raubenheimer, *Angew. Chem., Int. Ed.*, 2007, 46, 2497 .
6. D. Zopes, C. Hegemann, W. Tyrra and S. Mathur, *Chem. Commun.*, 2012, 48, 8805 RSC.
7. D.-A. Roşca, D. A. Smith, D. L. Hughes and M. Bochmann, *Angew. Chem., Int. Ed.*, 2012, 51, 10643 .
8. X.-G. Xiong and P. Pyykkö, *Chem. Commun.*, 2013, 49, 2103 RSC.
9. D.-A. Roşca, D. A. Smith and M. Bochmann, *Chem. Commun.*, 2012, 48, 7247 .
10. K. M.-C. Wong, L.-L. Hung, W. H. Lam, N. Zhu and V. W.-W. Yam, *J. Am. Chem. Soc.*, 2007, 129, 4350 .
11. For examples of electrochemical studies on (C^N^C) Au(III) pincer complexes see ref. 10 and (*a*) V. K.-M. Au, K. M.-C. Wong, N. Zhu and V. W.-W. Yam, *J. Am. Chem. Soc.*, 2009, 131, 9076 ; (*b*) V. K.-M. Au, K. M.-C. Wong, D. P.-K. Tsang, M.-Y. Chan, N. Zhu and V. W.-W. Yam, *J. Am. Chem. Soc.*, 2010, 132, 14273 ; (*c*) V. K.-M. Au, N. Zhu and V. W.-W. Yam, *Inorg. Chem.*, 2013, 52, 558 .
12. The reversible electrochemical oxidation of binuclear Au(I) chelate complexes to Au(II) compounds has been reported: M. Bardaj, N. G. Connelly, M. C. Gimeno, P. G. Jones, A. Laguna and M. Laguna, *J. Chem. Soc., Dalton Trans.*, 1995, 2245 .
13. A small, ill-defined reduction wave at −2.09 V can also be observed as a shoulder on the main wave and likely corresponds to some trace impurity arising during the synthesis of 1.
14. A. C. Testa and W. H. Reinmuth, *Anal. Chem.*, 1961, 33, 1320 .
15. D.-A. Roşca, J. A. Wright, D. L. Hughes and M. Bochmann, *Nat. Commun.*, 2013, 4, 2167, DOI:10.1038/ncomms3167.
16. The standard potentials differ from the observed reduction potentials due to the effect of the follow-up chemical steps, thus necessitating the use of digital mechanistic simulations in these more complicated cases.

CHAPTER 8

SIMULTANEOUS DETERMINATION OF COPPER, LEAD, AND CADMIUM AT HEXAGONAL MESOPOROUS SILICA IMMOBILIZED QUERCETIN MODIFIED CARBON PASTE ELECTRODE

Fangquan Xia, Xin Zhang, Changli Zhou, Danzi Sun, Yanmin Dong, and Zhen Liu

School of Chemistry and Chemical Engineering, University of Jinan, Jinan, 250022 Shandong, China

ABSTRACT

A new method was developed for simultaneous determination of copper, lead, and cadmium, based on their voltammetric response at a carbon paste electrode modified with hexagonal mesoporous silica (HMS) immobilized quercetin (HMS-Qu/CPE). Compared with quercetin modified carbon paste electrode (Qu/CPE) and quercetin ionic liquid modified carbon paste electrode (Qu-IL/CPE), the HMS-Qu/CPE exhibited improved selectivity and high sensitivity toward the detection of copper, lead, and cadmium. The properties of the HMS-Qu/CPE in 0.1 M HCOONa-HCl buffer solution (pH4.7) were investigated by adsorptive stripping voltammetry (ASV) and electrochemical impedance spectroscopy (EIS). The electrochemical behavior of copper, lead, and cadmium at the modified electrodes and factors affecting the preconcentration procedures were also investigated. Detection limits of 5.0, 0.8, 1.0 nM for copper, lead, and cadmium were obtained, respectively. The method is simple, fast, sensitive, and selective, and is successfully applied to soil sample.

INTRODUCTION

Heavy metal contamination in the environment is a major concern worldwide because of the toxicity of these metals and their potential threat to human health. Currently, quantification of heavy metals relies upon collection of liquid discrete samples for subsequent laboratory analysis using techniques such as ICP-MS, AAS, GC, HPLC, FT-IR, and GS/MS [1]. Electrochemical detection has several advantages over these methods in their simplicity, fast response, and suitability for the preparation of inexpensive and portable instrumentations [2]. Adsorptive stripping voltammetry (ASV) is not only an extremely sensitive electrochemical technique for measuring trace metals but also can carry out simultaneous detection of several metals in various matrices [3,4]. Chemically modified electrodes (CMEs), with surfaces designed for reacting and binding of target analyte, hold great promise for chemical sensing [5–7]. There are different directions by which CMEs can benefit for analytical application of environmental samples. These include acceleration of electron transfer reactions, preferential accumulation, or selective membrane permeation [8]. Such steps can impart higher selectivity, sensitivity, and stability on electrochemical devices.

As a member of the flavonoid family, quercetin (Qu) can form stable complexes with various metal cations. Many researches have focused on the application of Qu modified carbon paste electrode (CPE) [9, 10]. However, there are no

Citation: Fangquan Xia, Xin Zhang, Changli Zhou, Danzi Sun, Yanmin Dong, and Zhen Liu, "Simultaneous Determination of Copper, Lead, and Cadmium at Hexagonal Mesoporous Silica Immobilized Quercetin Modified Carbon Paste Electrode," Journal of Automated Methods and Management in Chemistry, vol. 2010, Article ID 824197, 6 pages, 2010. doi:10.1155/2010/824197.

reports about simultaneous determination of copper, lead, and cadmium by using Qu modified electrode. Ionic liquids (IL) are liquid electrolytes composed entirely of ions. In recent years, IL have emerged as a research frontier due to their unique chemical and physical properties [11–13]. IL, as the modifier, can offer improvements in the stability and reproducibility of electrode, and has attracted much attention in electrochemical field [14–16]. To our knowledge, there is no report about IL modified electrode to determine heavy metal ions.

Ordered mesoporous materials are highly unusual in their textural characteristics: uniform pore sizes, high surface areas, and long-range order of the packing of pores [17–20]. Hexagonal mesoporous silica (HMS), possessing hexagonal array of pores, is one of the most important mesoporous SiO^2 materials as adsorbents, gas storage materials, separation membranes, catalyst supports, and so on [21]. Yantasee [22–27] and coworkers have reported on simultaneous determination of copper, lead, and cadmium using a carbon paste electrode modified with mesoporous silica. In published documents, polymeric monolayers assembling on the surfaces of ordered mesoporous materials were the focus due to the superior stability and durability of electrode. However, selection of functional chelating agents (ligands) as the modifier may be restricted. Because of the micropore size of HMS limits the entry of large chelating agent leads to reduce the chelation of chelating agent toward metal ions. Otherwise, time-consuming synthesis procedures and preparation of the electrode are also difficult.

In this study, we developed a more simple and sensitive electrochemical method for simultaneous determination of copper, lead and cadmium utilizing the excellent properties of HMS and Qu. In this work, a new quercetin modified carbon paste electrode (Qu/CPE), quercetin ionic liquid modified carbon paste electrode (Qu-IL/CPE), and an HMS immobilized quercetin carbon paste electrode (HMS-Qu/CPE) were fabricated, respectively. The electrochemical responses of copper, lead and cadmium at CMEs were investigated in detail. The results show that HMS-Qu/CPE as a working electrode exhibits higher sensitivity and selectivity toward the detection of copper, lead and cadmium than Qu/CPE, and Qu-IL/CPE. A novel electrochemical method is developed for simultaneous determination of copper, lead, and cadmium, and applied to the analysis of soil samples with satisfying results.

EXPERIMENTAL

Reagents and Chemicals

Qu (purity >98.9%) was obtained from Shanghai Reagent Co. (Shanghai, China), and graphite powder (spectral pure) was obtained from Sinopharm Group Chemical Reagent Co. (Shanghai, China). Tetrafluoroborate 1-butyl-3-methylimidazolium ($[Bmim]BF_4$) was obtained from Shanghai Ling Feng Reagent Co. (Shanghai, China) and used without further purification. HMS was synthesized as described previously [28]. A stock standard solution of Cu(II) (0.01 M) was prepared by dissolving 0.1589 g of copper powder (Shanghai Reagent Company, China) in 4.0 M nitric acid 20 mL, then the resulting solution was moved into the 250 mL volumetric flask, added twice-distilled water to the mark and stirred. A stock standard solution of Pb(II) (0.01 M) was prepared by dissolving 0.5179 g lead powder (Tianjin Reagent Company, China) in 4.0 M nitric acid 20 mL, then the resulting solution was moved into the 250 mL volumetric flask, added twice-distilled water to the mark and stirred. A stock standard solution of Cd(II) (0.01 M) was prepared by dissolving 0.2810 g of cadmium powder (SHRC, China) in 3.0 M hydrochloric acid 20 mL, then the resulting solution was moved into the 250 mL volumetric flask, added twice-distilled water to the mark and stirred. An oxygen free nitrogen (OFN) gas (99.9995%, Jinan Gas Company, China) was used to purge the solution and to remove dissolved oxygen. All reagents were analytical grade materials. Twice-distilled water was used throughout the experiments.

Apparatus

Cyclic voltammetries (CV) and differential pulse voltammetries (DPV) were performed on electrochemical Analyzer (Chenhua Instrumental Co., China) CHI 842B controlled by a personal computer. Electrochemical impedance spectroscopy (EIS) was carried out by electrochemical Analyzer (ZAHNER Co., Germany) IM6ex and the frequency range was from 60 kHz to 0.1 Hz with an amplitude of 5 mV. A three-electrode system was employed, including a modified carbon paste electrode as the working electrode, a platinum wire as the auxiliary electrode, and an Ag/AgCl (3 M KCl) electrode as the reference electrode.

Electrode Preparation

The conventional CPE was prepared by mixing graphite powder and paraffin oil in a ratio of 5:2 (w/w) in a mortar. The resulting paste was packed firmly into the cavity (5 mm diameter) of a glass tube. A copper wire fixed to a graphite rod and inserted into the glass tube served to establish electrical contact with the analyzer. Qu-IL/CPE was prepared by dissolving 0.0240 g Qu in 50 mL ethanol and uniformly mixing with graphite powder. The ethanol was allowed to evaporate in a moment. 0.05 g paraffin oil and 0.6 mL $[Bmim]\ BF_4$ were added, and mixed until obtaining a uniformly wetted paste. A portion of the resulting paste was packed firmly into the cavity (5 mm diameter) of a glass tube. The electrical contact was established via a copper wire. Qu/CPE was prepared in the same fashion, but no $[Bmim]\ BF_4$ was added. HMS-Qu/CPE was fabricated by dissolving 0.0240 g Qu in 50 mL ethanol and mixed the solution with 0.1 g HMS. The ethanol was allowed to evaporate. The paraffin oil was mixed with graphite powder until a uniformly wetted paste was obtained. A portion of the resulting paste was packed firmly into the cavity (5 mm diameter) of a glass tube and the electrical contact was established by a copper wire. The surface of the electrodes was smoothed on a weighing paper and rinsed carefully with twice-distilled water prior to the each measurement.

Sample Solution Preparation

Soil samples were collected from Jinan Environmental Monitoring Central Station. The samples were dried, ground, and sifted by the 60 mesh nylon sieve. 0.5 g of powder was accurately weighed into the decomposing cell. 5.0 mL nitric acid, 5.0 mL hydrochloric acid, and 5.0 mL peroxide were added. The sample solution into decomposing cell is decomposed by microwave [29]. The resulting solution was moved into the 50 mL volumetric flask, and water was added to the mark and stirred. Placed 1 mL of sample solution was placed into 10 mL volumetric flask, and 0.1 M HCOONa-HCl buffer solution was added to the mark and stirred.

Voltammetric Analysis Procedure

Unless otherwise stated, 0.1 M HCOONa-HCl solution (pH4.7) was used as a supporting electrolyte for the determination. The multicomponent Cu(II)/Pb(II)/Cd(II) solution was deaerated with pure nitrogen for 10 min. The measurements were carried out after a preconcentration step, in which the solution was stirred during preconcentration time of 120 s at preconcentration potential of 0.6 V. After a rest period of 30 s, the response curve was recorded by scanning the potential in the negative direction with differential pulse voltammetry (DPV) technique.

RESULTS AND DISCUSSION

Electrochemical Characterization of Qu/CPE

The electrochemical behaviors of Qu/CPE and CPE were investigated in pH4.7 HCOONa-HCl buffer solution. Figure 1 shows the cyclic voltammograms of Qu/CPE (Figure 1, curve a) and CPE (Figure 1, curve b). A pair of well-defined redox peaks of Qu was observed at Qu/CPE, the anodic peak potential E_{pa} at 0.394 V, cathode peak potential E_{pc} at 0.284 V, and peak-to-peak separation (ΔE_p) of 0.11 V. The result indicates that Qu can undergo redox reaction at modified electrode. The redox peaks of Qu were also observed at Qu-IL/CPE and HMS-Qu/CPE.

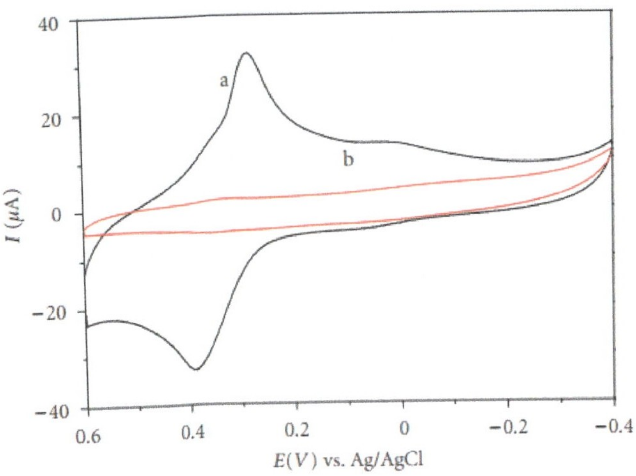

Figure 1: The cyclic voltammograms of Qu/CPE (a) and CPE (b) in 0.1 M HCOONa-HCl buffer (pH4.7), scan rate $0.1 V{\cdot}s^{-1}$.

The Electrochemical Behavior of Copper, Lead, and Cadmium at Different Modified Electrodes

Figure 2 shows the electrochemical responses of 1.0 μM (each) multicomponent Cu(II)/Pb(II)/Cd(II) solution at Qu/CPE, Qu-IL/CPE, and HMS-Qu/CPE. At the Qu/CPE (Figure 2, curve a) and Qu-IL/CPE (Figure 2, curve b), two reduction peaks are observed for copper and lead at −0.11 V and −0.48 V, respectively, and no obvious reduction peak of cadmium is observed. At the Qu-IL/CPE, the reduction peak currents of copper and lead are largely increased, indicating that the chemical and physical properties of IL improve the sensitivity of electrode. At the HMS-Qu/CPE (Figure 2, curve c), the reduction peak of cadmium appears at potential of −0.69 V and peak currents of 60.9 μA. The peak currents of copper and lead are also obviously increased from 31.8 μA to 33.6 μA and from 82.8 μA to 102.3 μA compared with that of Qu-IL/CPE. This tremendous increase in the current density for copper, lead, and cadmium with the HMS-Qu/CPE is apparently due to the cooperative effect of HMS and Qu. Meanwhile the reduction currents of Qu at different modified electrodes are remarkably lower in multicomponent Cu(II)/Pb(II)/Cd(II) solution than those in the blank solution. The reduction current change (ΔIp) of Qu at different modified electrodes was in a sequence of:

$$\begin{aligned} &\text{Qu/CPE}(\Delta Ip = 97.4\,\mu A) \\ &< \text{Qu-IL/CPE}(\Delta Ip = 117.4\,\mu A) \\ &< \text{HMS-Qu/CPE}(\Delta Ip = 153.8\,\mu A). \end{aligned} \quad (1)$$

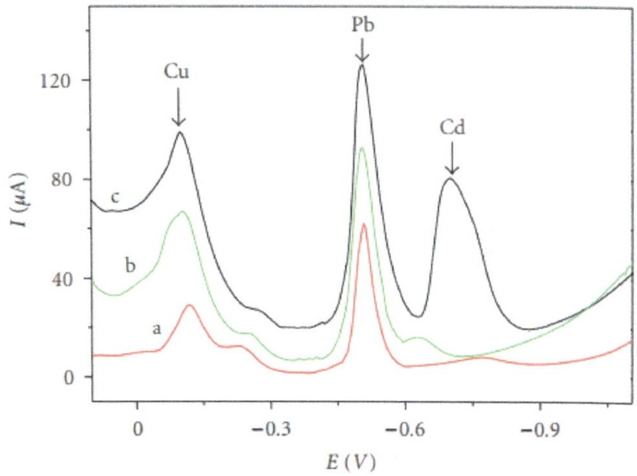

Figure 2: The differential pulse voltammograms of 1.0 μM (each) multicomponent Cu(II)/Pb(II)/Cd(II) solution at Qu/CPE (a), Qu-IL/CPE (b) and HMS-Qu/CPE (c). Condition: 0.1 M HCOONa-HCl (pH4.7), preconcentration potential 0.6 V, preconcentration time 120 s, potential scan range 0.6 V to −1.0 V.

According to above comparison, the concentration of free Qu at HMS-Qu/CPE is the lowest, suggesting that HMS-Qu/CPE can produce more complexes than other modified electrodes.

The chelation of Qu with metal ions in acidic solution has been studied in the previous literatures [9, 10]. When Qu was immobilized on the surface of HMS, effective reaction sites of Qu with metal ions on HMS are remarkably increased due to the high surface area and strong adsorption ability of HMS. Therefore, HMS-Qu/CPE exhibits better capability of adsorptive preconcentration, higher sensitivity and selectivity than Qu/CPE and Qu-IL/CPE. The result indicates that HMS-Qu/CPE is very satisfactory for the development of the simultaneous determination of copper, lead, and cadmium.

EIS experiments were carried out using different modified electrodes at the frequency range from 60 kHz to 0.1 Hz at the perturbation amplitude of 5 mV. Figure 3 shows that the semicircle in high frequency region was not obvious and the straight lines in low-frequency region were obvious in different modified electrodes. The result indicates that the electron transfer rate is fast and the process of electrode reaction is controlled mainly by adsorption and diffusion. The variety of linear slopes at different modified electrodes may be attributed to the different captance characteristic of electric double-layer at modified electrodes [30], which indicated that the surfaces of different electrodes had different specific adsorptions. The specific adsorption of HMS-Qu/CPE was the intensest, Qu/CPE and Qu-IL/CPE showed no obvious difference. These results suggested that HMS-Qu/CPE can effectively improve selectivity and sensitivity and achieved simultaneous determination of copper, lead, and cadmium by the strong chelation and adsorption of HMS-Qu/CPE with metal ions.

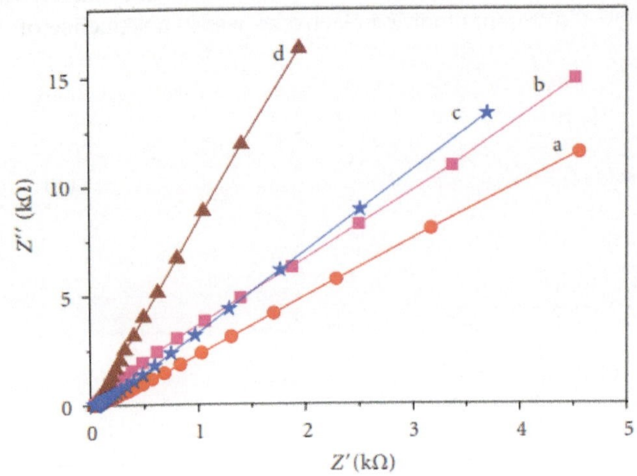

Figure 3: Nyquist plots for bare CPE (a), Qu/CPE (b), Qu-IL/CPE (c) and HMS-Qu/CPE (d) in 1.0 μM (each) multicomponent Cu(II)/Pb(II)/Cd(II) solution and 0.1 M HCOONa-HCl buffer (pH4.7) attached by the corresponding fitted plots (lines). Condition: perturbation amplitude 5 mV, the frequencies swept from 60 kHz to 0.1 Hz, initial potential 0.45 V.

Effect of Some Parameters on Voltammetric Analysis

Voltammetric behavior of the metal ions at HMS-Qu/CPE was evaluated in terms of the influence of stripping parameters such as effect of pH, preconcentration time, and preconcentration potential. The effect of pH was studied by using different types of supporting electrolytes for their suitability in the simultaneous detection of these metals. The HCOONa-HCl buffer solution was found to be more suitable for its better sensitivity. The effect of pH on peak currents was studied in HCOONa-HCl buffer solution. As shown in Figure 4, the peak currents of copper, lead, and cadmium is maximum at pH4.7. Preconcentration time has a significant influence on simultaneous determination. The effect of preconcentration time on the magnitude of the stripping peak current was examined in the range of 20–150 s, and shown in Figure 5. Taking into account the sensitivity and resolutions of adjacent peaks, preconcentration time is optimized to 120 s in the experiments. Preconcentration potential was also varied starting from 0.3 V to 0.8 V. The result showed that lower preconcentration potential might increase the possibility of interference between ions in this study, therefore, the preconcentration potential was applied at 0.6 V. At optimized experimental condition, three peaks are observed for copper, lead, and cadmium at −0.11 V, −0.48 V and −0.69 V, respectively.

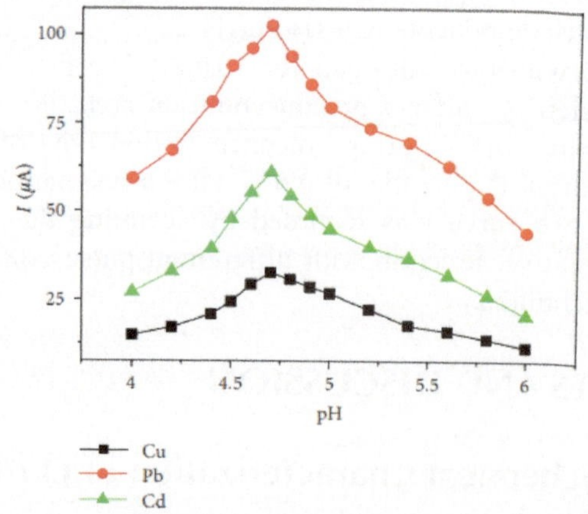

Figure 4: Effect of pH (4.0–6.0) on peak current at HMS-Qu/CPE in 1.0 μM (each) multicomponent Cu(II)/Pb(II)/Cd(II) solution and 0.1 M HCOONa-HCl buffers. Condition: preconcentration potential 0.6 V, preconcentration time 120 s, potential scan range 0.6 V to −1.0 V.

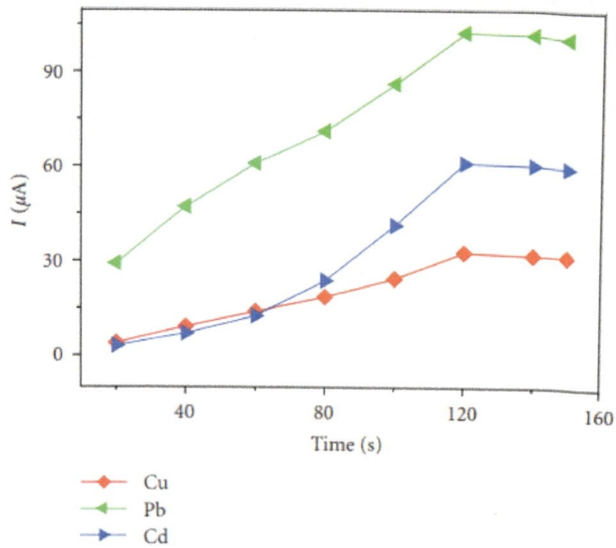

Figure 5: Effect of preconcentration time (20–150 s) on peak current at HMS-Qu/CPE in 1.0 μM (each) multicomponent Cu(II)/Pb(II)/Cd(II) solution and 0.1 M HCOONa-HCl buffer (pH4.7). Condition: preconcentration potential 0.6 V, potential scan range 0.6 V to −1.0 V.

In pH4.7 HCOONa-HCl buffer solution, the relationship between the peak current and concentration was studied by AdSV. Linear range and detection limits of copper, lead, and cadmium were shown in Table 1.

Table 1: The liner range, linear regression equation, and detection limit of copper, lead, and cadmium at HMS-Qu/CPE

Analyte	Linear range (μM)	Regression equation	Correlation coefficient ($n = 6$)	Detection limit (nM)
Cu(II)	0.02–6.0	$Ip(\mu A) = 29.81 + 0.68c\ (\mu M)$	0.9956	5.0
Pb(II)	0.01–8.0	$Ip(\mu A) = 96.61 + 1.37c\ (\mu M)$	0.9974	0.8
Cd(II)	0.004–2.0	$Ip(\mu A) = 58.06 + 0.85c\ (\mu M)$	0.9928	1.0

The interference experiment of some cations and anions was also investigated. Under the optimum conditions, relative error of determination was chosen in the order of 5% or smaller in 1.0 μM (each) multicomponent Cu(II)/Pb(II)/Cd(II) solution. The results showed that 1000-fold Al(III), K (I), sulfate, phosphate and nitrate, 200-fold Mg(II), Sc(III), Co(II) and Ni(II), 50-fold Mo(VI), V(V) and In(III), 25-fold Ag(I) and Cr(VI), did not interfere with the determination.

Under the optimum conditions, the stability and reproducibility of HMS-Qu/CPE were studied in 1.0 μM (each) multicomponent Cu(II)/Pb(II)/Cd(II) solution. After continuous determination 20 times, average stripping peak currents of copper, lead, and cadmium were 31.8 μA, 104.8 μA and 59.5 μA, respectively. The relative standard deviations (RSD) were 3.9%, 3.2% and 7.8%, respectively. In addition, after 30 days of storage in dry conditions, no significant change in current response was observed, suggesting that the modified electrode has an excellent reproducibility and stability.

Application to Real Samples

In order to ascertain the potential application, this newly developed method was employed to determine copper, lead, and cadmium in soil sample. The concentration of copper, lead, and cadmium was determined by the standard addition method, and the results were listed in Table 2. In order to testify the accuracy of this method, the concentration of copper, lead, and cadmium was also detected by FAAS [31]. The results obtained by two methods are in good agreement, indicating that this method is reliable and accurate.

Table 2: The results for determination of copper, lead, and cadmium in soil sample

Analyte	Sample	Found (n = 6, μg/g)	Average value (μg/g)	FAAS (μg/g)	RSD (%)
Cu(II)	Soil 1#	9.76, 10.03, 10.47, 9.88, 10.33, 9.69	10.026	10.064	3.15
	Soil 2#	24.57, 25.41, 25.67, 24.49, 24.63, 25.36	25.024	24.888	2.14
Pb(II)	Soil 1#	12.21, 12.67, 12.89, 12.32, 13.01, 12.28	12.563	12.518	2.71
	Soil 2#	34.29, 34.45, 34.21, 34.17, 34.33, 34.36	34.391	34.299	2.25
Cd(II)	Soil 1#	1.22, 1.19, 1.14, 1.36, 1.20, 1.38	1.247	1.27	7.73
	Soil 2#	1.27, 1.26, 1.46, 1.41, 1.36, 1.29	1.342	1.316	6.12

CONCLUSION

A series of new quercetin modified carbon paste electrodes were fabricated. The electrochemical responses of copper, lead, and cadmium at HMS-Qu/CPE were improved in comparison with Qu/CPE and Qu-IL/CPE. The HMS-Qu/CPE exhibited dramatically high sensitivity and selectivity due to high surface area, numerous active sites, and strong adsorption ability of HMS. After optimizing the parameters such as pH value, preconcentration time, and preconcentration potential, a novel electrochemical method is simple, fast, sensitive, and selective, and can simultaneously determine three metals with high accuracy and stability. Moreover, the HMS-Qu/CPE was successfully used as a sensor for simultaneous determination of copper, lead, and cadmium in soil sample.

ACKNOWLEDGMENTS

This work was supported by Jinan Environmental Monitoring Central Station in Shandong Province of China, the Natural Science Foundation of Shandong (no. ZR2009BM034) and the Key Subject Research Foundation of Shandong Province in China (no. XTD0705).

REFERENCES

1. S. Wang, H. Xu, and J. Li, "Environmental electroanalytical chemistry," Chinese Journal of Analytical Chemistry, vol. 30, no. 8, pp. 1005–1011, 2002.
2. M. Krausa and K. Schorb, "Trace detection of 2,4,6-trinitrotoluene in the gaseous phase by cyclic voltammetry," Journal of Electroanalytical Chemistry, vol. 461, no. 1-2, pp. 10–14, 1999.
3. C. M. G. Van den Berg, "Potentials and potentialities of cathodic stripping voltammetry of trace elements in natural waters," Analytica Chimica Acta, vol. 250, no. 1, pp. 265–276, 1991.
4. M. Paneli and A. Voulgaropoulos, "Applications of adsorptive stripping voltammetry in the determination of trace and ultratrace metals," Electroanalysis, vol. 5, no. 5-6, pp. 355–373, 1993.
5. A. R. Guadalupe and H. D. Abruna, "Electroanalysis with chemically modified electrodes," Analytical Chemistry, vol. 57, no. 1, pp. 142–149, 1985.
6. D. W. M. Arrigan, "Voltammetric determination of trace metals and organics after accumulation at modified electrodes," Analyst, vol. 119, no. 9, pp. 1953–1966, 1994.
7. P. Ugo and L. M. Moreto, "Ion-exchange voltammetry at polymer-coated electrodes: principles and analytical prospects," Electroanalysis, vol. 7, no. 12, pp. 1105–1113, 1994.
8. J. Wang, "Modified electrodes for electrochemical sensors," Electroanalysis, vol. 3, no. 4-5, pp. 255–259, 1991.
9. J. J. Fei, J. F. Wang, and J. N. Li, "Determination of copper in human serum by adsorptive stripping voltammetry at quercetin modified carbon paste electrode," Journal of Analytical Science, vol. 17, no. 5, pp. 375–378, 2001.
10. J. J. Fei, J. N. Li, and F. Y. Yi, "Determination of trace Lead by adsorptive stripping voltammetry at quercetin modified carbon paste electrode," Chinese Journal of Analytical Chemistry, vol. 29, no. 8, pp. 916–918, 2001.
11. M. C. Buzzeo, R. G. Evans, and R. G. Compton,

"Non-haloaluminate room-temperature ionic liquids in electrochemistry," ChemPhysChem, vol. 5, no. 8, pp. 1106–1120, 2004.

12. M. C. Buzzeo, C. Hardacre, and R. G. Compton, "Use of room temperature ionic liquids in gas sensor design," Analytical Chemistry, vol. 76, no. 15, pp. 4583–4588, 2004.

13. A. Lesniewski, J. Niedziolka, B. Palys, C. Rizzi, L. Gaillon, and M. Opallo, "Electrode modified with ionic liquid covalently bonded to silicate matrix for accumulation of electroactive anions," Electrochemistry Communications, vol. 9, no. 10, pp. 2580–2584, 2007.

14. C. Xiang, Y. Zou, L.-X. Sun, and F. Xu, "Direct electron transfer of cytochrome c and its biosensor based on gold nanoparticles/room temperature ionic liquid/carbon nanotubes composite film,"Electrochemistry Communications, vol. 10, no. 1, pp. 38–41, 2008.

15. Y. Liu, L. Huang, and S. Dong, "Electrochemical catalysis and thermal stability characterization of laccase-carbon nanotubes-ionic liquid nanocomposite modified graphite electrode," Biosensors and Bioelectronics, vol. 23, no. 1, pp. 35–41, 2007.

16. W. Sun, R.-F. Gao, R.-F. Bi, and K. Jiao, "Preparation and characteristics of room temperature ionic liquid N-butylpyridinium hexafluorophosphate modified carbon paste electrode," Chinese Journal of Analytical Chemistry, vol. 35, no. 4, pp. 567–570, 2007.

17. Y. Wan and D. Zhao, "On the controllable soft-templating approach to mesoporous silicates," Chemical Reviews, vol. 107, no. 7, pp. 2821–2860, 2007.

18. Y. Wan, Y. Shi, and D. Zhao, "Supramolecular aggregates as templates: ordered mesoporous polymers and carbons," Chemistry of Materials, vol. 20, no. 3, pp. 932–945, 2008.

19. A. Walcarius, E. Sibottier, M. Etienne, and J. Ghanbaja, "Electrochemically assisted self-assembly of mesoporous silica thin films," Nature Materials, vol. 6, no. 8, pp. 602–608, 2007.

20. Q. Zhang, L. Zhang, B. Liu, X. Lu, and J. Li, "Assembly of quantum dots-mesoporous silicate hybrid material for protein immobilization and direct electrochemistry," Biosensors and Bioelectronics, vol. 23, no. 5, pp. 695–700, 2007.

21. P. T. Tanev and T. J. Pinnavaia, "A neutral templating route to Mesoporous molecular sieves," Science, vol. 267, no. 5199, pp. 865–867, 1995.

22. W. Yantasee, Y. Lin, T. S. Zemanian, and G. E. Fryxell, "Voltammetric detection of lead(II) and mercury(II) using a carbon paste electrode modified with thiol self-assembled monolayer on mesoporous silica (SAMMS)," Analyst, vol. 128, no. 5, pp. 467–472, 2003.

23. W. Yantasee, L. A. Deibler, G. E. Fryxell, C. Timchalk, and Y. Lin, "Screen-printed electrodes modified with functionalized mesoporous silica for voltammetric analysis of toxic metal ions," Electrochemistry Communications, vol. 7, no. 11, pp. 1170–1176, 2005.

24. W. Yantasee, Y. Lin, X. Li, G. E. Fryxell, T. S. Zemanian, and V. V. Viswanathan, "Nanoengineered electrochemical sensor based on mesoporous silica thin-film functionalized with thiol-terminated monolayer," Analyst, vol. 128, no. 7, pp. 899–904, 2003.

25. W. Yantasee, C. Timchalk, G. E. Fryxell, B. P. Dockendorff, and Y. Lin, "Automated portable analyzer for lead(II) based on sequential flow injection and nanostructured electrochemical sensors," Talanta, vol. 68, no. 2, pp. 256–261, 2005.

26. W. Yantasee, Y. Lin, G. E. Fryxell, and B. J. Busche, "Simultaneous detection of cadmium, copper, and lead using a carbon paste electrode modified with carbamoylphosphonic acid self-assembled monolayer on mesoporous silica (SAMMS)," Analytica Chimica Acta, vol. 502, no. 2, pp. 207–212, 2004.

27. W. Yantasee, G. E. Fryxell, M. M. Conner, and Y. Lin, "Nanostructured electrochemical sensors based on functionalized nanoporous silica for voltammetric analysis of lead, mercury, and copper," Journal of Nanoscience and Nanotechnology, vol. 5, no. 9, pp. 1537–1540, 2005.

28. D. Li, J. Zhang, and M. Anpo, "Study on the fluorescence properties of benzopyrylium salt in Ti-HMS,"Dyes and Pigments, vol. 63, no. 1, pp. 71–76, 2004.

29. X. Wang, J. Cheng, and K. X. Duo, "Determination of lead in soil by graphite furnace atomic absorption spectrometry," Journal of Agro-Environment Science, vol. 23, pp. 74–76, 2006.

30. S. Liu and C. Cai, "Immobilization and characterization of alcohol dehydrogenase on single-walled carbon nanotubes and its application in sensing ethanol," Journal of Electroanalytical Chemistry, vol. 602, no. 1, pp. 103–114, 2007.

31. Environmental Protection Administration of China, Analysis Method of Water and Wastewater Monitoring, China Environmental Science Press, Beijing, China, 2002.

CHAPTER 9

ELECTRONIC STRUCTURES OF CYCLOMETALATED PALLADIUM COMPLEXES IN THE HIGHER OXIDATION STATES

Bao N. Nguyen[a], Luis A. Adrio[a], Tim Albrecht[a], Andrew J. P. White[a], Mark A. Newton[b], Maarten Nachtegaal[c], Santiago J. A. Figueroa[b] and King Kuok (Mimi) Hii[a]

[a]Department of Chemistry, Imperial College London, Exhibition Road, South Kensington, London SW7 2AZ, UK

[b]European Synchrotron Radiation Facility (ESRF), 6 rue Jules Horowitz, 38043, Grenoble, France

[c]Paul Scherer Institut, 5232 Villigen PSI, Switzerland

ABSTRACT

The electronic and redox properties of a series of cyclometalated Pd complexes with oxidation states of +2, +3 and +4 were examined using a range of currently available spectroscopic and electrochemical techniques. Changes in metal–ligand bond lengths were established by X-ray crystallography and correctly predicted by DFT calculations, from which the frontier orbitals and partial atomic charges can be obtained. X-ray absorption spectroscopy (XAS) revealed interesting XANES features that suggest a synergistic relationship between metal–ligand interactions. The electrochemical study of the Pd(II) dimer was found to contain two sequential oxidative potentials indicative of a weak metal–metal interaction.

INTRODUCTION

The direct activation of Ar–H bonds towards C–C and C–X bond forming reactions is of significant interest, as they have the potential to deliver greater step- and atom-economy than 'classical' cross-coupling reactions. Indeed, 'C–H activation of aromatics without the need for haloaromatics' was identified as one of 'the most aspirational reactions' and an 'unmet need', in a survey conducted by the ACS Green Chemistry Institute's Pharmaceutical Roundtable some years ago,[1] and still remains as one of the most significant challenges for synthesis in the 21st century. Palladium catalysis plays a critical role within this arena: many reactions can be performed simply by using $Pd(OAc)_2$ without extraneous ligands, in the presence of an appropriate oxidant.[2–5] Often referred to as 'oxidative coupling' reactions, it is believed that the reactions involve palladacycle intermediates of higher oxidation states (+3 and/or +4).[6–9] A classic example is the Pd-catalysed *ortho*-acetoxylation of 2-phenylpyridine (2-PhPy, Scheme 1) and its closely-related benzo[*h*]quinoline derivatives. The reaction proceeds in the presence of $Pd(OAc)_2$ and iodosobenzene diacetate, $PhI(OAc)_2$, which functions simultaneously as an oxidant and a source of the acetate.

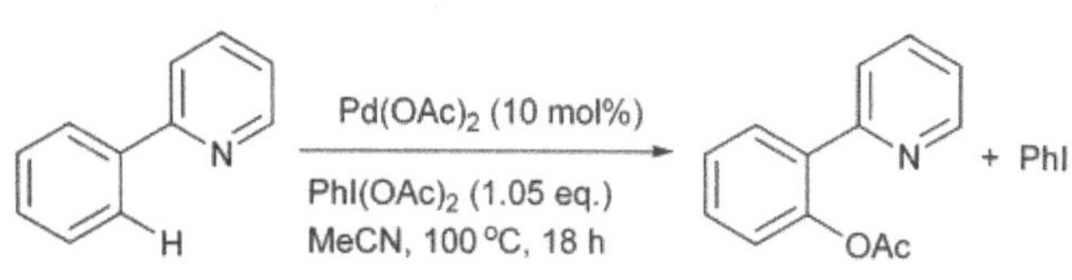

Scheme 1: Pd-catalysed *ortho*-acetoxylation of 2-phenylpyridine.

Citation: Bao N. Nguyen, Luis A. Adrio, Tim Albrecht, Andrew J. P. White, Mark A. Newton, Maarten Nachtegaal, Santiago J. A. Figueroa and King Kuok (Mimi) Hii, Electronic structures of cyclometalated palladium complexes in the higher oxidation states, DOI: 10.1039/C5DT02104A.

First reported by the group of Sanford,[10] the reaction was further investigated by Ritter and co-workers.[11] It is accepted that the reaction initiates with a chelated-assisted cyclometalation to form a stable diacetato-bridged palladacycle **1** (Scheme 2, step i), which undergoes oxidative addition with $PhI(OAc)_2$ to form a diamagnetic bimetallic Pd(III) intermediate **2** (step ii), prior to reductive elimination (step iii) to afford the *ortho*-acetoxylated product. An earlier proposal favoured the formation of Pd(IV) catalytic intermediates. Although a discreet monomeric complex was never isolated directly from the catalytic mixture, the viability of the +4 oxidation state was provided by the reaction of $PhI(OAc)_2$ with a bis-C,N-chelated complex (**3**) to give the octahedral Pd(IV) complex **4**, which undergoes reductive elimination at 60 °C spontaneously to afford the expected product (Scheme 3).

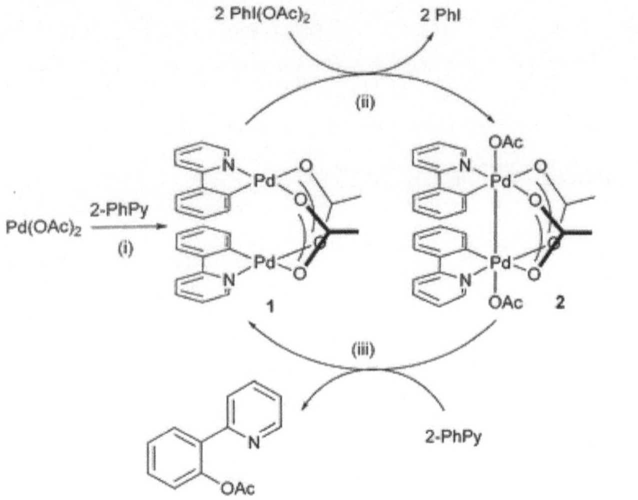

Scheme 2: Proposed catalytic steps of the Pd-catalyzed acetoxylation of 2-PhPy.

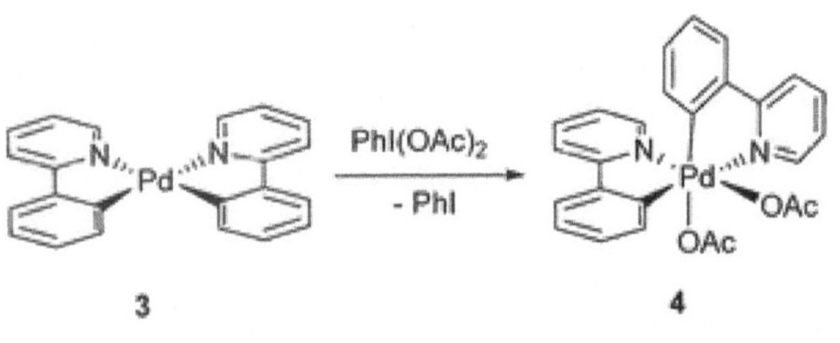

Scheme 3: Oxidative addition of $PhI(OAc)_2$ to complex **3**.

These studies provided a series of palladacycles **1–4** with formal oxidation states of +2, +3 or +4, with only 2-PhPy and acetate as ligands.[12] Thus far, characterisation of these complexes has been achieved by NMR spectroscopy, X-ray crystallography and, more recently, mass spectrometry.[13] While these techniques provide useful structural information, they do not offer direct information on the electronic properties of these cyclometalated complexes; particularly, electronic (re) distribution that occurs upon the increase in oxidation state that may account for their subsequent reactivity. The aim of this work is to combine spectroscopic measurements with theoretical methods to afford direct comparisons and insights into the electronic and redox properties of these closely-related palladacycles. As the extended π-system in 2-phenyl pyridine can also be redox active, we wish to determine the charge distribution and the extent of ligand participation[14] in these compounds.

RESULTS AND DISCUSSION

X-ray crystallography and DFT calculations

Solid-state structures for Pd(II) complexes **1**[15] and **2**[16] had been previously established by X-ray crystallography. In this study, crystals suitable for X-ray diffraction were also obtained for the bis-chelated complexes **3** and **4** (Fig. 1). The structure of **3** shows the expected twisted square planar coordination of two chelated 2-PhPy ligands to the palladium centre, the two [Pd,N,C] coordination planes being inclined to each other by *ca.* 24°. The coordination distances of the two ligands are very similar, with the Pd–N bond length being *ca.* 0.13 Å longer than its Pd–C counterpart in both instances. The X-ray crystal structure for complex **4** had been previously reported as a C2-symmetrical molecule (CCDC code: LUFQIN) with 50 : 50 disorder between the coordinated C and N atoms.[21] In this work, however, we were able to obtain an ordered crystal structure showing an asymmetric, all-*cis* geometry, which is commensurate with the solution state structure observed by NMR spectroscopy (more extensive discussion on the structure determination is provided in the ESI†).

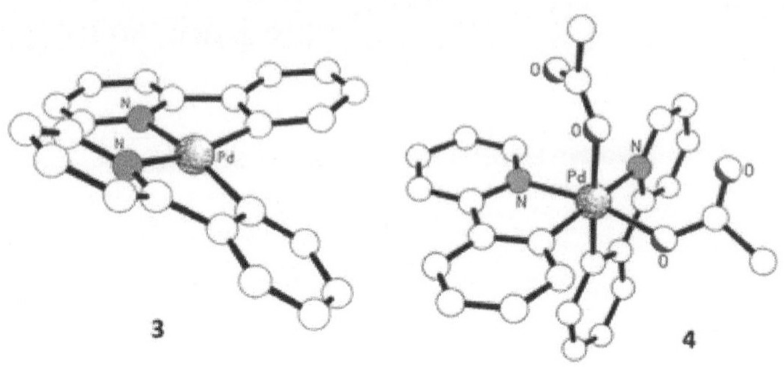

Figure 1: Molecular structures of complexes **3** and **4**.

As the oxidation state of the Pd increases from +2 to +3, the primary coordination sphere of its oxidative adduct might be expected to contract due to increased charge. However, a comparison of the structural parameters of complexes **1** and **2** showed that this only appears to affect the mutually *trans* Pd–O and Pd–C bonds, where a shortened Pd–O bond was counteracted by a corresponding lengthening of the Pd–C bond (by *ca.* 0.02 Å). In contrast, the mutually opposite Pd–N and Pd–O bonds remained essentially the same (Fig. 2). The same trends were also observed in complexes **3** and**4**. In the latter, the bond lengths in the primary coordination sphere of the highly asymmetric oxidative adduct are influenced by different *trans* effects; nevertheless, the equivalent Pd–N and Pd–C bonds (*trans* to C- and N-donor atoms, respectively) were found to be largely unaffected by the change in the formal oxidation state from +2 to +4. In terms of the covalent bonds, no statistically significant changes were observed in the bond lengths composing the chelate ring of the 2-PyPh ligand. Among the acetate ligands, the one positioned *trans* to the C(σ)-donor group in the Pd(IV) complex **4** is particularly notable by having practically identical C–O bond lengths of *ca.* 1.23 Å, while in all other instances the coordinated C–O bond is substantially longer.[17]

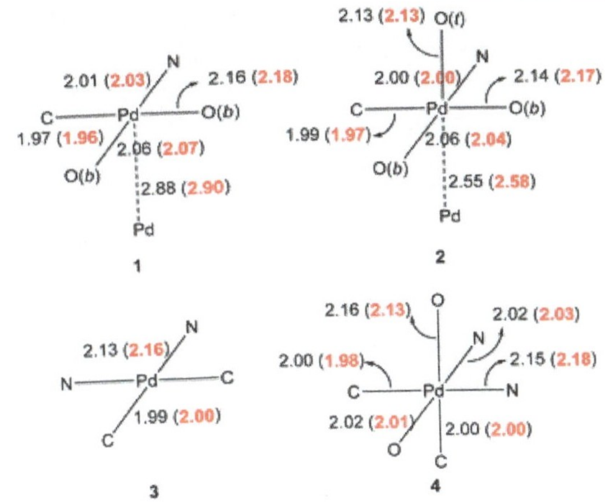

Figure 2: Experimental bond lengths (Å) of the primary coordination sphere (solid state structure). Calculated bond lengths are given in red in parenthesis (dichloroethane solvent field applied).

In order to rationalise these structural features, DFT calculations were performed, whereby the structures of all four complexes were optimised using the WB97XD functional[18] with a mixed Def2TZVP basis set[19] at Pd, 6-31G(d,p) basis set for all other atoms, and solvent correction (CPCM, dichloroethane). Taking into account the differences between solid and solution structures, the calculated bond lengths were in very good accord with the experimental values, with changes in the primary coordination spheres accurately predicted (Fig. 2), including the greater *trans* effects imposed by C- over N-donor atoms in the octahedral complex **4**.

Good agreement in the predicted bond lengths allows us to analyse the frontier orbitals of these molecules (Fig. 3). In terms of dimeric complexes **1** and **2**, it is interesting to note the change from antibonding and bonding interactions between the metal centres, thus supporting the development of M–M interactions during the oxidation process. With the exception of complex **1**, there is substantial ligand participation by both 2-phenylpyridyl and acetate, which are likely to work synergistically to stabilize the oxidative adducts. The participation of the acetate ligand *trans* to the C(σ)-donor group in complex **4** is particularly notable. Natural bond order analysis revealed nearly identical C–O bonds, which accounts for the shorter bond lengths observed previously.

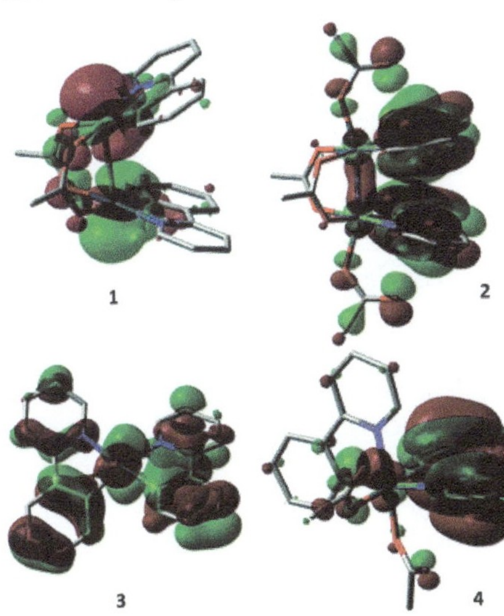

Figure 3: HOMO›s of compounds **1–4**.

The MO analysis was further supplemented by implementing Charge Model 5 (CM5), developed by Cramer and Truhlar[20] to afford prediction of partial atomic charges that are essentially independent of the level of theory and basis set (Fig. 4). As expected, positive charges of the Pd metal centre increased by 50 and 100% upon oxidation to Pd(III) and Pd(IV), respectively. Interestingly, changes in the ligand charges are not only dependent on the donor atom, but also their relative positions. In the e of Pd(III) complex **2**, the axial acetate carries a greater charge than its bridging counterpart. At the same time, only a slight reduction in the partial charge on the C(σ)-donor atom was observed (−0.14 to −0.12). In comparison, a more dramatic electronic redistribution occurred in the transformation of complex **3** to the Pd(IV) adduct **4**. In this e, a significant reduction of C(σ)-donor atom charges was observed (from −0.16 to *ca.* −0.1, irrespective of the *trans*-donor atom). The greater charge separation between the reactive centres, combined with the orbital contributions obtained from the DFT study, supports the experimental observations – the greater propensity for **4** to undergo reductive elimination to generate a C–O bond, compared to **2**.

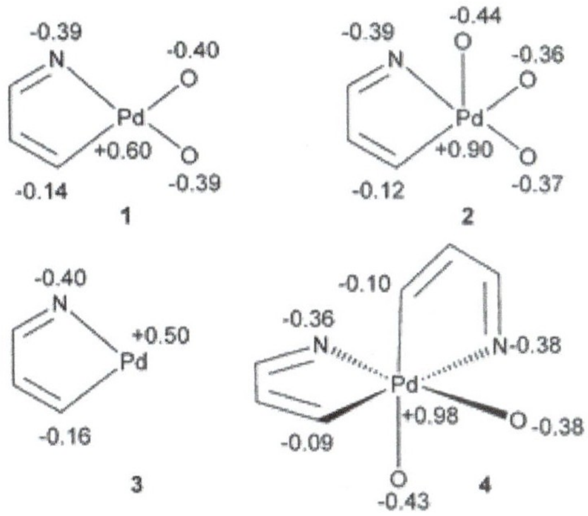

Figure 4: Partial atomic charges in complexes **1–4** obtained by CM5 analysis.

In the next part of our study, we attempted to obtain further experimental verification for the above analysis, in particular, whether changes in the metal›s electronic environment can be detected spectroscopically.

X-ray absorption studies

In a XAS measurement, an electron is excited from the metal's inner shell (in this e 1s, or 'K-edge') to its vacant valence orbitals (4d and 5p for Pd) or, when given more energy, to the continuum. This gives rise to an X-ray absorption spectrum corresponding to the local unoccupied density of states that can be directly correlated to the oxidation state. In the XAS spectrum of Pd, the dominant contribution arises from 1s → 5p transitions, giving rise to 'p peaks' ($\Delta L = 1$, generally accepted as the K-edge). Other transitions into empty orbitals, *i.e.* 1s → 4d transitions ('d peaks') give rise to pre-edge features in the XAS spectra of non-centrosymmetric complexes, due to mixing of (partially occupied) 4d orbitals with 5p-orbitals.[21]

Changes in the metal's oxidation state are expected to affect the positions of both the p- and d-peaks. For Pd complexes, an increase in the oxidation state of the metal can cause either an increase or decrease in the K-edge energy (p-peak) depending on whether the oxidation is metal- or ligand-based. For example, in a previous study of (diphosphine)Pd complexes, an increase in the oxidation state from Pd(0) to Pd(II) was reported to lower the K-edge energy.[22] Conversely, an increase in the K-edge energy was observed in the oxidation of a cyclometalated pentaphenylferrocene-oxazoline Pd(II) complex to a paramagnetic Pd(III) species.[23] In the present study, the relative positions of the d-peaks are also of interest,[24] as they are expected to be more sensitive to ligand effects (metal–ligand interactions operate primarily through the d-orbitals).

XAS spectra of complexes **1** and **3** were recorded at the Pd K-edge (24.35 keV), initially in the solid state (pressed discs), and then as solutions in 1,2-dichloroethane (DCE). The oxidative adducts **2**and **4** are unstable in the solid state and have limited stability in solution[25] at ambient temperature. For these compounds, $PhI(OAc)_2$ was added to the Pd(II) precursors, to generate these complexes *in situ*.[26] Due to the diluted sample, acquisition of multiple scans was required in order to improve the signal-to-noise ratio. The experiments were repeated with freshly prepared solution, such that spectra of sufficient quality can be extracted and inspected, before they were combined for analysis. The EXAFS regions of complexes **1–4** recorded as solution samples were compared with the structural parameters extracted from the X-ray diffraction studies. The goodness-of-fit of the primary coordination shell confirmed that the metal–ligand distances in the solid-state structures are maintained in solution (Fig. S4–S7, ESI†), although the poor resolution and the presence of multiple scatterers did not permit accurate determination of structural parameters beyond the primary coordination environment.

The expected increase in the absorption energy was observed in the XANES spectra for **3** → **4** (+2 → +4). However, the XANES regions for complexes **1** and **2** appeared to be identical (Fig. 5A). The XANES can be deconvoluted into p and d peaks by their transformation into negative second derivative plots (Fig. 5B), revealing three distinct features I–III.[27] Feature I at lower-energy is assigned as the d-peak, while the more energetic features II and III are the p-peaks. In the e of complexes **1** and **2**, the d peak remained unchanged, while the p peaks decreased and increased in energies, respectively. This suggests the presence of ligand contributions that can exert an opposite effect on the different energy states. In comparison, lowering in energy of the d-peak of complex **4** is accompanied by a rise in energy of both the p-peaks, relative to that of complex **3**, which is expected for a metal-based oxidation process.

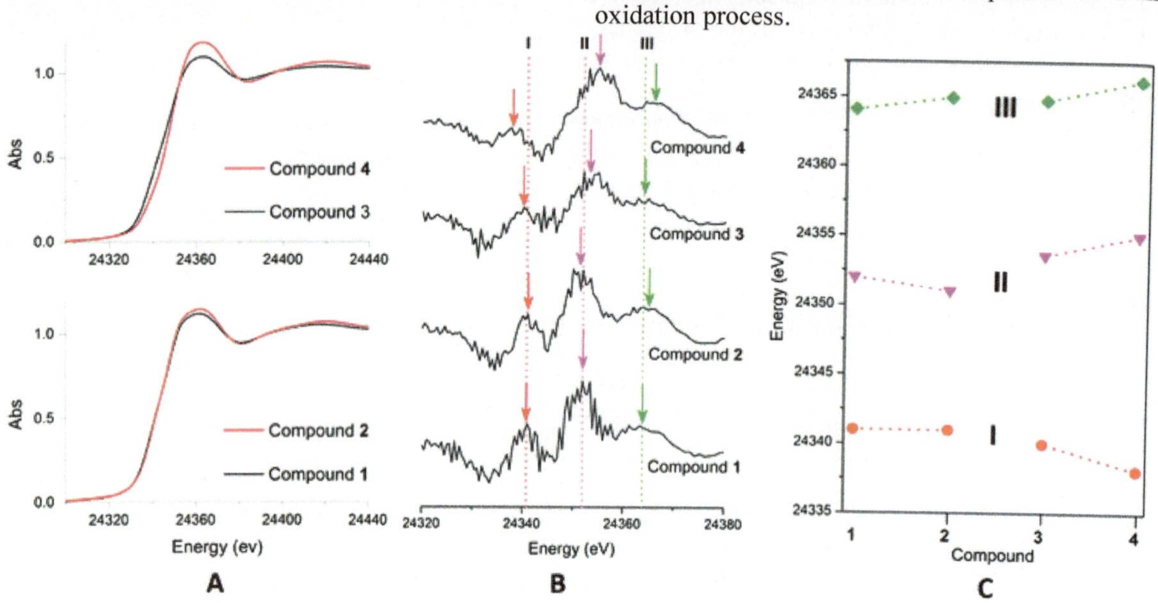

Figure 5: A = normalised Pd–K-edge XAS of complexes **1–4**; B = second-derivative XAS spectra of complexes **1–4**; C = energetic positions of the features corresponding to d- and p-peaks.

Electrochemistry

During the course of this work, the redox behaviour of the dinuclear complex **1** by cyclic voltammetry (CV) in acetonitrile (the reaction solvent – Scheme 1) was investigated. In a study by Bercaw and Green,[28] two irreversible 1-electron oxidative potentials at +0.43 and +1.4 V (*vs.* Fc/Fc+) were recorded in acetonitrile, which were attributed to $Pd^{II}–Pd^{II} \rightarrow Pd^{II}–Pd^{III}$ and $Pd^{II}–Pd^{III} \rightarrow Pd^{III}–Pd^{III}$couples, respectively. More recently, the electrochemical behaviour of complex **1** was re-examined by Dudkina *et al.*,[29] where two oxidative potentials were similarly observed at +0.88 and +1.82 V (*vs.*SCE). However, these were reported to be 2-electron transitions, leading the authors to reassign these to $Pd^{II}–Pd^{II}\rightarrow Pd^{III}–Pd^{III}$ and $Pd^{III}–Pd^{IV}\rightarrow Pd^{IV}–Pd^{IV}$ transitions, respectively.

Comparing these two reports, the assignment by Dudkina *et al.* is more consistent with the relatively low reductive potential of $PhI(OAc)_2$ (*vide infra*), which is able to oxidize complex **1** to **2** ($Pd^{II}–Pd^{II} \rightarrow Pd^{III}–Pd^{III}$), and also used as an oxidant in the catalytic process. If this is the e, it was speculated that the two-electron transition observed at the lower potential could be the result of two successive 1-electron oxidations, due to M···M interactions.[28,30] To pursue this further, the electrochemistry of complex **1** was revisited by duplicating the CV measurements under the reported conditions: using a glassy carbon (GC) electrode with $[Bu_4N][PF_6]$ as an electrolyte in acetonitrile. At a scan rate of 1 V s^{-1}, we were indeed able to resolve an additional, quasi-reversible transition at +0.28 V, in addition to the previously reported redox transition at ~+0.5 V (*vs.* Fc/Fc^+) (Fig. 6A). The two redox potentials were observed reproducibly using a Au(111) electrode, although the signal at higher potential appeared to be less prominent on the more reactive Au surface (S8, ESI†). The peak current ratio was found to depend strongly on the scan rate (Fig. 6B), changing from ~0.7 at 1.0 V s^{-1} to about ~0.15 at 0.1 V s^{-1}, which may explain why it was not observed in earlier experiments (recorded at a scan rate of 0.1 V s^{-1}). At the same time, the peak observed on the reversed scan at *ca.*+0.23 V also becomes more prominent at higher scan rates.

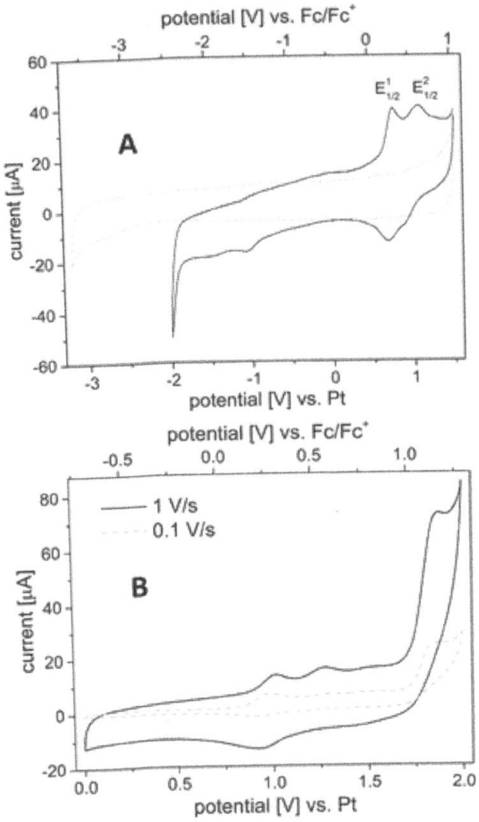

Figure 6: **A** (top) CV of complex **1** recorded using a GC electrode (0.1 M $[Bu_4N]PF_6$ in CH_3CN, 1 V s^{-1}) in the low potential region (grey dotted line = background scan). **B** (bottom) at different scan rates.

Hence, we have identified successive Pd^{II}–Pd^{II} → Pd^{II}–Pd^{III} and Pd^{II}–Pd^{III} → Pd^{III}–Pd^{III} transitions of complex **1** to occur in acetonitrile at redox potentials of +0.28 and ~+0.5 V (*vs.* Fc/Fc^+), respectively. The mixed-valent complex is highly unstable and can only be observed (as a reversible oxidation) at high scan rates. The revised oxidative potentials of complex **1** are commensurate with the use of $PhI(OAc)_2$ as an effective 2-electron oxidant. It is known that in the presence of an acid, the hypervalent iodonium reagent dissociates in acetonitrile to form phenylacetoxyiodonium ions that undergo irreversible 2-electron reduction at −0.91 V (*vs.* SCE) to the corresponding iodobenzene and acetate ion (Scheme 4),[31] *i.e.* the pre-dissociation of $PhI(OAc)_2$ generates nucleophilic acetate ions, as well as the active oxidant $[PhI(OAc)]^+$ with the requisite over-potential to effect the transformation of **1** to **2**, but not to Pd^{IV}.

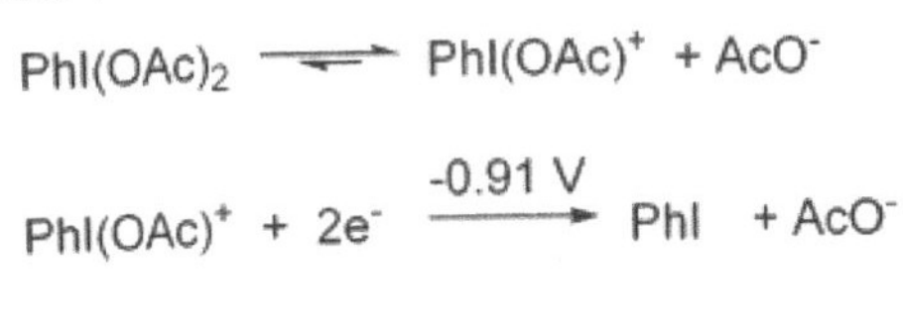

Scheme 4: Redox chemistry of $PhI(OAc)_2$.

CONCLUSIONS

A number of spectroscopic and theoretical techniques have been applied to examine the electronic features of two 2-PhPy–Pd complexes, as they undergo oxidative addition reactions with $PhI(OAc)_2$. Highly complementary information on the structure and reactivity of these complexes can be obtained from this study, offering a coherent picture of the electronic nature of these complexes: (1) a study of the solid-state structures of the complexes revealed some interesting features, with the ligands *trans*-effect being the overriding factor in M–L bond lengths, rather than changes in the formal oxidation states; (2) an appropriate method has been found for the DFT calculation, which correctly predicts the changes in M–L distances upon oxidative addition. The corresponding analysis of the frontier orbital gave insights into the participation of ligands in the stability of the oxidative adducts; (3) charge distributions in these metal complexes were revealed by CM5 calculations; (4) XAS studies revealed that the oxidation of a Pd(II) palladacycle **3** to Pd(IV) adduct **4** is predominantly a metal-based process (rise in K-edge), while an adjustment in the metal–ligand interaction was observed in the oxidative addition of a Pd(II) dimer to a Pd(III) dimer (opposite shifts in p-peaks); and (5) during the course of this work, the oxidative potentials for complex **1** in acetonitrile have been revised, commensurate with the use of $PhI(OAc)_2$ as an oxidant. This work provides a useful benchmark for future studies of similar palladacycles with different ligands, while highlighting additional challenges involve further refinement of these experiments to afford more accurate predictions/quantification of electronic properties that are relevant to redox catalysis. The practical and theoretical challenges of determining the structure of discreet molecular complexes (exacerbated by their limited solubility) are also highlighted in this work. This will require further design of sample environments/acquisition and detection methods to achieve better resolution of EXAFS spectra.

ACKNOWLEDGEMENTS

This work was supported by an EPSRC grant: 'Elucidate and Separate' (ELSEP, grant number: EP/G070172/1), with further contributions from the Pharmacat Consortium (AstraZeneca, GlaxoSmithKline and Pfizer). Travel to the SLS was supported by the European Commission's FP6 program (Strengthening the European Research Area, Research Infrastructures. No: RII3-CT-2004-506008). We thank Johnson Matthey PLC for an academic loan scheme, which provided Pd salts for this work. We are also grateful to ESRF (BM23) and SLS (SuperSAXS) for provision of beam time and access to their facilities.

NOTES AND REFERENCES

1. D. J. C. Constable, P. J. Dunn, J. D. Hayler, G. Humphrey, J. L. Leazer, R. J. Linderman, K. Lorenz, J. Manley, B. Pearlman, A. Wells, A. Zaks and T. Y. Zhang, *Green Chem.*, 2007, 9, 411 .
2. T. W. Lyons and M. S. Sanford, *Chem. Rev.*, 2010, 110, 1147.
3. K. M. Engle, T. S. Mei, M. Wasa and J. Q. Yu, *Acc. Chem. Res.*, 2012, 45, 788.
4. T. S. Mei, L. Kou, S. Ma, K. M. Engle and J. Q. Yu, *Synthesis*, 2012, 1778 .

5. S. R. Neufeldt and M. S. Sanford, *Acc. Chem. Res.*, 2012, 45, 936.

6. A. J. Canty, *Dalton Trans.*, 2009, 10409 .

7. K. Muniz, *Angew. Chem., Int. Ed.*, 2009, 48, 9412.

8. P. Sehnal, R. J. K. Taylor and I. J. S. Fairlamb, *Chem. Rev.*, 2010, 110, 824.

9. K. J. Bonney and F. Schoenebeck, *Chem. Soc. Rev.*, 2014, 43, 6609 .

10. (*a*) A. R. Dick, K. L. Hull and M. S. Sanford, *J. Am. Chem. Soc.*, 2004, 126, 2300; (*b*) D. Kalyani and M. S. Sanford, *Org. Lett.*, 2005, 7, 4149.

11. (*a*) D. C. Powers, M. A. L. Geibel, J. E. M. N. Klein and T. Ritter, *J. Am. Chem. Soc.*, 2009, 131, 17050; (*b*) D. C. Powers and T. Ritter, *Nat. Chem.*, 2009, 1, 419; (*c*) D. C. Powers and T. Ritter, *Nat. Chem.*, 2009, 1, 302.

12. Defined as the charge left on the metal after all ligands have been removed in their normal, close-shell configuration.

13. L. Guo, Y. Xu, X. Wang, W. Liu and D. Lu, *Organometallics*, 2013, 32, 3780 .

14. For example, see: (*a*) V. Lyaskovskyy and B. de Bruin, *ACS Catal.*, 2012, 2, 270 ; (*b*) V. K. K. Praneeth, M. R. Ringenberg and T. R. Ward, *Angew. Chem., Int. Ed.*, 2012, 51, 10228.

15. (*a*) M. Dincer, N. Ozdemir, M. E. Gunay and B. Cetinkaya, *Acta Crystallogr., Sect. E: Struct. Rep. Online*, 2008, 64, M381 ; (*b*) H. Y. Thu, W. Y. Yu and C. M. Che, *J. Am. Chem. Soc.*, 2006,128, 9048.

16. J. M. Racowski, A. R. Dick and M. S. Sanford, *J. Am. Chem. Soc.*, 2009, 131, 10974.

17. Experimental bond lengths for unbound acetate ions are 1.21 and 1.36 Å for C=O and C–O, respectively, see: M. Ibrahim and E. Koglin, *Acta Chim. Slov.*, 2004, 51, 453 .

18. J.-D. Chai and M. Head-Gordon, *Phys. Chem. Chem. Phys.*, 2008, 10, 6615 .

19. F. Weigend and R. Ahlrichs, *Phys. Chem. Chem. Phys.*, 2005, 7, 3297 .

20. A. V. Marenich, S. V. Jerome, C. J. Cramer and D. G. Truhlar, *J. Chem. Theory Comput.*, 2012, 8, 527 .

21. K. Getty, M. U. Delgado-Jaime and P. Kennepohl, *Inorg. Chim. Acta*, 2008, 361, 1059.

22. M. Tromp, J. A. van Bokhoven, M. Q. Slagt, R. J. M. K. Gebbink, G. van Koten, D. E. Ramaker and D. C. Koningsberger, *J. Am. Chem. Soc.*, 2004, 126, 4090.

23. S. H. Eitel, M. Bauer, D. Schweinfurth, N. Deibel, B. Sarkar, H. Kelm, H. J. Kruger, W. Frey and R. Peters, *J. Am. Chem. Soc.*, 2012, 134, 4683.

24. In theory, better resolved d-peaks can be obtained from XAS recorded at the Pd L3-edge. But it is not possible to acquire such low-energy spectra (*ca.* 3 keV) in this experiment as complexes 2and 4 can only be generated in solution.

25. Under these conditions, complexes 2 and 4 undergo slow decomposition over 3 h, where noticeable degradation of the spectra can be observed. The limited operational window prevents the application of other XAS techniques with long acquisition times, *e.g.* those based on florescence emissions.

26. The generation of the oxidative adduct under the conditions of XAS experiments were established by off-line NMR studies (see the ESI).

27. In principle, better resolution of the absorption edges may be obtained through high energy resolution fluorescence detection (HERFD) XAS. However, this is currently not feasible at such high energies (Pd K-edge 24.35 keV) at either SLS or ESRF, where these experiments were conducted.

28. J. E. Bercaw, A. C. Durrell, H. B. Gray, J. C. Green, N. Hazari, J. A. Labinger and J. R. Winkler, *Inorg. Chem.*, 2010, 49, 1801.

29. Y. B. Dudkina, D. Y. Mikhaylov, T. V. Gryaznova, A. I. Tufatullin, O. N. Kataeva, D. A. Vicic and Y. H. Budnikova, *Organometallics*, 2013, 32, 4785 .

30. I. O. Koshevoy, P. Lahuerta, M. Sanau, M. A. Ubeda and A. Domenech, *Dalton Trans.*, 2006, 5536 .

31. (*a*) M. Giffard, G. Mabon, E. Leclair, N. Mercier, M. Allain, A. Gorgues, P. Molinié, O. Neilands, P. Krief and V. Khodorkovsky, *J. Am. Chem. Soc.*, 2001, 123, 3852 ; (*b*) G. Kokkinidis, M. Papadopoulou and A. Varvoglis, *Electrochim. Acta*, 1989, 34, 133 .

CHAPTER 10

RECENT PROGRESS IN METAL–ORGANIC COMPLEXES FOR OPTOELECTRONIC APPLICATIONS

Hui Xu[ab], Runfeng Chen[bc], Qiang Sun[b], Wenyong Lai[c], Qianqian Su[b], Wei Huang[cd] and Xiaogang Liu[be]

[a]Key Laboratory of Functional Inorganic Material Chemistry, Ministry of Education and School of Chemistry and Materials, Heilongjiang University, 74 XueFu Road, Harbin 150080, China

[b]Department of Chemistry, Faculty of Science, National University of Singapore, 3 Science Drive 3, Singapore 117543, Singapore

[c]Key Laboratory for Organic Electronics & Information Displays, Institute of Advanced Materials, Nanjing University of Posts & Telecommunications, Nanjing 210023, China

[d]Singapore-Jiangsu Joint Research Center for Organic/Bio-electronics and Information Displays and Institute of Advanced Materials, Nanjing University of Technology, Nanjing 211816, China

[e]Institute of Materials Research and Engineering, Agency for Science, Technology and Research, 3 Research Link, Singapore 117602, Singapore

ABSTRACT

The design and characterization of metal–organic complexes for optoelectronic applications is an active area of research. The metal–organic complex offers unique optical and electronic properties arising from the interplay between the inorganic metal and the organic ligand. The ability to modify chemical structure through control over metal–ligand interaction on a molecular level could directly impact the properties of the complex. When deposited in thin film form, this class of materials enable the fabrication of a wide variety of low-cost electronic and optoelectronic devices. These include light emitting diodes, solar cells, photodetectors, field-effect transistors as well as chemical and biological sensors. Here we present an overview of recent development in metal–organic complexes with controlled molecular structures and tunable properties. Advances in extending the control of molecular structures to solid materials for energy conversion and information technology applications will be highlighted.

INTRODUCTION

Metal–organic complexes, combining facile synthesis with tunable optical properties, have attracted much interest over the past decades owing to their enormous potential application in optoelectronic devices. For example, with the right set of optoelectronic properties, metal–organic complexes may point the way for the next generation of solar energy conversion, replacing fragile, high-temperature processing semiconductor materials with durable, low-cost printed plastic electronics.

Fig. 1 shows important milestones in the development of metal–organic complexes for optoelectronic applications. Discovery of the first organic semiconductor copper phthalocyanine (CuPc) can be dated back to 1948.[1] In the late 1980s, Tang and Van Slyke at Kodak fabricated the first thin film light-emitting diode using tris(8-quinolinolato) aluminum(III) (Alq_3) as the emitting layer.[2] Research into organic light-emitting diodes culminated in 1998 with the work of Forrest

Citation: Hui Xu, Runfeng Chen, Qiang Sun, Wenyong Lai, Qianqian Su, Wei Huang and Xiaogang Liu, Recent progress in metal–organic complexes for optoelectronic applications, DOI: 10.1039/C3CS60449G.

and co-workers, who reported the discovery of organic light-emitting devices (OLEDs) employing phosphorescent metal–organic complexes.[3] These pioneering studies have truly revolutionized our understanding of the correlation between the chemical structure and optical properties of the complexes and also led to remarkable progress in developing efficient optoelectronic devices at a moderate price.

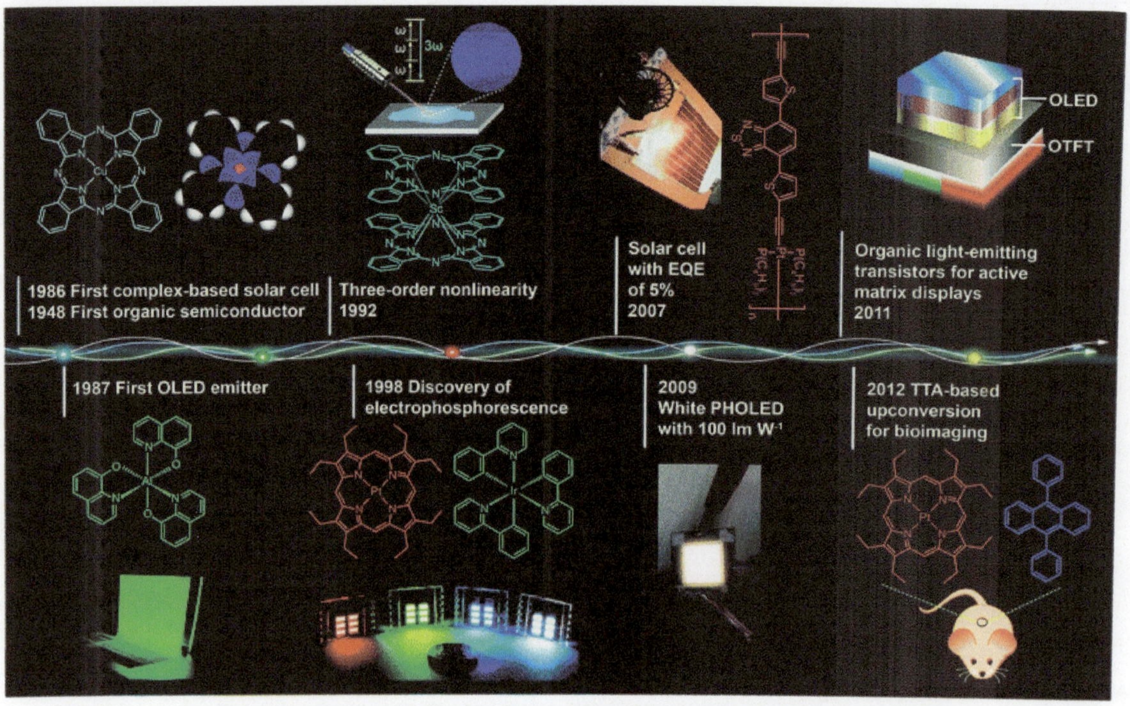

Figure 1: Milestones in the development of optoelectronic complexes. Eley firstly demonstrated semiconducting behavior of CuPc complexes in 1948.[1] The p-type CuPc was utilized by Tang to fabricate the first organic p–n junction solar cell with an efficiency of 1% in 1986.[4] In the following year, Tang and Van Slyke demonstrated the first OLED with practical luminance using another complex emitter Alq_3.[2] In 1992, a three-order nonlinear optical property was observed for phthalocyanine complex $ScPc_2$.[5] Baldo *et al.* in 1998 reported electrophosphorescence phenomenon using a red phosphorescent complex dye 2,3,7,8,12,13,17,18-octaethyl-21*H*,23*H*-porphine platinum(II) (PtOEP) as emitter.[3] In 2007, Wong *et al.* further improved the efficiency of complex-based solar cells to a record of 5% with a Pt(II) coordination polymer.[6] In 2009, with light out-coupling technology, Reineke *et al.* achieved a maximum power efficiency of 100 lm W^{-1} in phosphorescent OLEDs, which was comparable to that obtained from fluorescent tubes.[7] In 2011, McCarthy *et al.* demonstrated low-voltage driving active-matrix display pixels by integrating OLED cells into OTFT.[8] In 2012, the upconversion-emission nanoparticle containing triplet–triplet annihilation system was firstly applied in bio-imaging by Liu *et al.*[9]

This review article provides an update on recent advances in the development and implementation of metal–organic complexes for optoelectronic applications. We begin by describing the basic properties of the metal ion and organic molecule, which have a strong influence on the performance of an optoelectronic device. Next, a brief overview of typical luminescence processes occurring in organic molecules or metal–organic complexes will be presented, followed by thorough coverage of the latest research on OLEDs involving metal–organic complexes. The emphasis will be placed on the illustration of different types of fluorescent and phosphorescent emitters, and on the discussion of recent efforts in device configurations for high-efficiency OLEDs. The application of metal complex-based donors in organic photovoltaics will be covered in a subsequent chapter. The last chapter of this review highlights the emerging field of triplet–triplet annihilation upconversion through use of metal–organic complexes. A major goal of this review is to provide illustrative accounts on recent work and systematize our knowledge of the subject, extracting fundamental principles from diverse research topics.

BASIC PROPERTIES OF METAL–ORGANIC COMPLEXES

Metal ions

A key to the design of high performance metal–organic-based optoelectronic devices is the understanding of the pivotal role of metal ions involved in controlling the optoelectronic properties of metal–organic complexes.

Directing molecular coordination.

The metal centers in metal–organic complexes are endowed with tailor-made functional capabilities that the organic components may not be able to duplicate. For chemists, the availability of coordination sites on metal enables unique structural design in the assembly of organic molecules. The great versatility and directionality offered by metal–ligand coordinative bonding allow for the precise control over the shape and stereochemistry of the assemblies. With multidentate ligands, metal–organic frameworks and coordination polymers with high porosity and chemical stability can be realized. The ability to tailor coordination geometry of the complexes is crucial to the rational control over mechanical, thermodynamic, and optoelectronic properties of the resulting metal–organic materials.

Enhancing radiative transition.

For efficient optoelectronic devices, molecules with high radiative rates are generally selected as the emitters. As a rule of thumb, molecular rigidity reduces the possibility of non-radiative transitions by decreasing vibrations, thereby minimizing the intersystem crossing to triplet state and collision heat loss. The complexation of metal ions by conjugated organic molecules can improve structural rigidity and sustain many excitation–emission cycles. As a result, these metal–organic complexes can exhibit enhanced fluorescence. Furthermore, the involvement of metal orbitals in excited states is capable of inducing new photophysical behaviors in pure organic compounds, such as highly efficient phosphorescent emission. Indeed, the introduction of heavy metal ions, such as Pt(II) or Ir(III), in molecules favors the formation of the triplet states through rapid intersystem crossing and thus enhances spin-forbidden phosphorescence.

Improving charge transport characteristics.

The charge transport characteristics of organic semiconductors are key attributes that have important consequences for the device performance of these materials. For practical applications, it is important to realize high charge carrier mobilities within the organic semiconductor. Notably, the transport processes in these systems can be dramatically influenced by the addition of metal ions. Transition metals feature delocalized valence electrons and different oxidation states, which are beneficial for charge carrier injection and transportation.

Photoresponsive organic molecules

Owing to the weak electronic delocalization, photoresponsive organic materials have two distinctive features as compared to their inorganic counterparts. One is the existence of well-defined spin (singlet and triplet) states in isolated molecules. The luminescence processes in organic materials are typically associated with the excited states of molecules, while the luminescence spectra in inorganic materials are attributed to either defects or impurities in the host lattice or to the excited states of the isolated atom or ion. A second important difference arises from the fact that the size of excitations (Frenkel excitons) tends to be small in organic molecular crystals, dominated by the strong coulomb interaction between an electron and a hole. Thus, the molecular excitons with a considerable amount of binding energy (0.1–1 eV) can be located on one molecule.

Luminescence processes.

It must be emphasized that the luminescence spectrum is always investigated along with the absorption spectrum of a particular molecule. Fig. 2 shows the classic luminescence processes taking place in the π-electronic system of organic molecules. A complex is luminescent if it emits light upon absorption of a radiation. The absorption process, governed by quantum mechanical selection rules, occurs primarily between the ground state S_0 and the singlet states S_1, S_2, *etc.* Superimposed on each of the electronic levels is a set of sublevels associated with the vibrational and rotational energy of the molecule. The principal fluorescence emission is generally induced by the transition from the lowest excited singlet state S_1 to the ground state, irrespective of the initial state excited. This can be ascribed to the rapid nonradiative process of internal conversion between higher excited states S_2, S_3,..., *etc.* and the lowest excited state S_1. Nonradiative processes can also be observed in intersystem crossing from the singlet manifold to the triplet manifold and *vice versa*. These singlet–triplet transitions, albeit forbidden by quantum mechanics, will still occur but progress at significantly slower time scales than singlet–singlet transitions. The radiative decay from the excited triplet state back to a singlet state is known as phosphorescence. For a given molecule, the probability of nonradiative energy losses is much higher in the triplet state than in the singlet state because of the substantially longer lifetime of the triplet state. Phosphorescent molecules have the ability to store light energy and release it gradually. Pure organic compounds that exhibit room-temperature phosphorescence are very rare. However, the phosphorescence can be enhanced by the addition of metal ions that promote intersystem crossing *via* spin–orbital coupling.

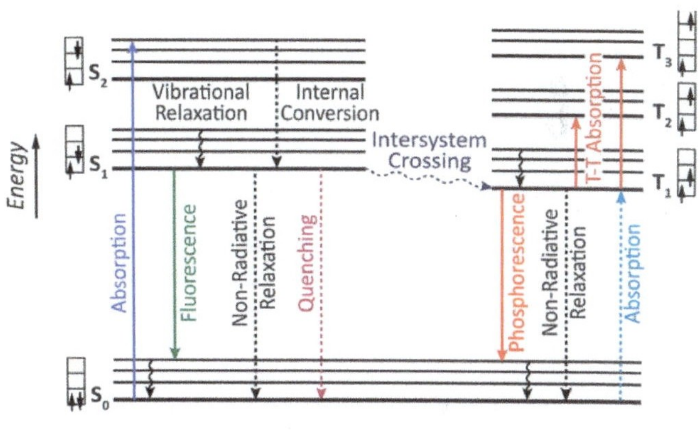

Figure 2: Jablonski energy level diagram showing principal luminescence processes in an organic molecule (left: singlet manifold; right: triplet manifold). The full and dotted arrows represent radiative and non-radiative processes.

The emphasis on the optoelectronic process can vary depending on specific applications. For organic photovoltaics, more attention is paid to the excitation process of organic molecules. The absorption spectra, corresponding to the excitation process, of the organic molecules should match well with incident solar energy to render high conversion efficiency. For light-emitting applications, the center of attention is given to the luminescence process which can be quantified by quantum efficiency. To achieve a high quantum efficiency, one has to boost the probability of radiative transition and suppress the energy loss transitions *via* vibrational relaxation or internal conversion. The relative photoluminescence quantum yield ( ) of organic compounds in dilute solutions can be measured using conventional dyes as references according to the equation shown below:

$$\Phi_s = \Phi_r \cdot \frac{A_r}{A_s} \cdot \frac{D_s}{D_r} \cdot \left(\frac{n_s}{n_r}\right)^2 \quad (1)$$

where A is the absorbance, D is the integrated area of the emission peaks, and n is the refractive index of the solvent.

The subscripts s and r represent the sample and reference, respectively. The absolute quantum yield is essentially the luminescence intensity integrated over the entire spectrum using an integrating sphere.

Energy transfer.

Two common mechanisms of energy transfer are the Förster or coulombic energy transfer mechanism and the Dexter or exchange energy transfer mechanism (Fig. 3). For highly efficient Förster resonance energy transfer (FRET) to occur, three primary conditions need to be met: (i) the donor and acceptor molecules must be in close proximity to one another, typically in the range of 1 to 10 nm; (ii) the absorption spectrum of the acceptor must overlap the emission spectrum of the donor; and (iii) the orientations of donor and acceptor transition dipoles must be approximately parallel.[10] The FRET efficiency (E) depends on the inverse sixth-distance between the donor and the acceptor as described in eqn (2):

$$E = \frac{R_0^6}{R_0^6 + r^6} \qquad (2)$$

where r is the actual distance between the donor and the acceptor and R_0 is the Förster distance at which half the energy is transferred. Note that R_0 is dependent on a number of factors as expressed byeqn (3):

$$R_0 = 0.211\left[\frac{\Gamma^2 n^{-4}\eta_D \int_0^\infty F_D(\lambda)\varepsilon_A \lambda^4\,d\lambda}{\int_0^\infty F_D(\lambda)d\lambda}\right]^{\frac{1}{6}} \qquad (3)$$

where ^{2} is the orientational factor, n is the refractive index, $_D$ is the fluorescence quantum yield of the donor in the absence of the acceptor, F_D is the normalized donor emission spectrum and $_A$ is the acceptor molar extinction coefficient.

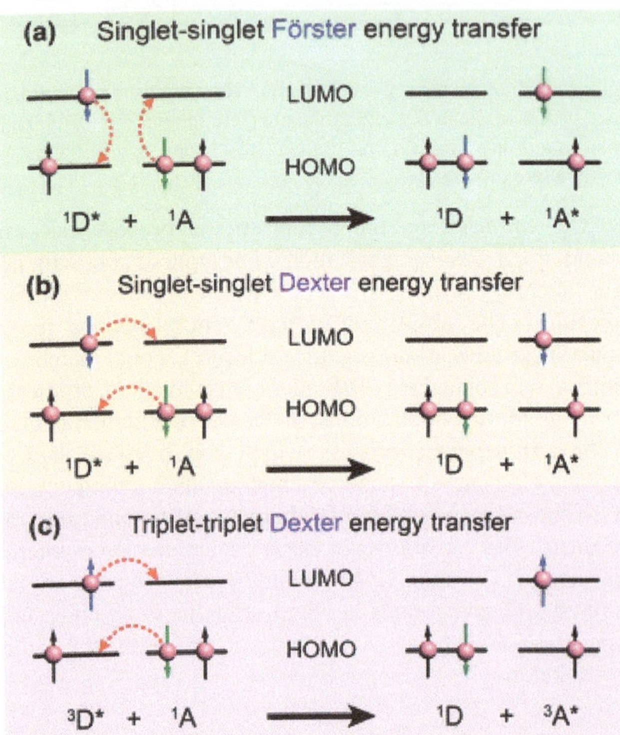

Figure 3: Schematic diagrams of (a) Förster energy transfer and (b, c) Dexter energy transfer. D and A denote donor and acceptor, respectively.

In contrast, Dexter energy transfer can travel only short distances (<1 nm) for both singlet and triplet excited states as the exchange interactions are dictated by the wavefunction overlap of electron clouds.[11] The rate constant of Dexter energy transfer is given by eqn (4):

$$k = CJ\exp\left[\frac{-2r}{L}\right] \qquad (4)$$

where C is an experimental constant, J is the normalized spectral overlap integral, r is the distance between the donor and the acceptor, and L is the sum of van der Waals radius.

The degree of spectral overlap between donor emission spectrum and acceptor absorption spectrum is clearly the dominant factor in determining the efficiency of energy transfer. To increase the probability of energy transfer, the excited energy gap between the donor and the acceptor generally needs to be controlled within the range of 2000–4000 cm^{-1}. Fortunately, a wealth of synthetic strategies and technologies for chemical synthesis allow for facile tailoring of the band gap and energy levels of organic molecules owing to their structure-dependent electronic properties.

METAL–ORGANIC COMPLEXES FOR OLEDS

An OLED is a thin, glowing component made of layers of organic semiconducting materials. A typical multilayer design is shown in Fig. 4. When a voltage is applied between two electrodes, the cathode pumps electrons into the dye-containing emission layer located between the two electrodes while the anode supplies electron holes. These holes can jump to the emission layer to form bound electron–hole pairs, namely the so-called excitons. The return of the excitons to the ground state, *i.e.* the recombination of holes and electrons, leads to a relaxation of the energy levels of the electrons in the form of light. The design of sandwiched layers facilitates charge injection and enhances the recombination rate of electrons and holes.

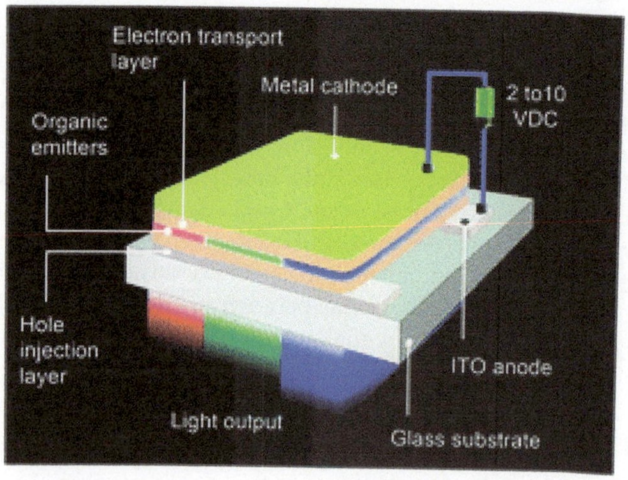

Figure 4: Device structures of prototype sandwich-structured OLEDs.

Presently, the power efficiency of three-color OLEDs, given in lumen per watt (lm W^{-1}), can reach up to 100 lm W^{-1} for green,[12] 30 lm W^{-1} for red,[13] and 40 lm W^{-1} for blue.[14] Similar to fluorescent tubes, white-emitting OLEDs can show remarkable efficiencies up to 100 lm W^{-1} through external out-coupling technologies that reduce light diffusion and scattering.[7] To suppress the concentration quenching effect of the emitters and improve device efficiency, the emission layer is often prepared by blending guest emitters into host matrices. The choice of the emitters is of paramount importance in the device design as they affect recombination and radiative probability, and consequently determine the device performance, including IV characteristics, brightness and power efficiency. On the basis of different excited states, the emitters can be classified into two categories: fluorescent and phosphorescent metal–organic complexes.

Electrofluorescent complexes

The ability of fluorescent complexes with different peripheral substituents to alter the emission profile of OLEDs is at the heart of their role as light-emitting centers. Since the pioneering work of Tang and Van Slyke on Alq_3,[2] significant progress has been made in developing highly efficient emitters or electron-transport materials containing Zn(II), Mg(II), Be(II), Al(III), and In(III) complexes.[15]

Electrofluorescent Zn(II) complexes.

The interest in Zn(II) complexes with Schiff bases and benzoheterocycles has been recently boosted due to their strong photoluminescence emissions in the blue region. However, organic electroluminescent devices with a high luminance of ~1000 cd m^{-2} and photostability were reported only through use of a limited number of Zn(II) complexes. Son*et al.* reported a series of light-emitting Zn(*N*,*O*-OPh^{OxZ}ArX)$_2$ complexes (**1–6**)[15b] in which the emission can be conveniently tuned from blue to green by introducing different substituents at the 4-position of the oxazolylphenolate ligand. They found that the electron-donating groups can induce a large red-shift of the emission, while the electron-withdrawing groups only slightly shift the emission wavelength. The authors attributed this notable observation to the different influence of the substituents on HOMOs of the complexes. Although these complexes showed very low quantum yield of less than 20% in solution, the non-doped devices containing the complex **3** gave rise to a high luminance of 1720 cd m^{-2} at 17 V with a moderate current efficiency of 0.2 cd A^{-1}.

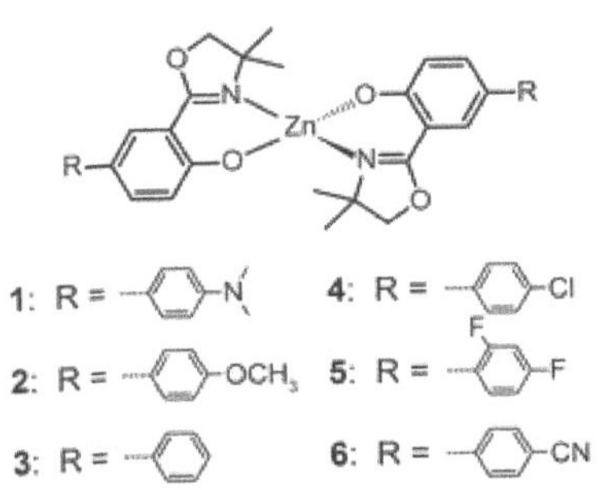

A subsequent study by Xu *et al.* reported the synthesis of blue-emitting Zn(II) complexes (**7–9**) using a phenylbenzoimidazole ligand.[15c] The authors found that the presence of a pendant *N*-hexylcarbazole group endows the complex **9** with a deep blue emission at 422 nm and a high quantum yield of 64%. Additionally, the carbazole substituent provides improved color purity and enhanced thermal stability with a decomposition temperature (T_d) of greater than 400 °C. This improvement is attributed to the largely elevated HOMO level (~0.6 eV) of the complex due to the existence of the peripheral carbazole moiety. These authors fabricated non-doped multilayer devices using the compound **9** and demonstrated much improved electroluminescence performance with a turn-on voltage of 4.5 V, a luminance of 2648 cd m^{-2} and a maximum current efficiency of 0.54 cd A^{-1}. A parallel development by the group of Li also reported the effects of the carbazole functional group on the emission and electrical properties of the Zn(II) complexes (**10** and **11**).[15d]

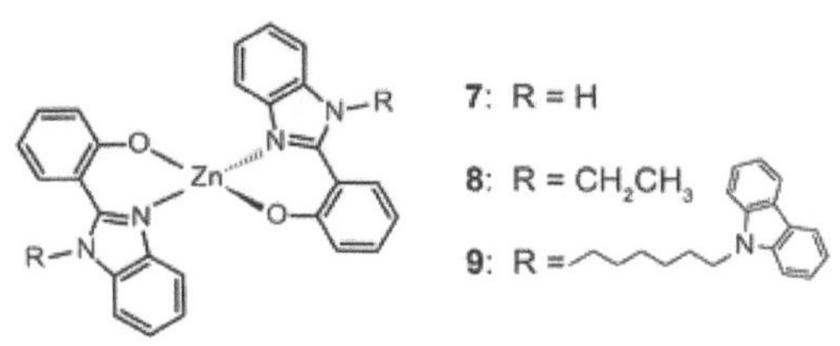

In 2009, Roh *et al.* investigated several green-emitting Zn(II) complexes (**12–15**) with benzothiazole and its derivatives to produce white-light emission.[15e] Interestingly, the substitution of an electron-donating methoxyl group in the Zn(II)-chelated complex **13** triggered a large red shift in emission from 495 ($_{max}$) to 524 nm in organic solvents. The incorporation of a highly conjugated fluorenyl group in **15** induced a further bathochromic shift of the fluorescence emission. The authors argued that the large Stokes shift is due to the intramolecular proton transfer at the excited state. In multilayer electroluminescent devices, these authors reported a high luminance (1 cd m^{-2} at 3.5 V, 10 400 cd m^{-2} at 14 V) for the fluorenyl-substituted Zn(II) complex **15**. This group also examined the electron transport property of the complexes and their suitability as electron-transporting layer materials as a replacement of widely used Alq_3.

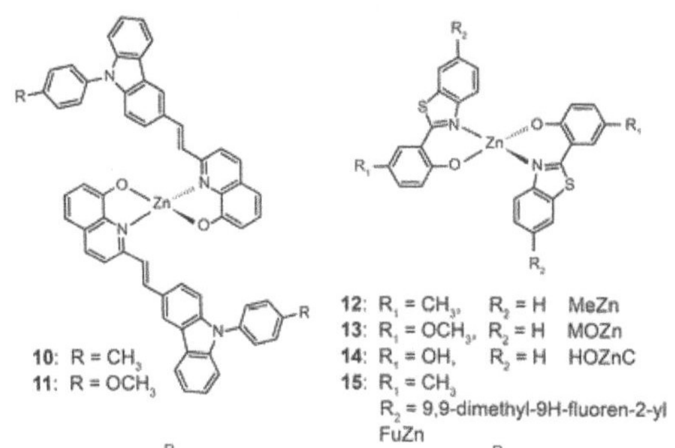

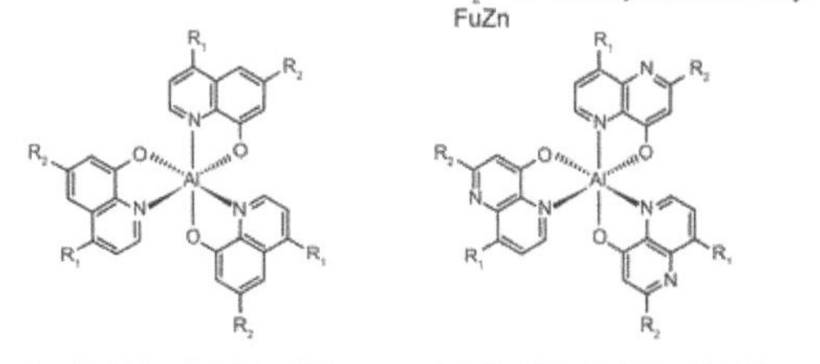

16: R_1 = H, R_2 = H Alq_3
17: R_1 = Me, R_2 = H $AlMeq_3$
18: R_1 = H, R_2 = F $AlFq_3$
19: R_1 = Me, R_2 = F $AlMeFq_3$
20: R_1 = Me, R_2 = CF_3 $AlMeCFq_3$
21: R_1 = Me, R_2 = CN $AlMeCNq_3$
22: R_1 = Me, R_2 = Ph $AlMePhq_3$

23: R_1 = Me, R_2 = H $AlmND_3$
24: R_1 = Me, R_2 = Me $AlmmND_3$
25: R_1 = Me, R_2 = Ph $AlmpND_3$

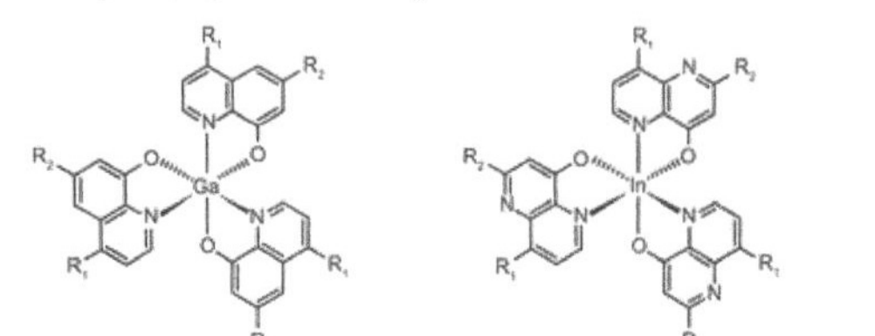

26: R_1 = Me, R_2 = H $GamND_3$
27: R_1 = Me, R_2 = Me $GammND_3$
28: R_1 = Me, R_2 = Ph $GampND_3$

29: R_1 = Me, R_2 = H $InmND_3$
30: R_1 = Me, R_2 = Me $InmmND_3$
31: R_1 = Me, R_2 = Ph $InmpND_3$

Electrofluorescent group III-type complexes.

Alq_3 (**16**) is one of the most stable and fluorescent solid-state materials with excellent electron-transport mobilities. It has been used as a green emitter and a common electron-transport layer in OLEDs. Its usual blue-green luminescence can be tuned through the addition of substitutional groups (**17–25**).[15f,16] For example, the presence of methyl substituents in complex **17** leads to a blue shift of ~20 nm, which can be ascribed to the increased LUMO levels. Further modification with F or CN groups in complexes **18–22** can reduce HOMO energy levels, resulting in a blue emission at ~480 nm. The steric hindrance caused by double substitutions in **18–22** also increases their quantum yields. The non-doped multilayer devices containing the Alq_3 complex **19** exhibited a low turn-on voltage of 3.5 V and a high luminance of 31 040 cd m^{-2}. Measured maximum external quantum and current efficiencies for the devices were 4.0% and 10.4 cd A^{-1}, respectively.

Liao *et al.* synthesized a series of group III (Al, Ga, and In) metal chelates (**23–31**) with hydroxynaphthyridine derivatives as chelating ligands.[16] These metal chelates exhibit deep blue emissions at 425–447 nm, wide band gaps and high thermal stability. The authors argued that the addition of an electron-donating group at the *para*-position of the pyridine ring promotes LUMO energy levels. Aluminium chelates **23–25** exhibited good fluorescence quantum yields around 45–47% in dichloromethane solution. Among the metal chelates, $AlmmND_3$**24** has been shown to be the most efficient non-doped blue emitter with a maximum emission at 432 nm (Fig. 5). For $AlmmND_3$-based OLEDs, these authors achieved a high current efficiency of 2.00 cd A^{-1} and an external quantum efficiency of 3.79%.

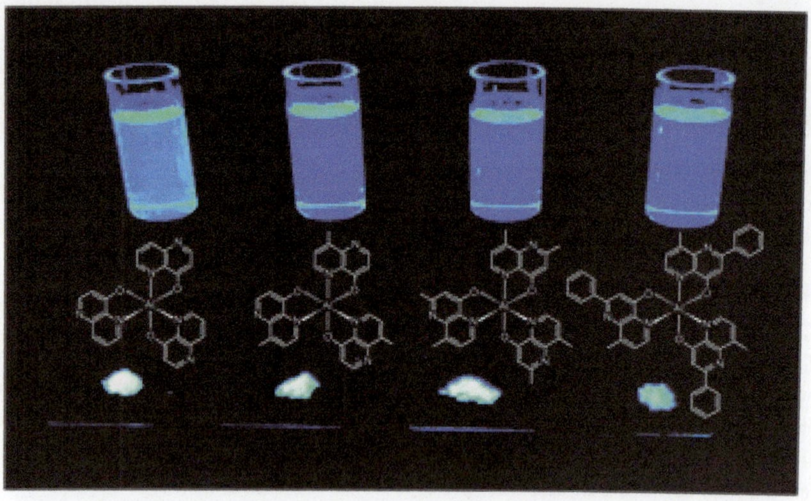

Figure 5: Emission images of $AlND_3$, $AlmND_3$**23**, $AlmmND_3$**24**, and $AlmpND_3$**25** recorded in solution and solid state under irradiation by UV light. (Reproduced with permission from ref. 16. Copyright 2009, American Chemical Society.)

It should be noted that most research activities have been focused on Al(III) complexes with three ligands symmetrically arranged around the metal center. This arrangement offers high thermal stability and low probability of exciton quenching. However, a recent study by Xu *et al.* on a mixed-ligand complex of Al(III) (Alq_2A; **32**) has revealed that an asymmetrical arrangement of ligands may offer improved electron-transport mobilities and high electroluminescence efficiency.[17] In contrast to OLEDs containing Alq_3 complex **16**, the devices utilizing complex **32** offer a larger luminance (15 650 cd m^{-2} at 12 V) and a higher power efficiency (3.03 lm W^{-1}). For performance improvement it is essential that the mixed ligands impart high electron mobility and show no interference of electron affinity. Another important contributing factor for high device efficiency is that the molecular structure of Alq_2A allows the facile formation of a uniform and smooth film. The physical properties and electroluminescence performance of representative fluorescent complexes is summarized in Table 1.

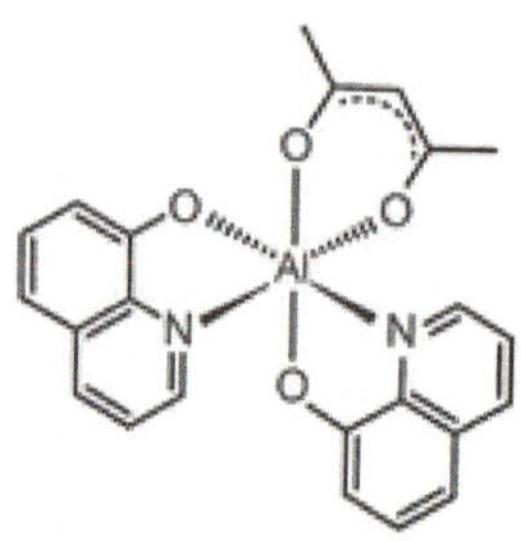

32

Table 1: Summary of physical properties and electroluminescence performance of fluorescent complexes

Emitter	Photophysical properties		HOMO/LUMO[b] (eV)	Device performance					Ref.
	Emission peak[a] (nm)	Φ (%)		Device structure[c]	Emission peak (nm)	Voltage (V)[d]	Max. brightness (cd m^{-2})	Max. efficiency[e]	
3	428, 463	8	−5.78/−2.45	ITO/NPB/**3**/BCP/Alq_3/LiF/Al	430	—	1720	0.3, —, —	15*b*
9	422, 422	—	−5.74/−2.49	ITO/NPB/**9**/BCP/Alq_3/LiF/Al	452	4.5	2648	0.54, —, 0.4	15*c*
12	—, 505	—	−5.33/−2.67	ITO/2-TNATA/NPB/**12**/Alq_3/LiF/Al	495	<4.2	~4000	1.84, 0.91	15*e*
13	—, 532	—	−5.31/−2.77	ITO/2-TNATA/NPB/**13**/Alq_3/LiF/Al	524	<4.8	~2000	2.10, 0.91	15*e*
15	—, 532	—	−5.81/−3.20	ITO/2-TNATA/NPB/**15**/Alq_3/LiF/Al	513, 551	<3.5	~10 000	1.90, 1.17	15*e*
16	526, 508	15	−5.62/−2.43	PEDOT:PSS/NPD/**16**/CsF/Al	—	5.5	52 300	4.6, —, 1.4	15*f*
17	501, 489	38	−5.53/−2.32	PEDOT:PSS/NPD/**17**/CsF/Al	—	3.5	15 300	5.2, —, 1.9	15*f*
18	499, 484	35	−5.82/−2.60	PEDOT:PSS/MoO_3/TPD/TCTA/**18**/TPBI/Yb/Ag	—	4.0	8570	0.6, —, 0.2	15*f*
19	478, 482	57	−5.81/−2.44	PEDOT:PSS/MoO_3/TPD/TCTA/**19**/Yb/Ag	—	3.5	31 040	10.4, —, 4.0	15*f*
20	482, 462	46	−5.78/−2.39	PEDOT:PSS/MoO_3/TPD/TCTA/**20**/TPBI/Yb/Ag	—	4.2	16 800	10.1, —, 4.1	15*f*
23	415, 431	45	−6.4/−3.0	ITO/NPB/CBP/**23**/$Bebq_2$/LiF/Al	458	—	7140	1.8, 1.0, 1.8	16
24	416, 419	45	—	ITO/NPB/CBP/**24**/Alq_3/LiF/Al	432	—	4444	2.0, 0.9, 3.8	16
32	—	—	−6.49/−3.63	ITO/2-TNATA/NPB/**32**/LiF/Al	513	—	15 650	4.35, 3.03, —	17

a **Measured in solution. *b* Measured according to the analysis of cyclic voltammetric data and optical energy gaps. *c* See the Abbreviations section for full names. *d* Turn-on voltage. *e* In the order of current efficiency (cd A^{-1}), power efficiency (lm W^{-1}) and external quantum efficiency (%).**

Electrophosphorescent complexes

The mechanism of phosphorescence involves intersystem crossing characterized by the nonradiative conversion of the initial excited state into another excited state of different multiplicity. The second state acts as an energy reservoir. Radiative decay back to the ground state can occur slowly if spin–orbital coupling causes a breakdown of the spin selection rule. As a result, the excited state of a phosphorescent complex may survive for microseconds or even longer. Theoretically, the internal quantum efficiency of phosphorescent OLEDs can approach 100% because of the contributions from both singlet and triplet excitons. Several excellent reviews on electrophosphorescent complexes and devices have appeared during the past few years.[18] Herein, we highlight recent advances in the development of high-performance OLEDs by means of phosphorescent complexes for multicolor displays and white light generation over a wide range of wavelengths.

Blue electrophosphorescent complexes.

Blue-emitting Ir(III) complexes.

An important example of phosphorescence is provided by blue-emitting Ir(III) complexes. These blue emitters are important for white light generation. One approach is to mix the blue phosphors with red and green phosphors, each exhibiting a relatively narrow emission spectrum, to create a spectral power distribution that appears white. Alternatively, the white light can be generated by the use of phosphors together with a blue emitter. When a yellow phosphor material is illuminated by the blue emitter, it renders yellow light having a fairly broad spectrum. The remaining blue light, when mixed with the yellow light, results in white light. However, it has been challenging to prepare blue phosphors with a high level of monochromaticity and photostability. For example, iridium(III) bis[(4,6-difluorophenyl)pyridinato-*N*,$C^{2'}$] picolinate (FIrpic), one of the most popular blue phosphors, emits at 475 nm with high electrophosphorescence, but is prone to quenching at high concentration and under large current density.[19]

The main approaches for achieving efficient blue phosphorescence include (i) increasing the energy gap by elevating LUMO level or lowering HOMO level; (ii) introducing ancillary ligands with strong field effects;[20] and (iii) shortening the effective conjugation length of molecules. For instance, Lee *et al.*[21] recently demonstrated a large reduction (0.33 eV) in the HOMO energy level of a phosphorescent Ir(III) complex (**33**) bearing fluorine-substituted bipyridine (dfpypy) through functional group modulation. The authors found that this complex exhibits intense blue emission at 438 nm with high color purity (CIE: $x = 0.14$, $y = 0.12$) and a higher photoluminescence quantum efficiency than that of FIrpic. The enhanced luminescence was thought to be due to improved metal–ligand charge-transfer. Density functional theory calculations of the complex lent support to this hypothesis.

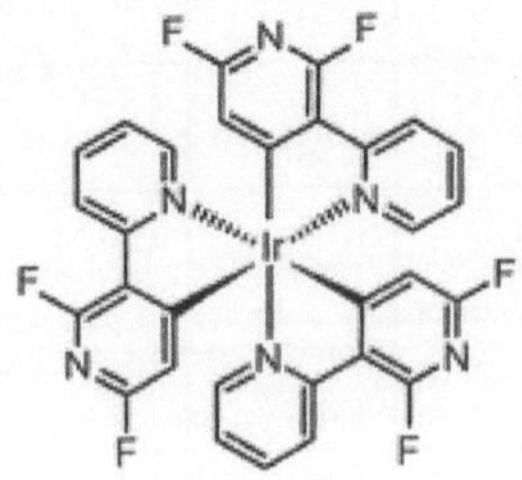

$Ir(dfpypy)_3$

33

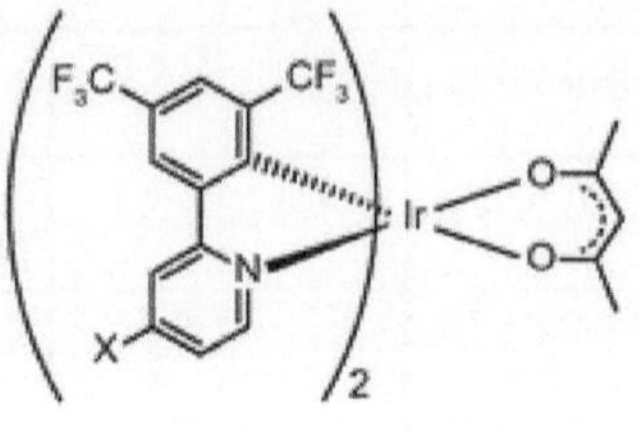

34: X = H

35: X = OMe

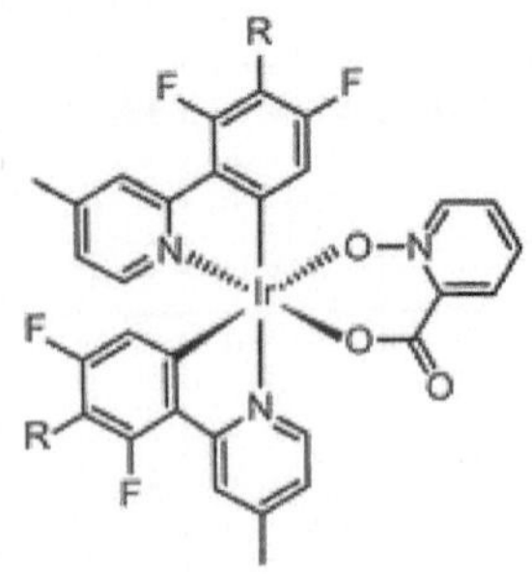

36: R = H

37: R = CF_3

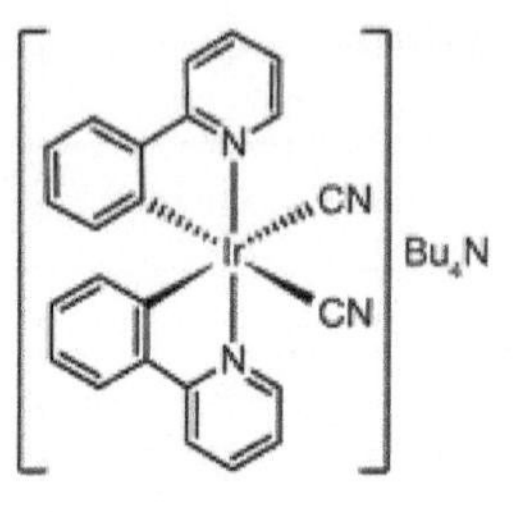

38

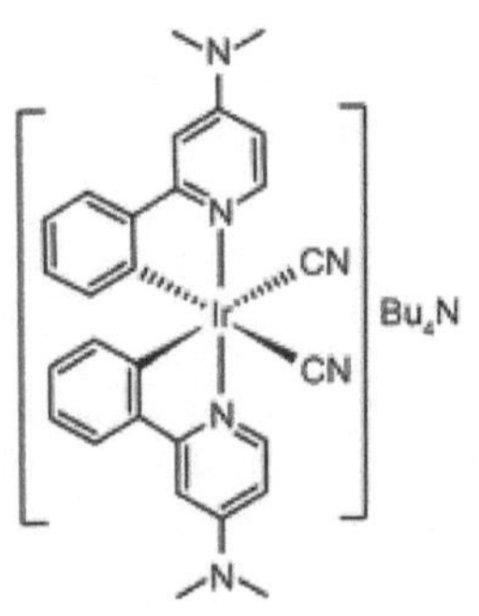

39

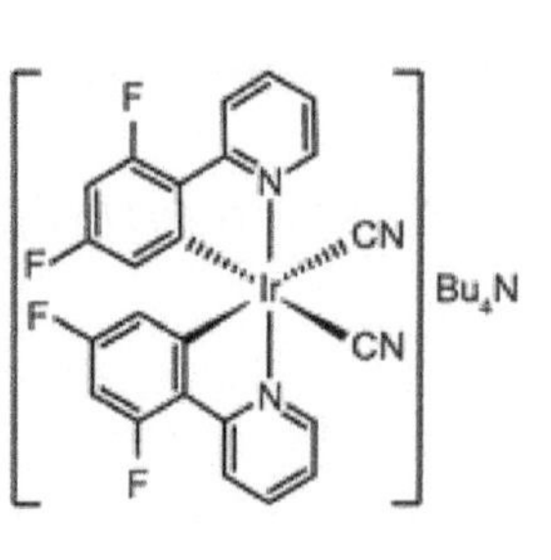

40

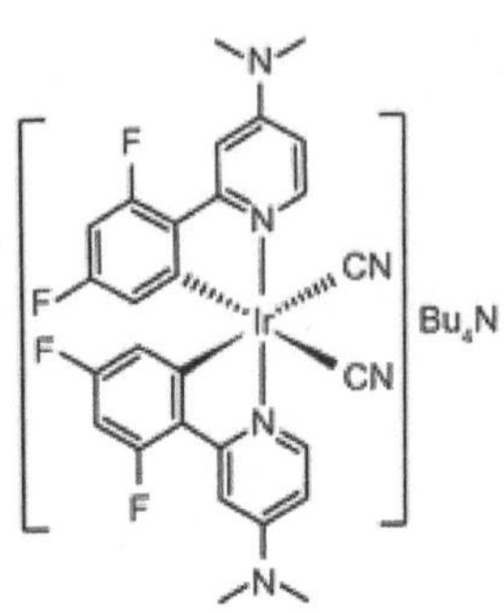

41

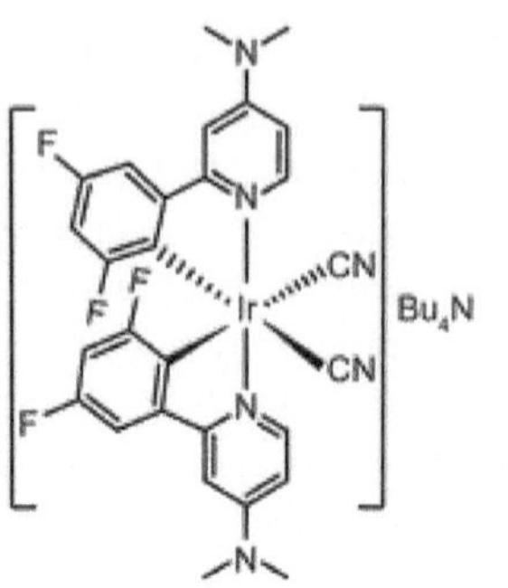

42

Lee and Kim demonstrated that the LUMO energy level of a phenylpyridine-based Ir(III) complex (**34**) can be destabilized by the addition of a methoxy group to the ligand.[22] The authors reported that the substitution of the electron-donating group in complex (**35**) results in a blue shift in emission. In a parallel investigation, Seo *et al.* reported a new series of highly efficient deep-blue phosphorescent Ir(III) complexes (**36** and **37**) bearing phenylpyridine-based ligands.[23] The authors found that the attachment of a trifluoromethyl (CF_3) group to the phenylpyridine ligand induces a bright blue emission at 454 nm due to stabilization of the HOMO level. The multilayer OLEDs based on complex **37** achieved a similar level of performance to that of FIrpic-based devices.

Ancillary ligands can interact strongly with metal ions, and the phosphorescent properties of the complex stem in large part from this interaction. Ancillary ligands with strong ligand field often induce hypochromatic shifts when they are present in a complex. This effect has been systematically investigated by Di Censo *et al.* using a series of cyano-stabilized Ir(III) complexes (**38–42**).[24]Replacement of one phenylpyridine ligand with two cyano groups leads to a blue shift (10 nm) in the emission spectrum of the resulting complex **38**.[24] In a separate study, the optical properties of complex (**43**) containing triphenylphosphine ligands was analyzed.[25] This complex was found to exhibit a similar hypochromatic shift to that observed for cyano-stabilized complex **38**. The spin-coated polymer light-emitting diodes comprising **43** revealed a moderate power efficiency of 0.86 lm W^{-1} and an external quantum efficiency of 4.0%. The research groups of Kim[26] and Ha[27] have pursued the development of blue-emitting phosphorescent Ir(III) complexes (**44–48**) by appending triphenylphosphine or triazolylpyridine as the ancillary ligand to the metal center.

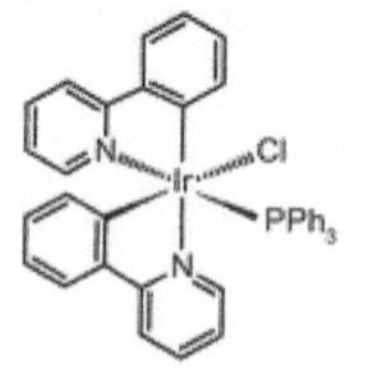

$Ir(ppy)_2(PPh_3)Cl$

43

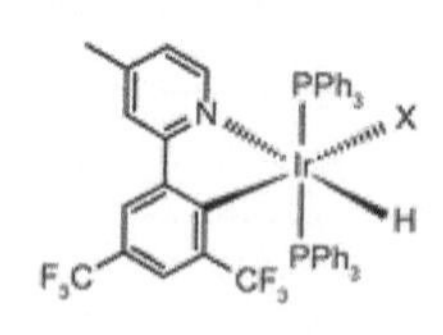

44: X = Cl $Ir[(CF_3)_2Meppy](PPh_3)_2(H)(Cl)$

45: X = CN $Ir[(CF_3)_2Meppy](PPh_3)_2(H)(CN)$

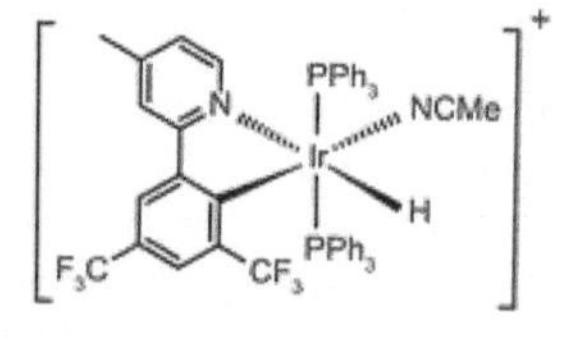

$Ir[(CF_3)_2Meppy](PPh_3)_2(H)(NCMe)]^+$

46

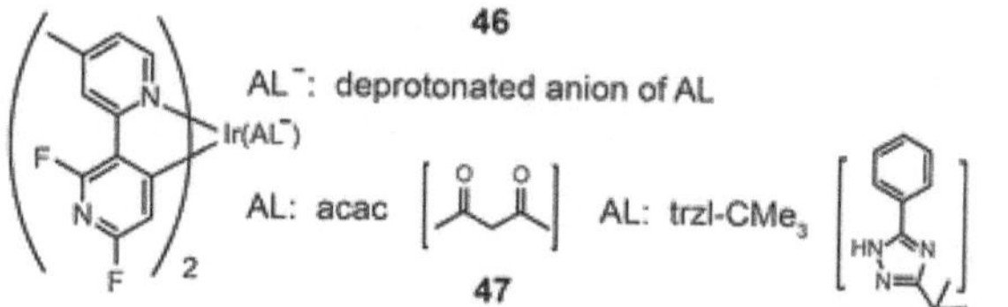

47

$Ir(4\text{-}Me\text{-}dfpypy)_2(trzl\text{-}CMe_3)$

48

The use of N-heterocyclic carbene ancillary ligands, equipped with very high triplet energy levels,[28] allows for facile construction of blue-emitting Ir(III) complexes.[29] For example, Chen and co-workers prepared $(fpmi)_2Ir(dmpypz)$ (**49**) and $(mpmi)_2Ir(dmpypz)$ (**50**) that emit pure blue emissions at 455 and 466 nm, respectively.[30] The multilayer devices based on **49** and **50** as the dopant emitter showed high external quantum efficiencies of 17.1% and 15.4% with commission internationale de l'éclairage (CIE) coordinates of (0.13, 0.16) and (0.13 and 0.18), respectively.

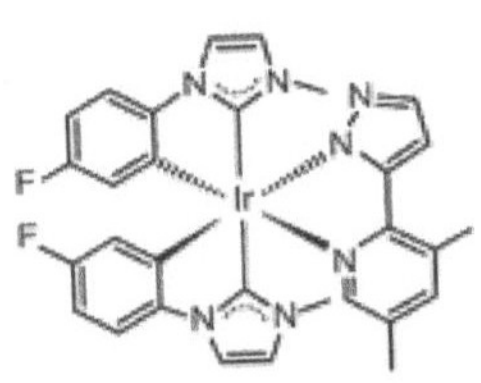

$(fpmi)_2Ir(dmpypz)$
49

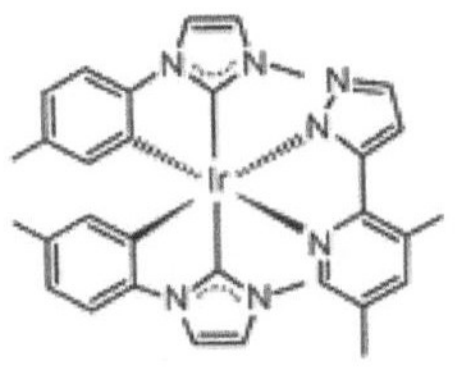

$(mpmi)_2Ir(dmpypz)$
50

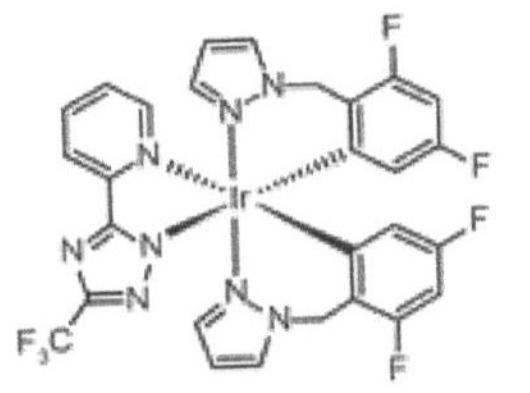

$Ir(dfb\text{-}pz)_2(fptz)$
51

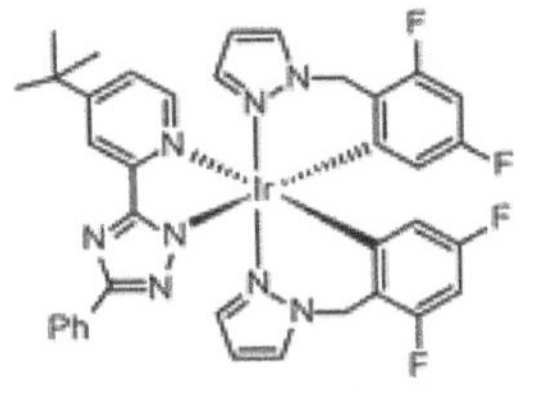

$Ir(dfb\text{-}pz)_2(pbptz)$
52

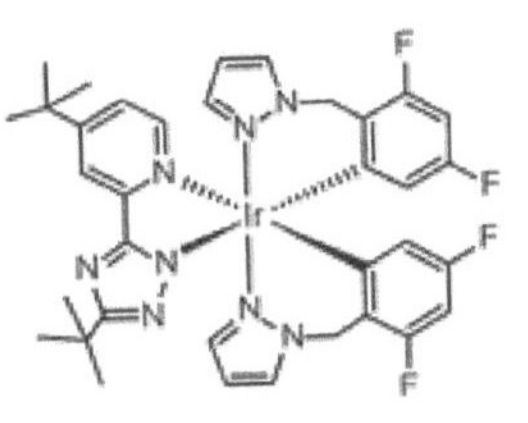

$Ir(dfb\text{-}pz)_2(bbptz)$
53

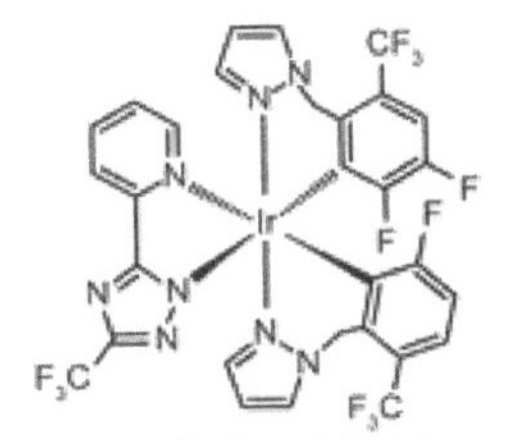

$Ir(tfmfb\text{-}pz)_2(fptz)$
54

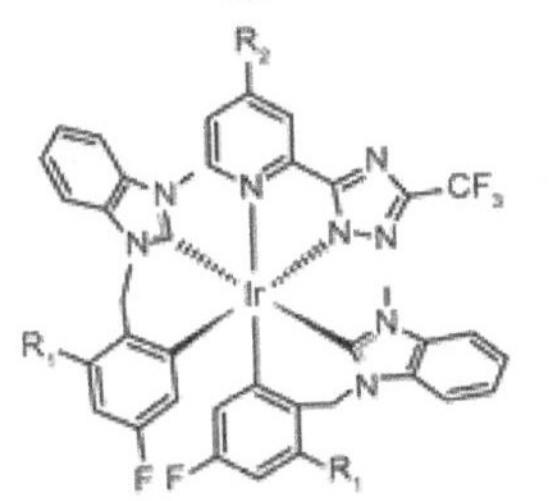

55: $R_1 = H, R_2 = {}^tBu$ $(fbmb)_2Ir(bptz)$
56: $R_1 = F, R_2 = H$ $(dfbmb)_2Ir(fptz)$

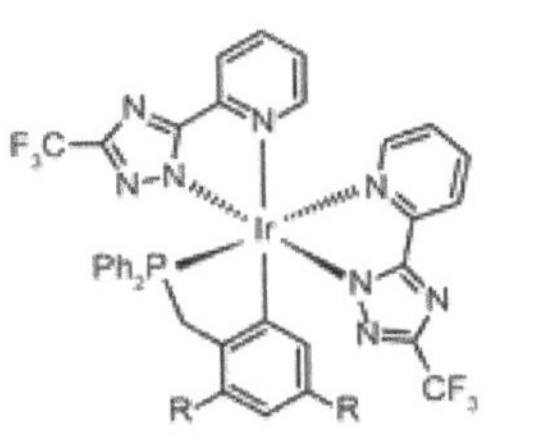

57: R = H $Ir(fppz)_2(bdp)$
58: R = F $Ir(fppz)_2(dfbdp)$

Complexes **51–54**, developed by Wu and co-workers,[31] contain nonconjugated *N*-benzylpyrazole ligands featuring a methylene spacer. This design feature allowed the synthesis of blue phosphorescent complexes with large energy gaps. The methylene spacer serves to effectively interrupt the π conjugation on reacting with a third chelating chromophore. The resulting complexes **51–54** have true blue emissions at 437, 464, 456 and 434 nm, respectively. It was suggested that conjugated ligands not only reduce energy gaps, but also partially offset the modification of HOMO and LUMO levels. However, the electroluminescence performance of these complexes was not satisfactory. Wu and co-workers[32] later developed chelating benzyl carbene ligands to alleviate much of this problem. The resulting two phosphors of $(fbmb)_2Ir(bptz)$ (**55**) and $(dfbmb)_2Ir(fptz)$ (**56**) showed bright blue emissions at 460 and 458 nm with photoluminescence quantum yields of 22% and 73%, respectively. They successfully fabricated vacuum-deposited, **56**-doped OLEDs, achieving pure blue emission with CIE coordinates of (0.16, 0.13). The electroluminescence efficiencies were also improved with a current efficiency of 6.3 cd A^{-1}, a power efficiency of 4.0 lm W^{-1}, and an external quantum efficiency of up to 6.0% photons per electron.

Replacing carbenes with high-field-strength chelating phosphine ligands in nonconjugated systems can further tune the emission color to the blue end. Chou and co-workers[33] have reported a breakthrough in the rational design of novel $Ir(fppz)_2(bdp)$ (**57**) and $Ir(fppz)_2(dfbdp)$ (**58**) phosphors, enabling the corresponding OLEDs to exhibit a true blue CIE chromaticity of (0.15, 0.11) and an external quantum efficiency of 12%. Concurrent increases in the current efficiency (~11 cd A^{-1}) and power efficiency (~8 lm W^{-1}) were observed, making these phosphors attractive for use in all phosphorescent displays and illumination devices.

In 2011, Zhu *et al.* published results on OLEDs containing phosphine oxide ligands.[34] The substitution of picolinate with phosphine oxide in FIrpic forms the blue-green emitting $Ir(dfppy)_2(tpip)$ (**59**). Although this does not present a significant improvement with respect to color saturation, the phosphorescent lifetime of **59** is reduced to 0.77 μs, indicating a short-lived excited state in favor of suppressed triplet–triplet annihilation (TTA). In addition to its carrier transport ability, the ancillary phosphine oxide group provides the metal complex with a LUMO level of −2.87 eV suitable for efficient electron injection. Accordingly, an OLED device prepared with **59** showed a current efficiency of 25.5 cd A^{-1}, a power efficiency of 23.5 lm W^{-1}, and a 6.58% efficiency roll-off from 100 to 1000 cd m^{-2}. A paper from Seo *et al.* shows that the luminescence performance of blue-emitting Ir(III) complexes may be achieved using hole-transporting carbazole moieties.[35] The incorporation of peripheral carbazole groups into complexes (**60–62**) can provide added intramolecular energy transfer to the metal core and facilitate hole injection and transportation. These complexes emit blue-green light with emission peaks located at about 480 nm and CIE coordinates of (0.21, 0.55). Devices doped with these complexes achieved a maximum current efficiency of 20.75 cd A^{-1}, a maximum power efficiency of 5.83 lm W^{-1}, and a peak external quantum efficiency of 7.16%.

The quenching effect caused by intermolecular interaction is perhaps one of the most serious problems for light-emitting phosphors. Although doping in active host matrices can mitigate the issues of TTA and concentration quenching to certain extent, it is generally believed that the potential phase separation may lead to device degradation and decreased performance over time. An intriguing solution is to covalently link the Ir(III) complex to the host matrix. Yang *et al.*[36] reported the use of polyhedral oligomeric silsesquioxane (POSS) in the preparation of monochromatic- and white-emitting devices. Notably, the blue-emitting OLEDs with POSS-modified Ir(III) complex (**63**) showed a brightness of 1000 cd m^{-2} and peak external quantum efficiencies in the range of 5–9% with a voltage potential less than 10 V. Despite synthetic challenges, the prospects of using POSS materials are exciting for the development of solution-processable electrophosphorescent devices.

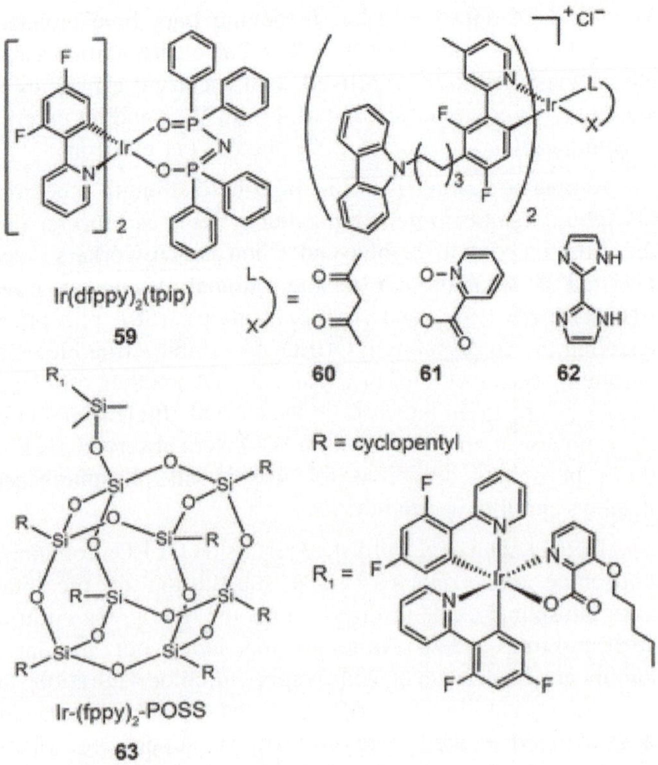

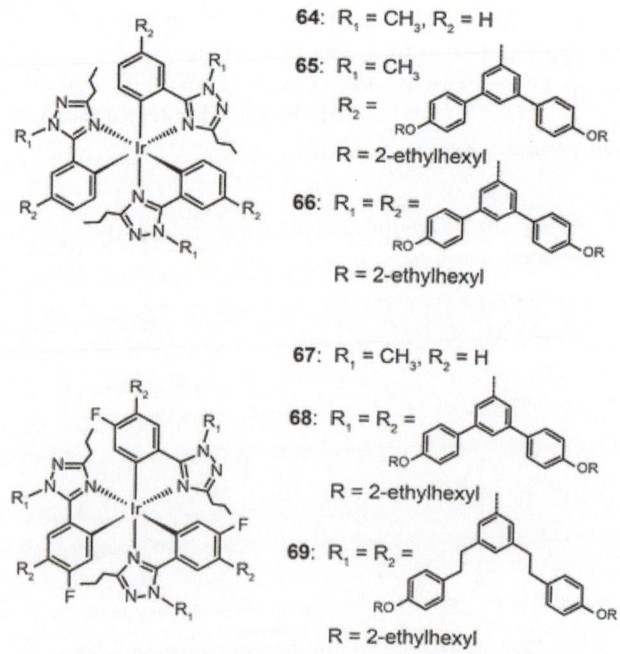

Concentration quenching can be avoided using dendrimers having highly branched molecular structures. An important property of dendrimers is the presence of dynamic internal cavities where luminescent ions or complexes can be comfortably embedded, resulting in a more protected metal center. In 2008, Lo *et al.* studied the effects caused on the luminescence properties of blue-emitting phenyltriazole Ir(III) complexes (**64–66**) by substitution of oligophenyl dendron units.[37] They showed that the dendron-modified molecules can host the metal in the interior site to suppress the quenching effect. For example, doubly dendronized complex (**66**) exhibited a blue emission at ~475 nm and a film photoluminescence quantum yield of 67%. It is particularly worth mentioning that the oligophenyl dendrons do not affect the electronic properties of the complexes. The spin-coated nondoped double-layer devices composed of **66** achieved a luminance of 142 cd m^{-2} at 3.8 V with an external quantum efficiency of 7.9%. One year later, the same group designed blue phosphorescent dendrimers (**67–70**) modified with both fluorine and methyl groups.[38] The involvement of fluorine atoms enables the emission tuning to a deep blue end, while the methyl group provides steric hindrance to mitigate the quenching effect induced by the fluorine atom. The doubly dendronized complex**70** showed good optical performance, with a high solution photoluminescence quantum yield of 94%, which is more than three times higher than was achievable by the parent complex core. The optimized double-layer devices showed deep blue emission with CIE coordinates of (0.16, 0.16) (Fig. 6).

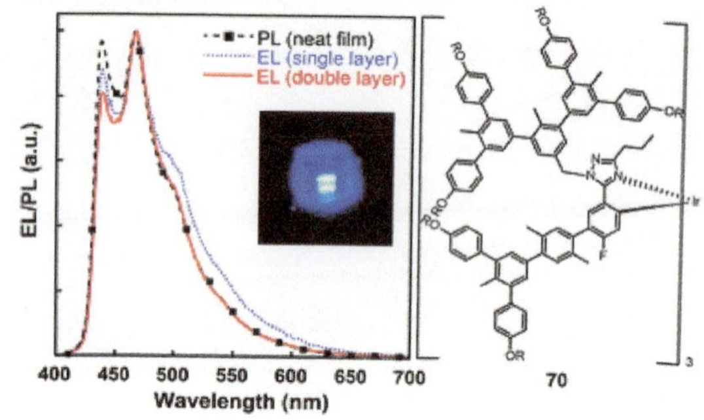

Figure 6: Photoluminescence (PL) and electroluminescence (EL) spectra of **70** (inset: the corresponding digital image of the device derived from **70**). (Reproduced with permission from ref. 38. Copyright 2009, American Chemical Society.)

Blue-emitting Pt(II) and Ce(III) complexes.

Platinum(II) complexes, characterized by a large variety of emissive excited states, are an important class of phosphorescent materials as dopants for fabricating blue-emitting OLEDs. However, when compared to Ir(III) complexes, Pt(II) complexes are much less explored as electrophosphorescent blue emitters for OLED applications. Yang *et al.* developed a blue-emitting Pt(II) complex (**71**) based on a fluorinated tridentate pyridine ligand.[39] This cyclometallated complex exhibited a narrow emission at 470 nm ($_{max}$) and a photoluminescence efficiency of 46%. To boost the color saturation, the authors utilized high-field-strength carbene moieties, combined with an ancillary ligand of acac, to tune the emission of Pt(II) complexes (**72–77**) to the spectrally pure blue region.[40] Complex **77** exhibited high optical performance, achieving a photoluminescence efficiency of 92%. The optimized devices employing **77** with a doping concentration of 12% showed a peak external quantum efficiency of 6.2%. Recently, the Che group has carried out considerable work on bis(carbene)-based Pt(II) complexes (**78–81**) featuring highly efficient deep blue emission (Fig. 7).[41] The OLED device doped with **80** achieved CIE coordinates of (0.16, 0.16) at 11 V and a maximum luminance of 1200 cd m^{-2} at a driving voltage of 15 V.

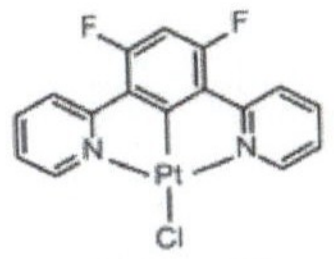

PtDFDPBC
71

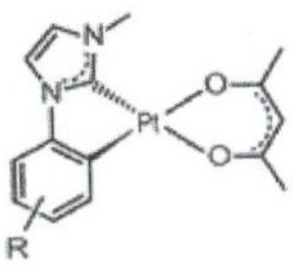

72: R = 4-H
73: R = 4-Br
74: R = 4-NO_2
75: R = 4-PMe
76: R = 4-Me
77: R = 2,3-OC_6H_4

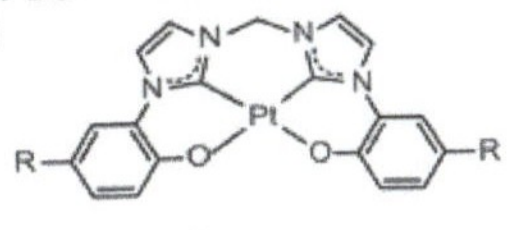

78: R = H
79: R = F
80: R = Me
81: R = ^{t}Bu

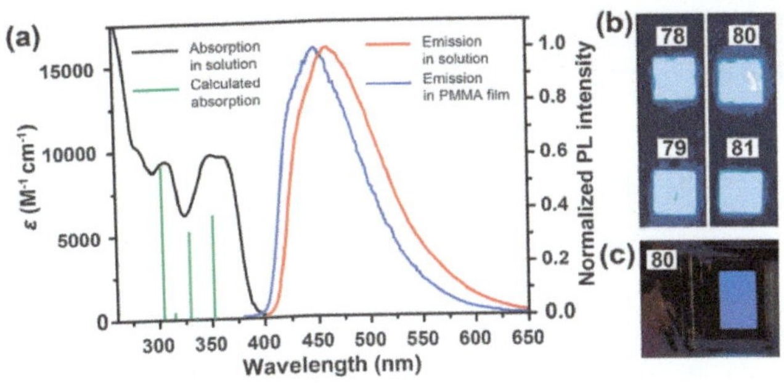

Figure 7: (a) Absorption and photoluminescence spectra of **78** and **79**. (b) Photographs showing electroluminescent devices derived from **78–81** under UV irradiation. (c) Photograph showing electroluminescence emission of the device based on **80**. (Reproduced with permission from ref. 41. Copyright 2011, Royal Society of Chemistry.)

Trivalent lanthanide complexes can also be designed as blue-emitting phosphors in OLED applications. Zheng *et al.* reported the luminescent cerium(III) complexes (**82–84**) with the metal ion encapsulated in the cavity formed by two face-to-face arranged tripodal ligands.[42] These complexes emitted blue light at ~440 nm. Complex **83** showed a high photoluminescence efficiency of 55%. The devices based on dopant emitter **83** achieved a current efficiency of 1.5 cd A^{-1} and a power efficiency of 0.52 lm W^{-1} with CIE coordinates of (0.18, 0.21). The authors argued that the blue emission is attributed to both metal-based 5d → 4f transition and the ligand sensitization. The ligand serves as a light-harvesting antenna that subsequently transfers the absorbed energy to the Ce(III) ion.

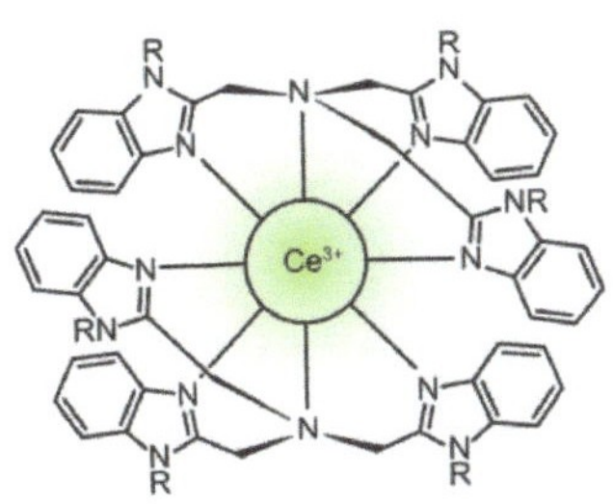

82: R = CH_3CH_2
83: R = $CH_3CH_2CH_2$
84: R = CH_2=$CHCH_2$

It is important to note that the development of phosphors for optoelectronics has been dominated by Ir(III) complexes since the discovery of electrophosphorescence. The main difference between Ir(III) and Pt(II) complexes is their different coordination modes. Compared with four-coordinated Pt(II) complexes, highly dimensional Ir(III) complexes with six-coordination mode are superior in controlling intermolecular interactions to suppress the quenching effects and in enhancing molecular rigidity to reduce vibronic progressions. In this regard, the Ir(III) complexes seem to be ideal as blue emitters. Furthermore, the yield of Pt(II) complexes containing ligands with strong coordination field is relatively low (<20%). To date, the highest external quantum efficiency realized by blue Pt(II) complexes is 15%, only half of that achieved by blue Ir(III) complexes.[43a,b] Nevertheless, the simpler coordination mode of Pt(II) complexes can benefit material synthesis and configuration design.

Phosphorescent emissions from Ir(III) and Pt(II) complexes are mainly contributed by ^{3}MLCT and ^{3}ILCT states. Therefore, the emission properties of these materials are significantly dictated by their ligands. Different from Ir(III) and Pt(II) analogues, the blue emission by Ce(III) complexes originates from $^4f \rightarrow {}^5d_j$ (j = 1–4) transitions of the Ce(III) ion. Strongly emissive Ce(III) complexes require efficient energy transfer and high recombination rates from the ligands under an electrical field. Owing to the relatively high excited state energy of Ce(III) ions, it is difficult to construct suitable ligands as sensitizers, which limits the development of highly efficient electroluminescent Ce(III) complexes.

Green electrophosphorescent complexes.

Ir(ppy)$_3$ is the first and also the most popular green-emitting phosphor used in OLEDs.[44] Although with a suitable host matrix the power efficiency of Ir(ppy)$_3$-based devices can exceed 100 lm W^{-1},[45a] there is still strong interest in developing high-performance green-emitting Ir(III) complexes with the aim of understanding the photophysics of these metal complexes and the relationship between molecular structures and device performance.[45b,c]

Small molecule-based green-emitting Ir(III) complexes.

The effect of ligand substitution has been thought to be the primary factor for modulating the emission property of metal complexes. Ir(dsippy)$_3$ (**85**) demonstrates that the attachment of a trimethylsilyl group can induce significant steric hindrance.[46] Importantly, the substitution of this bulky ligand results in improved photostability of the complex and a relatively narrow emission peak. Compared with the device doped with Ir(ppy)$_3$, the device comprising Ir(dsippy)$_3$ enables a lower driving voltage and more stable power efficiency. To assess the influence of electron-donating and electron-withdrawing groups on the emission properties of metal complexes, Zhou *et al.* systematically investigated a series of Ir(III)-based complexes (**86–92**).[47] All of these complexes exhibit high thermal and morphological stability with a T_d value of greater than 350 °C. The authors suggested that the substituents have strong effects on the charge distribution within the molecular orbitals in that the electron-donating group can increase the HOMO energy, while the electron-withdrawing group has the opposite directing effect. As a result, the emissions of these complexes can be tuned from 505 to 550 nm. As the inductive effect of the sulfonyl and phosphine oxide groups can simultaneously reduce the LUMO energy, the authors observed remarkable red-shifts in the emissions of complexes **86** and **90**. The devices using these Ir(III)-based complexes have shown moderate electroluminescence performance, of which the complex **88** showes a current efficiency of 35.02 cd A^{-1}, a power efficiency of 26.82 lm W^{-1}, and an external quantum efficiency of 11.05%.

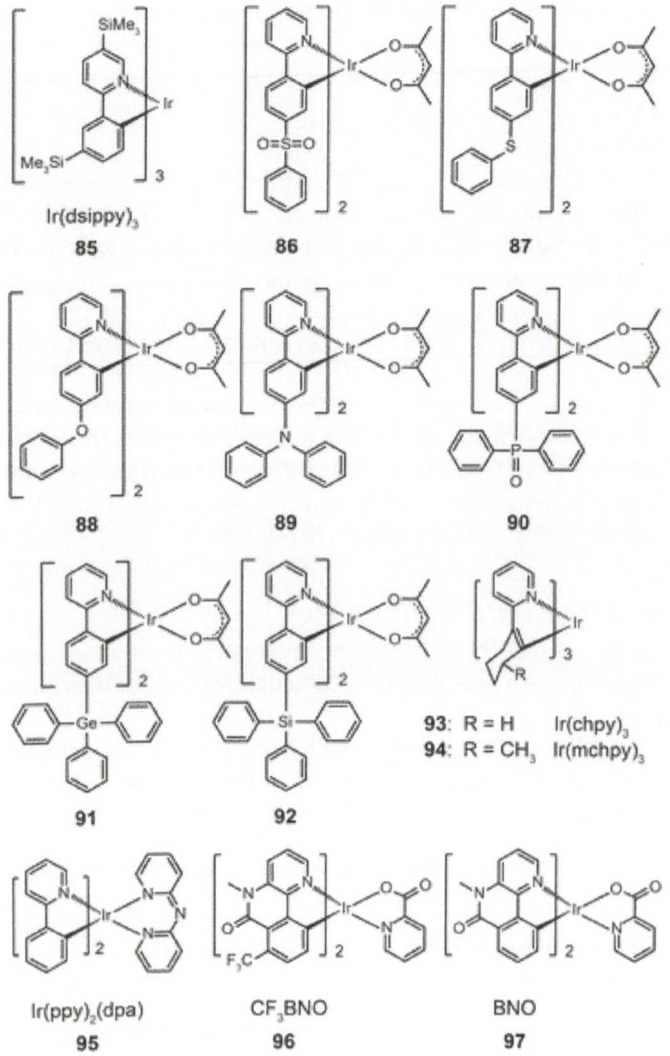

The use of cycloalkenyl (chpy or mchpy) group instead of the phenyl in $Ir(ppy)_3$ also can lead to increased steric hindrance.[48] The corresponding complexes, $Ir(chpy)_3$ (**93**) and $Ir(mchpy)_3$ (**94**), emit green yellowish color with the peaks centered around 530 nm. Their HOMO and LUMO energies were modified to −5.0 and −2.5 eV, matching the energy levels of carrier transporting layers. In 2011, Rai *et al.* reported an interesting dipyridylamido-supported cyclometalated Ir(III) complex, $Ir(ppy)_2(dpa)$ (**95**), which emits green electroluminescence with high current efficiency (123.5 cd A^{-1}) and power efficiency (43.2 lm W^{-1}) (Fig. 8).[49] Owing to increased charge transfer between the ligand and the metal, the lifetime of the complex was reduced to 0.17 μs. The authors observed a high photoluminescence quantum yield (87%) of complex **95** in a chloroform solution, which was believed to be due to minimized phosphorescence self-quenching by suppression of triplet–triplet annihilation. High-molecular-weight Ir(III) complexes (**96** and **97**) also can display relatively short excited-state lifetimes (0.30 and 0.32 μs, respectively) with improved phosphorescence quantum yields.[50] Notably, **96** and **97** are solution processible. The phosphorescent organic light-emitting diode (PHOLED) obtained from the spin-coated complex **96** emits blue-green color with a high current efficiency of 89.1 cd A^{-1} (69.8 lm W^{-1}) at 100 cd m^{-2}.

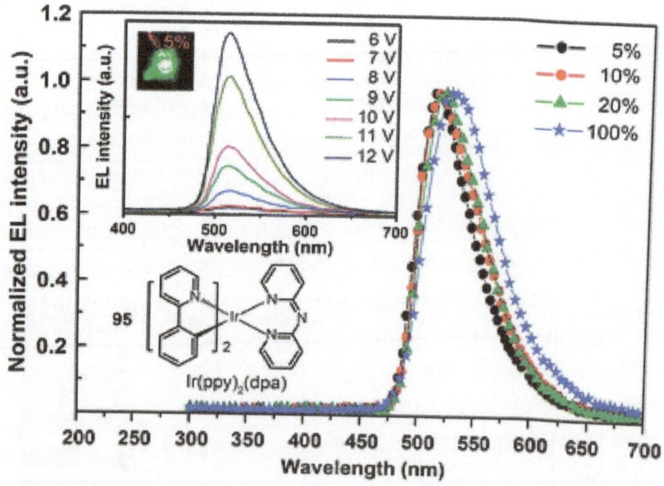

Figure 8: Electroluminescence spectra of PHOLEDs doped with **95** at different concentrations. (Reproduced with permission from ref. 49. Copyright 2011, Royal Society of Chemistry.)

Recently, the non-doped, solution-processed phosphorescent devices containing small-sized molecular phosphors have attracted considerable research interest because of their potential for facile, large-scale fabrication. Ir(III)-based complex (**98**) features a sterically hindered chelating amidinate ligand which has proven to be effective in solving the self-quenching problem (Fig. 9).[51] Weak intermolecular interaction of the complex and its short lifetime of 0.34 μs are believed to be the two main factors. The contribution of the amidinate ligand to the HOMO also improves the hole-injection ability of the complex. The non-doped device of **98** showed rather low driving voltages of 2.40 V for the onset voltage, 3.15 V at 100 cd m^{-2}, and 4.30 V at 1000 cd m^{-2}. The dramatically reduced self-quenching was further confirmed by high power efficiency (32.5 lm W^{-1}), in combination with the excellent stability of the device (28 and 22 lm W^{-1} at 100 and 1000 cd m^{-2}, respectively).

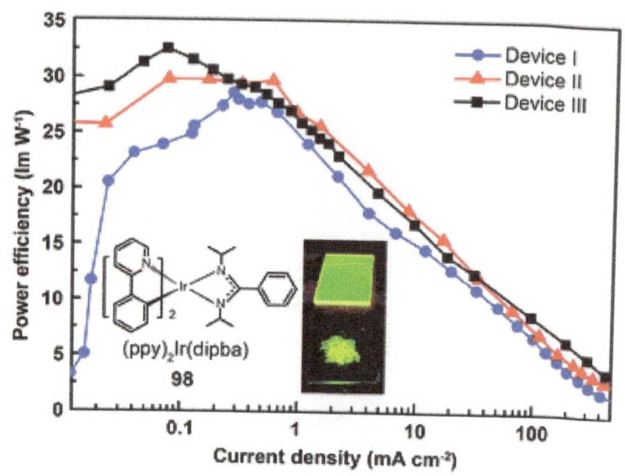

Figure 9: Power efficiency *vs.* current density curves of PHOLEDs using **98** as phosphor (inset: photographs showing phosphorescent emission of **98** in its film and powder states). (Reproduced with permission from ref. 51. Copyright 2009, Royal Society of Chemistry.)

Xu *et al.* utilized a combined modification of charge-transfer moieties and aliphatic chains to construct solution-processible phosphors (**99–101**).[52] The peripheral carbazolyl groups can facilitate the hole carrier injection and transport while the long hexyl chain improves the solubility of the complexes in common solvents. Furthermore, the surrounding rigid carbazolyl moieties are able to encapsulate the emitting metal center, protecting the latter from self-quenching. This design improved the phosphorescence quantum yield of the six-armed $Ir(Cz_2PhBI)_3$ **101** to ~70% without significantly altering its characteristic properties. A low turn-on voltage of 2.5 V and a high external quantum efficiency of 5.9% at 100 cd m^{-2} were achieved for the non-doped devices derived from complex **101**. However, the unbalanced carrier currents arising from asymmetric injection and transport characteristics of the carbazole groups allow rapid recombination of hole–electron pairs at the electrode, resulting in luminescence quenching. To balance the carrier injection and transport in metal complexes, the same group further prepared a series of Ir(III) complexes (**102–105**) having electron-transporting functional groups.[53] The space-filling model shows that substituted oxadiazole ligands at multiple positions in complex **105** are effective in segregating the metal centers. As anticipated, the authors observed a high photoluminescence performance with a balanced carrier injecting/transporting ability for complex **105** (Fig. 10). Its non-doped spin-coated devices achieved a current efficiency of 15.41 cd A^{-1} and a power efficiency of 8.08 lm W^{-1}.

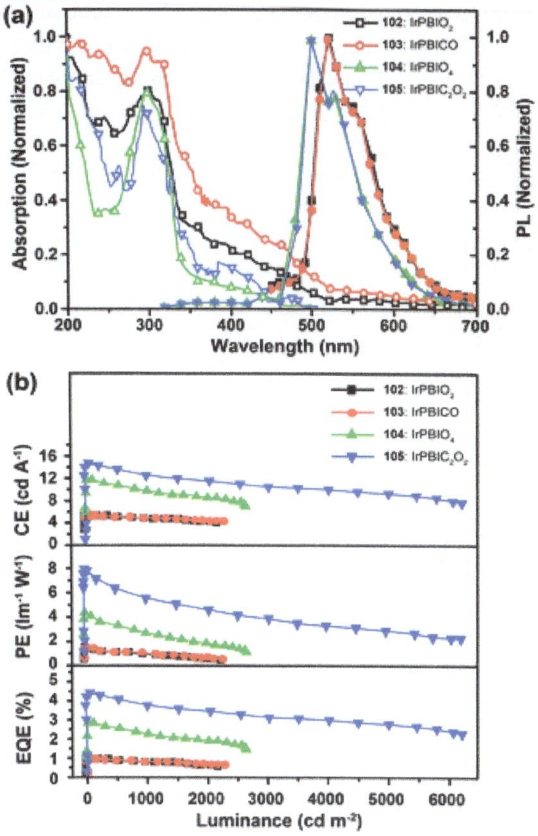

Figure 10: (a) Absorption and photoluminescence spectra of **102–105** dissolved in dilute CH_2Cl_2 solution. (b) The corresponding efficiency curves of the PHOLEDs derived from **102–105**. (Reproduced with permission from ref. 53. Copyright 2011, Royal Society of Chemistry.)

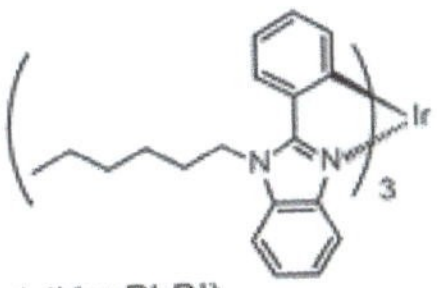

$Ir(HexPhBI)_3$
99

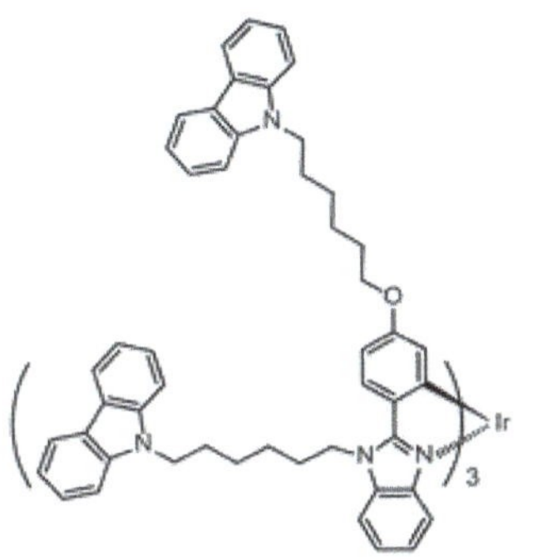

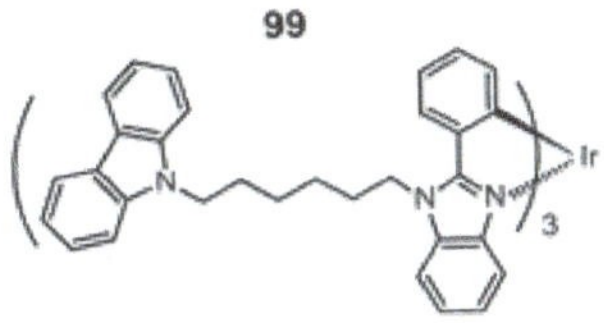

$Ir(CzPhBI)_3$
100

$Ir(Cz_2PhBI)_3$
101

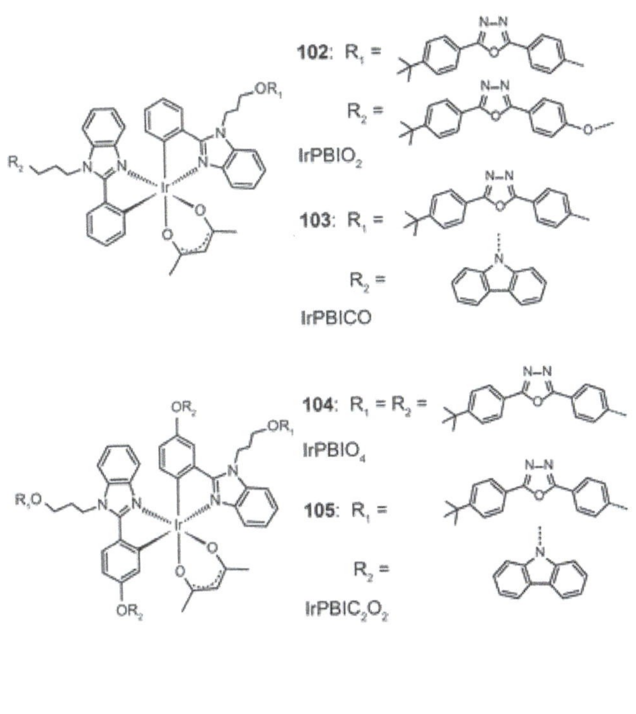

Recently, Chen *et al.* reported a similar study by employing Ir(III) complexes (**106–108**) comprising substituted oxadiazole ligands.[54] Through device optimization, complex **108** with four extended arms was used to realize a current efficiency of 23.4 cd A^{-1} and a power efficiency of 16.3 lm W^{-1}. These reports clearly demonstrate that the charge-transfer approach and the aliphatic chain strategy provide largely complementary benefits, which can be combined to construct high performance solution-processible phosphors with reduced quenching effects.

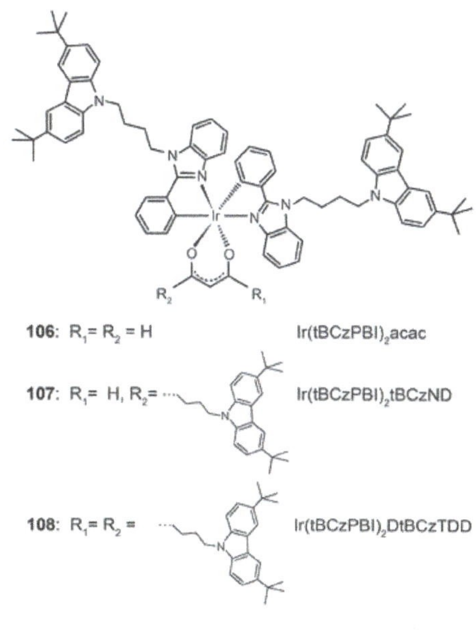

The attachment of oligomeric ligands to a metal ion often markedly influences the optical properties of the metal center. The Ir(III) complexes based on 2-carbazolylpyridine (**109–111**) or 3-carbazolylpyridine (**112–114**) have different emissions at 500 and 590 nm, respectively, which can be attributed to their difference in the length of π conjugation.[55] However, this difference is readily eliminated by introducing an oligofluorene linker between the carbazolyl and pyridine. The emission of both types of Ir(III) oligomers shifts to 550 nm with increasing numbers of the fluorenyl unit.

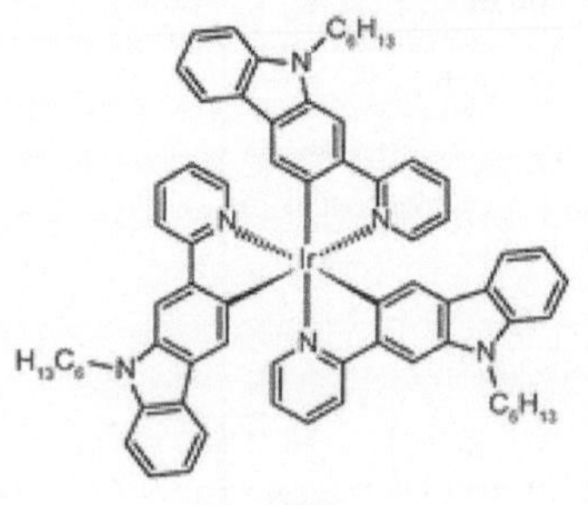

109

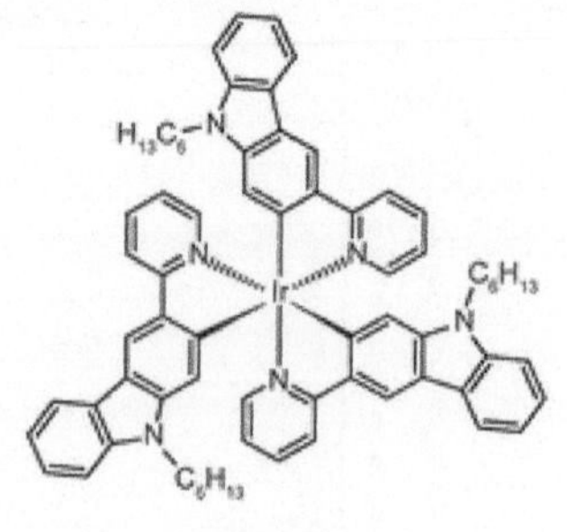

112

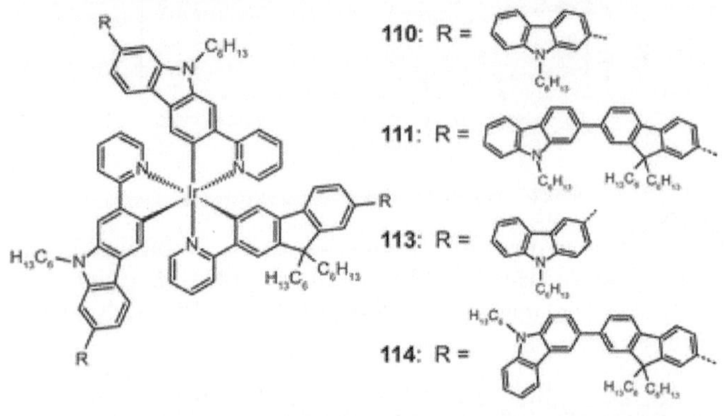

Bolink *et al.* in 2008 attached the Ir(ppy)$_3$ emitting unit to a solution processable dendrimer based on cyclic phosphazene to form complex **115**.[56] The cyclic phosphazene is an inert core used to link the dendronic groups for enhanced thermal and morphological stability. This dendrimer complex reveals optical properties similar to Ir(ppy)$_3$, but its emission peak shifts to 550 nm due to the extended conjugation by the diphenyl linker. The corresponding spin-coated devices derived from complex **115** can be turned on at 3 V, with a maximum current efficiency of 24.0 cd A^{-1}, a power efficiency of 16.7 lm W^{-1}, and an external quantum efficiency of 7.0%. The current efficiency *versus* luminance curves were rather flat as evident by the essentially unaltered current efficiency (19.4 cd A^{-1}) at maximum brightness (3362 cd m^{-2}). The authors ascribed this to the occurrence of balanced carrier injection and transport in emitting layers.

As the green-emitting dendrimers have been extensively studied, the research focus has recently shifted to developing effective strategies for dendronization. Many functional groups have shown strong effects on the optoelectronic performance of the dendrimers. As shown in Ir(III) complexes (**116** and **117**), the modification of the dendrons with functional groups further decreases the intermolecular interaction between the emitting cores, thereby resulting in improved luminescence performance.[57] On the other hand, increasing the density of substitution can boost the electroluminescence efficiency of the dendrons. Comparable electroluminescence performance for Ir(III) complexes with low density of high-generation dendrons can be achieved with high density of low-generation dendrons. Representative examples are Ir(III) complexes (**118** and **119**), which can be relatively easily synthesized.[58] Complex **119** with six carbazole groups endowed its nondoped spin-coated devices with a high current efficiency of 45.7 cd A^{-1} (with an external quantum efficiency of 13.4%), which is higher than that achieved with a similar Ir(III) complex modified with three second-generation dendrons (34.7 cd A^{-1}; 10.3%).

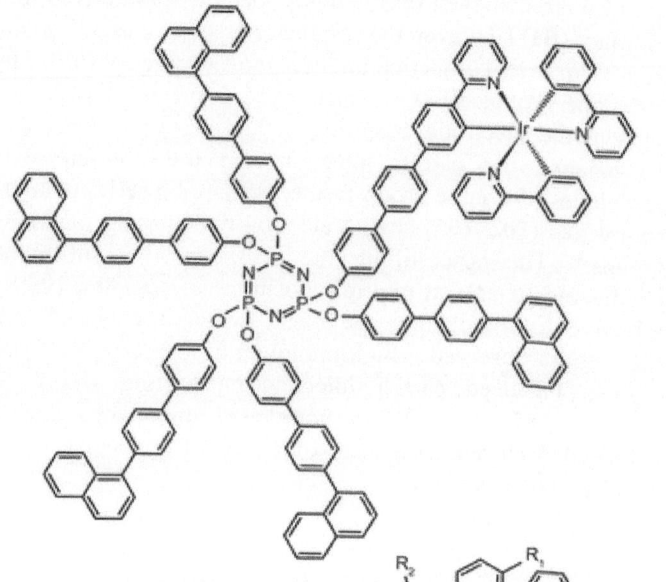

115

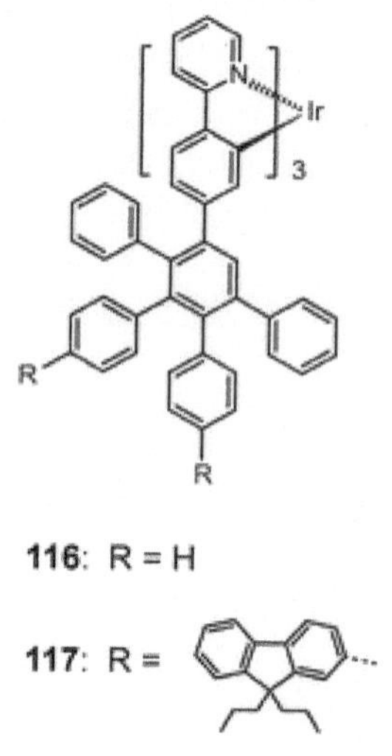

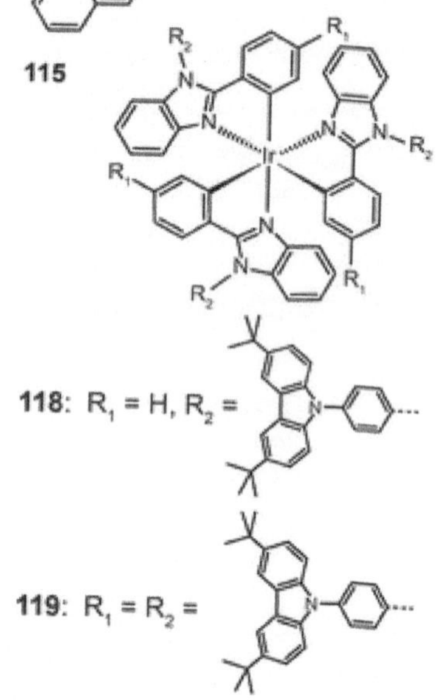

Müllen and co-workers reported in 2009 on a divergent approach to prepare high generation (up to fourth) polyphenylene dendrimers (**120–123**) with an Ir(III) core.[59] These hybranched Müllen-type dendrons can fill the empty space surrounding the core to the greatest extent. Complex **123**, which has nine phenyl groups spanning from the outer shell to the Ir(III) core, is perhaps the largest Ir(III) dendrimer reported thus far. The energy transfer between discrete Ir(III) cores can be efficiently suppressed due to the enormous size of the dendrons. The emission profiles of complex **123** in solid state and in solution were similar, with emission peaks at around 516 nm. The nondoped spin-coated devices incorporating complex **122** exhibited the best electroluminescence performance among the dendrimers under investigation, with a high current efficiency of 21.9 cd A^{-1} and an external quantum efficiency of 6.1%. The pioneering work of Müllen on Ir(III) dendrimers has provided a platform upon which new and emerging applications could be established.

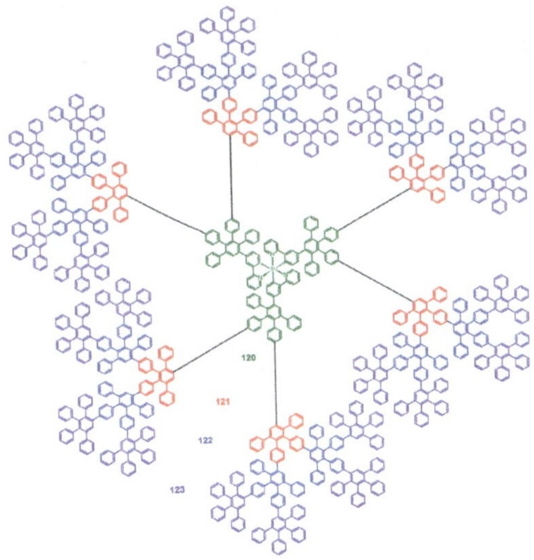

Green-emitting Cu(I) complexes.

Cu(I) complexes are important green-emitting metallophosphors because of their high reliability, low cost and rich luminescent properties. However, tetrahedral Cu(I) complexes often exhibit weak luminescence intensities caused by the distortion of excited states and consequently non-radiative decay. One notable solution to this problem is to enhance the structural rigidity of the complexes by adding bulky ligands. Deaton *et al.* in 2010 reported the fabrication of OLEDS derived from a binuclear Cu(I) complex (**124**), leading to a high external quantum efficiency of 8.7% with the emission maximum at 512 nm.[60] In the following year, Hashimoto *et al.* synthesized highly emissive three-coordinate Cu(I) complexes (**125–127**) with high degrees of symmetry.[61] These complexes showed intense green emissions with high photoluminescence quantum yields (57–71%) in amorphous films. The OLEDs containing **126** in the emitting layer exhibited a low turn-on voltage of 2.7 V, a high current efficiency of 65.3 cd A^{-1}, and a peak external quantum efficiency of 21.3%.

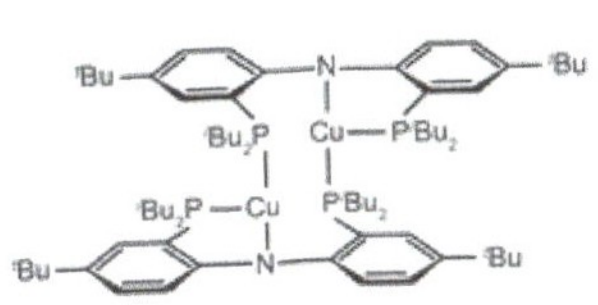

[Cu(PNP-ᵗBu)]₂
124

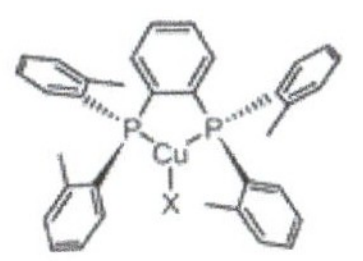

(dtpb)₂CuX
125: X = Cl
126: X = Br
127: X = I

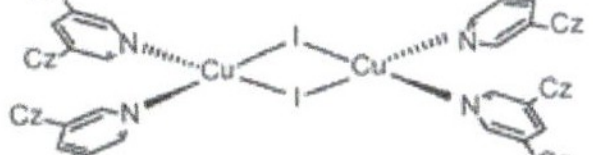

Cz = carbazolyl

[CuI(mCPy)₂]₂
128

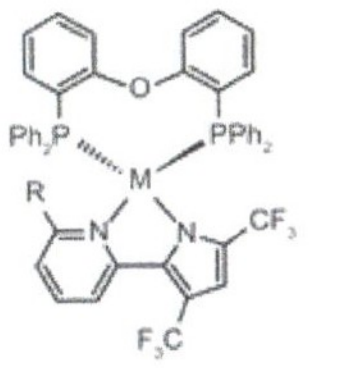

129: M = Cu, R = H
130: M = Cu, R = Ph
131: M = Ag, R = H

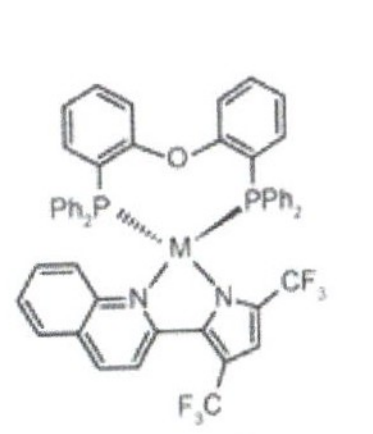

132: M = Cu
133: M = Ag

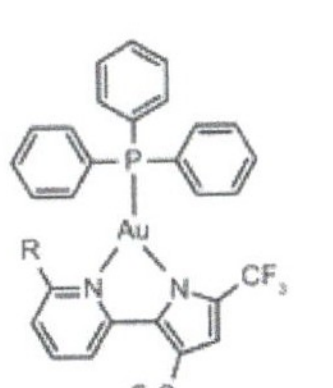

134: R = H
135: R = Ph

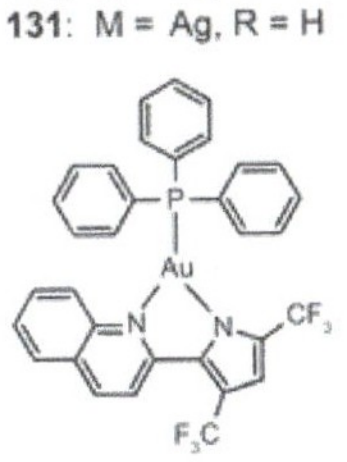

136

In 2011, Liu *et al.* developed a convenient route to high performance OLEDs based on green-emitting Cu(I) complexes as emissive layers (Fig. 11).[62] In their design, they co-deposited 3,5-bis(carbazol-9-yl)pyridine (mCPy) and CuI to form a CuI complex as emissive material in OLEDs. The pyridine-based mCPy ligand serves a dual role as both a ligand for stabilizing the metal and an accommodating host material for the complex, thereby resulting in the formation of a uniform film of the $[CuI(mCPy)_2]_2$ complex (**128**). With a simple three-layer device structure, the authors achieved pure green luminescence at 530 nm, a maximum luminance of 9700 cd m^{-2}, and a peak external quantum efficiency of 4.4%. In a separate note, Hsu *et al.* systematically investigated the influence of group 11 metal d-orbitals on the luminescence performance of the corresponding metal complexes (**129–136**).[63] They showed that Cu(I) complexes exhibit a substantially higher rate of intersystem crossing and more efficient phosphorescence than their Ag(I) and Au(I) analogues. Among the series of metal d^{10}-complexes studied, complex **129** exhibited the most intensive green emission with a photoluminescence quantum yield of 35% in the solid state. Although the maximum brightness of complex **129** is only ~500 cd m^{-2}, its doped devices yield peak electroluminescence efficiencies of 20.0 cd A^{-1}, 14.9 lm W^{-1}, and 6.6%.

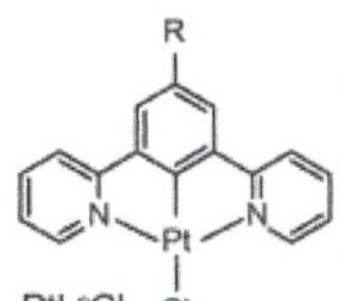

PtLⁿCl

137: L¹ R = H
138: L² R = C(O)OCH₃
139: L³ R = CH₃
140: L⁴ R = Py
141: L⁵ R = PhN(CH₃)₂
142: L⁶ R = mesityl

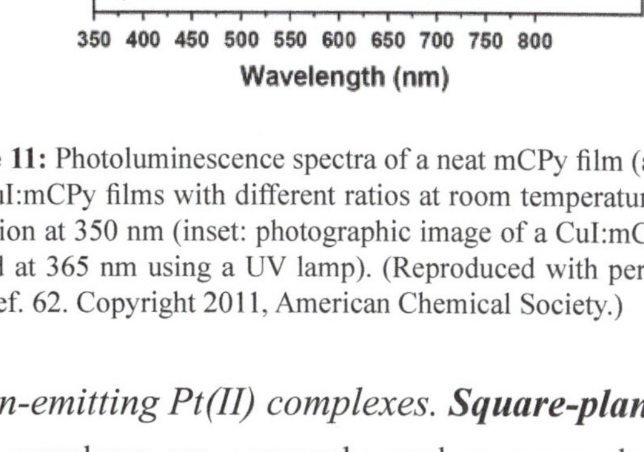

Figure 11: Photoluminescence spectra of a neat mCPy film (at 77 K) and CuI:mCPy films with different ratios at room temperature under excitation at 350 nm (inset: photographic image of a CuI:mCPy film excited at 365 nm using a UV lamp). (Reproduced with permission from ref. 62. Copyright 2011, American Chemical Society.)

Green-emitting Pt(II) complexes. ***Square-planar***

Pt(II) complexes are commonly used as green phosphors in OLEDs. Characteristic reactions for these complexes are ligand substitution and oxidative addition. In 2007, Cocchi *et al.* constructed a series of terdentate cyclometallated Pt(II) complexes (**137–142**) for electroluminescence applications.[64] The emissions of these complexes can be precisely tuned from green to yellow by varying the substituents on the N^C^N-coordinating ligand 1,3-di(2-pyridyl)benzene. The electron-donating substituents are able to considerably elevate HOMOs of the complexes and to facilitate hole injections. High efficiencies up to 16% for the peak external quantum efficiency and 40 cd A^{-1} for current efficiency were measured for OLEDs composed of Pt(II) complexes in a blended host matrix.

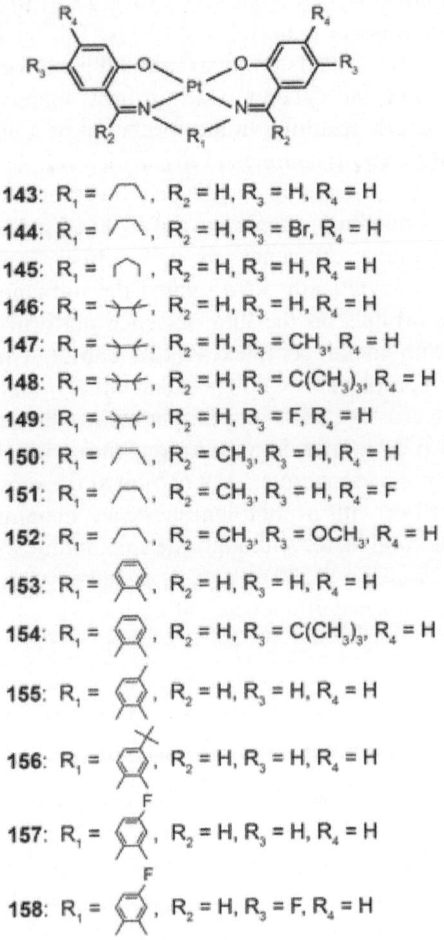

The Che group recently reported a series of highly luminescent Pt(II) complexes (**143–158**) with bidentate Schiff base ligands.[65] As a result of ligand rigidity, these complexes have high thermal stability with T_d values of more than 300 °C. The planar and conjugated structure of the Schiff base ligand benefits carrier injection and transport with HOMO and LUMO energies at −5.4 and −2.4 eV, respectively. The **146**-doped devices achieved a bright green-yellowish emission, with a maximum luminance of 23 000 cd m^{-2}, a turn-on voltage of 4 V, and a high current efficiency of 31 cd A^{-1}. To evaluate potential application in solution-processed devices, the same group further prepared a range of Pt(II) complexes (**159–177**) bearing tridentate ligands with fluorene oligomers.[66] The utilization of oligomeric organic ligands coordinated to the metal ions offers the prospect of low-cost, solution-processable optoelectronic devices as the fluorene moieties increase the solubility of the complexes in common solvents. The extended π-conjugation can also induce bathochromic shifts in the emission of the complexes. The emission spectra ($_{max}$ = 558–601 nm) of these complexes feature a vibronic structure partially resolved as usually observed for fluorene-based polymers. High emission quantum yields up to 76% were attained in degassed dichloromethane. All of the complexes showed excellent carrier-injection abilities with HOMO and LUMO levels of −5.2 and −3.0 eV, respectively. The **164**-doped devices achieved a maximum brightness of 27 000 cd m^{-2} with maximum current (14.7 cd A^{-1}) and luminous power efficiencies (9.2 lm W^{-1}).

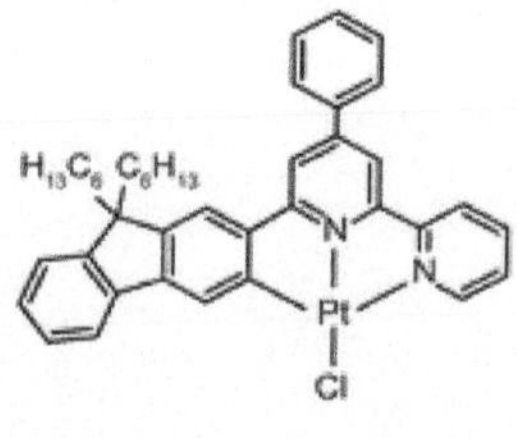

159

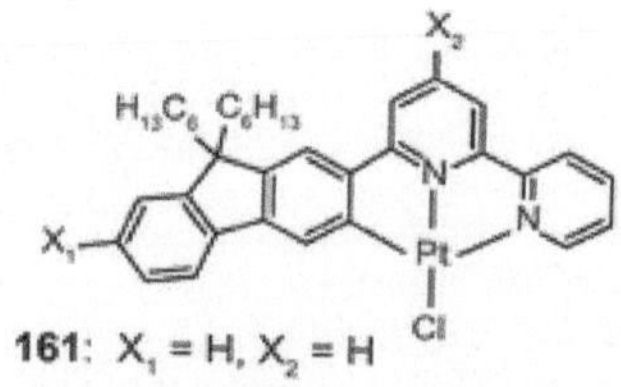

161: X_1 = H, X_2 = H

162: X_1 = 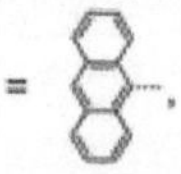, X_2 = H

163: X_1 = 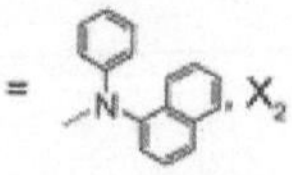, X_2 = H

164: X_1 = H, X_2 = Ph
165: X_1 = Br, X_2 = Ph
166: X_1 = H, X_2 = 3,5-^{t}BuPh

167: X_1 = 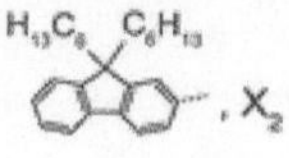, X_2 = Ph

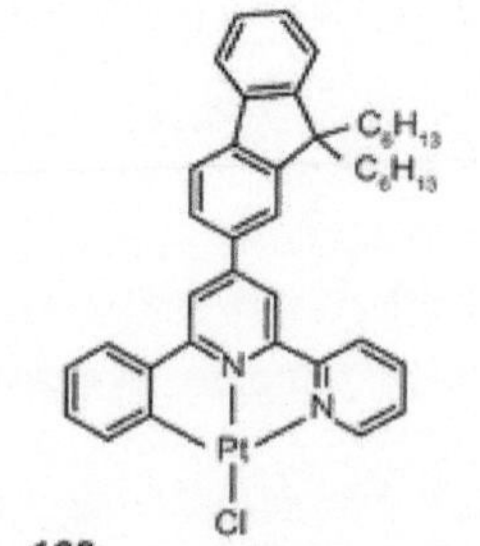

160

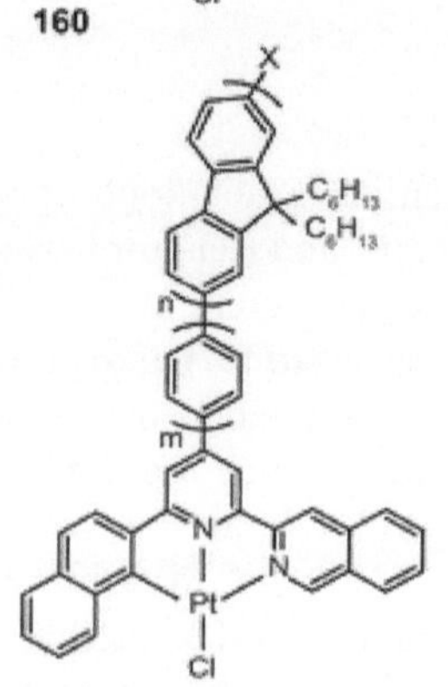

168: m = 0, n = 1, X = H
169: m = 0, n = 1, X = Br
170: m = 0, n = 2, X = H
171: m = 1, n = 1, X = H

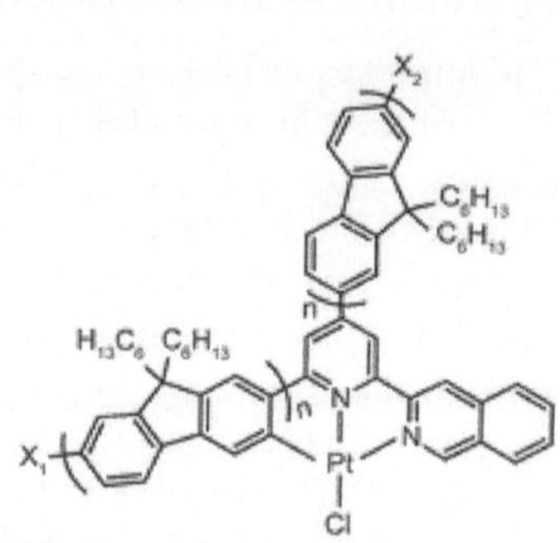

175: n = 1, X_1 = X_2 = H
176: n = 1, X_1 = X_2 = Br
177: n = 2, X_1 = X_2 = H

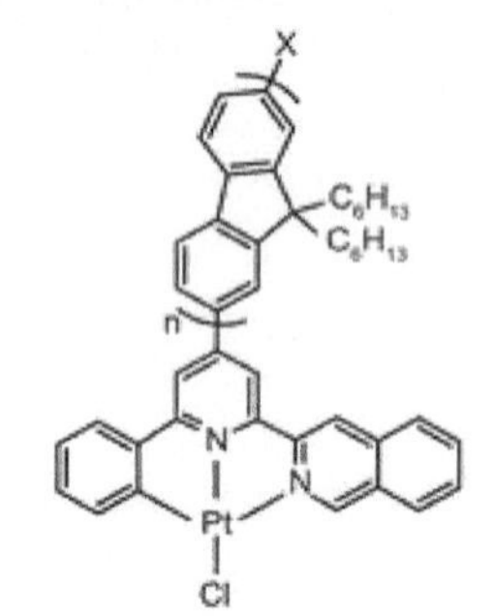

172: n = 1, X = H
173: n = 1, X = Br
174: n = 2, X = H

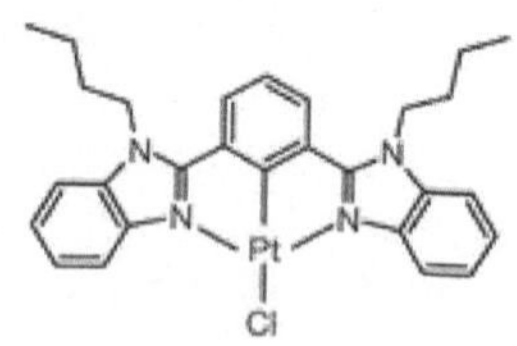

178

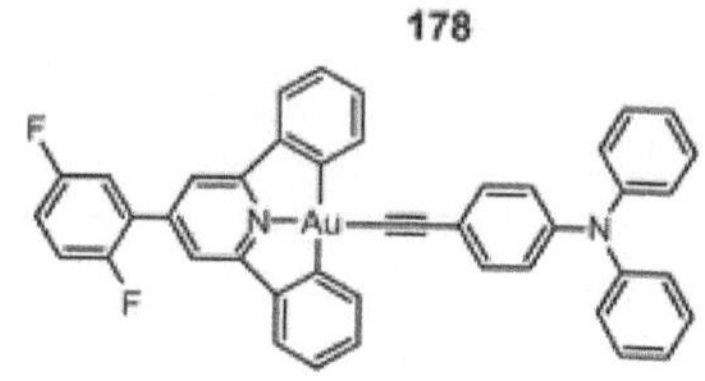

179

Recently, Tam *et al.* reported a dibenzimidazole-based Pt(II) complex (**178**).[67] Large d–d orbital splitting caused by the anionic phenyl ligand reduces nonradiative energy transfer processes and improves the quantum yields of the complex to 19% in solution and 45% in solid state. The **178**-doped devices yielded intensive green emission with high quantum (11.5%) and luminous power (27.2 lm W^{-1}) efficiencies. The authors also developed a novel alkynylgold(III) compound (**179**) with much improved luminescence.[68] The σ-donating alkynyl ligand strongly enhances the optical performance of the complex with a high quantum yield of 34% in poly(methyl methacrylate), and also provides main contribution to the HOMO energy level. As a result, a bright green electroluminescence with a peak at 528 nm was realized with high current (37.4 cd A^{-1}), luminous power (26.2 lm W^{-1}), and quantum efficiencies (11.5%).

Green-emitting Re(I) complexes.

To date, Re(I) complexes have not been subject to the same systematic treatment as other transition metal complexes, although the Re(I) complexes are known to exhibit bright photoluminescence. The main issue with the limited use of these complexes for device applications is their poor processability. Recently, Mauro *et al.* investigated a series of neutral, dinuclear, Re(I) complexes (**180–186**) as phosphorescent emitters in OLEDs.[69] Through functionalization of diazine-based ligands with alkyl groups at different positions, the emission color of these complexes can be tuned from yellowish-green to orange, with high quantum yields up to 22%. These complexes also exhibited high electrochemical stability and suitable HOMO and LUMO energy levels for efficient carrier injection. Notably, the OLEDs obtained with complex **186**, using spin-coating and vacuum evaporation, showed a turn-on voltage of 4.1 V and a maximum current efficiency of 11.0 cd A^{-1} at a luminance of 27 cd m^{-2}.

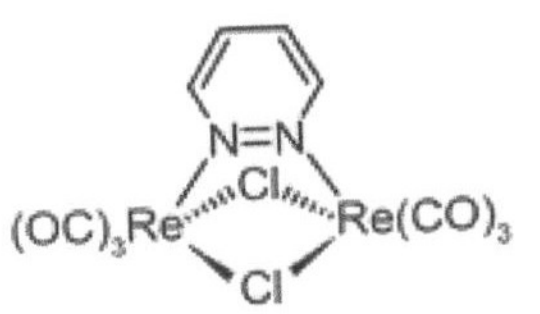

180

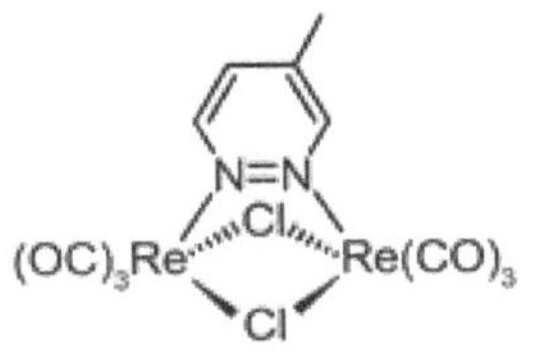

181

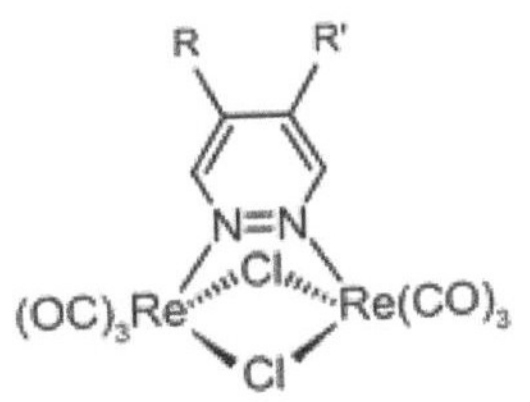

182: R = H, R' = n-hexyl
183: R = H, R' = ᵗBu
184: R = Me, R' = Me
185: R = Me, R' = n-Pr

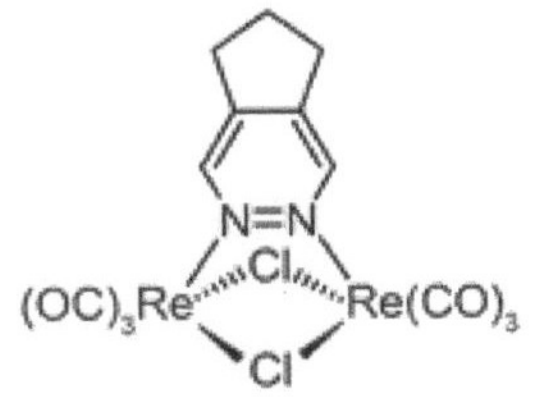

186

Yellow/orange electrophosphorescent complexes.

Yellow/orange emitters have recently gained significant interest for their use in the fabrication of white OLEDs. As described previously, a white OLED typically consists of a blue fluorophore, doped in a layer spatially separated from red and green phosphorescent layers. Alternatively, a combination of sky blue and yellow/orange phosphorescent dopants is able to generate white light. Considering the challenge and the high cost for designing deep-blue emitting phosphors, the strategy of using light sky blue phosphors is more desirable. In this context, yellow/orange phosphorescent emitters have become the second most important light sources apart from the three primary colors. The yellow/orange light can be readily obtained from $Ir(ppy)_3$-based complexes (**187–194**) through ligand modification. For example, the substitution of electron-donating methyl groups on 2-phenylpyridine can render a 35 nm red-shift in the emission of complex **187**.[70] This complex has a high quantum yield (47%) and a short lifetime (0.92 µs). The HOMO and LUMO energy levels of complex **187** are −5.32 and −2.84 eV, respectively, which are well aligned for double-carrier injection. The devices doped with **187** have a turn-on voltage of 4.2 V, a maximum current efficiency of 63.0 cd A^{-1}, and a peak power efficiency of 36.6 lm W^{-1}. In contrast, the functionalization of $(impy)_2Ir(acac)$ complex **188** with strong electron-withdrawing imide groups at the phenyl position can reduce both HOMO and LUMO energy levels to −5.8 and −3.5 eV, respectively.[71] The emission peak shifts to 560 nm. The devices doped with complex **188** (2–4%) exhibit intense orange electroluminescence at 560–572 nm, with high current (34.2–34.7 cd A^{-1}) and luminous power efficiencies (29.5–31.0 lm W^{-1}), corresponding to relatively high external quantum efficiencies (12.5–12.7%). The authors ascribed this to the suppressed quenching effect on intermolecular interaction exerted by the bulky iso-propyl group.

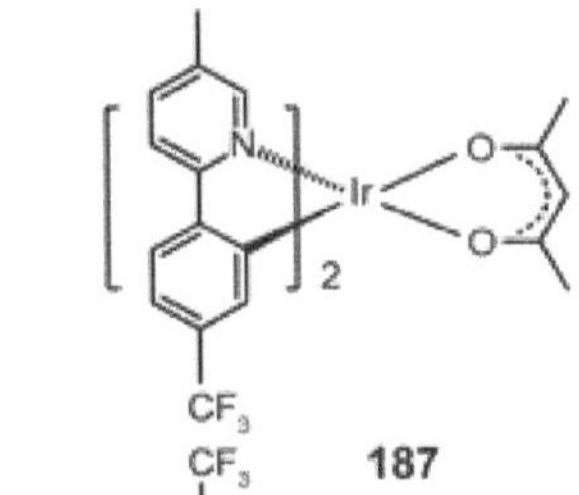

187

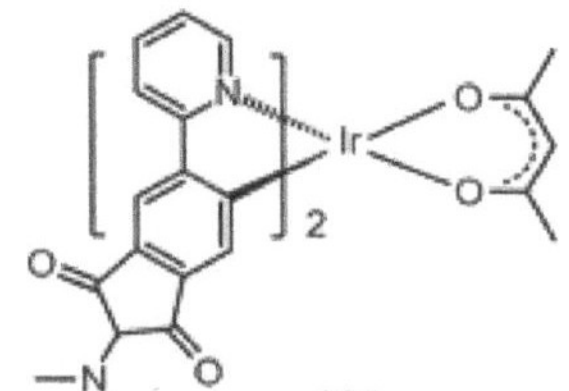

188

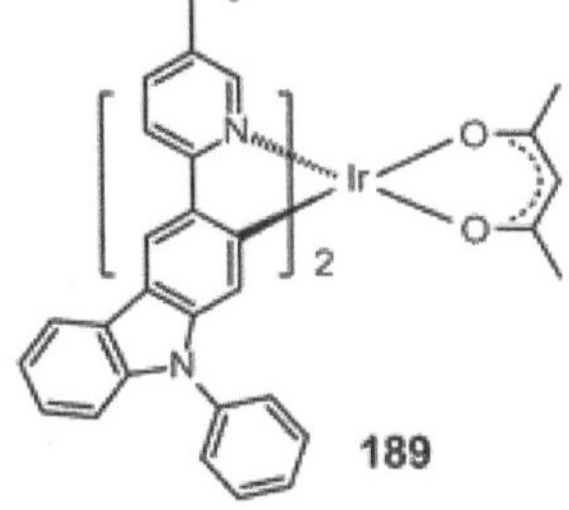

189

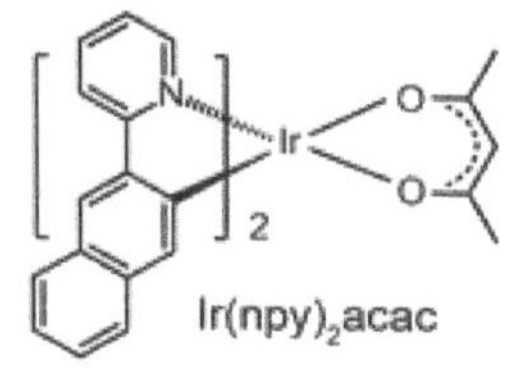

190

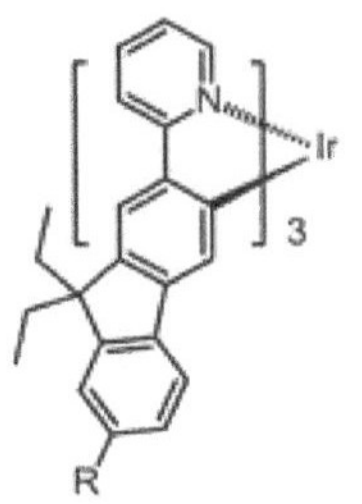

191: R = 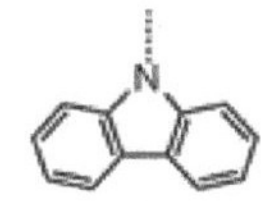

192: R = 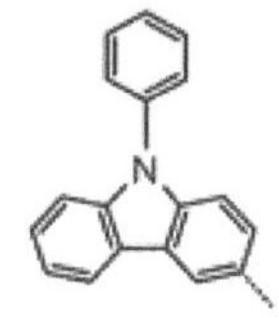

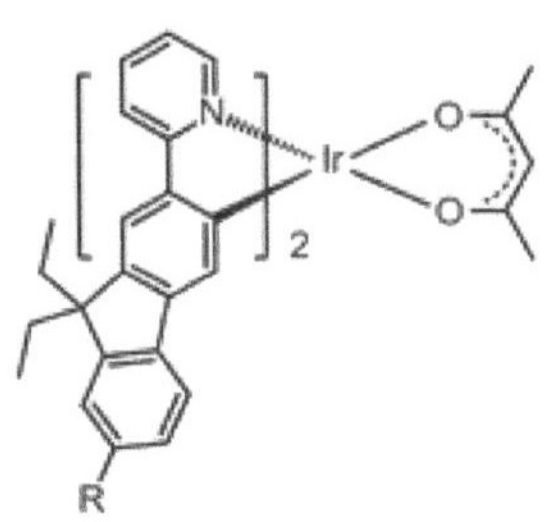

193: R = 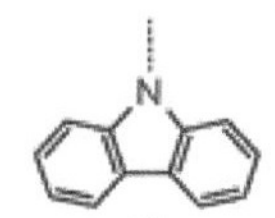

194: R = 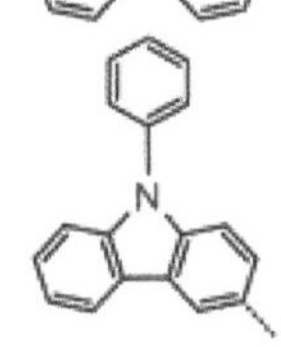

In 2008, Ho *et al.* developed a multifunctional Ir(III) complex **189**, which features the substitution of an electron-withdrawing CF_3 group on a phenylcarbazole-based ligand.[72] The authors argued that the incorporation of the carbazolyl moiety can promote hole transport, while the CF_3 group can suppress the self-quenching behavior and enhance the electron mobility. Remarkably, the CF_3 substituent induced a 50 nm red shift, leading to an emission at 567 nm with a quantum yield of 19% in solution and a short lifetime of 0.58 µs. The devices derived from complex **189** generated a maximum current efficiency of 40 cd A^{-1}, corresponding to an external quantum efficiency of 12% and a power efficiency of 24 lm W^{-1}.

Replacement of the phenyl group in the $Ir(ppy)_2acac$ complex with a naphthyl group results in complex

Ir(npy)$_2$acac **190**, which exhibits an emission at 550 nm.[73] Complex **190** has a high quantum yield of 22% in solution and suitable HOMO and LUMO energy levels (−5.1 and −2.8 eV, respectively), making it a promising candidate for OLED applications. Indeed, the **190**-doped devices achieved good quantum efficiency (10.5%) and luminous power efficiencies (21.8 lm W^{-1}). The substitution of the phenyl group with fluorene or its derivatives also can lead to yellow/orange emissions due to the extension of π-conjugation.[74] The resulting complexes **191–194** showed elevated HOMO levels for efficient carrier injection. In addition, the oligomeric ligand can also facilitate the formation of an amorphous state, enabling the fabrication of doped devices by wet-processing approaches. Among the series of fluorene-based complexes investigated, **192** supported its spin-coated doped devices with the highest electroluminescence efficiencies of 29.77 cd A^{-1}, 13.35 lm W^{-1}, and 9.58%.

Ir(bt)$_2$acac, which gives emission at 557 nm with a quantum yield of 26%, is one of the most popular orange-emitting complexes.[75] Substitution of the benzothiazole unit with a CF$_3$ group or a fluorine atom can shift the emission of (CF$_3$-bt)$_2$Ir(acac) (**195**) and (F-bt)$_2$Ir(acac) (**196**) complexes to 564 and 554 nm, respectively.[76] The devices prepared with these two complexes showed a low onset driving voltage of ~4 V, a driving voltage of ~10 V at brightness >60 000 cd m^{-2}, and high efficiencies (>20%, ~40 lm W^{-1}, 70 cd A^{-1}). In comparison, the substitution by an electron-donating naphthylphenylamine group resulted in much larger bathochromic shifts (>20 nm) of the emission band position for Ir(III) complexes (**197** and **198**).[77] The spectral shift mainly originates from the elevation of HOMO levels promoted by the aryl amine group. However, the quantum yields of the two complexes were lowered to 12% and 9%, presumably due to structural relaxation of the aryl amine group. The spin-coated doped devices of **197** and **198** gave moderate electroluminescence performance.

In 2008, Rehmann *et al.* reported a photo-crosslinkable Ir(III) complex (**199**; also known as x-emitter) containing oxetane moieties.[78] For a well-balanced charge transport, the authors utilized an additional electron-conducting layer on top of the emissive layer, leading to significantly improved device performance. The spin-coated doped devices of **199** realized a maximum current efficiency of 18.4 cd A^{-1} at an operating voltage of 5 V and a brightness of 100 cd m^{-2}. Notably, the maximum efficiency was improved by a factor of 30 relative to a device without the electron-transporting layer. Ir(bipz)$_3$ (**200**) and Ir(fipz)$_3$ (**201**) complexes were constructed by Chou and co-workers in 2010 according to a contrary strategy of utilizing high-field-strength ligands to induce hypochromatic shift in emission wavelength.[79] These two complexes showed bright yellow luminescence with the emission peaks at 567 and 545 nm, respectively. The **200**-doped devices exhibited higher efficiencies (14.6%, 26.1 lm W^{-1}, 34.8 cd A^{-1}) than those based on (dfpz)$_2$Ir(bipz) complex (**202**) (11.6%, 17.6 lm W^{-1}, 28.6 cd A^{-1}). The high efficiencies for complex **200** result from the balancing effect of the ancillary ligand on carrier injection and transport.

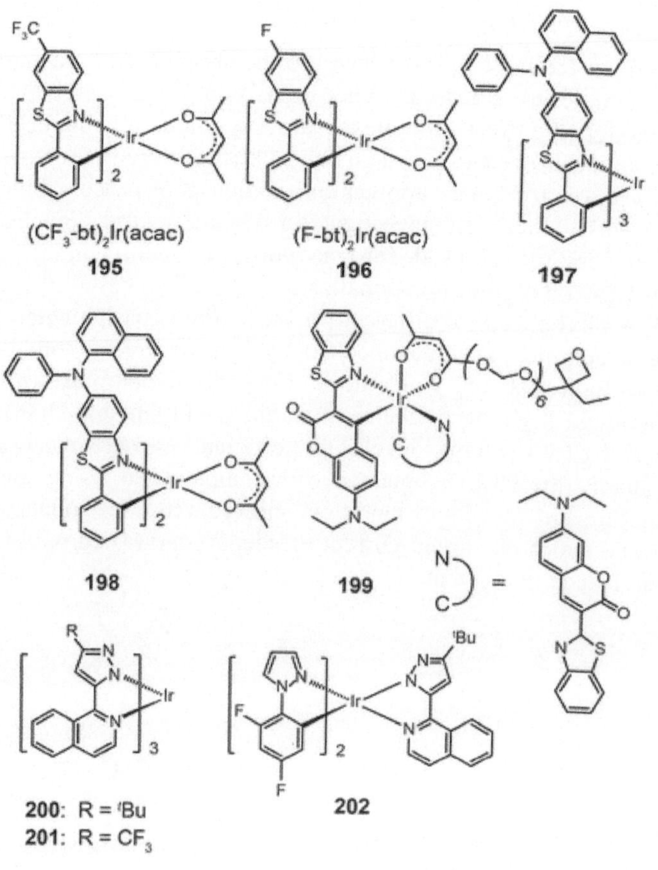

Red electrophosphorescent complexes.

Red-emitting Ir(III) complexes.

Owing to structural constraints and strong intermolecular interaction, red-emitting Ir(III) complexes often suffer from unbalanced carrier injection and transport, luminescence quenching, and poor device performance.[80] Introduction of electron-donating groups can effectively enhance the carrier injecting/transporting ability.[80] For instance, carbazole moieties in Ir(III) complexes (**203–206**) strongly elevate the HOMO levels of these complexes by coupling of the carbazole unit to the iridium isoquinoline core. They represent materials that emit in the red region of the optical spectrum with quantum yields up to 30%. These complexes also have excellent thermal and morphological stability (T_d > 340 °C and glass transition temperature T_g > 110 °C). Complex **205** exhibited a higher external quantum efficiency (11.76%) than **203** (10.87%), which could be attributed to the suppression of nonradiative transitions by the high field strength of the acac ligand in **205**. However, complex **203** displayed higher current (11.00 cd A^{-1}) and power efficiencies (6.16 lm W^{-1}) as compared with **205** (10.15 cd A^{-1}, 5.25 lm W^{-1}).

A similar effect of the carbazolyl group on quinoline-based Ir(III) complexes, (Et/EO-CVz-PhQ)$_2$Ir(pic-N–O), (Et/EO-CVz-PhQ)$_2$Ir(pic) and (Et/EO-CVz-PhQ)$_2$Ir(acac) (**207–212**), was demonstrated by Lee *et al.*[81] These complexes also have improved thermal and morphological stability with T_d of 385–344 °C and T_g of 150–287 °C. The ancillary ligands showed a strong effect on the optical properties of the complexes. The acac-based complexes displayed much lower photoluminescence quantum yields (3–5%) than other complexes, of which **210** gave the highest quantum yield (33%).

The emission peaks of the acac-based complexes are centered at around 623 nm, while the emissions of other complexes are at about 604 nm. Complex **208**-doped spin-coated devices achieved the best electroluminescence performance. The maximum efficiencies reached 8.89 cd A^{-1}, 3.41 lm W^{-1} and 5.51%. The carbazole group clearly improves the rigidity of the complexes and hence decreases the probability of nonradiative transition. The roll-offs of external quantum efficiency were less than 1% at 20 mA cm^{-2}.

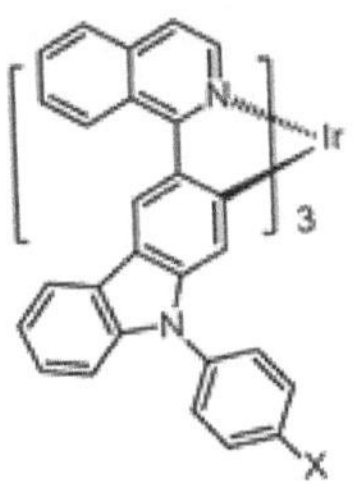

203: X = H
204: X = OMe

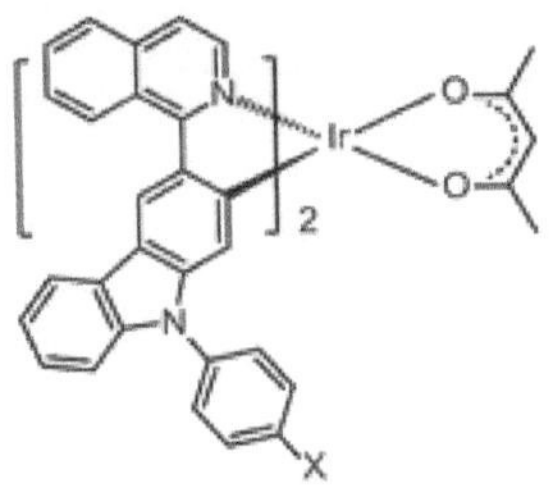

205: X = H
206: X = OMe

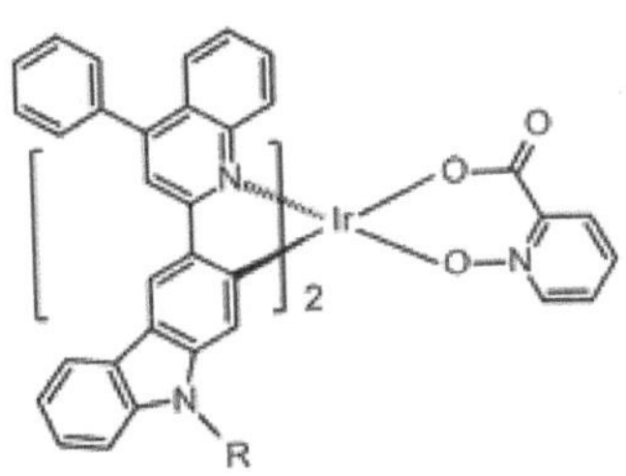

207: R = ethyl
208: R = 2-(2-methoxyethoxy)ethyl

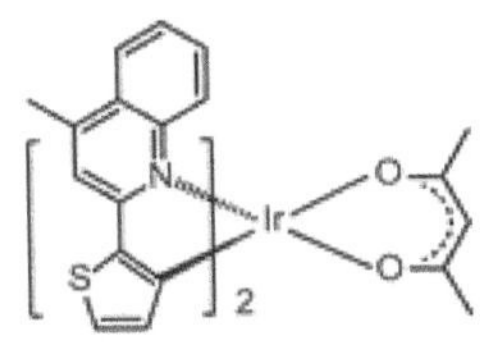

$(tmq)_2Ir(acac)$
213

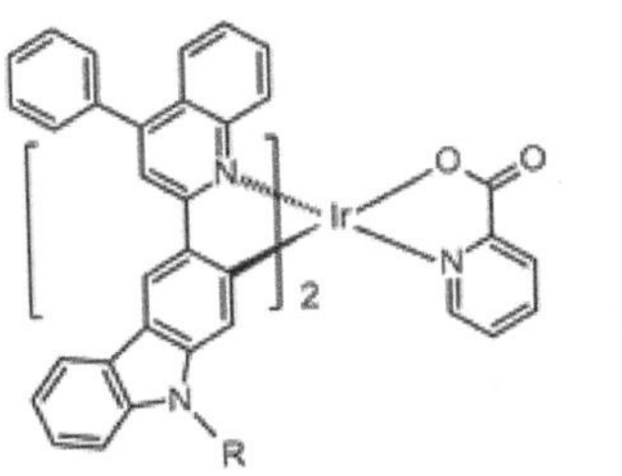

209: R = ethyl
210: R = 2-(2-methoxyethoxy)ethyl

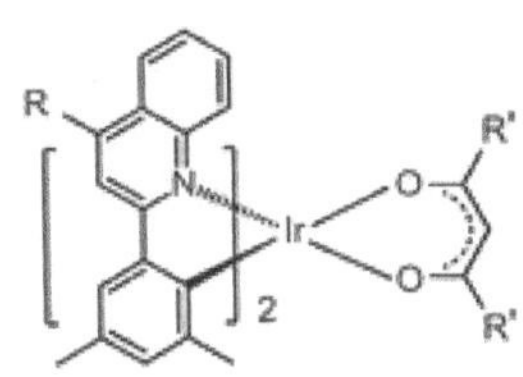

214: R = H, R' = Me
215: R = Me, R' = Me
216: R = H, R' = ^{t}Bu
217: R = Me, R' = ^{t}Bu

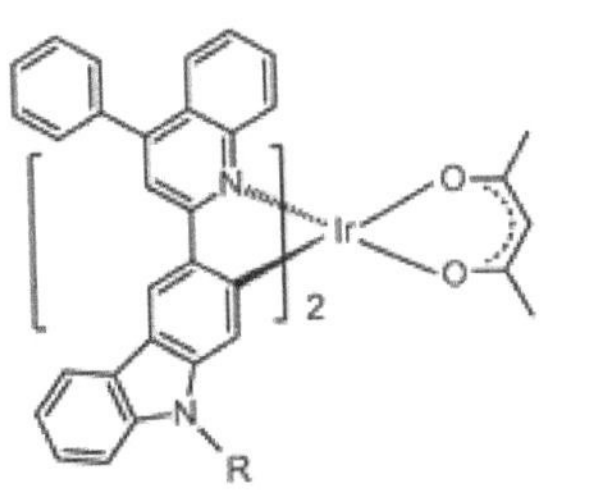

211: R = ethyl
212: R = 2-(2-methoxyethoxy)ethyl

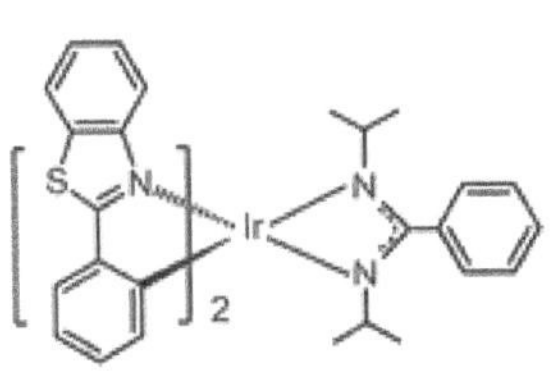

$(bt)_2Ir(dipba)$
218

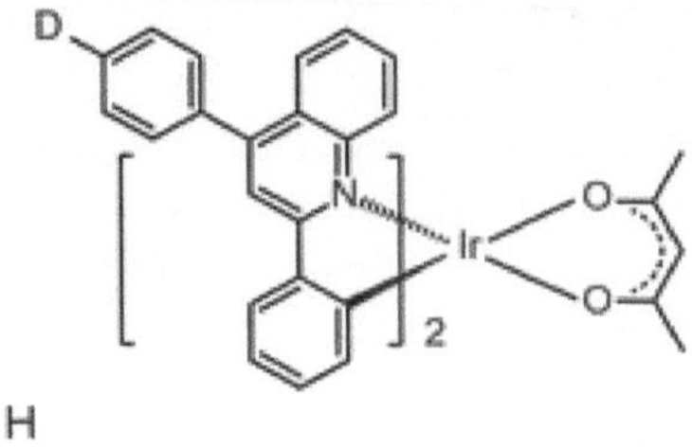

219: D = H

220: D =

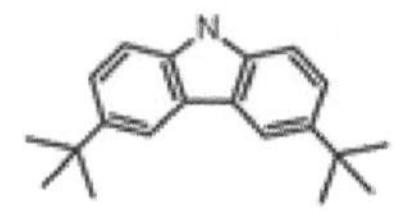

221: D =

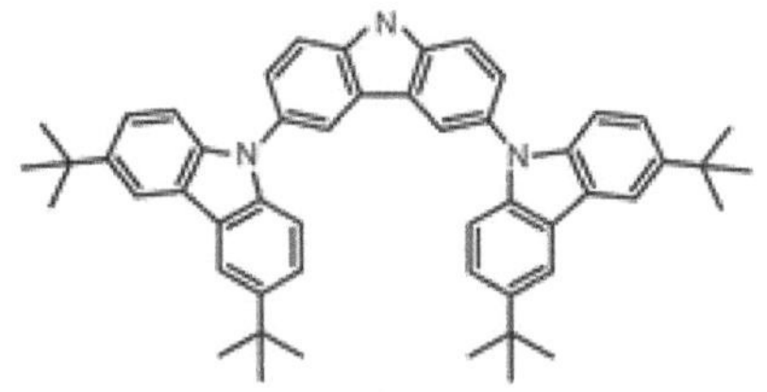

222: D =

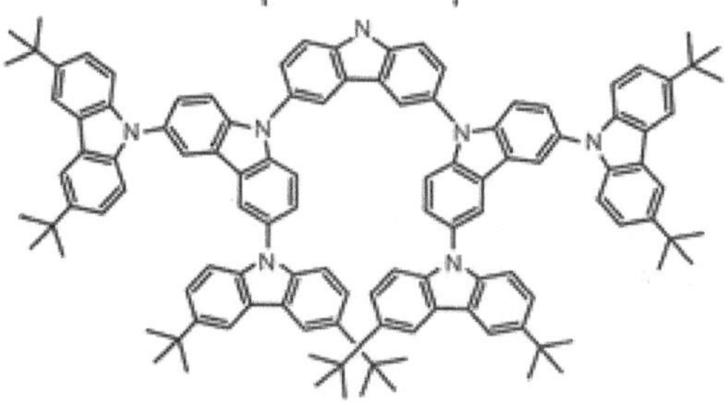

One common approach to suppress intermolecular interaction is to increase the steric hindrance of molecules by introducing bulky groups. A representative example was demonstrated in $(tmq)_2Ir(acac)$ complex (**213**) by substituting a methyl group at the 4-position of quinoline. Complex **213** emits at 611 nm with a full-width-at-half-maximum of 48 nm, which is much narrower than that of complex $(piq)_2Ir(acac)$.[13] With a conventional 4,4′-bis(9-carbazolyl)-2,2′-biphenyl (CBP) host, complex **213** realized a low turn-on voltage of 3.2 V and a high luminance of 54 508 cd m^{-2}, as well as high efficiencies of 21.5 cd A^{-1}, 19.3 lm W^{-1} and 15.1%. For (BIQS)/**213**-based devices, the best electroluminescence performance to date was obtained, including extremely high efficiencies of 37.3 cd A^{-1}, 32.9 lm W^{-1}, and 25.9%.[13] The increase in the number of methyl and *tert*-butyl substituents in Ir(III) complexes (**214–217**) can further increase the intermolecular steric hindrance.[82] The electron-donating methyl group induced the elevation of both HOMO and LUMO levels. The iso-propyl substituent in $(bt)_2Ir(dipba)$ complex (**218**) also revealed the ability to reduce intermolecular interaction.[83] It was found that the use of the amidinate ligand changed HOMO and LUMO levels to −4.88 and −2.81 eV. The suppressed quenching effect due to enhanced carrier injection in **218** was confirmed using its heavily doped

(30%) devices. Except for the low driving voltages of 2.5 V obtained for onset and 3.3 V at 100 cd m^{-2}, high efficiencies of 18.1 cd A^{-1}, 18.4 lm W^{-1}, and 15.4% were also realized. The nondoped devices of **218** also achieved high efficiencies of 8.4 cd A^{-1}, 6.9 lm W^{-1}, and 7.5% reported thus far. Another notable development was reported by Ding *et al.*, who showed that the use of dendritic protection, as shown in $Ir(phq)_2(acac)$-type dendrimers (**219–222**), can improve electroluminescence performance.[84] The efficiencies of nondoped devices based on **222**reached 3.3 cd A^{-1}, 1.3 lm W^{-1}, and 5.0%, while devices doped with 5% of **222** had the efficiencies of 13.0 cd A^{-1}, 7.2 lm W^{-1}, and 11.8%.

Red-emitting Ru(II) complexes.

For the oxidation states (II–IV) of ruthenium, there is great complexity as many complexes can undergo reversible oxidation and reduction reactions. Ru(II) complexes with nitrogen-based ligands now attract great attention for application in OLEDs. Zhu *et al.* recently reported an intriguing class of phosphorescent ruthenium(II) complexes (**223** and **224**) with peripheral carbazole substituents. These complexes exhibit intense deep red emission at ~660 nm, and can form high quality films using the electrochemical deposition technique.[85] Despite the encapsulation of the Ru(II) core by the peripheral carbazole group, the optical properties of **223** and **224** are essentially unaltered. However, the HOMO energy levels of **223** and **224** are raised by 0.1 eV, while their LUMOs are reduced by 0.4 eV. The device made with **223** achieved maximum luminous current (3.9 cd A^{-1}) and power efficiencies (1.1 lm W^{-1}) at a maximum brightness of 293 cd m^{-2}.

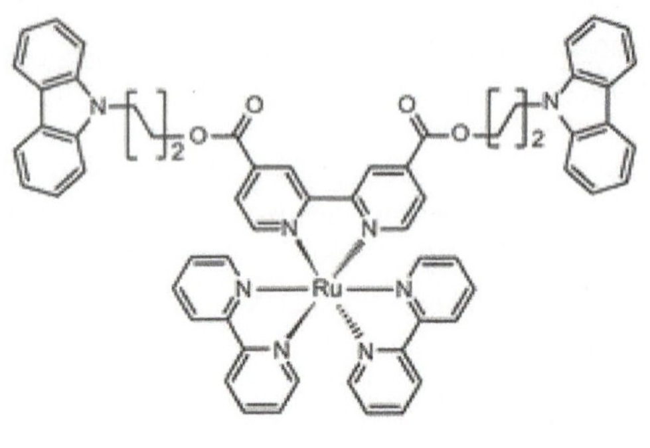

223

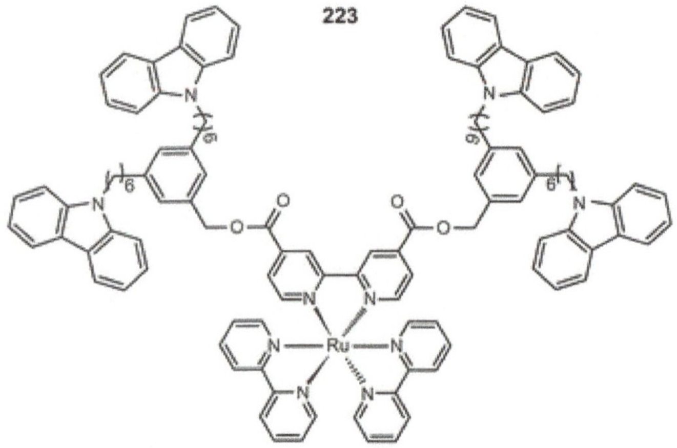

224

Red-emitting porphyrin-type complexes.

Tetrapyrrole macrocyclic rings, such as porphyrins, are of major biological and medical importance. But as emitters for OLED applications, these compounds are conventionally thought to be constrained by the quenching effect because of strong intermolecular interaction. The quenching effect can be circumvented using donor–acceptor (D–A) systems. To prevent the potential phase separation during the preparation of the D–A system, Montes *et al.* designed Alq_3-PtTPP-based systems (**225–228**), linked by oligofluorene bridges, for the investigation of triplet dynamics in D–A electroluminescent materials (Fig. 12).[86] Importantly, the linkers facilitate exothermic triplet energy transfer. The red emission at ~650 nm from PtTPP is sensitized by the Alq_3 moiety. The nondoped spin-coated devices based on **225** and **226** achieved external quantum efficiencies of 0.21% and 0.18%, respectively.

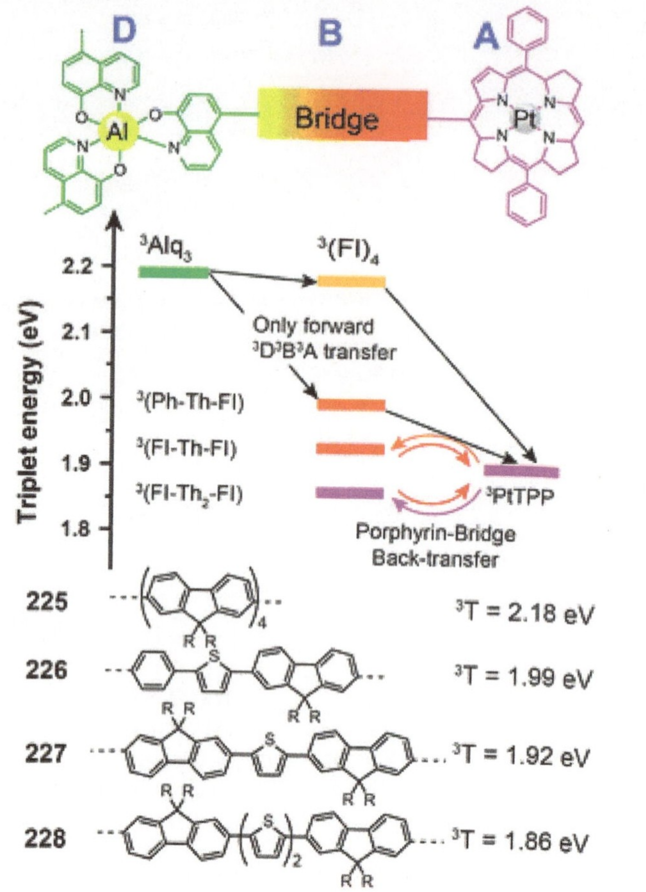

Figure 12: Intramolecular energy transfer involving compounds **225–228** (D, B and A refer to donor, bridge and acceptor, respectively). (Reproduced with permission from ref. 86. Copyright 2007, American Chemical Society.)

Red-emitting Eu(III) complexes.

Luminescent Eu(III) complexes are most famous for their intense monochromic red emissions at ~615 nm with high quantum yields.[87] As there are several excellent recent reviews on light-emitting lanthanide materials,[88] we herein only focus on recent development of functional phosphine oxide-based Eu(III) complexes. Tertiary phosphine oxides are a series of oxide ligands having the general formula R_3PO.[89] They are obtained by oxidation of trisubstituted phosphines, and constitute one of a few series of ligands in which electronic and steric properties can be precisely controlled by varying R.

The Huang group presented first reports on the synthesis of electroluminescent Eu(III) complexes, $Eu(TTA)_3DPEPO$ (**229**)[89a] and $Eu(TTA)_3NAPO$ (**230**),[89b,c] with phosphine oxide chelating ligands. The phosphine oxide ligands form more stable coordination bonds with lanthanide ions compared with N-heterocyclic ligands, such as pyridine. The phosphine oxide complexes thus have high thermal and morphological stability ($T_d > 300$ °C and $T_g \approx 100$ °C). Complex **229** reveals a very high quantum yield of 0.55 owing to the reduction in nonradiative loss. The compact and rigid structure and strong electron-transporting ability of the PO ligand endow **229**-coupled devices with very high efficiencies of 4.58 cd A^{-1}, 2.05 lm W^{-1}, and 2.89%.[89a]

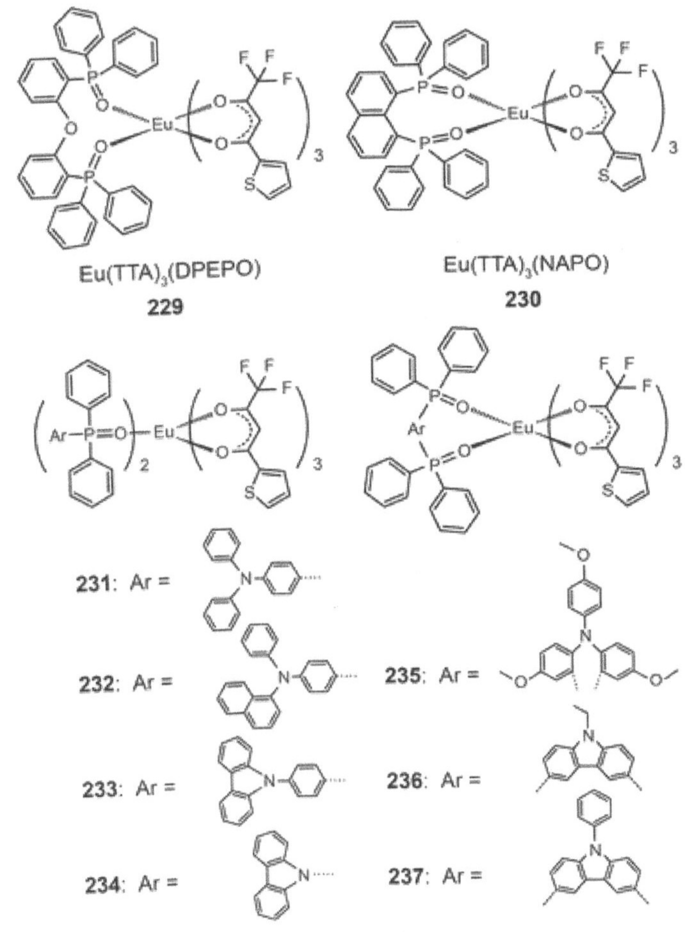

The same group further prepared $Eu(TTA)_3(TAPO)_2$, $Eu(TTA)_3(NADAPO)_2$, $Eu(TTA)_3(CPPO)_2$, and $Eu(TTA)_3(CPO)_2$ complexes (**231–234**) with hole-transporting groups incorporated into monodentate ligands in order to balance the carrier injection and transport in emissive layers.[90] These complexes have high quantum yields (up to 42%) that are attributable to the reduced quenching effect imparted by the bulky ligands. The hole-transporting groups enhanced the hole injection by elevating the HOMO level from −6.3 to about −5.5 eV. On the other hand, the LUMO levels of these complexes remained at −3.0 eV, which is similar to that of triphenylphosphine oxide-based Eu(III) complexes. The Eu(III) complex **232** with the bulkiest PO ligand in the series under investigation provided its nondoped devices with a high luminance of 59 cd m^{-1}. The electroluminescence performance of devices doped with **232** is among the best results reported thus far, with maximum efficiencies of 5.88 cd A^{-1}, 3.69 lm W^{-1} and 3.71%. In a following paper, the authors investigated the effect of different ligand structures on optoelectronic performance using D–A type Eu(III) complexes appended with PO ligands.[91] They discovered that the reductive effect of Eu(III) ions induces the migration of electron cloud from electron-rich hole-transporting groups to electron-deficient PO groups. Consequently, the abilities of both functional groups to inject or transport charge carriers are weakened.

Nevertheless, the insertion of a π-spacer between the donor and the acceptor group of a ligand does effectively diminish the inductive effect of Eu(III) ions on frontier molecular orbital electron cloud distributions of the ligand. Xu *et al.* duplicated this strategy in $Eu(TTA)_3(TMOADPO)$, $Eu(TTA)_3(EtCzDPO)$, and $Eu(TTA)_3(PhCzDPO)$ complexes (**235–237**).[92] The use of rigid bidentate ligands enhanced the thermal and morphological stability of the complexes and also reduced nonradiative transitions due to structural relaxation. The devices based on these complexes achieved a maximum luminance of >1000 cd m^{-2} and a high external quantum efficiency of >3.5%. By taking advantages of the strong chelating effect of the PO ligand toward lanthanide ions, the authors recently introduced PO-based Eu(III) complexes into polymeric light-emitting Eu(III) materials.[93]

Infrared electrophosphorescent complexes.

Infrared (IR) emitting materials are important with huge potential for application in optical communication, biosensing, and defence. However, the IR-emitting materials are scarce as the emission at around 1000 nm occurs only in a few pure organic compounds with strong electron-withdrawing groups and complexed structures.[94] For good measure, the emissions in the IR region are relatively readily achieved in porphyrin and phthalocyanine derivatives as well as organolanthanide complexes.

IR-emitting Pt(II) complexes.

In 2007, the Thompson group reported the first highly efficient metalloporphyrin near-infrared (NIR) dye, Pt(tpbp) (**238**).[95] The HOMO and LUMO levels of**238** are −4.9 and −2.5 eV, respectively, suitable for carrier injection. The strong emissive ability of this complex is manifested by its high quantum yield of 70%. Combined with an electron-transporting host Alq_3, **238** exhibited an excellent electroluminescence performance with a high maximum output of 750 μW cm^{-2} and a maximum external quantum efficiency of 6.0%. The efficiency roll-off was gradual and the lifetime of the devices was measured up to 1000 h. However, the emission wavelength (765 nm) of this material is considered to be too short for optical communication applications. Recently, Graham *et al.* reported on the development of IR emissive Pt–porphyrin derivatives (**239–246**), with emission wavelength stretched out to 1022 nm for Pt–TPTNP complex **245** and Pt–Ar_4TAP complex **246** by increasing the conjugation length from **238** (Fig. 13).[96] However, **237** exhibited the worst electroluminescence performance with a maximum efficiency of only 0.2%. The Pt–porphyrin derivatives of **238**, **239–244**, with limited conjugation lengths have relatively high quantum yields of ~45%. Their polymeric light-emitting devices achieved external quantum efficiencies of 2.07–3.2% using polyvinylcarbazole as the host. Their OLEDs made with Alq_3 as the host realized high external quantum efficiencies of 5.0–9.2%.

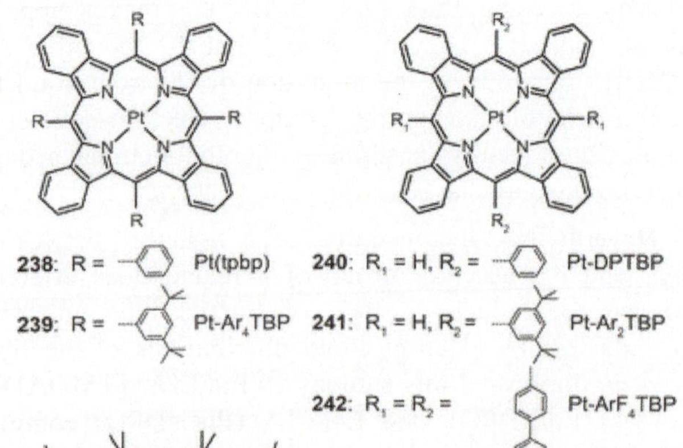

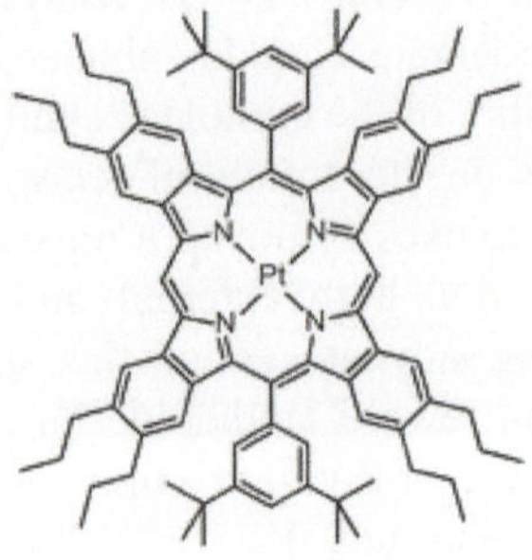

Pt-Ar$_2$OPrTBP
243

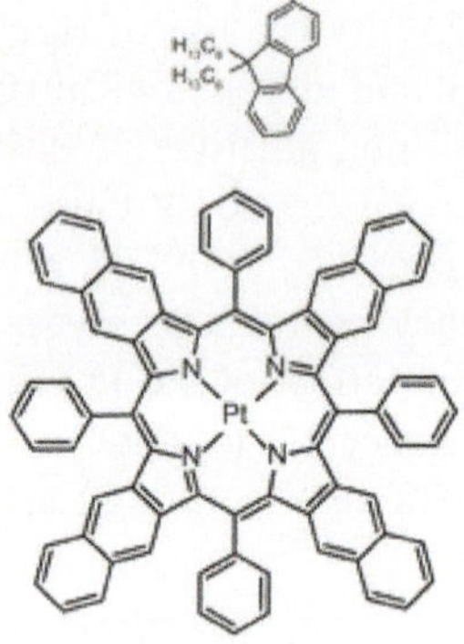

Pt-TPTNP
245

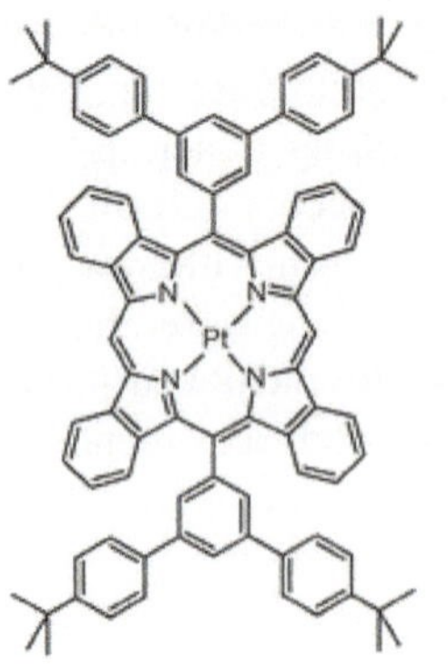

Pt-TAr$_2$TBP
244

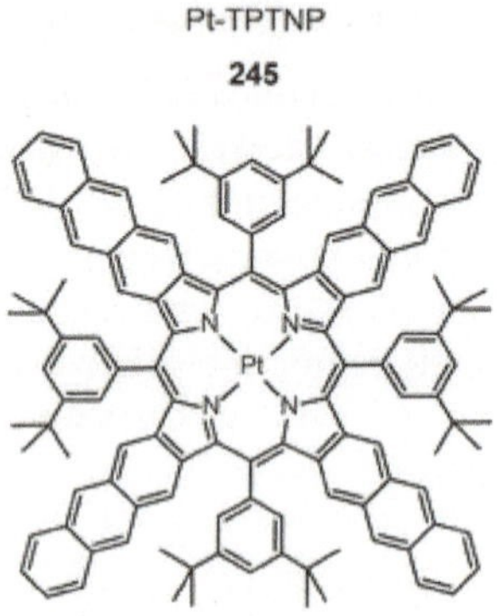

Pt-Ar$_4$TAP
246

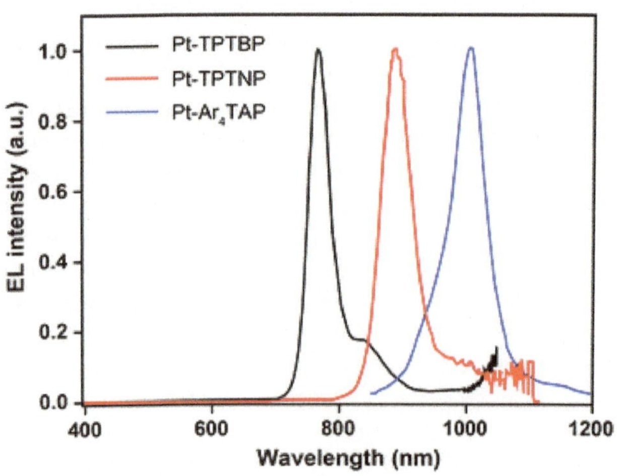

Figure 13: Electroluminescence spectra of PHOLEDs using **238**, **245** and **246** as the emitter, respectively, in the PVK:PBD host matrix. The bathochromic shifts in emission from 750 to 1000 nm are due to conjugation extension from phenyl to naphthyl to anthryl in their porphyrin rings. (Reproduced with permission from ref. 96. Copyright 2011, American Chemical Society.)

IR-emitting lanthanide complexes.

Lanthanide-doped inorganic materials are well-known for their successful optical applications as upconversion emitters and fiber amplifiers in optical communications. The Piguet group has developed a series of lanthanide–transition metal hybrid complexes to realize upconversion processes at the molecular level.[97] Common optical communication windows at 1550, 1300 and 980 nm correspond to the emission from Er(III), Nd(III) and Yb(III) complexes, respectively. Therefore, IR devices based on lanthanide complexes have been proposed. However, such developments are constrained by the strong absorption of IR radiation by C–H vibration and the inherent poor electrical properties of the complexes.[87a] A notable effort was demonstrated by Li *et al.*, who reported the synthesis of an $Er(PM)_3(TP)_2$ complex (**247**) based on pyrazolone and TPPO ligands.[98a] The two ligands possess an antenna effect that enhances the emission from Er(III) at 1540 nm. The nondoped devices based on **247** achieved a maximum output of 0.21 μW cm^{-2}.

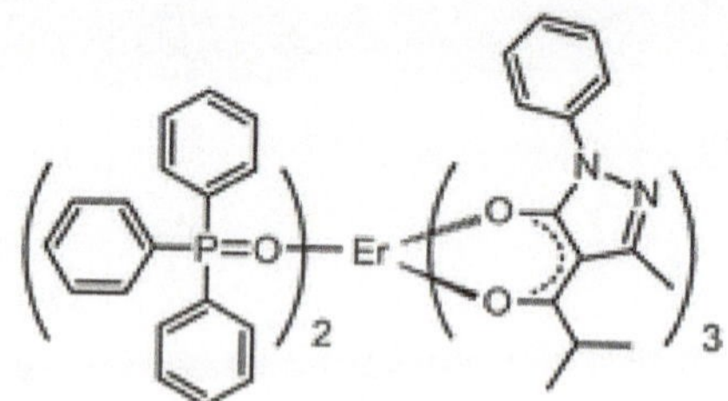

247

Katkova *et al.* prepared a range of NIR emitting Pr(III), Nd(III), Ho(III), Er(III), Tm(III), and Yb(III) complexes (**248–262**), sensitized by 2-(2-benzoimidazol-2-yl)phenolate, 2-(2-benzoxyazol-2-yl)phenolate or 2-(2-benzothiazol-2-yl) phenolate ligands.[98b] These complexes tend to form dimers comprising two lanthanide ions and six organic ligands. The nondoped devices of Nd(III) complexes**252** and **253** yielded strong light outputs at 1069 nm (74 and 29 μW cm^{-2}, respectively), while their Yb(III) counterparts **261** and **262** exhibited much higher light outputs at 982 nm (154 and 286 μW cm^{-2}, respectively).

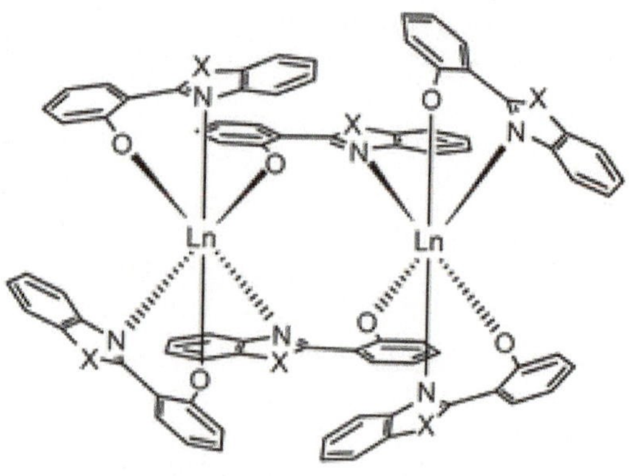

248: Ln = Pr, X = N
249: Ln = Pr, X = O
250: Ln = Pr, X = S
251: Ln = Nd, X = N
252: Ln = Nd, X = O
253: Ln = Nd, X = S
254: Ln = Ho, X = O
255: Ln = Ho, X = S
256: Ln = Er, X = O
257: Ln = Er, X = S
258: Ln = Tm, X = O
259: Ln = Tm, X = S
260: Ln = Yb, X = N
261: Ln = Yb, X = O
262: Ln = Yb, X = S

Full-color phosphorescent OLEDs are now realized with extremely high efficiencies using various metal–organic complexes. For instance, external quantum efficiencies approaching 25% have been achieved for blue, green and red emissions (Table 2). Nevertheless, it is worth noting that efficient blue phosphors are less abundant due to the difficulty in achieving high energy metal-to-ligand charge-transfer (^{3}MLCT) and ligand-to-ligand charge-transfer (^{3}LLCT) states. Furthermore, relatively low performances of OLEDs are typically observed for red phosphors as they are prone to quenching. This stimulates continuing efforts to develop novel red-emitting complexes with low polarity, weak intermolecular interaction, and short emission lifetime.

Table 2: Physical properties and electroluminescence performance of typical phosphorescent complexes

Emitter	Photophysical properties		HOMO/LUMO[c] (eV)	Device performance			Ref.
	Emission peak[a] (nm)	Φ (%)		Voltage (V)[d]	Max. brightness (cd m^{-2})	Max. efficiency[e]	
a **Measured in solution. The values in parentheses indicate the shoulder peak.** ***b*** **Measured in solid matrices.** ***c*** **Measured according to the analysis of cyclic voltammetric data and energy bandgaps.** ***d*** **Turn-on voltage.** ***e*** **In the order of current efficiency (cd A^{-1}), power efficiency (lm W^{-1}) and external quantum efficiency (%).**							
37	454(482), —	—	−6.2/−3.2	~6	~2000	36.1, 17.3, 23.3	23
49	455, —	—	5.2/3.2	3.2	20 649	22.3, 19.8, 15.1	30
56	458, —	—	—	4	—	6.3, 4.0, 6.0	32
57	430(458), —	0.11	—	5.4	1817	11.4, 7.9, 11.9	33
58	428(455), —	0.4	—	4.4	4044	11.3, 8.6, 11.7	33
59	485(515), 510	3.81	−5.51/−2.87	3.2	38 963	25.45, 23.52, —	34
66	468(495), —	76	—	<3.8	—	17, 14, 7.9	37
70	435(465), 438(466)	94	−5.4/−2.2	—	—	5.4, 3.4, 3.9	38
71	—	46	—	—	—	20.6, 6.4, 12.6	39
80	460, —	7	—	—	1200	0.5, —, —	41
84	441, —	55	—	—	—	1.5, 0.52, —	42
88	505, —	40	−5.22/−2.00	3.3	48 295	35.02, 26.82, 11.05	47
93	536, —	—	−5.0/−2.5	3.4		69.0, 62.0, 18.7	48
94	535, —	—	−5.1/−2.6	3.7		62.5, 53.1, 17.1	48
95	480(510), 530	87	—	3.5	46 808	123.5, 43.2, —	49
96	—	95	−5.1/−2.6	—	—	89.2, 69.8, 23.6	50
98	543, 553	30	−4.78/−2.29	2.4	—	—, 32.5, —	51
101	491(517), —	72	−5.13/−2.9	2.5	6150	—, —, 5.9	52
105	493(523), 520(528)	37	−5.36/−2.49	4.9	6255	15.41, 8.08, 4.68	53
115	—	—	—	3	3362	24.0, 16.7, 7.0	56
119	517, 521	83	−5.06/−2.63	—	19 000	45.7, 37.8, 13.4	58
128	—, 495(528)	—	—	3.6	9700	10.2, 13.8, 4.4	62
137	491, —	60	—	—	—	40, —, 16	64
178	506(505[b]), —	19	−6.32/−3.86	—	—	38.9, 27.2, 11.5	67
179	669(537[b]), 590	0.85	−5.6/−3.4	—	—	37.4, 26.2, 11.5	68
188	560(595), —	—	−5.8/−3.5	3.3	—	34.2, 31.0, 12.7	71
192	562(605), 561(610)	12	−4.96, −2.54	6.7	21 570	29.77, 13.35, 9.58	74
195	564(600), —	36		4.0	80 190	76.0, 39.8, 27.2	76
203	620, 608(658)	19	−4.96/−2.43	4.6	5846	11.0, 6.16, 10.87	80
213	611, —	55	—	3.1	58 688	37.3, 32.9, 25.9	13
218	609–616	—	−4.88/−2.81	2.5	15 701	8.4, 6.9, 7.5	83
232	611, 615	36.1	−5.29/−3.08	4.8	1158	5.88, 3.69, 3.71	90
236	612, 615	36.3	−5.56/−3.21	7.6	1276	5.60, 2.26, 3.54	92
238	765, —	—	−4.9/−2.5	—	740(μW cm^{-2})	—, —, 6.3	95

Recent progress on metal complex-based OLEDs

Since the invention of OLEDs, the device fabrication has been inseparable from the synthesis of organic materials. The first two-layer sandwich-type configuration designed by Tang and Van Slyke[2] has led to a remarkable improvement in the maximum brightness of OLEDs and a significant reduction in the driving voltage from hundreds of volts for bulk single-crystal devices to just a few volts. This achievement clearly undeores the paramount importance of device fabrication in improving the overall electroluminescence performance. Till the present time, there are three major types of organic light-emitting devices, namely organic light-emitting diodes,[18d,99] light-emitting electrochemical cells (LECs),[100] and organic light-emitting transistors (OLETs).[101]

New approaches to modifying electrodes.

Experimental results have shown that the electrode of a device not only determines charge-carrier transport and injection, but also influences the stability, processability and lifetime of the device.[102] The large barrier between the low LUMO level of organic semiconductors and the high work function metal (*e.g.*, Al) cathode often prevents the injection of electrons. Low work function metal (*e.g.*, Ca and Ba) cathodes can improve the carrier injection. However, they also facilitate the formation of bipolarons at metal–polymer interfaces and enhance nonradiative decay. By inserting an inert layer of BaF_2 with a thickness of ~3 nm between an emissive layer and the overlying Ca layer, the nonradiative decay can be effectively minimized. A reaction between the BaF_2 layer and the Ca layer typically takes place during the deposition to generate a mixed region of BaF_2, Ba, CaF_2, and Ca in favor of electron injection. The trilayer cathode of BaF_2:Ba/CaF_2/Ca can prolong the device lifetime two times longer than that achieved with a similar bilayer cathode.

Rapid patterning of metal cathodes onto different substrates is an important issue for high throughput, large-scale production of OLEDs.[103] Lee[102] and Chae[103] developed a simple but practical technology named selective metal transfer (SMT) (Fig. 14). This technology obviates the need for a mask or mold typically required in conventional techniques involving either vacuum evaporation or screen printing. The adhesion strength of the Al metal strongly depends on the surface properties of the polymer used. Polyvinylcarbazole (PVK) and poly[2-methoxy-5-(2′-ethyl-hexyloxy)-1,4-phenylene vinylene] (MEH-PPV) have a much stronger adhesion strength with Al surface than hydrophilic poly(vinyl pyrrolidone) (PVP). This difference can be utilized to selectively transfer Al onto PVK-containing emissive layers.

Indium tin oxide (ITO) is a common anode material for OLEDs due to the close match of its high work function to the HOMO level of organic semiconductors and its high transmittance to visible light. However, the high cost and difficult processing procedure of ITO limit its commercial applications in OLEDs. Therefore, solution-processable, conductive carbon-based anode materials become an attractive candidate for the replacement of ITO.

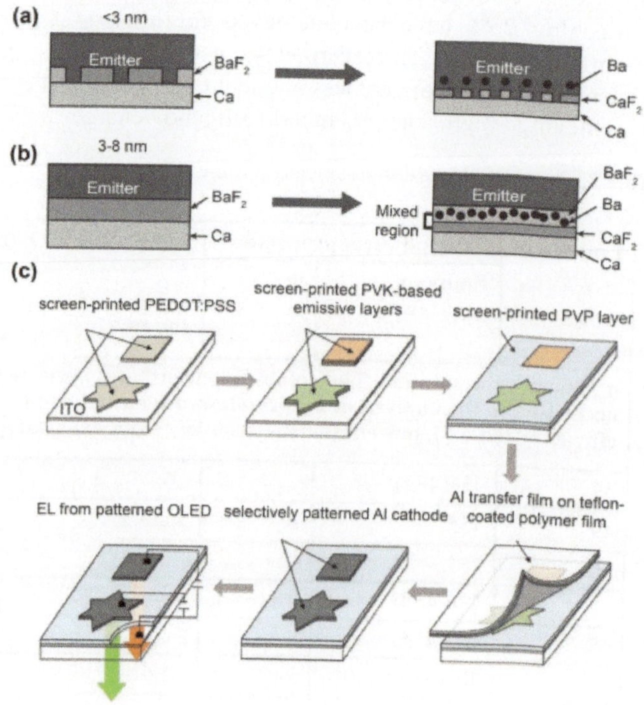

Figure 14: (a and b) Formation of a triple-layer cathode of BaF_2:Ba/CaF_2/Ca adjacent to an emitting layer by the reaction of BaF_2 with Ca layers. (a) With a thickness of BaF_2 less than 3 nm, direct contact between Ca and emitting layers facilitates nonradiative decay of excitons. (b) With a BaF_2thickness of greater than 3 nm, a complete trilayer cathode can be formed to realize both efficient electron injection and the separation of Ca and emitting layers. (c) Schematic showing the fabrication procedure of patterned Al cathodes on the surface of PVK-containing emitting layers by a selective metal transfer technique. (Reproduced with permission from ref. 102 and 103. Copyright 2009 and 2011, Wiley-VCH Verlag GmbH & Co. KGaA.)

Ou *et al.* used a PEDOT:PSS composite (PS^C) to modify the morphology of carbon nanotubes (CNTs) (Fig. 15a and b).[104] When treated with PS^C, the CNT thin film anode exhibited a 40–70% reduction in roughness. The thin film anode also showed an excellent tensile strength as compared with ITO-based anodes. The utilization of CNTs remarkably improved the electroluminescence performance of the devices with a high luminance of over 8000 cd m^{-2} and a maximum efficiency of 10.7 cd A^{-1}. Relative to CNT, graphene has a high carrier mobility and a comparable resistance and light outcoupling properties. The sheet resistance (R_{sh}) of a graphene film can be estimated according to eqn (5):

$$R_{sh} = \frac{1}{e\mu N_i N} = \frac{62.4\,\Omega}{N} \quad (5)$$

where μ is charge carrier mobility (about 10^5 $cm^2V^{-1}s^{-1}$), N_i refers to doping concentration (approximately 10^{12} cm^{-2}), and N is the number of graphene layers.

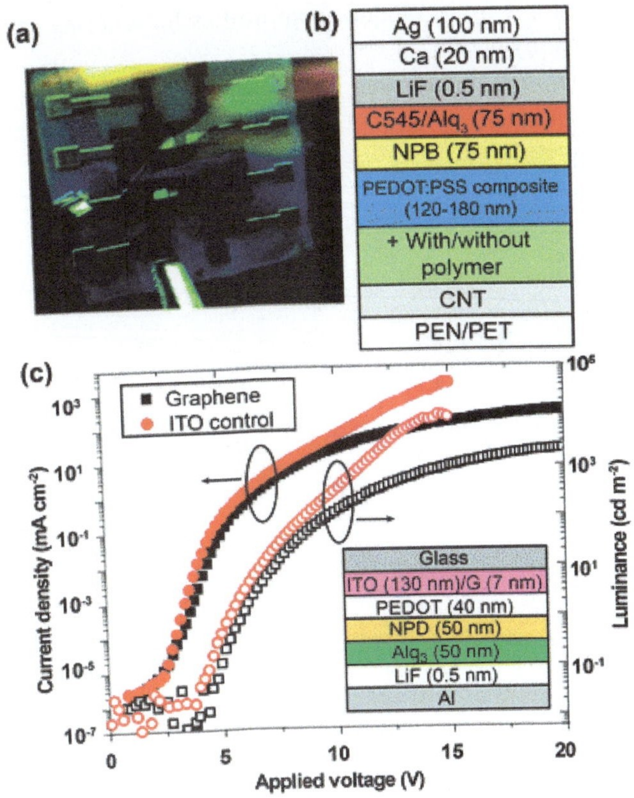

Figure 15: (a) Photographic image of devices made with a carbon nanotube anode. (b) The corresponding device configuration. (c) The brightness–current density–voltage curves of the devices using graphene as the anode (inset: device configuration). (Reproduced with permission from ref. 104 and 105. Copyright 2009, American Chemical Society.)

Wu *et al.* prepared graphene films with a thickness of ~7 nm and a surface roughness less than 3 nm (Fig. 15c).[105] The driving voltages of the graphene-based devices, obtained at 4.5 V for onset and 11.7 V for 100 cd m^{-2}, are comparable to those of ITO-based devices with 3.8 and 9.9 V for onset and 100 cd m^{-2}, respectively. However, the efficiencies of the graphene-based devices were slightly lower than those of ITO-based devices because of the higher sheet resistance and mismatched work function of the graphene film. Matyba *et al.* further replaced the metal cathode and the ITO anode with graphene and PEDOT:PSS, respectively, to construct a solution-processible LEC based on full carbon materials.[106] They constructed the emitting layer with a blend of a light-emitting polymer, a 'super yellow' dye, and an electrolyte. The authors reasoned that the ions generated by dissolution of the electrolyte can rearrange at the interface between the emitting layer and the electrode and facilitate carrier injection, thus providing a solution to the high-work-function issue (4.2–4.6 eV) of the graphene. As a result, this type of devices achieved a low turn-on voltage of 2.8 V. Both transparent electrodes of graphene and PEDOT:PSS supported the double-side emissions with similar maximum brightness (~1000 cd m^{-2} for each side).

New structures of tandem OLEDs.

A tandem OLED design involves the stacking of two or more OLED units that are connected by special charge-generating interlayers. In contrast to single-cell devices, the tandem OLED device configuration offers improved luminance and efficiencies and allows facile tuning of color emission. In this regard, the Ma group has devoted significant efforts over the years.[107] In 2007, they used Al/WO_3/Au (2/3/16 nm) as the interconnecting layer.[107d] It should be noted that WO_3 is perhaps the most active layer used for improving device efficiency. The authors obtained current efficiencies for the double-unit device (Fig. 16a) twice as high as that of the single-unit counterparts (Fig. 16b). Recently, the Ma group reported the fabrication of devices comprising a series of thiophene–naphthalene derivatives (NaT*n*, **263–267**) (Fig. 16c and d).[108] They used C_{60} as the charge-generating layer. It was found that the two contributing factors to the high efficiency of the tandem device are the high carrier mobility of the thiophene–naphthalene derivatives and the energy match between the HOMO of NaT*n* and the LUMO of C_{60}.

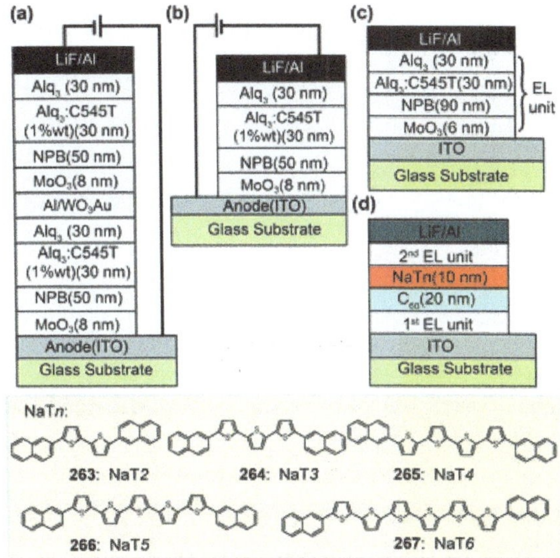

Figure 16: (a and b) Device configurations of a double-unit tandem OLED and a single-unit OLED. (c and d) Device configurations of a single-unit OLED and a double-unit tandem OLED modified with a charge-generating bilayer of NaT*n*/C_{60}. (Reproduced with permission from ref. 107*d* and 108. Copyright 2007 and 2011, respectively, American Institute of Physics and Royal Society of Chemistry.)

Hamwi *et al.* reported the utility of an intriguing design, based on an interconnecting unit that consists of a $CsCO_3$-doped 4,7-diphenyl-1,10-phenanthroline (BPhen) layer and a WO_3 film, for remarkably improved electroluminescence performance.[109] They used 4,4′,4″-tris(*N*-carbazolyl)-triphenylamine (TCTA) and 1,3,5-tri(1-phenyl-1*H*-benzo[*d*]imidazol-2-yl)phenyl (TPBI) as the hole-transporting layer and the electron-transporting layer, respectively. The charge generation and separation mainly occurred at the interface between the WO_3 film and the hole-transporting layer. The $CsCO_3$-doped BPhen layer enables rapid electron transport into the TPBI layer (Fig. 17a). Cesium azide (CsN_3) was also doped in the electron-transporting layer because of its high chemical stability in air and its low evaporation temperature.[110] The tandem OLEDs containing the interconnector of MoO_3/BPhen:CsN_3 realized a very high current efficiency over 80 cd A^{-1} (Fig. 17b). Recently, Perumal*et al.*[111] utilized the same concept of doped carrier transporting layers in tandem OLEDs (Fig. 17c). In their studies, the charge carriers were generated in carrier transporting layers under alternating current rather than injection through external electrodes. The maximum brightness (~1000 cd m^{-2}) of this type of devices indicates their potential use for display and lighting applications.

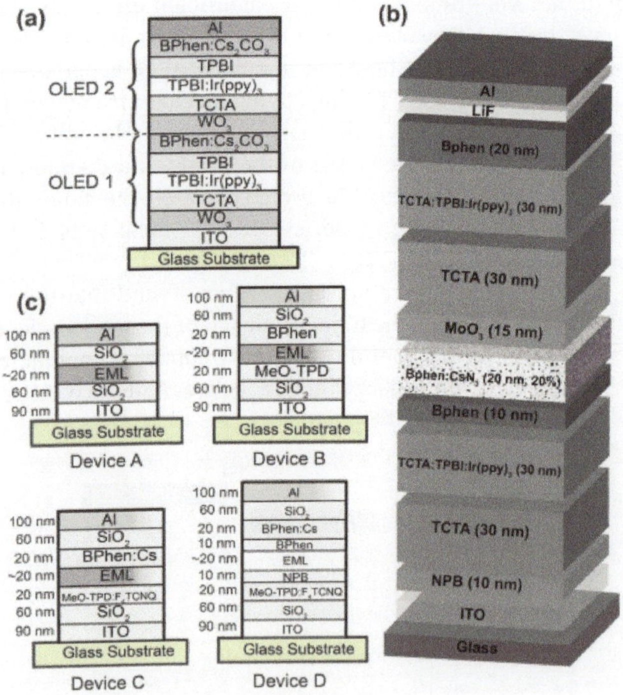

Figure 17: Device configurations of tandem OLEDs using organic–inorganic hybrids, including (a) BPhen:Cs_2CO_3, (b) BPhen:CsN_3 and (c) BPhen:Cs, as the charge generation layer. (Reproduced with permission from ref. 109–111. Copyright 2010 and 2012, Wiley-VCH Verlag GmbH & Co. KGaA.)

Solution-processible multilayer OLEDs through cross-linking technology.

The cross-linking technology in polymer research, widely adopted for making negative photoresists in semiconductor industry,[112a] plays a critical role in achieving multilayer OLEDs with enhanced device performance. If the cross-linking bond is introduced between adjacent molecular chains, the thermal stability of the polymers could be dramatically improved and the cross-linked molecular chains would be much more resistant to flow when stress is applied.[112] Many research groups, including Huang,[113] Fréchet[114] and Meerholz,[78] have developed a large array of charge-transporting materials and emitters with thermal- or photo-crosslinkable groups. In a recent review, Zuniga *et al.* highlighted the progress in the field of cross-linking technology for OLEDs.[115] Owing to the difference in the solubility of the cross-linkable materials before and after crosslinking, solution-processed multilayer devices can also be realized. Ma *et al.* demonstrated the fabrication of an all-solution processed multi-layer PHOLED (Fig. 18).[116] They first used an $Ir(ppz)_3$-based cross-linkable monomer (**268**) to form a hole-transporting and electron-blocking film made of polymeric Ir(III) complex (**272**). Phosphorescent monomers **269–271** were then spin-coated onto the top layer of **272** and subsequently crosslinked to form light-emitting films of **273** (blue), **274** (green) and **275** (red). Electron-transporting layers were formed by spin-coating of poly-OXA (**276**) onto these different color-emitting films. The three-layer OLEDs achieved moderate external quantum efficiencies (5–10%). Although the reported electroluminescence performance of these spin-coated multi-layer devices is far from ideal, the superiority of cross-linking technology in terms of low-cost and scalability offers unique opportunities for tailoring carrier fluxes in polymeric OLEDs.

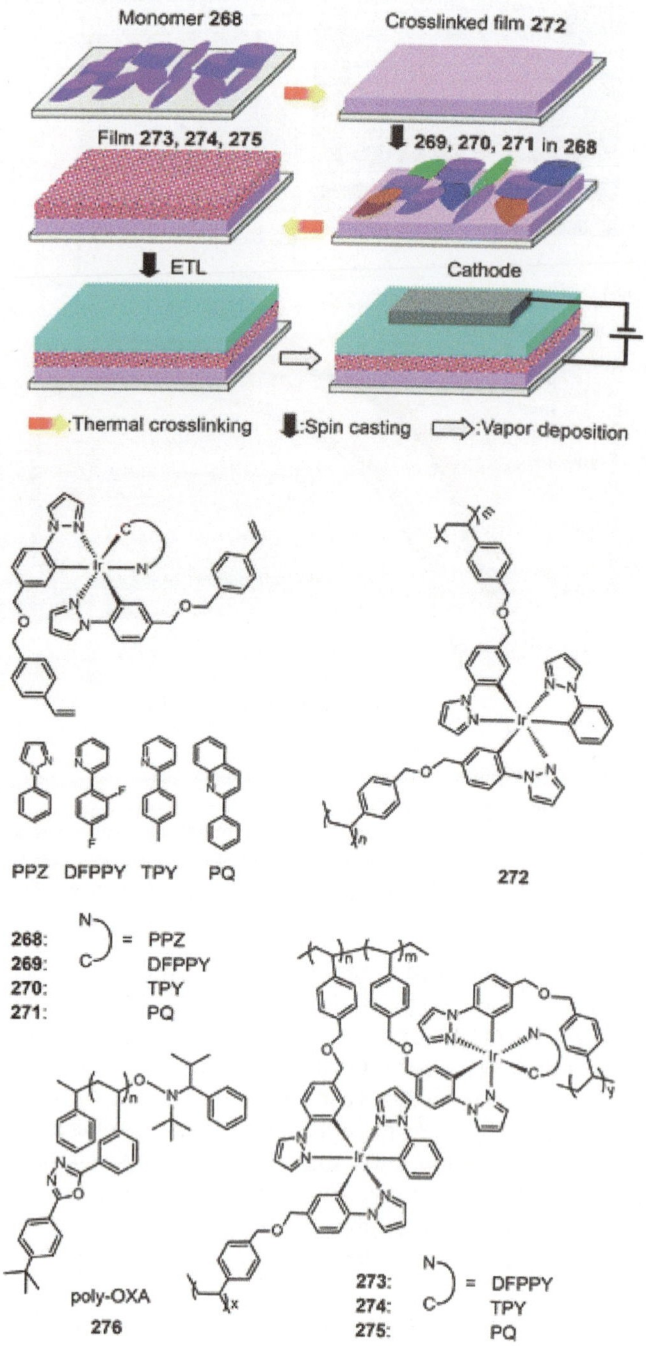

Figure 18: procedure of solution-processed multi-layer devices on the basis of crosslinkable monomers **268–271** and the corresponding crosslinked polymers **272–275**. (Reproduced with permission from ref. 116. Copyright 2009, Wiley-VCH Verlag GmbH & Co. KGaA.)

Enhancing out-coupling efficiency by morphological modification.

Low refractive indices of organic materials, ITO and glass substrates are the main reason for the poor light extraction efficiency (~0.2) recorded for prototype OLEDs. In this regard, the out-coupling technology with the aim of enhancing light extraction efficiency is crucial for practical applications of OLEDs. A simple and cheap technology was recently developed by Wang and co-workers, who used Ta_2O_5 as the coating material instead of expensive high-index substrates to enhance the out-coupling efficiency (Fig. 19a).[117a] With this new design, a green PHOLED with a conventional multilayer configuration achieved an extremely high external quantum efficiency of 63% (Fig. 19b and c). To solve the dependence of light extraction on wavelength, Koo *et al.*[117b] developed an interesting quasi-periodic buckling structure to substitute conventional periodic Bragg diffraction gratings (Fig. 19d and e). The corresponding Alq_3-based OLEDs exhibited a two-fold increase in electroluminescence intensity with essentially unaltered spectral shifts.

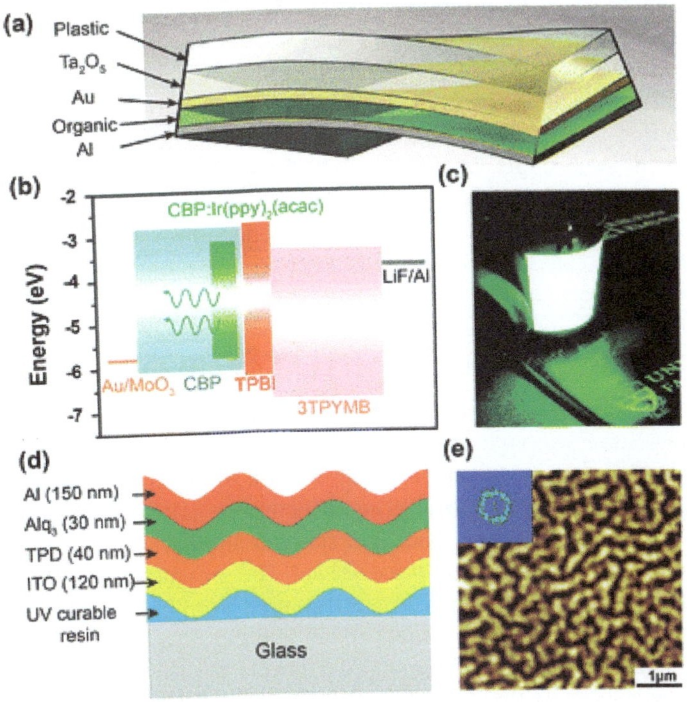

Figure 19: (a) Device configuration of a PHOLED using Ta_2O_5 as the light out-coupling layer. (b) The energy level diagram of active layers used in PHOLEDs. (c) The photographic image of the device in operation. (d) A quasi-periodic buckling structure designed for enhancing light out-coupling. (e) Topographic AFM image of buckling patterns formed by a 10 nm-thick aluminium layer. (Reproduced with permission from ref. 117*a* and *b*. Copyright 2011 and 2010, Nature Publishing Group.)

Stretchable or polarizable displays.

Stretchable organic electronics are attractive because of their potential applications in flexible displays, energy conversion devices, smart sensors and actuators. Previous studies on conductive rubbers have actually paved the way for the development of stretchable optoelectronic devices.[118] Stretchable OLED panels could redefine the user experience with flexible displays for mobile devices and curved displays for televisions. Printable elastic conductors with conductivity over 100 S cm^{-1} and a stretchability of 100% have been developed by blending single-wall carbon nanotubes and a fluorinated rubber.[119] This elastic conductor can be used as a substrate to integrate 16 × 16 active-matrix OLEDs into a tensile display with a stretchability of 30–50% (Fig. 20a).[119a] The Pei group also used a similar strategy to realize stretchable polymeric light-emitting cells that can sustain strain over 120%.[119b]

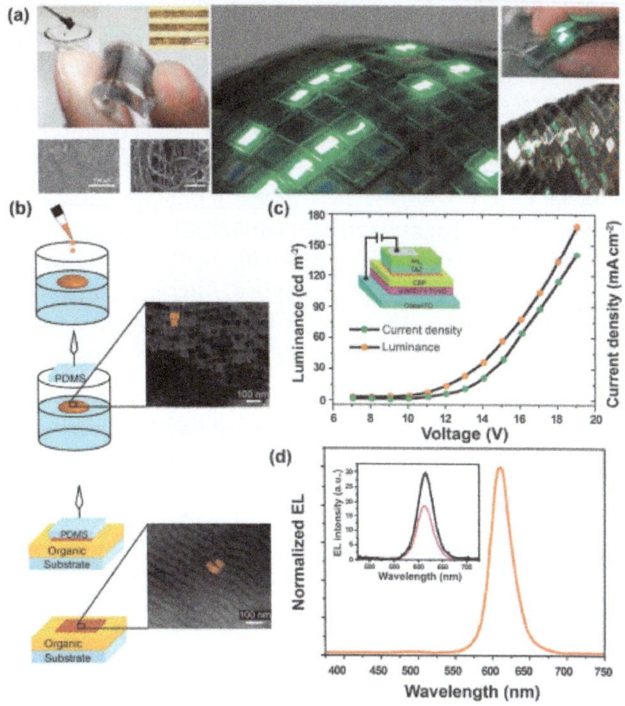

Figure 20: (a) Characterization of stretchable PHOLEDs derived from printable elastic conductors comprising single-walled carbon nanotubes (SWNTs) uniformly dispersed in a fluorinated rubber. (b) Fabrication procedure for multilayer inorganic/organic polarized OLED devices with aligned nanorods as confirmed by TEM characterization. (c) Brightness–*J*-voltage curves and (d) electroluminescence spectrum of the nanorod-based OLEDs (insets show device configuration and the electroluminescence spectra in two orthogonal directions, respectively). (Reproduced with permission from ref. 119*a* and 120. Copyright 2009, Nature Publishing Group and American Chemical Society, respectively.)

Another noteworthy development is the fabrication of polarized LED devices that would largely benefit general illumination with an improved contrast and minimize eye discomfort. In 2009, Rizzo*et al.* reported an interesting organic/inorganic polarized LED device based on ordered arrays of semiconducting nanorods (Fig. 20b–d).[120] Using a solution-based processing method, the authors prepared CdSe@CdS core–shell nanorod-based films aligned in smectic or nematic phases on water surface. The films were subsequently transferred to LED devices by a contact printing technique. Charge injection and transportation in these devices were modified with organic carrier transporting layers. Polarized emissions along the nanorod alignment direction were observed from the nanorods with a maximum brightness of 170 cd m^{-2} and a maximum current efficiency of 0.24 cd A^{-1}.

Highly efficient OLETs.

An OLET is an alternative form of light-emitting source with transistor characteristics.[101b] In this sense, an OLET actually differs from an OLED in that an active matrix can be made entirely of an OLET. In comparison, a traditional OLED needs to be combined with a thin-film transistor (TFT) where the recombination of electrons and holes occurs to emit light. Despite this advantage, single-layer OLET displays are often constrained by unbalanced electron and hole injection from drain and source.

A substantial improvement over emission performance can be achieved using an OLET with a stacked trilayer configuration, in which hole- and electron-transporting layers are inserted between source/drain, the dielectric layer and emission layers to balance the flux of carriers. A representative example was demonstrated by Capelli *et al.*, who fabricated a trilayer OLET with the maximum luminance greater than 200 nW and the peak external quantum efficiency exceeding 4%.[121] In their design, Alq_3 doped with 4-(dicyanomethylene)-2-methyl-6-(*p*-dimethylaminostyryl)-4*H*-pyran (DCM) was used as the emission layer, while thiophene derivatives, DH-4T (**277**) and DFH-4T (**278**), were used as hole- and electron-transporting layers, respectively (Fig. 21a and b). However, the requirement of a relatively high gate voltage for the OLETs is still a problem for their application.

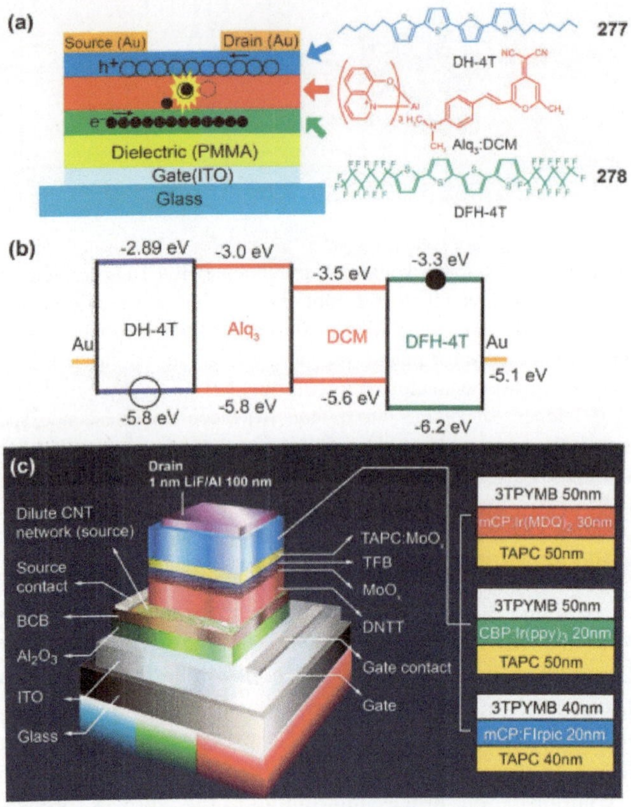

Figure 21: (a and b) Device configuration and energy level scheme of a trilayer OLET using Alq_3:DCM, DH-4T and DFH-4T as emitting, hole-transporting, and electron-transporting layers, respectively. (c) Device structures of a trilayer OLET with direct integration of red, green and blue OLED and OTFT. (Reproduced with permission from ref. 8 and 121. Copyright 2011 and 2010, respectively, American Association for the Advancement of Science and Nature Publishing Group.)

McCarthy *et al.*[8] recently reported active matrix displays using highly efficient OLETs with low driving voltage and low power dissipation (Fig. 21c). The key design is the use of a thin film of single-wall carbon nanotubes that permits the organic semiconductors to drive the high currents needed by OLED pixels, but at much lower operating voltages. The resulting OLETs achieved a high luminance of ~10 000 cd m^{-2} at a driving voltage of less than 10 V. Efficiencies of the devices obtained are comparable to those of conventional PHOLEDs, with ~13 cd A^{-1} for red, ~50 cd A^{-1} for green and ~25 cd A^{-1} for blue. Significantly, the on/off states of these OLET displays can be readily switched by applying negative and positive gate voltages, thus offering potential applications in the fabrication of simplified pixels for active matrix displays.

In general, the low efficiency of electroluminescent devices is the most important limiting factor that stands in the way of using OLEDs for practical applications. The development of outcoupling technology and integrated device configurations, such as tandem OLEDs and OLETs, would be likely to satisfy the constraints associated with current electroluminescent devices. For outcoupling technology, a key issue that needs to be addressed is to improve the performance of thin film-type coupling materials. Tandem OLEDs offer several advantages in terms of technological feasibility for enhancing efficiencies and realizing white emissions. However, this kind of devices suffer from repeatability problems as defects present even in a single layer would lead to significantly reduced performance. We believe that OLETs are more promising to be commercialized because of their reduced fabrication costs and new functionalities as well as easy integration in different substrates.

PHOTOVOLTAIC COMPLEXES FOR OPVs

Most of the energy available to us on this planet comes from the Sun. Although the amount of solar energy reaching the Earth is enormous, only a small fraction of the energy is used to support life. Almost 30% of incoming solar radiation is immediately reflected back into space as short-wave radiation, while nearly half is converted to heat. The growing demand for sustainable energy supplies has triggered a rapid development of photovoltaic devices that can tap into the massive source of solar energy by converting sunlight into electrical current. These devices can be divided into five categories based on inorganic semiconductors (InGaN, GaN, GaAs, *etc.*), silicon (single crystal, polycrystalline, amorphous, *etc.*), dye-sensitizers (Ru-based complexes, quantum dots, *etc.*), organic compounds, and recently reported perovskite materials. In particular, solution-processable all-solid organic photovoltaics (OPVs) work well in all light conditions and have much greater flexibility.[122]Over the past two years, we have seen an impressive acceleration in the field of OPVs. In 2011, Mitsubishi Chemical reportedly set a record conversion efficiency of 9.2%. In 2013, the Yang group updated the conversion efficiency over 10% using tandem device structures.[123] However, the technical challenge remains for material synthesis and device fabrication, particularly for organic solar cells requiring low batch-to-batch variation in conversion efficiency.[124]

The layout of active layers, adopting either bilayer, grade, bulk, inverted,[125a] or tandem[125b,c] cell structures, significantly impacts the overall performance of the resulting OPVs (Fig. 22).[126] When selecting organic materials for photovoltaic devices, one needs to consider the following criteria: (i) organic semiconductors with high absorption coefficients for efficiently harvesting sunlight; (ii) suitable heterojunctions for separating photogenerated excitons; (iii) compatible HOMO and LUMO levels and high charge mobility for efficient transfer of the separated charge carries to the electrodes; (iv) high stability of the materials and long operating lifetimes of the devices; (v) good solution processability; and (vi) good scalability of the materials and devices for large area applications.

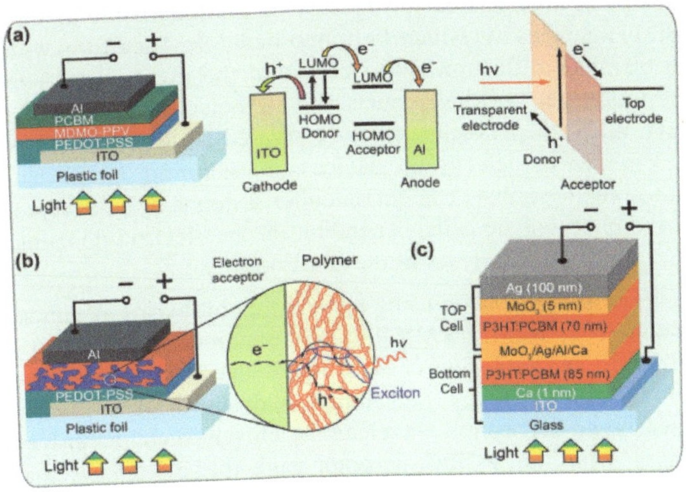

Figure 22: General device designs of OPVs in (a) bilayer, (b) bulk, and (c) tandem heterojunction solar cells. (Reproduced with permission from ref. 126*a–c*. Copyright 2007, 2013 and 2010, respectively, American Chemical Society, In-Tech Open Access Publisher and American Institute of Physics.)

It is imperative that a bicontinuous network, characterized by a domain width of ~20 nm (approximately twice that of the exciton diffusion length) and a high donor/acceptor interfacial area,[112d] favors the dissociation of excitons and the transport of the separated charge carriers to the collecting electrodes.[124] To achieve high efficiencies, the difference between the LUMO levels of the donors and acceptors needs to be within ~0.3 eV to facilitate the dissociation of the excitons, and a bandgap in the range of 1.2–1.7 eV is preferred to allow maximal absorption of sunlight. This would correspond to donor HOMO levels of −5.2 to −5.7 eV to provide a large open-circuit voltage (V_{oc}), when [6,6]-phenyl-C_{61}-butyric acid methyl ester (PCBM) with a LUMO level of approximately −4.3 eV is used as the acceptor (Fig. 23).[127] In addition, the material should possess a charge transfer channel with high charge carrier mobility and thus high short-circuit current density (J_{sc}) for high-efficiency carrier separation and collection at a given distance from the electrodes.[128] It should be noticed that morphology adjustment of the active composite layer plays a very important role too.[112a] Bad morphology of the active layer will lead to a low fill factor (FF). Therefore, an ideal material used for PCBM-based bulk OPV devices should have the following attributes: a HOMO level lower than −5.20 eV, a bandgap in the range of 1.30–1.90 eV, and a hole mobility greater than 1 × 10^{-3} $cm^2\ V^{-1}\ s^{-1}$ (Fig. 24).[129]

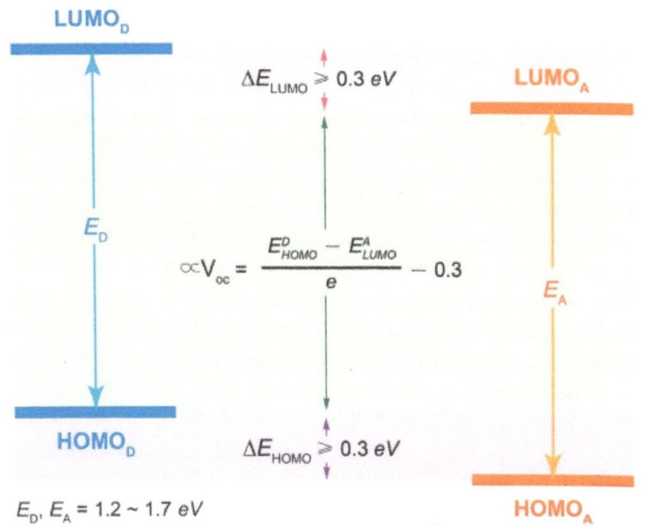

Figure 23: Schematic drawing of the donor and acceptor energy levels for the OPV devices.

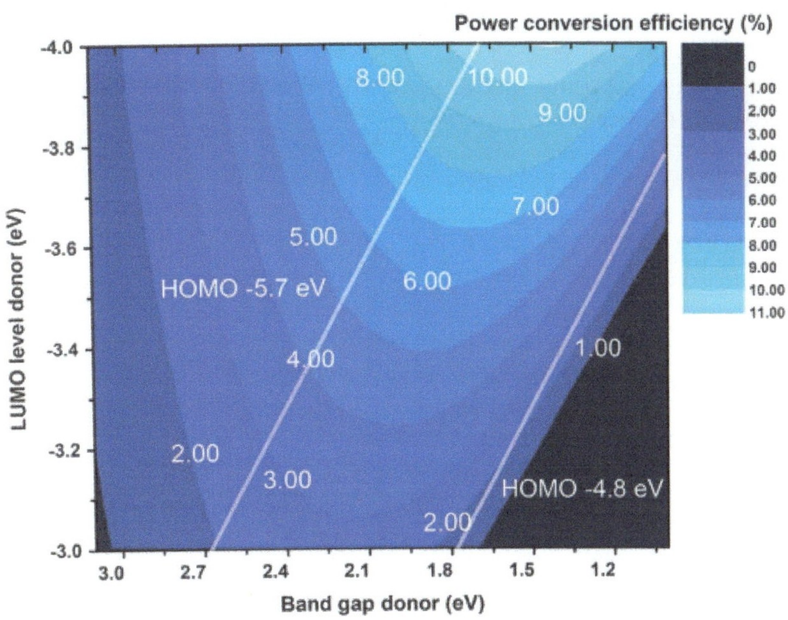

Figure 24: Theoretical relations between the power conversion efficiency and the HOMO level of the donors in PCBM-based bulk OPVs. (Reproduced with permission from ref. 129. Copyright 2006, Wiley-VCH Verlag GmbH & Co. KGaA.)

Several methods have been developed to minimize the energy loss of OPV devices by fine-tuning the bandgap and frontier orbital energy levels of organic semiconducting materials. These include D–A hybrid architecture, copolymerization, molecular topology, and supramolecular alignment.[130] Despite their usefulness in tuning the HOMO and LUMO levels, most of the organic molecules reported to date do not fulfill the stringent requirements as ideal donors for OPVs. Fig. 25 lists some widely used building blocks for donor molecules, which have been designed and assessed in bulk-heterojunction OPV devices.[131] These building blocks are usually covalently attached to newly designed molecules to prepare desirable OPV materials.[132]

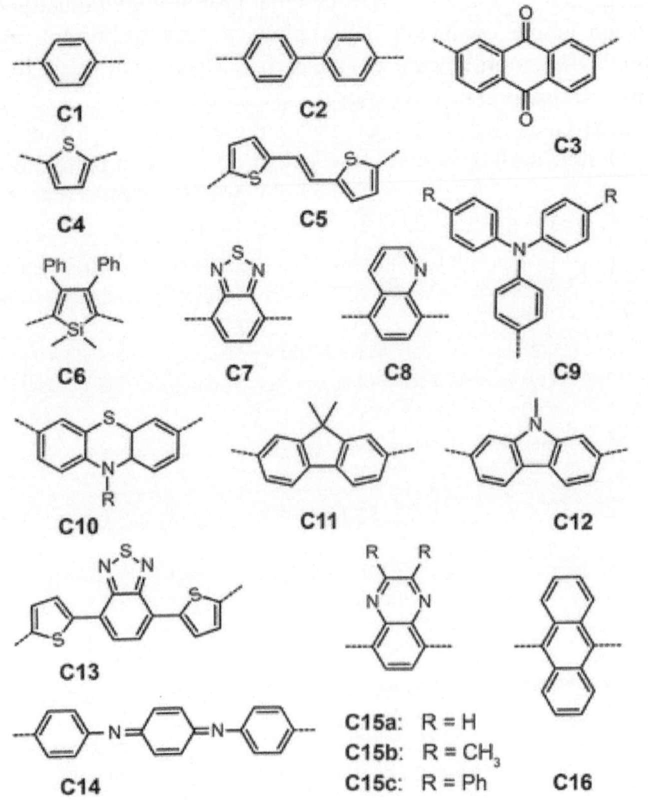

Figure 25: Conventional building blocks for OPVs.

Although organic oligomers and polymers, such as polythiophenes, dominate the field of OPVs,[133] their metal-containing derivatives represent another important class of organic semiconductors.[134,135] Metal ions may provide redox-active and paramagnetic centers to generate active species for charge transport and thus greatly alter the electronic and optical properties of the organic π-systems. As already discussed in previous sections, the introduction of heavy metals into the organic system promotes the formation of triplet excited states. The lifetime of triplet excited states is typically in the microsecond regime, which is three orders of magnitude longer than the nanosecond decays typically observed for singlet excited states in conjugated molecules. Given that a significant increase in lifetime should lead to a longer exciton diffusion length, the use of metal–organic complexes would be attractive for developing highly efficient photovoltaic devices.

Porphyrins and phthalocyanines

Porphyrins and phthalocyanines are highly conjugated macrocyclic systems (26 π electrons) that exhibit excellent thermal and chemical stability.[136] These macrocycles also possess tunable flexibility over their electronic and optical properties either by varying the functional group at the peripheral position of the molecules or by introducing an appropriate metal atom (Cu, Zn, Al, and Ti as shown in complexes **279–282**) into the interior region of the macrocycles.[136a] In view of their large extinction coefficients and strong absorbing ability in the NIR range, porphyrins and phthalocyanines have been investigated for decades as photodynamic therapy (PDT) agencies to convert light into heat.[136b,c] Therefore, wide chemical modification approaches and the great impacts on the molecular packing models and optoelectronic properties have made them intensively investigated as the active layer in organic optoelectronic devices. In addition, porphyrins and phthalocyanines can act as either a donor or an acceptor in the photovoltaic cells, depending on the electron-donating/accepting characteristics of the substitutions.

The first report on using metal-free porphyrins as donors in solar cells was in 1992 with a conversion efficiency of only 0.03%.[137] In 2007, Sun *et al.*[138] found that liquid crystalline porphyrin derivatives, substituted with long alkyl groups, have broad absorption wavelengths, closely matched energy levels with the PCBM acceptor and the electrodes, and unique homeotropical alignment in deposited films. The bulk heterojunction OPV cells showed conversion efficiencies of up to 0.78%. In 2011, Ryuzaki *et al.*[139] took one step further and synthesized a zinc-coordinated porphyrin complex (**283**). The authors carefully examined the structural effect of **283** used as a donor molecule on the conversion efficiency of OPV cells [ITO/**283**/C_{60}/Al]. They found that a crystalline film of**283** favors intermolecular charge transfer when compared to its amorphous counterpart, leading to increased mobility of charge carriers and thus enhanced conversion efficiencies. It should be noted that metal–porphyrin complexes are mainly used as sensitizers in dye-sensitized solar cells. These complexes are rarely used as active layers in heterojunction solar cells, largely due to their poor charge transport properties.[140]

In contrast to porphyrins, phthalocyanines exhibit excellent charge transport characteristics, thereby enabling their promising application in organic optoelectronics. In particular, metal–phthalocyanine complexes have gained increasing attention. CuPc (**284**) is the first OPV material, reported by Tang in 1986,[4] with 1% conversion efficiency obtained from its two-layer structured device under simulated air-mass2 illumination. In a recent development, Roy and co-workers[141a] reported bulk heterojunction OPV cells, based on tetramethyl-substituted Cu(II) phthalocyanine complex (**285**). Complex **285** displayed excellent photovoltaic performance with a J_{sc} of 16.3 mA cm^{-2}, a larger V_{oc} of 0.58 V, a FF of 0.56, and a conversion efficiency of 5.3%.[141b]

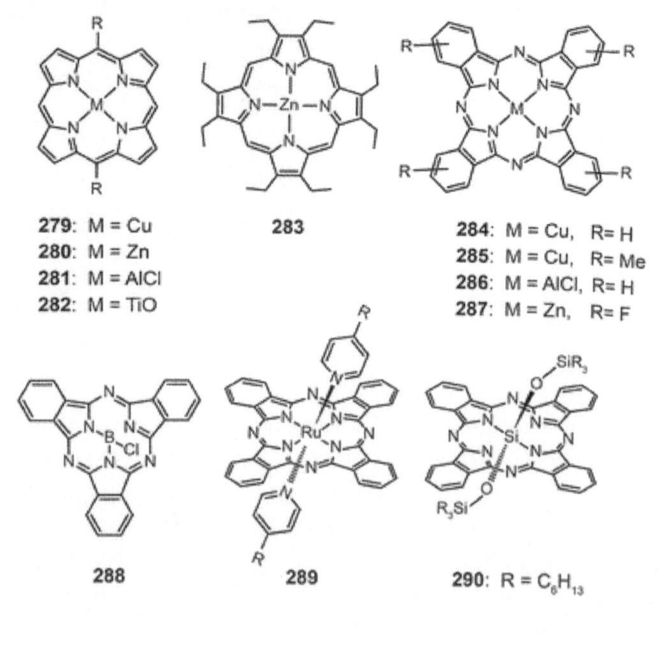

When the copper in phthalocyanine complexes is replaced with aluminium chloride as shown in Al(III) complex (**286**), a non-planar phthalocyanine structure forms as evident by a significant absorption peak at 755 nm.[142] A systematic investigation of the interfacial properties of ClAlPc–C_{60} organic heterojunctions in an inverted device configuration[143] showed that the complex **286** adopts a lying configuration after being vacuum-deposited along with C_{60} onto an ITO electrode. This structure promoted charge transport in heterojunction-based solar cell devices and resulted in a relatively large V_{oc} of 0.67 V, in contrast to 0.37 V obtained in the **284**/C_{60}-based device. Consequently, the complex exhibited high device performance with conversion efficiencies of >4% (Fig. 26).[144]

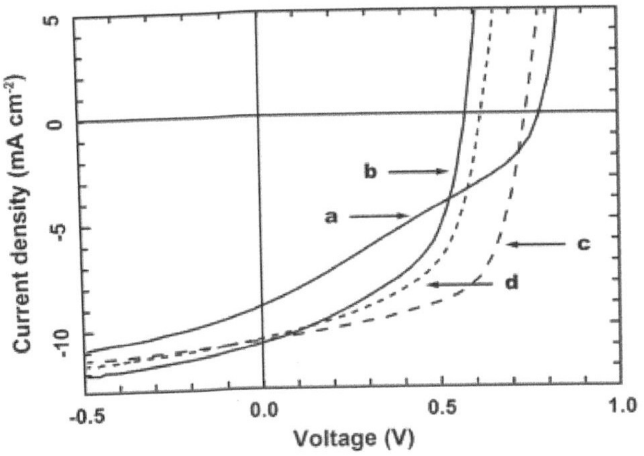

Figure 26: Current density *vs.* voltage (*J*–*V*) curves under 100 mW cm^{-2} air mass 1.5G simulated solar illumination, of (a) ITO/MoO_3/**286**:C_{60} (105 °C)/C_{60}/BCP/Ag; (b) ITO/MoO_3/FDTS/**286**:C_{60} (105 °C)/C_{60}/BCP/Ag; (c) ITO/MoO_3/**286** (105 °C)/C_{60}/BCP/Ag; and (d) ITO/MoO_3/FDTS/**286** (105 °C)/C_{60}/BCP/Ag. (Reproduced with permission from ref. 144. Copyright 2010, Elsevier B.V.)

Like copper and aluminium, zinc also shows good device performance when incorporated in phthalocyanine. Meiss *et al.*[145] prepared an efficient single bulk heterojunction organic solar cell with a power conversion efficiency of 4.6%. The solar cell was based on a blend of a fluorinated zinc phthalocyanine (**287**, M = Zn) as an electron donor and fullerene C_{60} as an electron acceptor. When boron was introduced into phthalocyanine, the researchers still observed fairly good OPV performance. Pandey *et al.*[146] made both bilayer and bulk OPV devices with boron subphthalocyanine chloride (**288**) and C_{60}. The conversion efficiencies of the devices reached 3.05 and 3.44%, respectively. Additional information about using boron subphthalocyanines as an emerging class of high performance organoelectronic materials can be found in a recent review.[147]

Fischer *et al.*[148] reported a highly soluble ruthenium phthalocyanine complex (**289**) using pyridine-functionalized dendritic oligothiophenes as axial ligands. The solution-processed bulk heterojunction solar cells derived from **289** showed a high conversion efficiency (~1.6%). Additionally, rare-earth metals can be incorporated into the phthalocyanine ligand. For example, Wang *et al.*[149] used Sm^{3+} to prepare a layered structure of phthalocyanine. They found that the hybrid solar cells possess the capability to harvest light with a broad wavelength range from UV-visible to NIR. In other attempts, Honda *et al.*[150] synthesized a silicon phthalocyanine derivative (**290**) and incorporated it into a bulk solar cell of poly(3-hexylthiophene) (P3HT) and PCBM. They observed an increased short-circuit current density and improved overall conversion efficiency by 20%.

Because of the low solubility of most phthalocyanine-based molecules in typical organic solvents, their devices were mostly fabricated by vacuum-deposited techniques. This significantly limits their application in solution-processed heterojunction solar cells. By introducing soluble long alkyl substitutes, Hori *et al.*[151] prepared a soluble, mesogenic phthalocyanine derivative of 1,4,8,11,15,18,22,25-octahexylphthalocyanine (C6PcH_2). Bulk heterojunction solar cells, fabricated by spin-coating of a mixed solution of C6PcH_2 and PCBM, showed an energy conversion efficiency of 3.2%. For researchers working in the field of porphyrin- and phthalocyanine-based photovoltaic cells, we suggest further reading of the recent literature on state-of-the-art methodology.[152]

Pt(II) complexes

Pt(II) complexes are famous for their anticancer activity. They also possess excellent optoelectronic properties that are suitable for application in OPVs, mainly due to their square planar structure and relatively strong Pt–Pt interaction.[153] Thus far, Pt(II)-containing polyethynylene (**291**),[6] cyclometalated Pt(II) (**292**),[154] and bis(8-hydroxyquinolinato) Pt(II) (**293**)[155] complexes are among the most extensively investigated Pt(II) complexes for OPVs. When platinum ions are conjugated with alkyne units to form a one-dimensional polymer chain, the d-orbitals of the platinum ions overlap with the p-orbitals of the alkyne units, leading to an enhancement of both π-electron delocalization and intra-chain charge transport along the polymer chain. This orbital overlap enhances the spin–orbit coupling and facilitates intersystem crossing from the singlet to triplet manifolds, thereby enhancing the formation of triplet excitons with extended diffusion lengths.[156] Another key advantage of using Pt(II) complexes is their good solubility in common solvents enabling film fabrication *via* solution processing.[157]

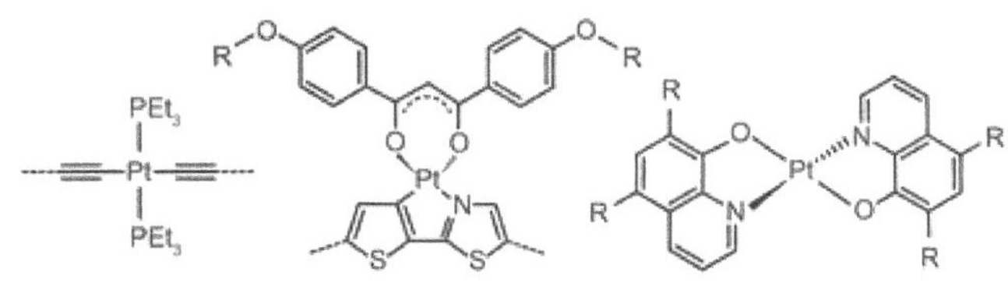

291 **292** **293**

The first report on the photovoltaic effect in a wide bandgap Pt(II) polyyne was by Köhler *et al.*[158] in 1994, with a quantum yield up to 0.6%. Two years later, the same group improved the quantum yield up to ~2% in an OPV device that contains 7 wt% of C_{60} to serve as an electron acceptor and assist in the dissociation of excitons.[159] For the most part, research into the electron transfer from the Pt(II)–polyyne donor to the C_{60} acceptor explained away the efficiency enhancement. The authors attributed it to ionization of the triplet exciton in the Pt(II)–polyyne complex.

In 2007, Wong *et al.*[6] reported an interesting platinum metallopolyyne polymer (**294**) with a low-bandgap π-conjugation by employing a D–A molecular design (Fig. 27). Based on ninety OPV cells fabricated, the authors obtained remarkably high, albeit debatable,[160] conversion efficiencies (up to 5%).[6] Several years later, Wang *et al.*[161] further developed a new solution-processable Pt(II) metallopolyyne polymer (**295**) functionalized with both triphenylamine and 2,1,3-benzothiadiazole in a donor–π bridge–acceptor–π bridge–donor (D–π–A–π–D) motif. The best conversion efficiency of 1.61% was achieved with V_{oc} of 0.77 V, J_{sc} of 4.94 mA cm^{-2}, and FF of 0.39 when illuminated with an air mass 1.5 solar cell simulator. Recently, they extended the use of the D–A motif to prepare a series of soluble Pt(II) metallopolyynes, for example complex **296**, containing Zn(II) porphyrin chromophores.[162] These metallopolymers blended with PCBM showed a maximum conversion efficiency of 1.04%, with V_{oc} of 0.77 V, J_{sc} of 3.42 mA cm^{-2}, and FF of 0.39. OPV devices with further improved conversion efficiencies were fabricated by Wong and co-workers using Pt(II)-containing triphenylamine-based pseudo-3D polymers[163] or bis(aryleneethynylene)-derived compounds (**297**).[164]

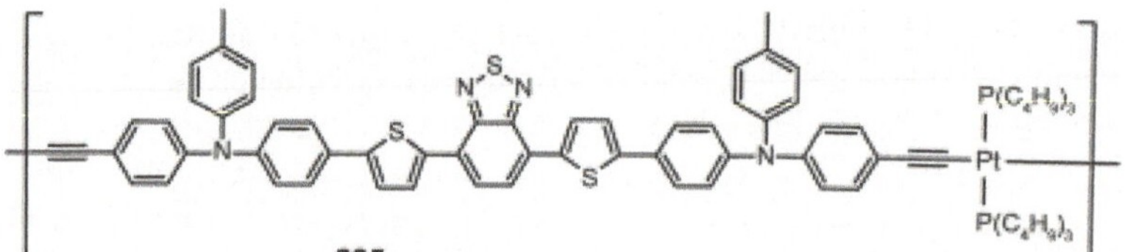

295

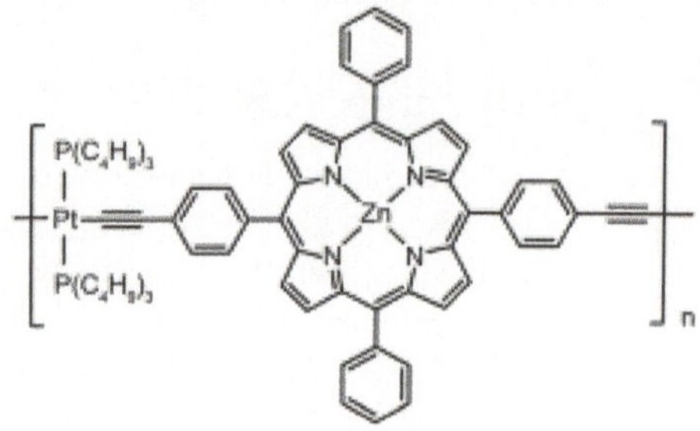

296

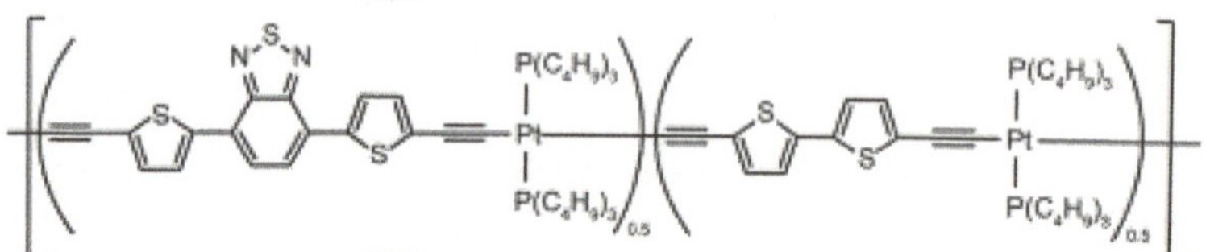

297

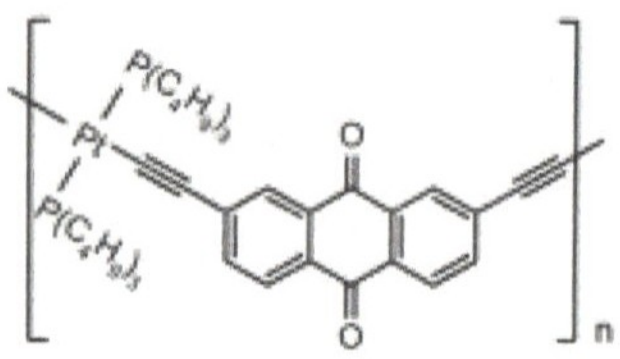

298

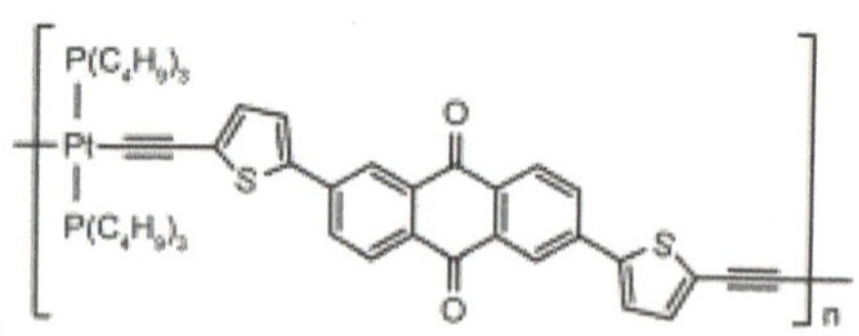

299

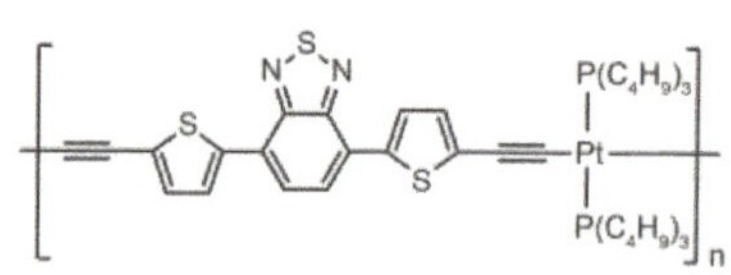

294

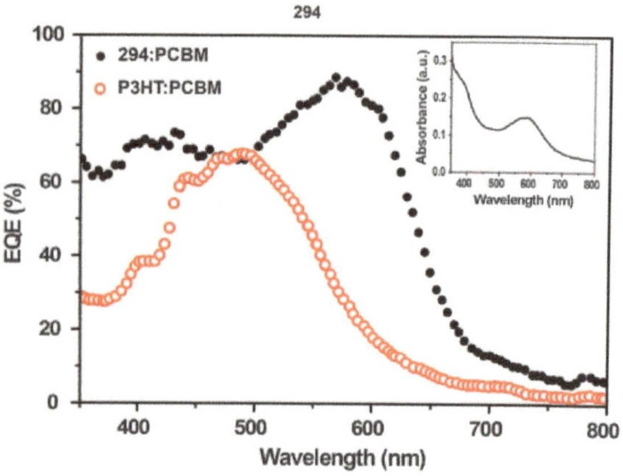

Figure 27: Absorption wavelength-dependent external quantum efficiencies of an OPV device containing Pt(II)-based organometallic D–A conjugated polymers (inset: absorption spectra of**294**). (Reproduced with permission from ref. 6. Copyright 2007, Nature Publishing Group.)

Several other groups have also made significant contributions to development of Pt(II)-based complexes for OPVs. Li *et al.*[165] prepared a solution-processable acetylide polymer (**298**) functionalized with an electron-deficient anthraquinone spacer. Optical spectroscopy and electrochemical data revealed that **298** had a narrow bandgap and intra-molecular charge transfer occurred along the polymer backbone through involvement of the D–A structural motif. Li *et al.*[165] also found that the insertion of an anthraquinone unit between thiophene fragments, as shown in complex **299**, can expand the width of the absorption band for sunlight harvesting. Wu *et al.*[166] prepared Pt(II)-bridged organometallic D–A conjugated polymers (**300** and **301**) by incorporating various electron acceptors *via*Sonogashira reaction. The OPV device made of **300** reached a high conversion efficiency of 2.4%, which is in good agreement with the value of 2.2% theoretically predicted by Janssen and co-workers.[160] Clem *et al.*[154] reported the test of a series of low-bandgap cyclometalated Pt(II) polymer complexes (**302** and **303**) for OPVs. Their corresponding photovoltaic devices yielded conversion efficiencies up to 1.3%.

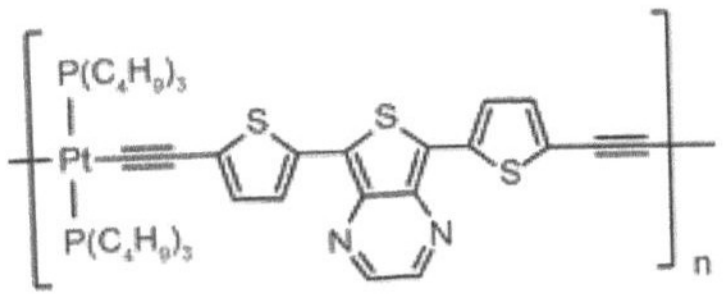

300

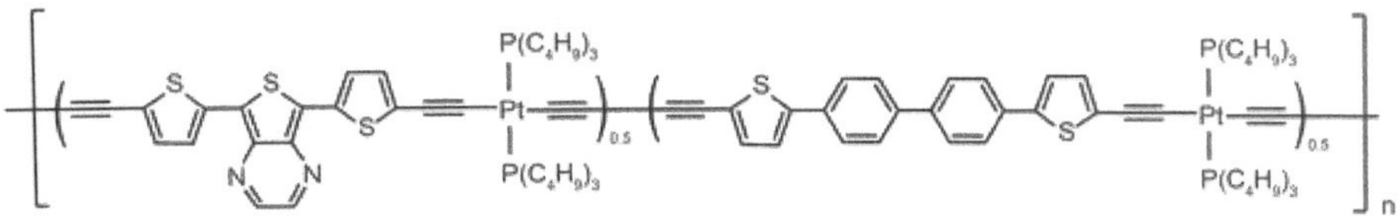

301

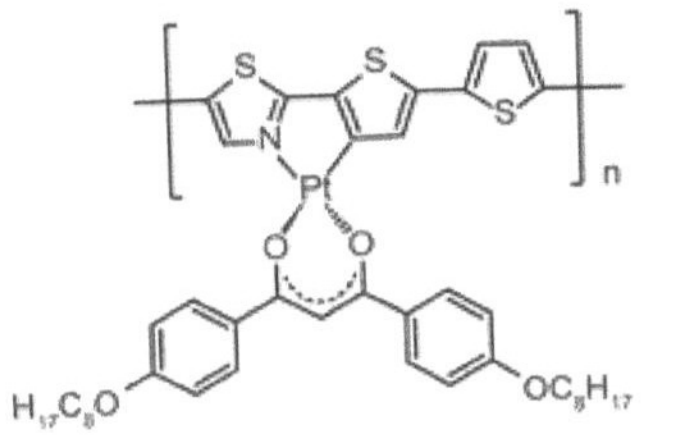

302

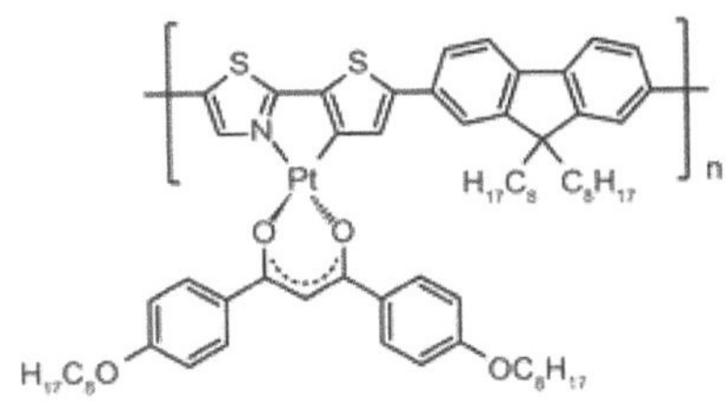

303

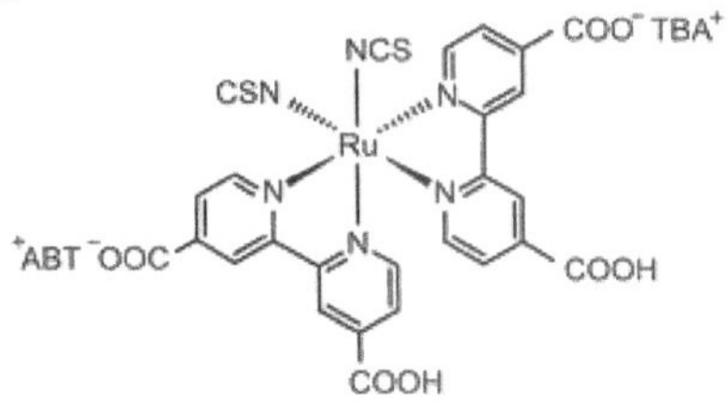

N719

304

Other metal complexes

Despite the dominant use of Pt(II) complexes for OPVs over the past two decades, other metal complexes, however, are gaining momentum. For instance, Ru(II) complexes, especially N719 dye (**304**),[167] have been demonstrated to produce high short-circuit current in dye-sensitized solar cells (DSSCs). These Ru(II) complexes feature broad UV-visible absorption bands and relatively long excited-state lifetimes, resulting in respectable conversion efficiencies (higher than 10%). Wang *et al.*[168a] introduced a $[Ru(bpy)_3]^{2+}$ coordination complex (**305**) to a polyimide to prepare photovoltaic materials. The open-circuit voltages of the corresponding OPV devices were determined within the range of 0.19–0.52 V. Low *et al.*[155] also reported the replacement of Pt with Ni and Cu metals chelated to the 5,7-dimethyl-8-hydroxy-quinoline (HMe_2q) ligand. However, no photovoltaic phenomena were observed for the devices derived from $[Ni(Me_2q)_2]$ (**306**) and $[Cu(Me_2q)_2]$ (**307**) complexes. A fairly good power conversion efficiency of 2.8% was achieved by Fleetham *et al.*[168b] using a bilayer device made from a cyclometalated Ir(III) complex (**308**) and C_{60}. Despite having an estimated exciton energy of only 1.55 eV, **308** rendered its device a dramatically improved V_{oc} (1 V).

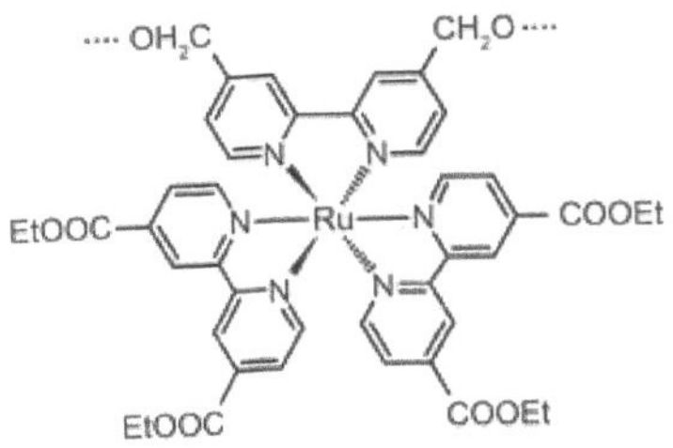

305

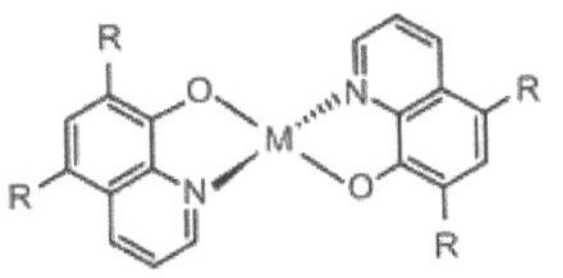

306: M = Ni
307: M = Cu

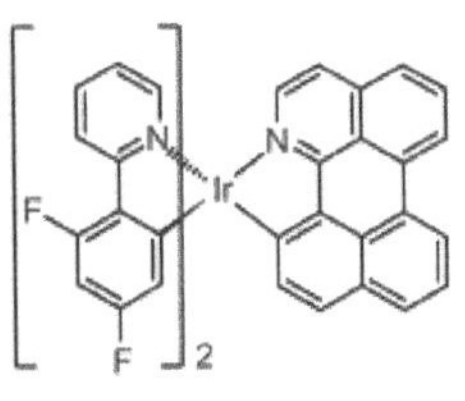

308

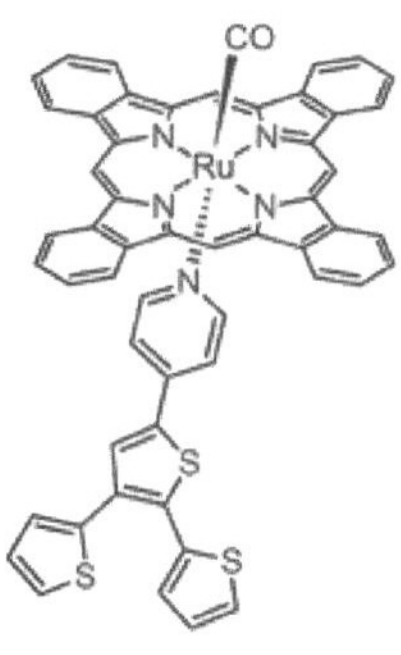

309

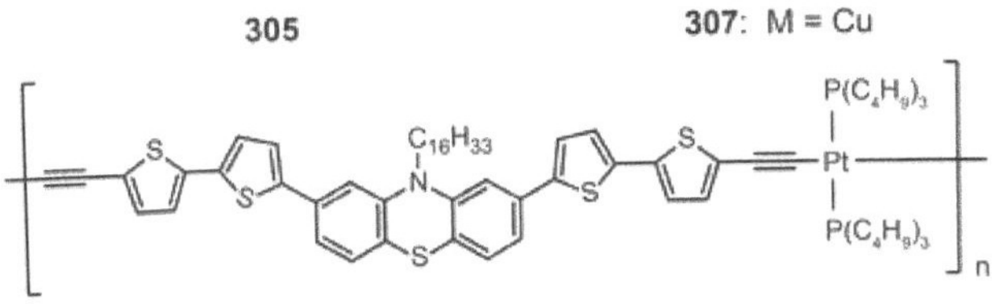

310

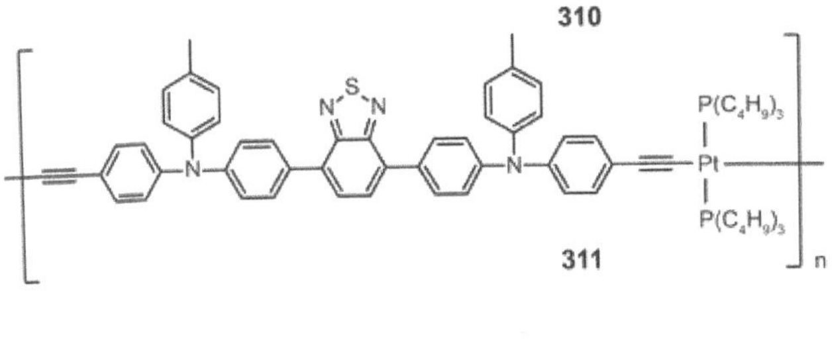

311

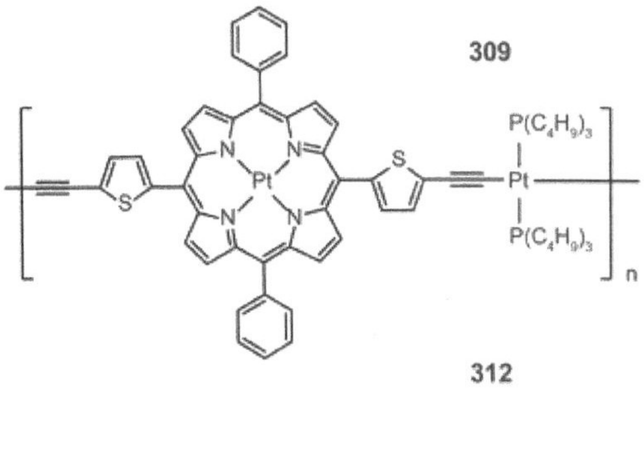

312

In 2013, a record conversion efficiency of 15.4% was achieved by perovskite solar cells using a metal–organic complex, namely $CH_3NH_3Pb(I,Cl)_3$, as the photoactive layer.[169a] The perovskite solar cells are one kind of DSSCs with a typical heterojunction configuration. The perovskite with the same structure as calcium titanium dioxide shows a broad absorption edge beyond 1000 nm, enabling improved harvesting of solar energy.[169b] Furthermore, the special metal–organic structure endows $CH_3NH_3Pb(I,Cl)_3$ with ambipolar characteristics to afford excellent electrical performance. Therefore, high charge mobility, long lifetime, and large diffusion distance (0.1–1 μm) of the generated carriers should be the main factors responsible for the high conversion efficiency of the perovskite solar cells.[169c,d]

Table 3 shows the electronic properties and device performance for a wide range of metal complexes used as donors in OPVs. As discussed previously, an ideal donor for OPVs should have a broad absorption band in the solar spectrum to ensure an effective harvesting of the solar photons, and a high charge carrier mobility for hole and electron transport. The donor should also have a low-lying HOMO energy level for a large V_{oc} and a suitable LUMO energy level to provide enough offset for charge separation. Furthermore, the energy levels of the polymer complexes must match those of the fullerides (*e.g.*, PCBM) used in bulk heterojunction solar cells. It is clear that the complexes based on phthalocyanines and Pt(II) arylene ethynylenes meet the abovementioned criteria to improve the performance of OPVs. Apart from these two candidates, much remains to be learned from exploring other organometallic heterojunction materials for OPVs. In comparison with the low-bandgap polymers used in metal-free OPVs, the metal complexes as those described above generally allow facile control over absorption characteristics, energy levels, and charge-carrier mobilities *via* ligand substitution and modification. Through optimum design of devices and subtle control over the interaction between an organic chelating ligand and a suitable metal, synergistic effects can be harnessed to drastically boost the efficiency of the corresponding solar cells.

Table 3: Electronic properties and device performance of various donors and sensitizers for OPVs

Emitter	Building blocks[a]			Abs. (nm)	HOMO	LUMO	Acceptor	V_{oc} (V)	J_{sc} (mA cm^{-2})	PCE[d] (%)	Ref.
	B1	B2	B3								
294	291	C13			−5.37	−3.14	PCBM	0.82	15.4	4.93	6
286	286						C_{60}	0.74	10.7	4.80	144
287	285						C_{60}	0.69	12.1	4.60	145
288	287						C_{60}	1.10	7.10	3.44	146
294	291	C13		380, 572	−5.15	−3.02	$PC_{71}BM$	0.77	9.65	2.41	166
293	293						C_{60}	0.42	14.8	2.4	155
295	291	C9	C7	380, 539	−5.78	−3.46	PCBM	0.78	4.94	1.61	161
309	289	C9		348, 651	−5.13	−3.58	PCBM	0.56	8.3	1.60	148
310	291	C10	C4	264, 430	−5.51	−3.00	PCBM	0.79	4.06	1.29	170
302	292	C11		375, 520	−5.40	−3.75	PCBM	0.65	5.3	1.29	154
311	291	C9	C3	363, 484	−6.09	−3.31	PCBM	0.80	4.00	1.09	161
312	291	C5a	C4		−5.53	−3.71	PCBM	0.77	3.42	1.04	162
284	284						Perylene	0.45	2.3	0.95	4
297	291	C13	C4	398, 546	−5.20	−3.17	$PC_{71}BM$	0.68	4.21	0.71	166
296	291	C5a	C1		−5.58	−3.64	PCBM	0.72	2.74	0.68	162
303	292	C4		335, 610	−5.60	−3.50	PCBM	0.38	3.5	0.40	154
300	291	C15a		415, 674	−4.82	−3.11	$PC_{71}BM$	0.52	2.71	0.36	166
299	291	C3	C4		−5.96	−3.78	PCBM	0.78	1.40	0.35	165
301	291	C13	C11	402, 540	−5.23	−3.09	$PC_{71}BM$	0.64	2.35	0.31	166
298	291	C3			−6.06	−3.76	PCBM	0.70	0.14	0.03	165

a **The first (B1), second (B2), and/or third (B3) building blocks of donors. *b* Open-circuit voltage (V_{oc}) in V. *c* Short-circuit current density (J_{sc}) in mA cm^{-2}. *d* Power conversion efficiency (PCE).**

TTA-BASED UPCONVERSION EMISSION MATERIALS

Organic materials with fluorescence emission usually follow the famous principle of the Stokes' law, meaning that output photon energy is weaker than input photon energy. Anti-Stokes emission by two-photon absorption (TPA) that converts two low-energy pump photons into a high energy emitting photon was first observed in 1961 by Kaiser and Garrett in a CaF_2:Eu(II) crystal sample.[171] Anti-Stokes emission or photon upconversion is a distinct optical process characterized by the emission of photons with energies 10–100 times kT greater than the excitation energy.[172]

Theory of TTA-based upconversion

Over the past decades, significant efforts have been devoted to the research of lanthanide-based materials that are essential to many aspects of our life.[172] The lanthanides are a family of fifteen highly electropositive metals in the f-block that falls between the s- and d-block. In contrast to the wide variation in properties across each series of transition metals, the chemical properties of the lanthanides are remarkably uniform because of the shielding by the 5s- and 5p-electrons. The electron transitions within the $4f^n$ configuration of the lanthanides made possible the discovery of many fascinating optical properties, such as photon upconversion first introduced in the mid-1960s by Auzel, Ovsyankin, and Feofilov.[172] When compared to conventional anti-Stokes processes such as TPA, photon upconversion emission is based on physically existing energy states, allowing for frequency conversion with much higher efficiency.

Unlike the well-established upconversion involving lanthanide-doped phosphors or nanocrystals,[173–182] upconversion based on organic molecules has not been widely reported, much of which is due to the high excitation intensity (on the order of MW cm^{-2}) required for organic molecules.[183] Recently, a new approach has been proposed involving the TTA of excitons in organic molecules where an upconversion-induced delayed fluorescence occurs. This approach requires two molecular entities that can interact in a triplet state to produce one molecular entity in an excited singlet state and another in its ground singlet state. This bimolecule-based upconversion process does not need a coherence excitation and the required excitation power density (a few mW cm^{-2}) is significantly lower than that for TPA.[184] The latter characteristic of TTA is particularly attractive for OPVs given that direct use of normal sunlight (power density of AM1.5G: ~100 mW cm^{-2}) has an intuitive plausibility.

The upconversion-induced delayed fluorescence supported by bimolecular TTA generally involves a three-step process (Fig. 28). First, a phosphorescent donor molecule is excited from its ground state to a singlet excited state, followed by intersystem crossing to an excited triplet state. The second step is characterized by the triplet–triplet energy transfer occurring between the donor molecule and a nearby emitting acceptor. Subsequent collision of the acceptor molecules at the excited triplet state results in an energy level twice that of the lowest triplet energy gap.

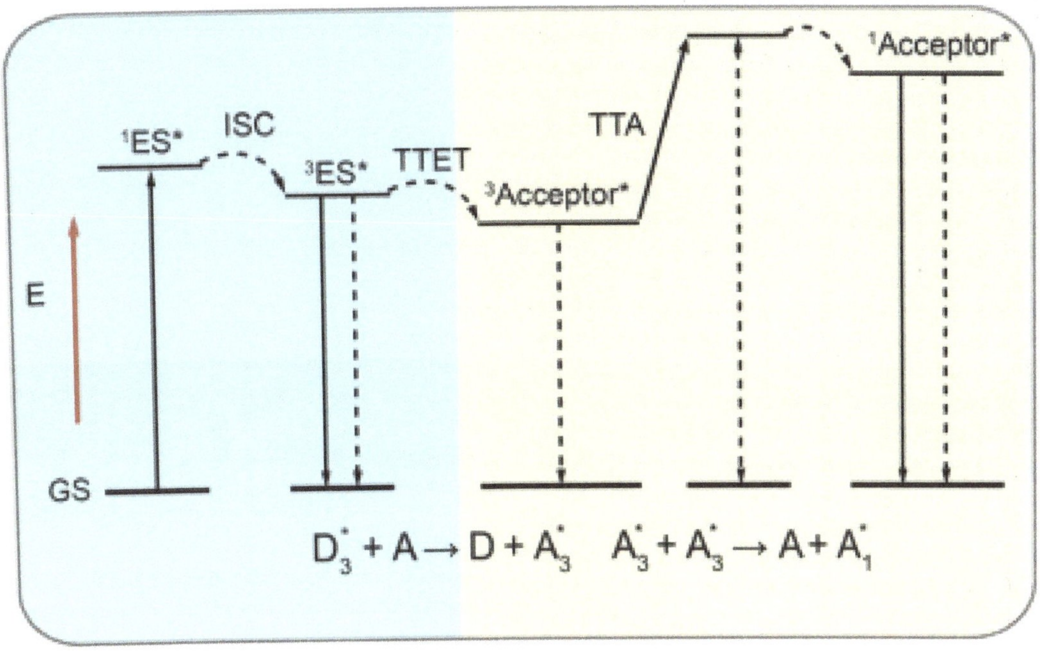

Figure 28: Upconversion emission through radiative decay from the singlet excited state of an acceptor (A_1*) *via* TTA. The ground state (GS) of a donor molecule is excited to its singlet excited state (¹ES*), followed by intersystem crossing (ISC) to its triplet excited state (³ES*). Subsequently, the triplet–triplet energy transfer (TTET) between the donor and an acceptor molecule populates the triplet excited state of the acceptor. The singlet excited state of the acceptor is generated by collision of the acceptor molecules at the triplet excited state (A_3*).

For TTA-based upconversion, the triplet state of the donor molecule should have a relatively long lifetime to ensure energy transfer to the triplet state of the acceptor molecule. In recent years, many groups[185] have demonstrated that the incorporation of metal–organic chromophores can markedly enhance upconversion yields because the metal center enables the population of the triplet state of the sensitizers with high efficiency following absorption of single photons.

Pd(II) complex-based sensitizers

In a study published in 2007,[186] a team led by Stanislav Baluschev looked into upconversion of excitation at ~700 nm (1 W cm^{-2}) to blue-green emission (480–580 nm) using a sensitizer–emitter couple based on a Pd(II)–porphyrin complex (**313** or **314**) and bis(tetracene) (**315**) (Fig. 29). The external yield of this upconversion system was estimated to be more than 4%. The anti-Stokes shift between the excitation and the emission for the sensitizer–emitter couple was estimated to be as large as 0.7 eV.

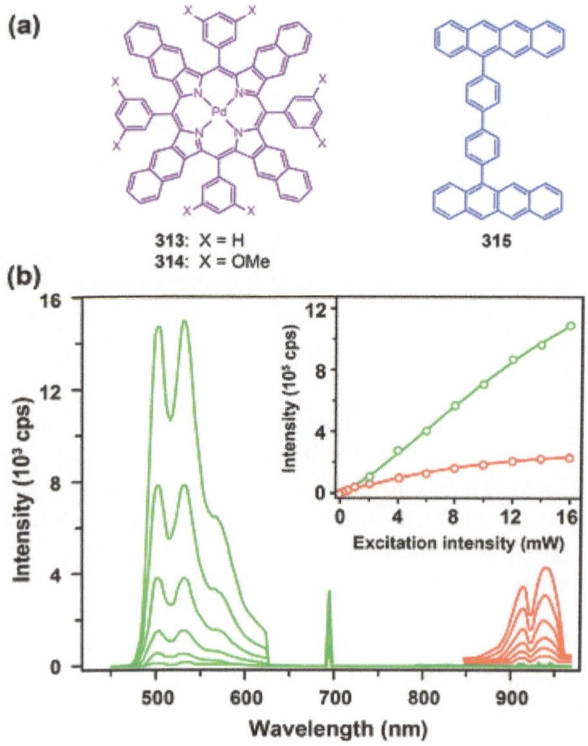

Figure 29 (a) TTA systems based on the coupling of a Pd(II)–porphyrin complex **313** (or **314**) with bis(tetracene) **315**. (b) Emission spectra of **313** (red lines) and **313**:**315** (green lines) in toluene by a single-mode CW diode laser excitation at 695 nm with different intensities (inset: integral intensity comparison of emission from **313** (red circles) and **313**:**315** (green circles)). (Reproduced with permission from ref. 186. Copyright 2007, Wiley-VCH Verlag GmbH & Co. KGaA.)

Noncoherent TTA-based upconversion is typically observed in solution. Islangulov *et al.*[187] have taken one step further and reported noncoherent low-power upconversion in solid polymer films by blending high concentrations of Pd(II) octaethylporphyrin (PdOEP: **316**) sensitizer and 9,10-diphenylanthracene (DPA: **317**) emitter within a rubbery host polymer. The upconversion through delayed singlet fluorescence of **317** was confirmed by quadratic incident power dependence studies and time-resolved emission data. These measurements indicate that the triplet sensitization of **317** can take place by selective excitation of **316**. This work also provides evidence that the use of a host polymer is quite effective for achieving TTA-based upconversion in the solid state.

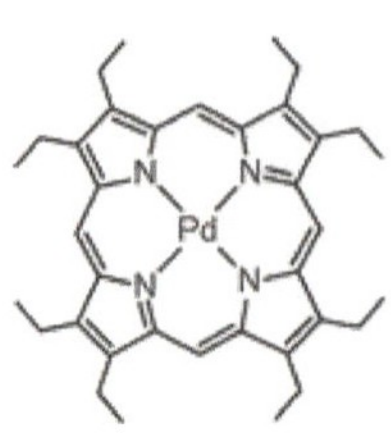

PdOEP

316

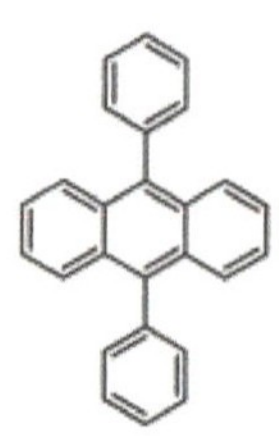

DPA

317

Ru(II) complex-based sensitizers

In 2011, Ji *et al.*[188] reported a series of pyrene-containing Ru(II) polyimine complexes (**318–322**) with long-lived intraligand (^{3}IL) excited states. The lifetime of the Ru(II) complex can be substantially prolonged by switching the emissive state from its usual ^{3}MLCT to ^{3}IL excited states, or by establishing an equilibrium between ^{3}MLCT and ^{3}IL excited states. $[Ru(bpy)_2Phen]^{2+}$**319** shows a typical^{3}MLCT emission with a lifetime of 0.45 μs. $[Ru(bpy)_2PEPhen]^{2+}$**322** has an ^{3}IL excited state with a lifetime of 58.4 μs, which is much longer than that of **319** (Fig. 30). An equilibrium between the^{3}MLCT and ^{3}IL excited states of $[Ru(bpy)_2PPhen]^{2+}$**321** results in a lifetime of 9.22 μs. The upconversion was studied using **317** as an acceptor. The anti-Stokes shift obtained is 0.48 eV, from 473 nm of the laser excitation to 400 nm in emission. $[Ru(bpy)_2PEPhen]^{2+}$**322** showed the strongest upconverted fluorescence, while **319** gave rise to a very week upconverted fluorescence. These data suggest that ^{3}IL excited states are much more efficient in sensitizing the TTA-based upconversion than ^{3}MLCT excited states.

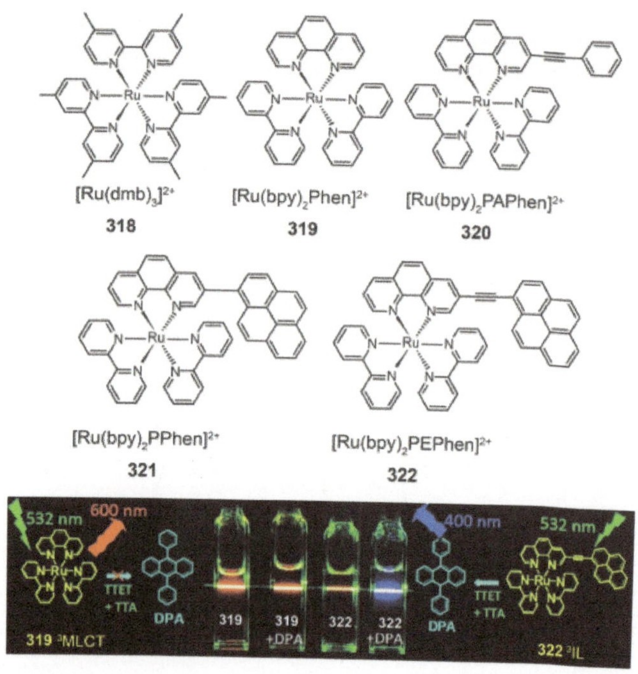

Figure 30: Molecular structures of Ru(II) sensitizers **318–322** and photographic images of upconversion emissions by **319** and **322**. (Reproduced with permission from ref. 188. Copyright 2011, Wiley-VCH Verlag GmbH & Co. KGaA.)

Another interesting investigation was reported by Singh-Rachford *et al.*[189] using a conjugated supermolecule Pyr_1RuPZn_2 (**323**) with appreciably long lifetime of excited states (> μs) as the sensitizer and *N,N*-bis(ethylpropyl) perylene-3,4,9,10-tetracarboxylicdiimide (PDI; **324**) as the acceptor. The selective NIR excitation at 780 nm of **323** in a solution containing **324** gives rise to a large anti-Stokes shift with an energy gain of 0.7 eV, resulting in the generation of yellow fluorescence at 541 nm.

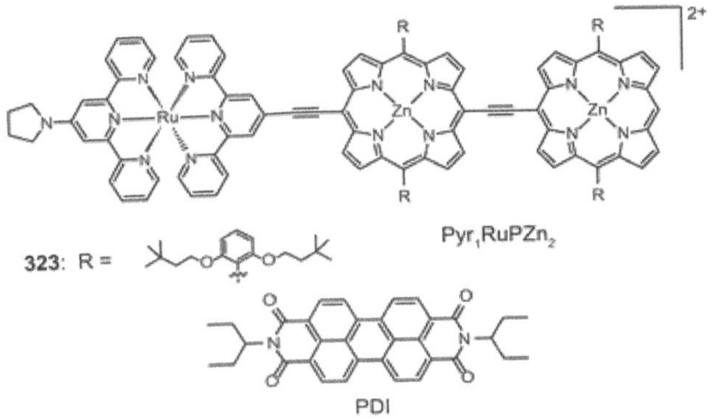

323: R = Pyr$_1$RuPZn$_2$

PDI

324

BODIPY-type sensitizers

In addition to heavy metal-containing organic complexes, molecules based on a single chromophore of boron-dipyrromethene (BODIPY) also can be used as triplet sensitizers for TTA upconversion.[190a] Wu *et al.*[190b] reported a two BODIPY-derived sensitizers (**325** and **326**) with absorption in the range of 510–629 nm and long-lived triplet excited states of up to 66.3 μs. Upconversion emission with ${}_{UC}$ up to 6.1% was observed for the samples dissolved in solution or dispersed in polymer films using 1-chloro-9,10-bis(phenylethynyl) anthracene (CBPEA; **327**) as an acceptor.

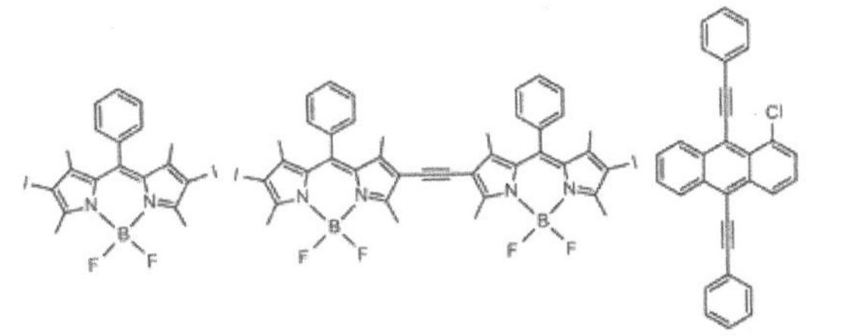

8-Phenyl-BODIPY-I$_2$ **325** E(8-Phenyl-BODIPY-I)$_2$ **326** 1CBPEA **327**

In summary, TTA-based upconversion offers tunable excitation and emission wavelengths by independent selection of sensitizers and annihilators. The high absorption coefficient (~10^{-17} cm^{-2}) of the available sensitizers enables the realization of high upconversion quantum yield under a low-threshold excitation. When compared to photon upconversion involving lanthanide doping, TTA-based upconversion seems to be more efficient with quantum yields (>5%) that are generally higher than those of lanthanide-doped NaYF$_4$ upconversion nanoparticles (<3%). Another notable advantage of TTA-based upconversion is the ability to precisely control excitation and emission combinations, such as NIR-to-green[186] and red-to-blue,[191] while there is relatively limited access to tunable emission wavelengths from upconversion nanoparticles due to the shortage of efficient lanthanide activators. Nevertheless, upconversion nanoparticles can realize larger anti-Stokes shifts owing to the involvement of three, four and even five photon upconversion processes. Furthermore, as a bimolecular system, thc TTA upconversion material may face constraints in dynamic systems and *in vivo*applications because the intermolecular interactions between the donor and the acceptor would be strongly influenced by environmental factors.

CONCLUSIONS AND PERSPECTIVES

The increasing awareness of the benefit of metal–organic complexes for optoelectronic applications has fuelled a growing demand for new approaches to prepare more complex molecular structures with tunable optical and electronic properties. The performance in certain aspects of these judiciously designed metal–organic complexes has even exceeded that of pure organic compounds. This improvement signifies the importance of metal coordination in controlling photophysical properties of the complexes. We anticipate that new developments, which may shape the future of optoelectronic technology, will be on the rise in the following areas.

(i) *Multifunctionalization*. The limited availability of photofunctional complexes will drive innovative strategies for molecular multifunctionalization.[192–196] The post-functionalization approach through direct modification of existing metal–organic complexes can simplify material design and render the complexes with well-defined optoelectronic properties and much improved device performance.

(ii) *New theories and viewpoints*. Will our classical theoretical bases always be intact when applied to grasp the mechanisms underlying the beneficial effects of metal–organic coordination? Perhaps this will not be the e. New models and challenging views, including the design of blue-emitting phosphors and elongation of the lifetime of ^{3}IL excited states for TTA, could have a profound effect on the development of highly efficient optoelectronic materials.[197–202]

(iii) *Improvement on device compatibility*. There are important challenges to be tackled with the rise of flexible devices, in large part owing to the differences between the properties of inorganic and organic compounds. New device configurations and fabrication technologies, such as roll-to-roll technology, will take place in developing stretchable and biocompatible optoelectronic devices.[119b]

ACKNOWLEDGEMENTS

H.X. acknowledges financial support from the National Natural Science Fund of China (NSFC, 61176020 and 51373050), the New Century Excellent Talents Supporting Program of MOE (NCET-12-0706), and the Key Project of MOE (212039). W.H. acknowledges financial support from the National Basic Research Program of China (973 Program, 2009CB930601), NSFC (BZ2010043, 21274065, 20974046, 20774043, 51173081, 50428303, 61136003) and the Natural Science Foundation of Jiangsu Province (BK2008053, BK2009025, BK2011751, 10KJB510013). X.L. acknowledges the National Research Foundation and the Economic Development Board (Singapore-Peking-Oxford Research Enterprise, COY-15-EWI-RCFSA/N197-1), the Ministry of Education (MOE2010-T2-1-083), and the Agency for Science, Technology and Research (A*STAR) for providing support during the time this review was written.

REFERENCES

1. D. D. Eley, *Nature*, 1948, 162, 819.
2. C. W. Tang and S. A. Van Slyke, *Appl. Phys. Lett.*, 1987, 51, 913–915.
3. M. A. Baldo, D. F. O'Brien, Y. You, A. Shoustikov, S.

Sibley, M. E. Thompson and S. R. Forrest, *Nature*, 1998, 395, 151–154.

4. C. W. Tang, *Appl. Phys. Lett.*, 1986, 48, 183–185.
5. J. S. Shirk, J. R. Lindle, F. J. Bartoli and M. E. Boyle, *J. Phys. Chem.*, 1992, 96, 5847–5852.
6. W.-Y. Wong, X.-Z. Wang, Z. He, A. B. Djurisic, C.-T. Yip, K.-Y. Cheung, H. Wang, C. S. K. Mak and W.-K. Chan, *Nat. Mater.*, 2007, 6, 521–527.
7. S. Reineke, F. Lindner, G. Schwartz, N. Seidler, K. Walzer, B. Lussem and K. Leo, *Nature*, 2009, 459, 234–238.
8. M. A. McCarthy, B. Liu, E. P. Donoghue, I. Kravchenko, D. Y. Kim, F. So and A. G. Rinzler, *Science*, 2011, 332, 570–573.
9. Q. Liu, T. Yang, W. Feng and F. Li, *J. Am. Chem. Soc.*, 2012, 134, 5390–5397.
10. T. Förster, *Ann. Phys.*, 1948, 437, 55–75 .
11. D. L. Dexter, *J. Chem. Phys.*, 1953, 21, 836–850.
12. S. J. Su, C. Cai and J. Kido, *Chem. Mater.*, 2011, 23, 274–284.
13. C.-H. Fan, P. Sun, T.-H. Su and C.-H. Cheng, *Adv. Mater.*, 2011, 23, 2981–2985.
14. H.-H. Chou and C.-H. Cheng, *Adv. Mater.*, 2010, 22, 2468–2471.
15. (*a*) L. S. Sapochak, A. Padmaperuma, N. Washton, F. Endrino, G. T. Schmett, J. Marshall, D. Fogarty, P. E. Burrows and S. R. Forrest, *J. Am. Chem. Soc.*, 2001, 123, 6300–6307; (*b*) H.-J. Son, W.-S. Han, J.-Y. Chun, B.-K. Kang, S.-N. Kwon, J. Ko, S. J. Han, C. Lee, S. J. Kim and S. O. Kang, *Inorg. Chem.*, 2008, 47, 5666–5676; (*c*) H. Xu, Z.-F. Xu, Z.-Y. Yue, P.-F. Yan, B. Wang, L.-W. Jia, G.-M. Li, W.-B. Sun and J.-W. Zhang, *J. Phys. Chem. C*, 2008, 112, 15517–15525; (*d*) H.-P. Zeng, G.-R. Wang, G.-C. Zeng and J. Li, *Dyes Pigm.*, 2009, 83, 155–161; (*e*) S.-G. Roh, Y.-H. Kim, K. D. Seo, D. H. Lee, H. K. Kim, Y.-I. Park, J.-W. Park and J.-H. Lee,*Adv. Funct. Mater.*, 2009, 19, 1663–1671; (*f*) C. Perez-Bolivar, S.-y. Takizawa, G. Nishimura, V. A. Montes and P. Anzenbacher, *Chem.–Eur. J.*, 2011, 17, 9076–9082.
16. S.-H. Liao, J.-R. Shiu, S.-W. Liu, S.-J. Yeh, Y.-H. Chen, C.-T. Chen, T. J. Chow and C.-I. Wu, *J. Am. Chem. Soc.*, 2009, 131, 763–777.
17. B. Xu, L. Chen, X. Liu, H. Zhou, H. Xu, X. Fang and Y. Wang, *Appl. Phys. Lett.*, 2008, 92, 103305.
18. (*a*) P.-T. Chou and Y. Chi, *Chem.–Eur. J.*, 2007, 13, 380–395; (*b*) L. Xiao, Z. Chen, B. Qu, J. Luo, S. Kong, Q. Gong and J. Kido, *Adv. Mater.*, 2011, 23, 926–952; (*c*) Y. Chi and P.-T. Chou, *Chem. Soc. Rev.*, 2010, 39, 638–655 ; (*d*) S.-H. Hwang, C. N. Moorefield and G. R. Newkome, *Chem. Soc. Rev.*, 2008, 37, 2543–2557 .
19. R. J. Holmes, S. R. Forrest, Y.-J. Tung, R. C. Kwong, J. J. Brown, S. Garon and M. E. Thompson, *Appl. Phys. Lett.*, 2003, 82, 2422–2424.
20. A. B. Tamayo, S. Garon, T. Sajoto, P. I. Djurovich, I. M. Tsyba, R. Bau and M. E. Thompson, *Inorg. Chem.*, 2005, 44, 8723–8732.
21. S. J. Lee, K.-M. Park, K. Yang and Y. Kang, *Inorg. Chem.*, 2009, 48, 1030–1037.
22. S.-C. Lee and Y. S. Kim, *Mol. Cryst. Liq. Cryst.*, 2008, 491, 209–216.
23. H.-J. Seo, K.-M. Yoo, M. Song, J. S. Park, S.-H. Jin, Y. I. Kim and J.-J. Kim, *Org. Electron.*, 2010, 11, 564–572.
24. D. Di Censo, S. Fantacci, F. De Angelis, C. Klein, N. Evans, K. Kalyanasundaram, H. J. Bolink, M. Gratzel and M. K. Nazeeruddin, *Inorg. Chem.*, 2008, 47, 980–989.
25. Y.-M. Wang, F. Teng, L.-H. Gan, H.-M. Liu, X.-H. Zhang, W.-F. Fu, Y.-S. Wang and X.-R. Xu, *J. Phys. Chem. C*, 2008, 112, 4743–4747 .
26. H. W. Ham, I. J. Kim and Y. S. Kim, *Mol. Cryst. Liq. Cryst.*, 2009, 505, 130/[368]–138/[376] Search .
27. S. Y. Ahn and Y. Ha, *Mol. Cryst. Liq. Cryst.*, 2010, 530, 116/[272]–122/[278] Search .
28. T. Sajoto, P. I. Djurovich, A. Tamayo, M. Yousufuddin, R. Bau, M. E. Thompson, R. J. Holmes and S. R. Forrest, *Inorg. Chem.*, 2005, 44, 7992–8003.
29. R. J. Holmes, S. R. Forrest, T. Sajoto, A. Tamayo, P. I. Djurovich, M. E. Thompson, J. Brooks, Y.-J. Tung, B. W. D'Andrade, M. S. Weaver, R. C. Kwong and J. J. Brown, *Appl. Phys. Lett.*, 2005,87, 243507.
30. K.-Y. Lu, H.-H. Chou, C.-H. Hsieh, Y.-H. O. Yang, H.-R. Tsai, H.-Y. Tsai, L.-C. Hsu, C.-Y. Chen, I. C. Chen and C.-H. Cheng, *Adv. Mater.*, 2011, 23, 4933–4937.
31. Y.-H. Song, Y.-C. Chiu, Y. Chi, Y.-M. Cheng, C.-H. Lai, P.-T. Chou, K.-T. Wong, M.-H. Tsai and C.-C. Wu, *Chem.–Eur. J.*, 2008, 14, 5423–5434.
32. C.-F. Chang, Y.-M. Cheng, Y. Chi, Y.-C. Chiu, C.-C. Lin, G.-H. Lee, P.-T. Chou, C.-C. Chen, C.-H. Chang and C.-C. Wu, *Angew. Chem., Int. Ed.*, 2008, 47, 4542–4545.
33. Y.-C. Chiu, J.-Y. Hung, Y. Chi, C.-C. Chen, C.-H. Chang, C.-C. Wu, Y.-M. Cheng, Y.-C. Yu, G.-H. Lee and P.-T. Chou, *Adv. Mater.*, 2009, 21, 2221–2225.
34. Y.-C. Zhu, L. Zhou, H.-Y. Li, Q.-L. Xu, M.-Y. Teng, Y.-X. Zheng, J.-L. Zuo, H.-J. Zhang and X.-Z. You, *Adv. Mater.*, 2011, 23, 4041–4046.
35. H.-J. Seo, M. Song, S.-H. Jin, J. H. Choi, S.-J. Yun and Y.-I. Kim, *Adv.*, 2011, 1, 755–757 .
36. X. Yang, J. D. Froehlich, H. S. Chae, S. Li, A. Mochizuki and G. E. Jabbour, *Adv. Funct. Mater.*, 2009, 19, 2623–2629.
37. S.-C. Lo, R. N. Bera, R. E. Harding, P. L. Burn and I. D. W. Samuel, *Adv. Funct. Mater.*, 2008, 18, 3080–3090.
38. S.-C. Lo, R. E. Harding, C. P. Shipley, S. G. Stevenson, P. L. Burn and I. D. W. Samuel, *J. Am. Chem. Soc.*, 2009, 131, 16681–16688.
39. X. Yang, Z. Wang, S. Madakuni, J. Li and G. E. Jabbour, *Adv. Mater.*, 2008, 20, 2405–2409.
40. Y. Unger, D. Meyer, O. Molt, C. Schildknecht, I. Münster, G. Wagenblast and T. Strassner, *Angew. Chem., Int. Ed.*, 2010, 49, 10214–10216.
41. K. Li, X. Guan, C.-W. Ma, W. Lu, Y. Chen and C.-M. Che, *Chem. Commun.*, 2011, 47, 9075–9077 .
42. X.-L. Zheng, Y. Liu, M. Pan, X.-Q. Lu, J.-Y. Zhang, C.-

Y. Zhao, Y.-X. Tong and C.-Y. Su, *Angew. Chem., Int. Ed.*, 2007, 46, 7399–7403.

43. (*a*) Z. M. Hudson, C. Sun, M. G. Helander, Y.-L. Chang, Z.-H. Lu and S. Wang, *J. Am. Chem. Soc.*, 2012, 134, 13930–13933; (*b*) C.-W. Lee and J.-Y. Lee, *Adv. Mater.*, 2013,25, 5450–5454.
44. M. A. Baldo, S. Lamansky, P. E. Burrows, M. E. Thompson and S. R. Forrest, *Appl. Phys. Lett.*, 1999, 75, 4–6.
45. (*a*) S. J. Su, T. Chiba, T. Takeda and J. Kido, *Adv. Mater.*, 2008, 20, 2125–2130; (*b*) C. Wu, H.-F. Chen, K.-T. Wong and M. E. Thompson, *J. Am. Chem. Soc.*, 2010, 132, 3133–3139; (*c*) L.-H. Xie, R. Zhu, Y. Qian, R.-R. Liu, S.-F. Chen, J. Lin and W. Huang, *J. Phys. Chem. Lett.*, 2010, 1, 272–276.
46. S. O. Jung, Q. Zhao, J.-W. Park, S. O. Kim, Y.-H. Kim, H.-Y. Oh, J. Kim, S.-K. Kwon and Y. Kang, *Org. Electron.*, 2009, 10, 1066–1073.
47. G. Zhou, C.-L. Ho, W.-Y. Wong, Q. Wang, D. Ma, L. Wang, Z. Lin, T. B. Marder and A. Beeby, *Adv. Funct. Mater.*, 2008, 18, 499–511.
48. D. M. Kang, J.-W. Kang, J. W. Park, S. O. Jung, S.-H. Lee, H.-D. Park, Y.-H. Kim, S. C. Shin, J.-J. Kim and S.-K. Kwon, *Adv. Mater.*, 2008, 20, 2003–2007.
49. V. K. Rai, M. Nishiura, M. Takimoto and Z. Hou, *Chem. Commun.*, 2011, 47, 5726–5728 .
50. J.-H. Jou, M.-F. Hsu, W.-B. Wang, C.-L. Chin, Y.-C. Chung, C.-T. Chen, J.-J. Shyue, S.-M. Shen, M.-H. Wu, W.-C. Chang, C.-P. Liu, S.-Z. Chen and H.-Y. Chen, *Chem. Mater.*, 2009, 21, 2565–2567.
51. Y. Liu, K. Ye, Y. Fan, W. Song, Y. Wang and Z. Hou, *Chem. Commun.*, 2009, 3699–3701 .
52. H. Xu, D.-H. Yu, L.-L. Liu, P.-F. Yan, L.-W. Jia, G.-M. Li and Z.-Y. Yue, *J. Phys. Chem. B*, 2009, 114, 141–150.
53. J.-X. Cai, T.-L. Ye, X.-F. Fan, C.-M. Han, H. Xu, L.-L. Wang, D.-G. Ma, Y. Lin and P.-F. Yan, *J. Mater. Chem.*, 2011, 21, 15405–15416 .
54. L. Chen, Z. Ma, J. Ding, L. Wang, X. Jing and F. Wang, *Chem. Commun.*, 2011, 47, 9519–9521 .
55. S. Bettington, M. Tavasli, M. R. Bryce, A. Beeby, H. Al-Attar and A. P. Monkman, *Chem.–Eur. J.*, 2007, 13, 1423–1431.
56. H. J. Bolink, S. G. Santamaria, S. Sudhakar, C. Zhen and A. Sellinger, *Chem. Commun.*, 2008, 618–620 .
57. R. N. Bera, N. Cumpstey, P. L. Burn and I. D. W. Samuel, *Adv. Funct. Mater.*, 2007, 17, 1149–1152.
58. J. Ding, B. Wang, Z. Yue, B. Yao, Z. Xie, Y. Cheng, L. Wang, X. Jing and F. Wang, *Angew. Chem., Int. Ed.*, 2009, 48, 6664–6666.
59. T. Qin, J. Ding, L. Wang, M. Baumgarten, G. Zhou and K. Müllen, *J. Am. Chem. Soc.*, 2009, 131, 14329–14336.
60. J. C. Deaton, S. C. Switalski, D. Y. Kondakov, R. H. Young, T. D. Pawlik, D. J. Giesen, S. B. Harkins, A. J. M. Miller, S. F. Mickenberg and J. C. Peters, *J. Am. Chem. Soc.*, 2010, 132, 9499–9508.
61. M. Hashimoto, S. Igawa, M. Yashima, I. Kawata, M. Hoshino and M. Osawa, *J. Am. Chem. Soc.*, 2011, 133, 10348–10351.
62. Z. Liu, M. F. Qayyum, C. Wu, M. T. Whited, P. I. Djurovich, K. O. Hodgson, B. Hedman, E. I. Solomon and M. E. Thompson, *J. Am. Chem. Soc.*, 2011, 133, 3700–3703.
63. C.-W. Hsu, C.-C. Lin, M.-W. Chung, Y. Chi, G.-H. Lee, P.-T. Chou, C.-H. Chang and P.-Y. Chen, *J. Am. Chem. Soc.*, 2011, 133, 12085–12099.
64. M. Cocchi, D. Virgili, V. Fattori, D. L. Rochester and J. A. G. Williams, *Adv. Funct. Mater.*, 2007, 17, 285–289.
65. C.-M. Che, C.-C. Kwok, S.-W. Lai, A. F. Rausch, W. J. Finkenzeller, N. Zhu and H. Yersin, *Chem.–Eur. J.*, 2010, 16, 233–247.
66. M.-Y. Yuen, S. C. F. Kui, K.-H. Low, C.-C. Kwok, S. S.-Y. Chui, C.-W. Ma, N. Zhu and C.-M. Che, *Chem.–Eur. J.*, 2010, 16, 14131–14141.
67. A. Y.-Y. Tam, D. P.-K. Tsang, M.-Y. Chan, N. Zhu and V. W.-W. Yam, *Chem. Commun.*, 2011, 47, 3383–3385 .
68. V. K.-M. Au, K. M.-C. Wong, D. P.-K. Tsang, M.-Y. Chan, N. Zhu and V. W.-W. Yam, *J. Am. Chem. Soc.*, 2010, 132, 14273–14278.
69. M. Mauro, P. E. Quartapelle, Y. Sun, C.-H. Chien, D. Donghi, M. Panigati, P. Mercandelli, P. Mussini, G. D'Alfonso and C. L. De, *Adv. Funct. Mater.*, 2009, 19, 2607–2614.
70. S. Chen, G. Tan, W.-Y. Wong and H.-S. Kwok, *Adv. Funct. Mater.*, 2011, 21, 3785–3793.
71. D.-S. Leem, S. O. Jung, S.-O. Kim, J.-W. Park, J. W. Kim, Y.-S. Park, Y.-H. Kim, S.-K. Kwon and J.-J. Kim, *J. Mater. Chem.*, 2009, 19, 8824–8828 .
72. C.-L. Ho, W.-Y. Wong, Q. Wang, D. Ma, L. Wang and Z. Lin, *Adv. Funct. Mater.*, 2008, 18, 928–937.
73. S.-L. Lai, S.-L. Tao, M.-Y. Chan, M.-F. Lo, T.-W. Ng, S.-T. Lee, W.-M. Zhao and C.-S. Lee, *J. Mater. Chem.*, 2011, 21, 4983–4988 .
74. C.-L. Ho, W.-Y. Wong, G.-J. Zhou, B. Yao, Z. Xie and L. Wang, *Adv. Funct. Mater.*, 2007, 17, 2925–2936.
75. S. Lamansky, P. Djurovich, D. Murphy, F. Abdel-Razzaq, H.-E. Lee, C. Adachi, P. E. Burrows, S. R. Forrest and M. E. Thompson, *J. Am. Chem. Soc.*, 2001, 123, 4304–4312.
76. R. Wang, D. Liu, H. Ren, T. Zhang, H. Yin, G. Liu and J. Li, *Adv. Mater.*, 2011, 23, 2823–2827.
77. R. Wang, D. Liu, R. Zhang, L. Deng and J. Li, *J. Mater. Chem.*, 2012, 22, 1411–1417 .
78. N. Rehmann, C. Ulbricht, A. Koehnen, P. Zacharias, M. C. Gather, D. Hertel, E. Holder, K. Meerholz and U. S. Schubert, *Adv. Mater.*, 2008, 20, 129–133.
79. K. Chen, C.-H. Yang, Y. Chi, C.-S. Liu, C.-H. Chang, C.-C. Chen, C.-C. Wu, M.-W. Chung, Y.-M. Cheng, G.-H. Lee and P.-T. Chou, *Chem.–Eur. J.*, 2010, 16, 4315–4327.
80. (*a*) B. X. Mi, P. F. Wang, Z. Q. Gao, C. S. Lee, S. T. Lee, H. L. Hong, X. M. Chen, M. S. Wong, P. F. Xia, K. W. Cheah, C. H. Chen and W. Huang, *Adv. Mater.*, 2009, 21, 339–343; (*b*) L.-X. Wang, Q.-B. Mei, F. Yan, B. Tian, J.-N. Weng, B. Zhang and W. Huang, *Acta Phys.-Chim.*

Sin., 2012, 28, 1556–1569 ; (*c*) C.-L. Ho, W.-Y. Wong, Z.-Q. Gao, C.-H. Chen, K.-W. Cheah, B. Yao, Z. Xie, Q. Wang, D. Ma, L. Wang, X.-M. Yu, H.-S. Kwok and Z. Lin, *Adv. Funct. Mater.*, 2008, 18, 319–331; (*d*) S.-J. Liu, N.-N. Song, J.-X. Wang, Y.-Q. Huang, Q. Zhao, X.-M. Liu, S. Sun and W. Huang, *Phys. Chem. Chem. Phys.*, 2011, 13, 18497–18506 ; (*e*) Q.-L. Xu, H.-Y. Li, C.-C. Wang, S. Zhang, T.-Y. Li, Y.-M. Jing, Y.-X. Zheng, W. Huang, J.-L. Zuo and X.-Z. You, *Inorg. Chim. Acta*, 2012, 391, 50–57.

81. S.-J. Lee, J.-S. Park, M. Song, I. A. Shin, Y.-I. Kim, J. W. Lee, J.-W. Kang, Y.-S. Gal, S. Kang, J. Y. Lee, S.-H. Jung, H.-S. Kim, M.-Y. Chae and S.-H. Jin, *Adv. Funct. Mater.*, 2009, 19, 2205–2212.

82. D. H. Kim, N. S. Cho, H.-Y. Oh, J. H. Yang, W. S. Jeon, J. S. Park, M. C. Suh and J. H. Kwon, *Adv. Mater.*, 2011, 23, 2721–2726.

83. T. Peng, H. Bi, Y. Liu, Y. Fan, H. Gao, Y. Wang and Z. Hou, *J. Mater. Chem.*, 2009, 19, 8072–8074 .

84. J. Ding, J. Lu, Y. Cheng, Z. Xie, L. Wang, X. Jing and F. Wang, *Adv. Funct. Mater.*, 2008, 18, 2754–2762.

85. Y. Zhu, C. Gu, S. Tang, T. Fei, X. Gu, H. Wang, Z. Wang, F. Wang, D. Lu and Y. Ma, *J. Mater. Chem.*, 2009, 19, 3941–3949 .

86. V. A. Montes, C. Perez-Bolivar, L. A. Estrada, J. Shinar and P. Anzenbacher Jr., *J. Am. Chem. Soc.*, 2007, 129, 12598–12599.

87. (*a*) H.-Y. Li, J. Wu, W. Huang, Y.-H. Zhou, H.-R. Li, Y.-X. Zheng and J.-L. Zuo, *J. Photochem. Photobiol., A*, 2009, 208, 110–116; (*b*) H. You, J. Fang, L. Wang, X. Zhu, W. Huang and D. Ma, *Opt. Mater.*, 2007, 29, 1514–1517.

88. (*a*) S. V. Eliseeva and J.-C. G. Bunzli, *Chem. Soc. Rev.*, 2010, 39, 189–227 ; (*b*) K. Binnemans, *Chem. Rev.*, 2009, 109, 4283–4374.

89. (*a*) H. Xu, L. H. Wang, X. H. Zhu, K. Yin, G. Y. Zhong, X. Y. Hou and W. Huang, *J. Phys. Chem. B*, 2006, 110, 3023–3029; (*b*) H. Xu, K. Yin, L. Wang and W. Huang, *Thin Solid Films*, 2008, 516, 8487–8492; (*c*) H. Xu, Y. Wei, B. Zhao and W. Huang, *J. Rare Earths*, 2010, 28, 666–670; (*d*) H. Xu, K. Yin and W. Huang, *Synth. Met.*, 2010, 160, 2197–2202; (*e*) H. Xu and W. Huang, *J. Photochem. Photobiol., A*, 2011, 217, 213–218; (*f*) K. Yin, H. Xu, G. Zhong, G. Ni and W. Huang, *Appl. Phys. A: Mater. Sci. Process.*, 2009, 95, 595–600.

90. H. Xu, K. Yin and W. Huang, *Chem.–Eur. J.*, 2007, 13, 10281–10293.

91. H. Xu, K. Yin and W. Huang, *ChemPhysChem*, 2008, 9, 1752–1760.

92. H. Xu, K. Yin and W. Huang, *J. Phys. Chem. C*, 2010, 114, 1674–1683 .

93. (*a*) H. Xu, R. Zhu, P. Zhao and W. Huang, *J. Phys. Chem. C*, 2011, 115, 15627–15638; (*b*) H. Xu, R. Zhu, P. Zhao, L.-H. Xie and W. Huang, *Polymer*, 2011, 52, 804–813.

94. (*a*) G. Qian, Z. Zhong, M. Luo, D. Yu, Z. Zhang, Z. Y. Wang and D. Ma, *Adv. Mater.*, 2009, 21, 111–116; (*b*) G. Qian, B. Dai, M. Luo, D. Yu, J. Zhan, Z. Zhang, D. Ma and Z. Y. Wang, *Chem. Mater.*, 2008, 20, 6208–6216.

95. C. Borek, K. Hanson, P. I. Djurovich, M. E. Thompson, K. Aznavour, R. Bau, Y. Sun, S. R. Forrest, J. Brooks, L. Michalski and J. Brown, *Angew. Chem., Int. Ed.*, 2007, 46, 1109–1112.

96. K. R. Graham, Y. Yang, J. R. Sommer, A. H. Shelton, K. S. Schanze, J. Xue and J. R. Reynolds, *Chem. Mater.*, 2011, 23, 5305–5312.

97. L. Aboshyan-Sorgho, M. Cantuel, S. Petoud, A. Hauser and C. Piguet, *Coord. Chem. Rev.*, 2012, 256, 1644–1663.

98. (*a*) Z. Li, J. Yu, L. Zhou, H. Zhang, R. Deng and Z. Guo, *Org. Electron.*, 2008, 9, 487–494; (*b*) M. A. Katkova, A. P. Pushkarev, T. V. Balashova, A. N. Konev, G. K. Fukin, S. Y. Ketkov and M. N. Bochkarev, *J. Mater. Chem.*, 2011, 21, 16611–16620 .

99. (*a*) S. Chen, L. Deng, J. Xie, L. Peng, L. Xie, Q. Fan and W. Huang, *Adv. Mater.*, 2010, 22, 5227–5239; (*b*) G. M. Farinola and R. Ragni, *Chem. Soc. Rev.*, 2011, 40, 3467–3482 ; (*c*) H. Wu, L. Ying, W. Yang and Y. Cao, *Chem. Soc. Rev.*, 2009, 38, 3391–3400 ; (*d*) X. Zhao and X. Zhan, *Chem. Soc. Rev.*, 2011, 40, 3728–3743 ; (*e*) X.-H. Zhu, J. Peng, Y. Cao and J. Roncali, *Chem. Soc. Rev.*, 2011, 40, 3509–3524 .

100. (*a*) D. Dini, *Chem. Mater.*, 2005, 17, 1933–1945; (*b*) C. Ulbricht, B. Beyer, C. Friebe, A. Winter and U. S. Schubert, *Adv. Mater.*, 2009, 21, 4418–4441; (*c*) R. D. Costa, E. Orti and H. J. Bolink, *Pure Appl. Chem.*, 2011, 83, 2115–2128; (*d*) A. Duarte, K.-Y. Pu, B. Liu and G. C. Bazan, *Chem. Mater.*, 2011, 23, 501–515; (*e*) W.-J. Xu, S.-J. Liu, T.-C. Ma, Q. Zhao, A. Pertegas, D. Tordera, H. J. Bolink, S.-H. Ye, X.-M. Liu, S. Sun and W. Huang, *J. Mater. Chem.*, 2011, 21, 13999–14007 .

101. (*a*) C. Wang, H. Dong, W. Hu, Y. Liu and D. Zhu, *Chem. Rev.*, 2011, 112, 2208–2267; (*b*) C. Santato, F. Cicoira and R. Martel, *Nat. Photonics*, 2011, 5, 392–393; (*c*) M. Mas-Torrent and C. Rovira, *Chem. Soc. Rev.*, 2008, 37, 827–838 .

102. T.-W. Lee, M.-G. Kim, S. H. Park, S. Y. Kim, O. Kwon, T. Noh, J.-J. Park, T.-L. Choi, J. H. Park and B. D. Chin, *Adv. Funct. Mater.*, 2009, 19, 1863–1868.

103. D.-H. Lee, H.-C. Shin, H. Chae and S. M. Cho, *Adv. Mater.*, 2011, 23, 1851–1854.

104. E. C. W. Ou, L. Hu, G. C. R. Raymond, O. K. Soo, J. Pan, Z. Zheng, Y. Park, D. Hecht, G. Irvin, P. Drzaic and G. Gruner, *ACS Nano*, 2009, 3, 2258–2264.

105. J. Wu, M. Agrawal, H. c. A. Becerril, Z. Bao, Z. Liu, Y. Chen and P. Peumans, *ACS Nano*, 2009, 4, 43–48.

106. P. Matyba, H. Yamaguchi, G. Eda, M. Chhowalla, L. Edman and N. D. Robinson, *ACS Nano*, 2010, 4, 637–642.

107. (*a*) H. M. Zhang, Y. F. Dai and D. G. Ma, *J. Phys. D: Appl. Phys.*, 2008, 41, 102006 ; (*b*) Q. Wang, J. Q. Ding, Z. Q. Zhang, D. G. Ma, Y. X. Cheng, L. X. Wang and F. S. Wang, *J. Appl. Phys.*, 2009, 105, 076101; (*c*) Y. H. Chen, J. S. Chen, D. G. Ma, D. H. Yan and L. X. Wang, *J. Appl. Phys.*, 2011, 110, 074504; (*d*) H. M. Zhang, Y. F. Dai and D. G. Ma, *Appl. Phys. Lett.*, 2007, 91, 123504; (*e*) F. W. Guo and D. G. Ma, *Appl. Phys. Lett.*, 2005, 87, 173510; (*f*) Y. H. Chen, J. S. Chen, D. G. Ma, D. H. Yan,

L. X. Wang and F. R. Zhu, *Appl. Phys. Lett.*, 2011, 98, 243309; (*g*) Y. H. Chen, J. S. Chen, D. G. Ma, D. H. Yan and L. X. Wang, *Appl. Phys. Lett.*, 2011, 99, 103304.

108. Y. Chen, H. Tian, Y. Geng, J. Chen, D. Ma, D. Yan and L. Wang, *J. Mater. Chem.*, 2011, 21, 15332–15336 .

109. S. Hamwi, J. Meyer, M. Kröger, T. Winkler, M. Witte, T. Riedl, A. Kahn and W. Kowalsky, *Adv. Funct. Mater.*, 2010, 20, 1762–1766.

110. K. S. Yook, S. O. Jeon, S.-Y. Min, J. Y. Lee, H.-J. Yang, T. Noh, S.-K. Kang and T.-W. Lee, *Adv. Funct. Mater.*, 2010, 20, 1797–1802.

111. A. Perumal, M. Fröbel, S. Gorantla, T. Gemming, B. Lüssem, J. Eckert and K. Leo, *Adv. Funct. Mater.*, 2011, 22, 210–217 .

112. (*a*) R.-Q. Png, P.-J. Chia, J.-C. Tang, B. Liu, S. Sivaramakrishnan, M. Zhou, S.-H. Khong, H. S. O. Chan, J. H. Burroughes, L.-L. Chua, R. H. Friend and P. K. H. Ho, *Nat. Mater.*, 2010, 9, 152–158; (*b*) S. H. Khong, S. Sivaramakrishnan, R. Q. Png, L. Y. Wong, P. J. Chia, L. L. Chua and P. K. H. Ho, *Adv. Funct. Mater.*, 2007, 17, 2490–2499; (*c*) G. Winroth, G. Latini, D. Credgington, L.-Y. Wong, L.-L. Chua, P. K.-H. Ho and F. Cacialli, *Appl. Phys. Lett.*, 2008, 92, 103308; (*d*) B. Liu, R.-Q. Png, L.-H. Zhao, L.-L. Chua, R. H. Friend and P. K. H. Ho, *Nat. Commun.*, 2012, 3, 1321.

113. (*a*) D. F. Tang, G. A. Wen, X. Y. Qi, H. Y. Wang, B. Peng, W. Wei and W. Huang, *Polymer*, 2007, 48, 4412–4418; (*b*) G. A. Wen, Y. Xin, X. R. Zhu, W. J. Zeng, R. Zhu, J. C. Feng, Y. Cao, L. Zhao, L. H. Wang, W. Wei, B. Peng and W. Huang, *Polymer*, 2007, 48, 1824–1829; (*c*) G. A. Wen, X. R. Zhu, L. H. Wang, J. C. Feng, R. Zhu, W. Wei, B. Peng, Q. B. Pei and W. Huang, *J. Polym. Sci., Part A: Polym. Chem.*, 2007, 45, 388–394; (*d*) D. F. Tang, G. A. Wen, W. Wei and W. Huang, *Polym. Int.*, 2008, 57, 1235–1241.

114. B. Ma, F. Lauterwasser, L. Deng, C. S. Zonte, B. J. Kim, J. M. J. Fréchet, C. Borek and M. E. Thompson, *Chem. Mater.*, 2007, 19, 4827–4832.

115. C. A. Zuniga, S. Barlow and S. R. Marder, *Chem. Mater.*, 2011, 23, 658–681.

116. B. Ma, B. J. Kim, D. A. Poulsen, S. J. Pastine and J. M. J. Fréchet, *Adv. Funct. Mater.*, 2009, 19, 1024–1031.

117. (*a*) Z. B. Wang, M. G. Helander, J. Qiu, D. P. Puzzo, M. T. Greiner, Z. M. Hudson, S. Wang, Z. W. Liu and Z. H. Lu, *Nat. Photonics*, 2011, 5, 753–757; (*b*) W. H. Koo, S. M. Jeong, F. Araoka, K. Ishikawa, S. Nishimura, T. Toyooka and H. Takezoe, *Nat. Photonics*, 2010, 4, 222–226.

118. (*a*) M. Park, J. Im, M. Shin, Y. Min, J. Park, H. Cho, S. Park, M.-B. Shim, S. Jeon, D.-Y. Chung, J. Bae, J. Park, U. Jeong and K. Kim, *Nat. Nanotechnol.*, 2012, 7, 803–809; (*b*) J. Ge, H.-B. Yao, X. Wang, Y.-D. Ye, J.-L. Wang, Z.-Y. Wu, J.-W. Liu, F.-J. Fan, H.-L. Gao, C.-L. Zhang and S.-H. Yu, *Angew. Chem., Int. Ed.*, 2013, 52, 1654–1659.

119. (*a*) T. Sekitani, H. Nakajima, H. Maeda, T. Fukushima, T. Aida, K. Hata and T. Someya, *Nat. Mater.*, 2009, 8, 494–499; (*b*) J. Liang, L. Li, X. Niu, Z. Yu and Q. Pei, *Nat. Photonics*, 2013, 7, 817–824.

120. A. Rizzo, C. Nobile, M. Mazzeo, M. D. Giorgi, A. Fiore, L. Carbone, R. Cingolani, L. Manna and G. Gigli, *ACS Nano*, 2009, 3, 1506–1512.

121. R. Capelli, S. Toffanin, G. Generali, H. Usta, A. Facchetti and M. Muccini, *Nat. Mater.*, 2010, 9, 496–503.

122. A. J. Heeger, *Chem. Soc. Rev.*, 2010, 39, 2354–2371 .

123. (*a*) J. You, C.-C. Chen, Z. Hong, K. Yoshimura, K. Ohya, R. Xu, S. Ye, J. Gao, G. Li and Y. Yang, *Adv. Mater.*, 2013, 25, 3973–3978; (*b*) J. You, L. Dou, K. Yoshimura, T. Kato, K. Ohya, T. Moriarty, K. Emery, C.-C. Chen, J. Gao, G. Li and Y. Yang, *Nat. Commun.*, 2013, 4, 1446.

124. Y. Liang, Z. Xu, J. Xia, S.-T. Tsai, Y. Wu, G. Li, C. Ray and L. Yu, *Adv. Mater.*, 2010, 22, E135–E138.

125. (*a*) T. Ameri, G. Dennler, C. Waldauf, H. Azimi, A. Seemann, K. Forberich, J. Hauch, M. Scharber, K. Hingerl and C. J. Brabec, *Adv. Funct. Mater.*, 2010, 20, 1592–1598; (*b*) J. Y. Kim, K. Lee, N. E. Coates, D. Moses, T. Q. Nguyen, M. Dante and A. J. Heeger, *Science*, 2007, 317, 222–225; (*c*) X.-Y. Zhao, Z.-G. Li, T.-J. Zhu, B.-X. Mi, Z.-Q. Gao and W. Huang, *J. Phys. D: Appl. Phys.*, 2013, 46, 195105.

126. (*a*) S. Gunes, H. Neugebauer and N. S. Sariciftci, *Chem. Rev.*, 2007, 107, 1324–1338; (*b*) R. Rugescu, *Application of Solar Energy*, In-Tech Publications, 2013 Search ; (*c*) X. W. Sun, D. W. Zhao, L. Ke, A. K. K. Kyaw, G. Q. Lo and D. L. Kwong, *Appl. Phys. Lett.*, 2010, 97, 053303.

127. G. Dennler, M. C. Scharber and C. J. Brabec, *Adv. Mater.*, 2009, 21, 1323–1338.

128. C. Deibel, T. Strobel and V. Dyakonov, *Adv. Mater.*, 2010, 22, 4097–4111.

129. M. C. Scharber, D. Mühlbacher, M. Koppe, P. Denk, C. Waldauf, A. J. Heeger and C. J. Brabec, *Adv. Mater.*, 2006, 18, 789–794.

130. R. R. Lunt, T. P. Osedach, P. R. Brown, J. A. Rowehl and V. Bulovi, *Adv. Mater.*, 2011, 23, 5712–5727.

131. N. M. O Boyle, C. M. Campbell and G. R. Hutchison, *J. Phys. Chem. C*, 2011, 115, 16200–16210 .

132. S. R. Cowan, N. Banerji, W. L. Leong and A. J. Heeger, *Adv. Funct. Mater.*, 2012, 22, 1116–1128.

133. W. Cai, X. Gong and Y. Cao, *Sol. Energy Mater. Sol. Cells*, 2010, 94, 114–127.

134. W.-Y. Wong, *J. Organomet. Chem.*, 2009, 694, 2644–2647.

135. C.-L. Ho and W.-Y. Wong, *Coord. Chem. Rev.*, 2011, 255, 2469–2502.

136. (*a*) D. Qi and J. Jiang, *J. Phys. Chem. A*, 2011, 115, 13811–13820; (*b*) R. Bonnett, *Chem. Soc. Rev.*, 1995, 24, 19–33 ; (*c*) E. D. Sternberg, D. Dolphin and C. Brückner,*Tetrahedron*, 1998, 54, 4151–4202.

137. S. Gunster, S. Siebentritt, J. Elbe, L. Kreienhoop, B. Tennigkeit, D. Wohrle, R. Memming and D. Meissner, *Mol. Cryst. Liq. Cryst.*, 1992, 216, 117–121 .

138. Q. Sun, L. Dai, X. Zhou, L. Li and Q. Li, *Appl. Phys. Lett.*, 2007, 91, 253505.

139. S. Ryuzaki, T. Kai, Y. Toda, S. Adachi and J. Onoe, *J. Phys. D: Appl. Phys.*, 2011, 44, 145103 .

140. A. Yella, H. W. Lee, H. N. Tsao, C. Y. Yi, A. K. Chandiran,

M. K. Nazeeruddin, E. W.-G. Diau, C.-Y. Yeh, S. M. Zakeeruddin and M. Grätzel, *Science*, 2011, 334, 629–634.

141. (*a*) Z.-X. Xu, V. A. L. Roy, K.-H. Low and C.-M. Che, *Chem. Commun.*, 2011, 47, 9654–9656 ; (*b*) X. Li, Y. Chen, J. Sang, B.-X. Mi, D.-H. Wu, Z.-G. Li, H. Zhang, Z.-Q. Gao and W. Huang, *Org. Electron.*, 2013, 14, 250–254.

142. R. F. Bailey-Salzman, B. P. Rand and S. R. Forrest, *Appl. Phys. Lett.*, 2007, 91, 013508.

143. S. Zhong, J. Q. Zhong, X. Z. Wang, M. Y. Huang, D. C. Qi, Z. K. Chen and W. Chen, *J. Phys. Chem. C*, 2012, 116, 2521–2526 .

144. B. Verreet, R. Muller, B. P. Rand, K. Vasseur and P. Heremans, *Org. Electron.*, 2011, 12, 2131–2139.

145. J. Meiss, A. Merten, M. Hein, C. Schuenemann, S. Schaefer, M. Tietze, C. Uhrich, M. Pfeiffer, K. Leo and M. Riede, *Adv. Funct. Mater.*, 2012, 22, 405–414.

146. R. Pandey, A. A. Gunawan, K. A. Mkhoyan and R. J. Holmes, *Adv. Funct. Mater.*, 2012, 22, 617–624.

147. G. E. Morse and T. P. Bender, *ACS Appl. Mater. Interfaces*, 2012, 4, 5055–5068 .

148. M. K. Fischer, I. Lopez-Duarte, M. M. Wienk, M. V. Martinez-Diaz, R. A. Janssen, P. Bauerle and T. Torres, *J. Am. Chem. Soc.*, 2009, 131, 8669–8676.

149. Q. Wang, Y. Li, X. Yan, M. Rathi, M. Ropp, D. Galipeau and J. Jiang, *Appl. Phys. Lett.*, 2008, 93, 073303.

150. S. Honda, T. Nogami, H. Ohkita, H. Benten and S. Ito, *ACS Appl. Mater. Interfaces*, 2009, 1, 804–810 .

151. (*a*) T. Hori, N. Fukuoka, T. Masuda, Y. Miyake, H. Yoshida, A. Fujii, Y. Shimizu and M. Ozaki, *Sol. Energy Mater. Sol. Cells*, 2011, 95, 3087–3092 ; (*b*) T. Hori, Y. Miyake, N. Yamasaki, H. Yoshida, A. Fujii, Y. Shimizu and M. Ozaki, *Appl. Phys. Express*, 2010, 3, 101602 .

152. (*a*) A. W. Hains, Z. Liang, M. A. Woodhouse and B. A. Gregg, *Chem. Rev.*, 2010, 110, 6689–6735; (*b*) M. V. Martínez-Díaz, G. de la Torrea and T. Torres, *Chem. Commun.*, 2010, 46, 7090–7108 .

153. S. Archer and J. A. Weinstein, *Coord. Chem. Rev.*, 2012, 256, 2530–2561.

154. T. A. Clem, D. F. J. Kavulak, E. J. Westling and J. M. J. Frechet, *Chem. Mater.*, 2009, 22, 1977–1987 .

155. K.-H. Low, Z.-X. Xu, H.-F. Xiang, S. S.-Y. Chui, V. A. L. Roy and C.-M. Che, *Chem.–Asian J.*, 2011, 6, 3223–3229.

156. W.-Y. Wong and P. D. Harvey, *Macromol. Rapid Commun.*, 2010, 31, 671–713.

157. W.-Y. Wong and C.-L. Ho, *Acc. Chem. Res.*, 2010, 43, 1246–1256.

158. A. Köhler, H. F. Wittmann, R. H. Friend, M. S. Khan and J. Lewis, *Synth. Met.*, 1994, 67, 245–249 .

159. A. Kohler, H. F. Wittmann, R. H. Friend, M. S. Khan and J. Lewis, *Synth. Met.*, 1996, 77, 147–150 .

160. J. Gilot, M. M. Wienk and R. A. J. Janssen, *Nat. Mater.*, 2007, 6, 704–705.

161. Q. Wang and W.-Y. Wong, *Polym. Chem.*, 2011, 2, 432–440 .

162. H. Zhan, S. Lamare, A. Ng, T. Kenny, H. Guernon, W.-K. Chan, A. B. Djurišić, P. D. Harvey and W.-Y. Wong, *Macromolecules*, 2011, 44, 5155–5167.

163. Q. Wang, Z. He, A. Wild, H. Wu, Y. Cao, U. S. Schubert, C.-H. Chui and W.-Y. Wong, *Chem.–Asian J.*, 2011, 6, 1766–1777.

164. F.-R. Dai, H.-M. Zhan, Q. Liu, Y.-Y. Fu, J.-H. Li, Q.-W. Wang, Z. Xie, L. Wang, F. Yan and W.-Y. Wong, *Chem.–Eur. J.*, 2012, 18, 1502–1511.

165. L. Li, W.-C. Chow, W.-Y. Wong, C.-H. Chui and R. S.-M. Wong, *J. Organomet. Chem.*, 2011, 696, 1189–1197.

166. P. T. Wu, T. Bull, F. S. Kim, C. K. Luscombe and S. A. Jenekhe, *Macromolecules*, 2009, 42, 671–681.

167. C. Y. Chen, J. G. Chen, S. Wu, J. Y. Li, C. G. Wu and K. C. Ho, *Angew. Chem., Int. Ed.*, 2008, 47, 7342–7345.

168. (*a*) Y. Wang, S. Xu, T. Chen, H. Guo, Q. Liu, B. Ye, Z. Zhang, Z. He and S. Cao, *Polym. Chem.*, 2010, 1, 1048–1055 ; (*b*) T. B. Fleetham, Z. Wang and J. Li, *Inorg. Chem.*, 2013, 52, 7338–7343.

169. (*a*) J. Buhka, N. Pellet, S.-J. Moon, R. H. Baker, P. Gao, M. K. Nazeeruddin and M. Grätzel, *Nature*, 2013, 499, 316–319; (*b*) C. C. Stoumpos, C. D. Malliakas and M. G. Kanatzidis, *Inorg. Chem.*, 2013, 52, 9019–9038; (*c*) S. D. Stranks, G. E. Eperon, G. Grancini, C. Menelaou, M. J. Alcocer, T. Leijtens, L. M. Herz, A. Petrozza and H. J. Snaith, *Science*, 2013, 342, 341–344; (*d*) G. Xing, N. Mathews, S. Sun, S. S. Lim, Y. M. Lam, M. Gratzel, S. Mhaisalkar and T. C. Sum, *Science*, 2013, 342, 344–347.

170. W.-Y. Wong, W.-C. Chow, K.-Y. Cheung, M.-K. Fung, A. B. Djurišić and W.-K. Chan, *J. Organomet. Chem.*, 2009, 694, 2717–2726.

171. W. Kaiser and C. G. B. Garrett, *Phys. Rev. Lett.*, 1961, 7, 229–231.

172. F. Auzel, *Chem. Rev.*, 2004, 104, 139–174.

173. (*a*) F. Wang and X. Liu, *Chem. Soc. Rev.*, 2009, 38, 976–989 ; (*b*) J. Zhou, Z. Liu and F. Li, *Chem. Soc. Rev.*, 2012, 41, 1323–1349 ; (*c*) M. Haase and H. Schäfer, *Angew. Chem., Int. Ed.*, 2011, 50, 5808–5829; (*d*) Y. Liu, D. Tu, H. Zhu and X. Chen, *Chem. Soc. Rev.*, 2013, 42, 6924–6958 .

174. (*a*) F. Wang and X. Liu, *J. Am. Chem. Soc.*, 2008, 130, 5642–5643; (*b*) F. Wang, Y. Han, C. S. Lim, Y. Lu, J. Wang, J. Xu, H. Chen, C. Zhang, M. Hong and X. Liu, *Nature*, 2010, 463, 1061–1065; (*c*) F. Wang, R. Deng, J. Wang, Q. Wang, Y. Han, H. Zhu, X. Chen and X. Liu, *Nat. Mater.*, 2011, 10, 968–973; (*d*) Q. Su, S. Han, X. Xie, H. Zhu, H. Chen, C. K. Chen, R. S. Liu, X. Chen, F. Wang and X. Liu, *J. Am. Chem. Soc.*, 2012, 134, 20849–20857.

175. (*a*) F. Vetrone, R. Naccache, V. Mahalingam, C. G. Morgan and J. A. Capobianco, *Adv. Funct. Mater.*, 2009, 19, 2924–2929; (*b*) K. A. Abel, J. C. Boyer and F. C. van Veggel, *J. Am. Chem. Soc.*, 2009, 131, 14644–14645; (*c*) J. Wang, F. Wang, C. Wang, Z. Liu and X. G. Liu, *Angew. Chem., Int. Ed.*, 2011, 50, 10369–10372; (*d*) R. Deng, X. Xie, M. Vendrell, Y.-T. Chang and X. Liu, *J. Am. Chem. Soc.*, 2011, 133, 20168–20171; (*e*) J. Wang, R. Deng, M. A. MacDonald, B. Chen, J. Yuan, F.

Wang, D. Chi, T. S. A. Hor, P. Zhang, G. Liu, Y. Han and X. Liu, *Nat. Mater.*, 2014, 13, 157–162.

176. (*a*) N. M. Idris, M. K. Gnanasammandhan, J. Zhang, P. C. Ho, R. Mahendran and Y. Zhang, *Nat. Med.*, 2012, 18, 1580–1585; (*b*) L. Cheng, K. Yang, S. Zhang, M. W. Shao, S. T. Lee and Z. Liu, *Nano Res.*, 2010, 3, 722–732; (*c*) S. Jeong, N. Won, J. Lee, J. Bang, J. H. Yoo, S. G. Kim, J. A. Chang, J. Kim and S. Kim, *Chem. Commun.*, 2011, 47, 8022–8024 .

177. (*a*) V. Mahalingam, C. Hazra, R. Naccache, F. Vetrone and J. A Capobianco, *J. Mater. Chem. C*, 2013, 1, 6536–6540 ; (*b*) N. J. Johnson, A. Korinek, C. Dong and F. C. van Veggel, *J. Am. Chem. Soc.*, 2012, 134, 11068–11071.

178. (*a*) T. Zheng, L. Sun, J. C. Zhou, W. Feng, C. Zhang and C. H. Yan, *Chem. Commun.*, 2013, 49, 5799–5801 ; (*b*) D. Yang, C. Li, G. Li, M. Shang, X. Kang and J. Lin, *J. Mater. Chem.*, 2011,21, 5923–5927 .

179. (*a*) X. Huang, S. Han, W. Huang and X. Liu, *Chem. Soc. Rev.*, 2013, 42, 173–201 ; (*b*) B. M. Van der Ende, L. Aarts and A. Meijerink, *Phys. Chem. Chem. Phys.*, 2009, 11, 11081–11095 ; (*c*) W. Zou, C. Visser, J. A. Maduro, M. S. Pshenichnikov and J. C. Hummelen, *Nat. Photonics*, 2012, 6, 560–564; (*d*) X. Xie and X. Liu, *Nat. Mater.*, 2012, 11, 842–843.

180. (*a*) R. B. Liebherr, T. Soukka, O. S. Wolfbeis and H. H. Gorris, *Nanotechnology*, 2012, 23, 485103; (*b*) J. Shen, G. Chen, T. Y. Ohulchanskyy, S. J. Kesseli, S. Buchholz, Z. Li, P. N. Prasad and G. Han, *Small*, 2013, 9, 3213–3217 .

181. (*a*) X. Ye, J. E. Collins, Y. Kang, J. Chen, D. T. Chen, A. G. Yodh and C. B. Murray, *Proc. Natl. Acad. Sci. U. S. A.*, 2010, 107, 22430–22435; (*b*) F. Zhang, Y. Wan, T. Yu, F. Q. Zhang, Y. F. Shi, S. H. Xie, Y. G. Li, L. Xu, B. Tu and D. Y. Zhao, *Angew. Chem., Int. Ed.*, 2007, 46, 7976–7979; (*c*) H. T. Wong, H. L. W. Chan and J. H. Hao, *Opt. Express*, 2010, 18, 6123–6130; (*d*) E. M. Chan, G. Han, J. D. Goldberg, D. J. Gargas, A. D. Ostrowski, P. J. Schuck, B. E. Cohen and D. J. Milliron, *Nano Lett.*, 2012, 12, 3839–3845; (*e*) F. Chen, W. Bu, S. Zhang, J. Liu, W. Fan, L. Zhou, W. Peng and J. Shi, *Adv. Funct. Mater.*, 2013, 23, 298–307; (*f*) J. Zhao, D. Jin, E. P. Schartner, Y. Lu, Y. Liu, A. V. Zvyagin, L. Zhang, J. M. Dawes, P. Xi, J. A. Piper, E. M. Goldys and T. M. Monro, *Nat. Nanotechnol.*, 2013, 8, 729–734; (*g*) Y. Zhang and X. Liu, *Nat. Nanotechnol.*, 2013, 8, 702–703.

182. (*a*) X. Xie, N. Gao, R. Deng, Q. Sun, Q. Xu and X. Liu, *J. Am. Chem. Soc.*, 2013, 135, 12608–12611; (*b*) J. Shen, G. Chen, A.-M. Vu, W. Fan, O. S. Bilsel, C.-C. Chang and G. Han, *Adv. Opt. Mater.*, 2013, 1, 644–650 ; (*c*) Y. Wang, G. Liu, L. Sun, J. Xiao, J. Zhou and C. H. Yan, *ACS Nano*, 2013, 7, 7200–7206; (*d*) H. Wen, H. Zhu, X. Chen, T. F. Hung, B. Wang, G. Zhu, S. F. Yu and F. Wang, *Angew. Chem., Int. Ed.*, 2013, 52, 13419–13423; (*e*) R. Deng and X. Liu, *Nat. Photonics*, 2014, 8, 10–12.

183. Y. R. Shen, *The Principles of Nonlinear Optics*, Wiley-VCH, Weinheim, 2002 Search .

184. (*a*) M. M. Fejer, G. A. Magel, D. H. Jundt and R. L. Byer, *IEEE J. Quantum Electron.*, 1992, 28, 2631–2654 ; (*b*) J.-C. Boyer, L. A. Cuccia and J. A. Capobianco, *Nano Lett.*, 2007, 7, 847–852; (*c*) S. Baluschev, T. Miteva, V. Yakutkin, G. Nelles, A. Yasuda and G. Wegner, *Phys. Rev. Lett.*, 2006, 97, 143903; (*d*) X.-L. Zhang, H.-R. Yang, H.-B. Sun, S.-J. Liu, Q. Zhao and W. Huang, *Prog. Chem.*, 2012, 24, 1880–1889 .

185. (*a*) D. V. Kozlov and F. N. tellano, *Chem. Commun.*, 2004, 2860–2861 ; (*b*) R. R. Islangulov, D. V. Kozlov and F. N. tellano, *Chem. Commun.*, 2005, 3776–3778 ; (*c*) P. E. Keivanidis, S. Baluschev, T. Miteva, G. Nelles, U. Scherf, A. Yasuda and G. Wegner, *Adv. Mater.*, 2003, 15, 2095–2098; (*d*) S. Baluschev, F. Yu, T. Miteva, S. Ahl, A. Yasuda, G. Nelles, W. Knoll and G. Wegner, *Nano Lett.*, 2005, 5, 2482–2484.

186. S. Baluschev, V. Yakutkin, T. Miteva, Y. Avlasevich, S. Chernov, S. Aleshchenkov, G. Nelles, A. Cheprakov, A. Yasuda, K. Müllen and G. Wegner, *Angew. Chem., Int. Ed.*, 2007, 46, 7693–7696.

187. R. R. Islangulov, J. Lott, C. Weder and F. N. tellano, *J. Am. Chem. Soc.*, 2007, 129, 12652–12653.

188. S. Ji, W. Wu, W. Wu, H. Guo and J. Zhao, *Angew. Chem., Int. Ed.*, 2011, 50, 1626–1629.

189. T. N. Singh-Rachford, A. Nayak, M. L. Muro-Small, S. Goeb, M. J. Therien and F. N. tellano, *J. Am. Chem. Soc.*, 2010, 132, 14203–14211.

190. (*a*) T. N. Singh-Rachford, A. Haefele, R. Ziessel and F. N. tellano, *J. Am. Chem. Soc.*, 2008, 130, 16164–16165; (*b*) W. Wu, H. Guo, W. Wu, S. Ji and J. Zhao, *J. Org. Chem.*, 2011, 76, 7056–7064.

191. G. J. Wilson, A. Launikonis, W. H. F. Sasse and A. W.-H. Mau, *J. Phys. Chem. A*, 1997, 101, 4860–4866.

192. E. S.-H. Lam, D. P.-K. Tsang, W. H. Lam, A. Y.-Y. Tam, M.-Y. Chan, W.-T. Wong and V. W.-W. Yam, *Chem.–Eur. J.*, 2013, 19, 6385–6397.

193. T. Giridhar, W. Cho, J. Park, J.-S. Park, Y.-S. Gal, S. Kang, J. Y. Lee and S.-H. Jin, *J. Mater. Chem. C*, 2013, 1, 2368–2378 .

194. C. Shi, H. Sun, Q. Jiang, Q. Zhao, J. Wang, W. Huang and H. Yan, *Chem. Commun.*, 2013, 49, 4746–4748 .

195. T. Peng, G. Li, K. Ye, C. Wang, S. Zhao, Y. Liu, Z. Hou and Y. Wang, *J. Mater. Chem. C*, 2013, 1, 2920–2926 .

196. C. Yang, J. Xu, Y. Zhang, Y. Li, J. Zheng, L. Liang and M. Lu, *J. Mater. Chem. C*, 2013, 1, 4885–4901 .

197. C.-H. Chang, C.-L. Ho, Y.-S. Chang, I. C. Lien, C.-H. Lin, Y.-W. Yang, J.-L. Liao and Y. Chi, *J. Mater. Chem. C*, 2013, 1, 2639–2647 .

198. F. Kessler, Y. Watanabe, H. Sasabe, H. Katagiri, M. K. Nazeeruddin, M. Gratzel and J. Kido, *J. Mater. Chem. C*, 2013, 1, 1070–1075 .

199. S. Lee, S.-O. Kim, H. Shin, H.-J. Yun, K. Yang, S.-K. Kwon, J.-J. Kim and Y.-H. Kim, *J. Am. Chem. Soc.*, 2013, 135, 14321–14328.

200. X.-L. Chen, R. Yu, Q.-K. Zhang, L.-J. Zhou, X.-Y. Wu, Q. Zhang and C.-Z. Lu, *Chem. Mater.*, 2013, 25, 3910–3920.

201. G. Cheng, P.-K. Chow, S. C. F. Kui, C.-C. Kwok and C.-M. Che, *Adv. Mater.*, 2013, 25, 6765–6770.
202. X.-C. Hang, T. Fleetham, E. Turner, J. Brooks and J. Li, *Angew. Chem., Int. Ed.*, 2013, 52, 6753–6756.

CHAPTER 11

A SURFACE CONFINED YTTRIUM(III) BIS-PHTHALOCYANINATO COMPLEX: A COLOURFUL SWITCH CONTROLLED BY ELECTRONS

I. Alcón, M. Gonidec, M. R. Ajayakumar, M. Mas-Torrent and J. Veciana

Institut de Ciència de Materials de Barcelona (ICMAB-CSIC), Networking Re Center on Bioengineering, Biomaterials and Nanomedicine (CIBER-BBN), Campus de la UAB, 08193, Bellaterra, Spain

ABSTRACT

SAMs of a Y(III) double-decker complex on ITO have been prepared and their electrical and optical properties explored, exhibiting three accessible stable redox states with characteristic absorption bands in the visible spectra, corresponding to three complementary colors (*i.e.*, green, blue and red). These absorption bands are exploited as output signals of this robust ternary electrochemical switch, behaving hence as an electrochromic molecular-based device.

INTRODUCTION

In the last 20 years huge efforts have been devoted to studying and developing organic molecules for electronic applications. Molecules can be, in principle, synthesized in mass production at a relatively low cost and, by chemical design, their properties can be tuned. In order to fabricate devices, molecules are typically supported on inorganic substrates (mainly metals or metal-oxides), facilitating their manipulation and the possibility to direct the application of external stimuli on them. One common route is the fabrication of self-assembled monolayers (SAMs) which is focused on the use of molecules with a specific functional group that spontaneously bonds to the surface.[1,2] Electrochemical molecular switches are a particularly appealing class of molecular devices where electroactive molecules are switched reversibly between different redox states triggered by an electrical signal.[3] Optical, magnetic, electrical or chemical outputs can be used to read the state of the switch.[4–12] Most of the reported examples are based on bi-stable molecules where the two accessible redox states can be visualized as 1's or 0's mimicking the terminology employed in the binary logic system which is the basis of current memory devices. However, it is known that the fabrication of devices with a higher number of states would facilitate the processing of higher memory densities.[13,14]Despite this interest, only a few examples based on electroactive SAMs that can present three or more states have been reported to date.[13–15] Most of these systems take advantage of the different optical absorption levels

Citation: I. Alcón, M. Gonidec, M. R. Ajayakumar, M. Mas-Torrent and J. Veciana, A surface confined yttrium(III) bis-phthalocyaninato complex: a colourful switch controlled by electrons, DOI: 10.1039/C6SC00443A.

that the distinct redox states exhibit at determined wavelengths as the readout mechanism.[16–18] Double and triple-decker phthalocyanine lanthanide complexes are potential building blocks for the fabrication of electrochemical switches owing to their rich electrochemistry that allows an easy access to a range of oxidation states centred on the ligands.[19–21] Lindsey and Bocian demonstrated with Eu phthalocyanine triple-decker SAMs that four available redox states could be accessed electrochemically.[4] In solution and in thin films it is widely known that the reduction and oxidation processes in these materials are accompanied by significant changes in their optical absorption spectra.[22–24] This prompted us to explore the possibility to exploit this property as the output of a surface confined switch based on a double-decker phthalocyanine lanthanide complex. In this work, a ternary switchable SAM of a bis-phthalocyaninato–Y(III) complex has been prepared. By the application of a low bias voltage, three redox states have been accessed and clearly identified using optical absorption spectroscopy. Remarkably, each state shows characteristic absorption bands giving complementary colours. The SAMs are revealed to be very robust and stable upon the application of more than 100 switching cycles.

RESULTS AND DISCUSSION

Synthetic procedures

The yttrium(III) double-decker complex **2** (Scheme 1) was especially designed to form SAMs on indium-tin oxide (ITO). ITO was chosen as the supporting substrate due to its excellent properties for performing spectro-electrochemistry experiments since it is a transparent and electrical conductor oxide. Compound **2** bears a triethoxysilane moiety as a surface anchoring group. Scheme 1 shows the route followed to synthesize it. The first step consisted of the synthesis of compound **1**, a bis-phthalocyaninato yttrium(III) complex bearing a long alkyl chain with an end-vinyl group, which was carried out in one step adapting methods from the literature.[25] Firstly, a solution of acetyl-acetonate–Y(III) (1 eq.), phthalonitrile (6.34 eq.), alkene phthalonitrile (1.6 eq.), a catalytic amount of potassium acetate and diaza-bicycloundecene (DBU, 4 eq.) in hexanol was heated at 165 °C for 14 h. The solid product was extensively purified using a silica-gel chromatography column and a size-exclusion chromatography column for separating **1** from the non-alkylated double-decker complex and the multi-alkylated ones. Compound **1** was obtained in a 12% yield.

Compound **2** was synthesized by dissolving **1** in distilled triethoxy-silane, adding a catalytic amount of Karstedt catalyst (Pt(0)-1,1,3,3-tetramethyldisiloxane 2% solution in xylene) and stirring the solution for 24 h at room temperature. After purification using silica-gel chromatography, compound **2** was obtained in a 35% yield. All the experimental and characterization details of compounds **1**and **2** are provided in the ESI.

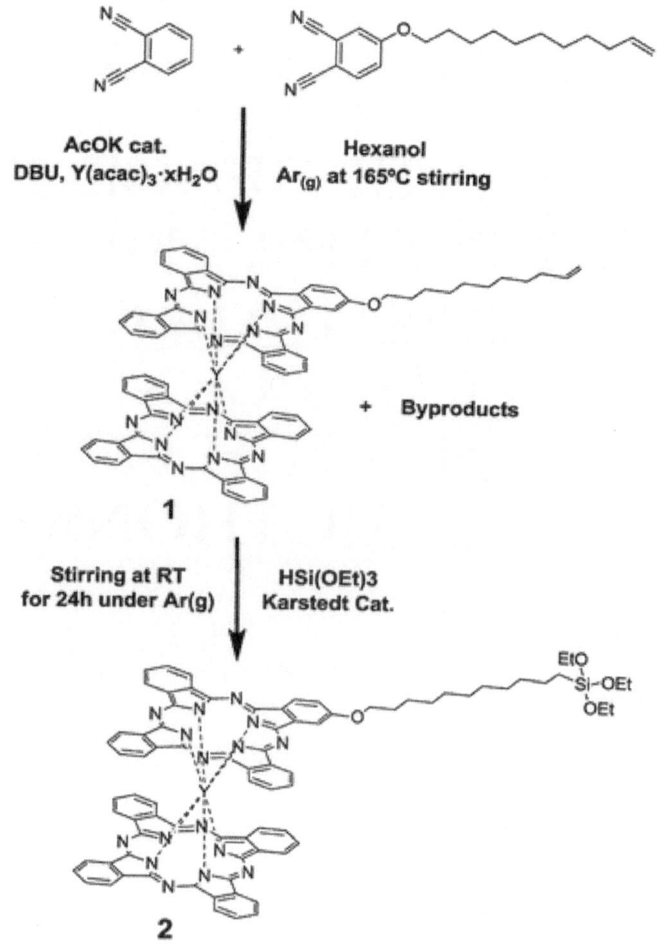

Scheme 1: Synthesis of compounds **1** and **2**.

Electro-optical characterization of 2

The cyclic voltammetry (CV) of **2** in solution (see ESI†) showed two redox waves at low potential bias corresponding to a one electron reduction process ($\mathbf{2^0} \rightarrow \mathbf{2^-}$) at −0.18 V and to a one electron oxidation process ($\mathbf{2^0} \rightarrow \mathbf{2^+}$) at 0.35 V (*vs.* Ag(s)). Next, UV-vis spectro-electrochemistry experiments were performed to follow the redox inter-conversion between the 3 redox states ($\mathbf{2^+}$, $\mathbf{2^0}$ and $\mathbf{2^-}$). A specially designed UV-vis cuvette for performing electrochemistry was used for this purpose with a Pt–Rh net as the working electrode (WE) and Ag and Pt wires as the reference (RE) and counter electrodes (CE), respectively. As an electrolyte, a deoxygenated 50 mM $TBAPF_6$ solution in 1,2-dichlorobenzene was employed ($TBAPF_6$ = tetrabutylammonium hexafluorophosphate). The electrochemical potentials used to perform the chronoamperometry were −1 V ($\mathbf{2^0} \rightarrow \mathbf{2^-}$), +0.25 V ($\mathbf{2^-} \rightarrow \mathbf{2^0}$), +1 V ($\mathbf{2^0} \rightarrow \mathbf{2^+}$) and −0.1 V ($\mathbf{2^+} \rightarrow \mathbf{2^0}$) *vs.* Ag(s).

Fig. 1 illustrates the UV-vis spectra recorded during the experiments where several clear isosbestic points can be observed demonstrating the reversibility of the redox processes. The UV-vis absorption spectrum of the neutral compound showed the typical absorption peaks of this family of compounds: the Soret bands at 334 and 371 nm, the π-radical band at 474 nm, the vibronic bands around 600 nm and the Q-band at 663 nm. Upon reduction, the Soret bands of the neutral complex fuse into a single absorption band at 358 nm. The π-radical band gradually disappears (corroborating the closed-shell nature of the reduced species) and the neutral Q-band at 663 nm splits into two bands (625 and 690 nm). Upon oxidation, the Soret band is blue-shifted. Also, the π-radical band at 474 nm shifts to higher wavelengths and increases in intensity (in accordance with the bi-radical character of the oxidized species). The vibronic bands of the neutral state (600 nm) disappear and, finally, the Q-band at 663 nm moves to lower energies and decreases in intensity. Remarkably, with the naked eye one can observe that the green solution of **2** (c = 0.1 mM) becomes blue upon reduction and red when it is oxidized (see Fig. 1, below). Such colour complementarity between the different redox states in solution is highly appealing for its exploitation in electro-optical molecular switches.

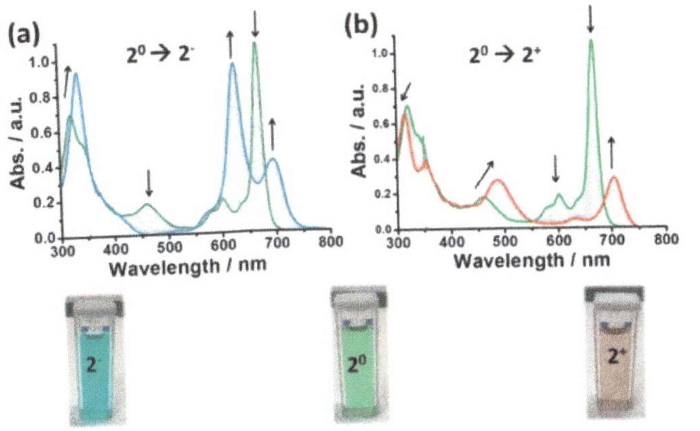

Figure 1: Spectro-electrochemistry experiments for (a) 1e-reduction ($2^0 \rightarrow 2^-$) and (b) 1e-oxidation ($2^0 \rightarrow 2^+$) processes of a solution of 0.1 mM of **2** using a 50 mM $TBAPF_6$ solution in 1,2-dichlorobenzene as the electrolyte, a Pt–Rh net as the working electrode and Pt and Ag wires as the counter and reference electrodes, respectively. On the bottom, the solutions of **2** at each redox state are shown.

2-SAM preparation

Once the excellent electro-optical properties of **2** for the purpose of this work had been corroborated, the following step was the anchoring of the active molecule on a solid support, in this e ITO. The ITO double coated-glass substrates were chemically activated by being immersed into an oxidant solution $NH_4OH : H_2O_2 : H_2O$ (1 : 1 : 5), rinsed with water and acetone, and dried with a N_2 flow. Immediately, the substrates were immersed in a 0.4 mM solution of **2** in toluene. The solution was heated at 75 °C for 1 hour and then maintained at 45 °C for 14 hours under an argon atmosphere. After that time the ITO slides were removed from the organic solution, rinsed with toluene to remove the physisorbed material and dried with a N_2 stream to give the **2**-SAM (Fig. 2a). X-ray photoelectron spectroscopy (XPS) and time-of-flight secondary ion mass spectrometry (ToF-SIMS) characterization was carried out confirming the formation of the monolayer (ESI†).

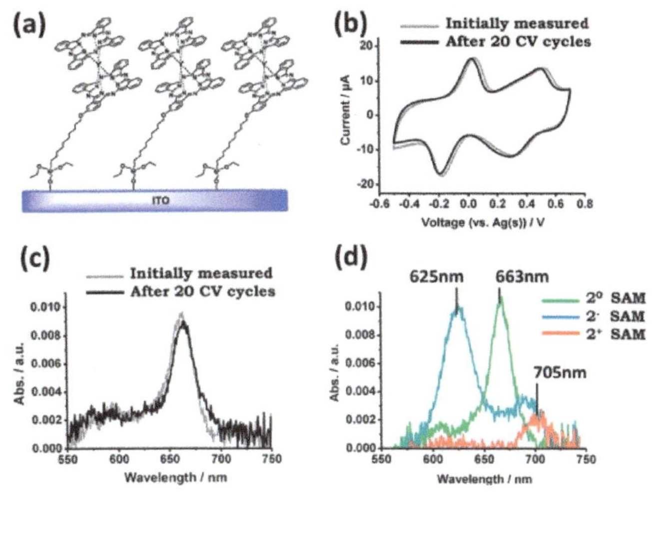

Figure 2: (a) Scheme of **2**-SAM. (b) CV of **2**-SAM used as the WE and Pt and Ag wires as the CE and RE, respectively, in a 50 mM $TBAPF_6$ solution in 1,2-dichlorobenzene at a scan rate of 0.3 V s^{-1}. (c) Vis absorption spectra of **2**-SAM. The grey and black lines in (b) and (c) are the results before and after applying 20 CV cycles, respectively. (d) Vis absorption spectra of **2**-SAM in the neutral,2^0, oxidised, 2^+, and reduced, 2^-, states.

Electro-optical characterization of 2-SAM

Besides the elemental surface characterization, the electrochemical and optical properties of **2**-SAM were also studied. CV experiments were performed using the functionalized substrate as the WE and Ag and Pt wires as the RE and CE, respectively, in a 50 mM $TBAPF_6$ solution in 1,2-dichlorobenzene as the electrolyte medium. As can be seen in Fig. 2b, two redox processes appeared in the CV at similar potential values as previously observed for this molecule in solution: the reduction process (2^0-SAM → 2^--SAM) appeared at −0.07 V (*vs.* Ag(s)) and the oxidation process (2^0-SAM → 2^+-SAM) at +0.42 V (*vs.* Ag(s)). These results demonstrate that the electro-active nature of **2** is preserved after being surface bonded. It is worth noting that the wide peak observed in the CV for the oxidation of **2**-SAM could be caused by the presence of molecules with different chemical environments.[26] The current intensity of both waves linearly increased with the applied scan rates (*i.e.*, 0.05, 0.1, 0.2, 0.3 and 0.4 V s^{-1}), which is characteristic of surface-confined electroactive species (ESI†). Further, for testing the robustness of the **2**-SAM upon electrical stress, 20 cycles between −0.7 and +0.8 V were applied without observing any loss in the current intensity (Fig. 2b).

Visible absorption characterization was also performed for **2**-SAM before and after applying the 20 CV cycles (Fig. 2c). Since it is known that the ITO functionalized substrates exhibit a broad band in the region of 400–500 nm,[2] we focused on the absorption characteristics above this region. In the spectrum, the typical Q-band absorption (= 663 nm) corresponding to the neutral state is observed. The initial recorded spectrum perfectly matches the one measured after the application of 20 CV cycles, pointing out, again, the good stability of the **2**-SAM.

Switch-ability of 2-SAM

The next step was to explore the possibility to employ the optical properties to follow the electrochemically triggered changes in **2**-SAM. For this, the absorption spectra were registered in the 550–750 nm range while sequentially applying pulses of different voltages over 2 minutes for generating each accessible redox state in the **2**-SAM (*i.e.*, +0 V for $\mathbf{2^0}$-SAM, −0.4 V for $\mathbf{2^-}$-SAM and +0.6 V for $\mathbf{2^+}$-SAM). In Fig. 2d the absorption spectra of the three redox states of **2**-SAM are shown, observing clear fingerprints for each of them. Indeed, the maximum absorption bands at 625, 663 and 705 nm can be employed to identify unambiguously the states $\mathbf{2^-}$-SAM, $\mathbf{2^0}$-SAM, and $\mathbf{2^+}$-SAM, respectively. That is, these absorption bands can be used as output signals.

The evolution of the absorption of **2**-SAM at the three characteristic wavelengths of the maximum absorption bands (*i.e.*, 625, 663 and 705 nm) was investigated when switching the surface between the different states by applying the corresponding voltage value for a 3 s pulse (Fig. 3). It was observed that the three wavelengths can be exploited to track the state of the switch since at each wavelength $\mathbf{2^0}$-SAM, $\mathbf{2^-}$-SAM and $\mathbf{2^+}$-SAM exhibit different levels of absorption. However, a wavelength of 663 nm seems to be more suitable since the three states have more separated absorption intensities. Table 1 summarizes the **2**-SAM switching behaviour, where high absorption values are labelled as "H", medium as "M" and low as "L". This molecular device was switched between the three states for more than 100 cycles without signal loss, unambiguously demonstrating the robustness of the system.

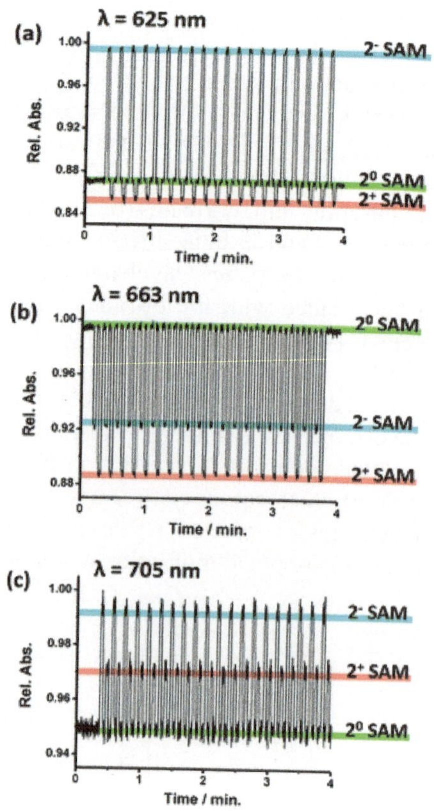

Figure 3: Application of 20 electrochemical switching cycles to the **2**-SAM while measuring the absorbance at each of the three characteristic wavelengths: (a) 625 nm, (b) 663 nm and (c) 705 nm. Each potential pulse was applied for 3 s. The different absorption plateaus correspond to the different states of the switch, highlighted with the corresponding colours.

Table 1: Truth table for the **2**-SAM switch

Input (V *vs.* Ag(s))	**Logic state**	**Redox state**	**Output^a I (625 nm)**	**Output^a II (663 nm)**	**Output^a III (705 nm)**
***a* The letters L, M and H in the table body indicate low, medium and high absorption intensity levels, respectively.**					
0	0	2^0-SAM	M	H	L
−0.4	1	2^--SAM-	H	M	H
+0.6	2	2^+-SAM	L	L	M

We have found that the time required for the complete switching between the different redox states of the **2**-SAM is 0.3–0.4 s (see ESI†). This value is in accordance with other reported electro-active SAMs on ITO.[5]

Stability of the 2-SAM switch states

The preservation of oxidized and reduced states without the application of an external stimulus was also tested. Voltage pulses were applied to electro-generate the corresponding redox state and then the system was left at open voltage. Each state was followed by recording the absorption of its characteristic band (Fig. 4). After 2 minutes, the reduced $\mathbf{2^-}$-SAM state started to show a decrease in its absorption band, while the oxidized $\mathbf{2^+}$-SAM did not reveal any sign of deterioration for more than 7 minutes. The lower stability of the $\mathbf{2^-}$-SAM state compared with $\mathbf{2^0}$-SAM and $\mathbf{2^+}$-SAM is in accordance with the behaviour of reduced double-decker derivatives in solution.[27]

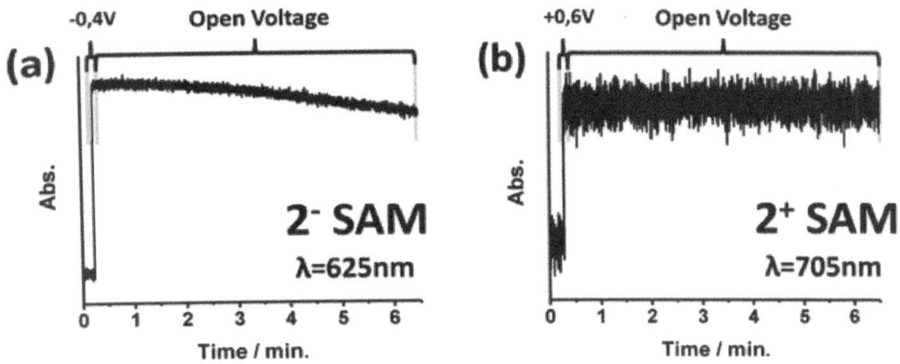

Figure 4: Stability of the electro-generated **2**$^{-}$-SAM and **2**$^{+}$-SAM states followed by the UV-vis absorption of the surface at 625 and 705 nm, respectively. Potential pulses of 10 s were applied before leaving the system at open voltage.

Finally, the shelf-stability of the **2**-SAM was also explored when the functionalized substrate was kept at room temperature and in environmental conditions. Using absorption spectroscopy, it was demonstrated that after one month and a half there was no signal loss (ESI†).

CONCLUSIONS

In summary, a novel phthalocyanine double-decker Y(III) complex has been designed and synthesized to be grafted on oxide substrates. SAMs of this molecule on ITO, a conducting and transparent substrate, have been prepared and their electrochemical and optical properties explored. This system exhibits three accessible redox states at relatively low voltages and behaves as a stable ternary electrochemical switch. Outstandingly, each state reveals clear characteristic absorption bands in the visible spectra, which are in fact responsible for the three complementary colours that the solutions of each redox state of this compound show (*i.e.*, green, blue and red). Such absorption bands in the SAM were exploited as output signals of the system, behaving hence as an electrochromic molecular-based device. The high robustness and stability that the system revealed point towards the high potential that phthalocyanine lanthanide complexes have as molecular electro-optical switches.

ACKNOWLEDGEMENTS

This work was funded by ERC StG 2012-306826 e-GAMES, the EU project ITN iSwitch (GA no. 642196), the Networking Re Center on Bioengineering, Biomaterials and Nanomedicine (CIBER-BBN), DGI (Spain) BE-WELL CTQ2013-40480-R, and Generalitat de Catalunya 2014-SGR-17. I. A. acknowledges CIBER-BBN for a grant.

REFERENCES

1. D. Vuillaume, in *Oxford handbook of nanoscience and nanotechnology. Frontiers and advances*, ed. A. Narlikar and Y. Fu, Oxford university press, Oxford, 2009, pp. 312–3342 .
2. B. K. Pathem, S. A. Claridge, Y. B. Zheng and P. S. Weiss, *Annu. Rev. Phys. Chem.*, 2013, 64, 605–630.
3. N. Fuentes, A. Martín-Lasanta, L. Álvarez de Cienfuegos, M. Ribagorda, A. Parra and J. M. Cuerva, *Nanoscale*, 2011, 3, 4003–4014 .
4. Z. Liu, A. A. Yasseri, J. S. Lindsey and D. F. Bocian, *Science*, 2003, 302, 1543–1545.
5. C. Simão, M. Mas-Torrent, N. Crivillers, V. Lloveras, J. M. Artés, P. Gorostiza, J. Veciana and C. Rovira, *Nat. Chem.*, 2011, 3, 359–364 .
6. G. de Ruiter and M. E. van der Boom, *Angew. Chem., Int. Ed.*, 2012, 51, 8598–8601.
7. J. Lahann, S. Mitragotri, T.-N. Tran, H. Kaido, J. Sundaram, I. S. Choi, S. Hoffer, G. A. Somorjai and R. Langer, *Science*, 2003, 299, 371–374.
8. E. Marchante, N. Crivillers, M. Buhl, J. Veciana and M. Mas-Torrent, *Angew. Chem., Int. Ed.*, 2016, 55, 368–372.
9. D. I. Gittins, D. Bethell, D. J. Schiffrin and R. J. Nichols, *Nature*, 2000, 408, 67–69.
10. I. V. Pobelov, Z. Li and T. Wandlowski, *J. Am. Chem. Soc.*, 2008, 130, 16045–16054 .
11. E. H. van Dijk, D. J. T. Myles, M. H. van der Veen and J. C. Hummelen, *Org. Lett.*, 2006, 8, 2333–2336.
12. S. S. Jang, Y. H. Jang, Y.-H. Kim, W. A. Goddard III, A. H. Flood, B. W. Laursen, H.-R. Tseng, J. F. Stoddart, J. O. Jeppesen, J. W. Choi, D. W. Steuerman, E. Delonno and J. R. Heath, *J. Am. Chem. Soc.*, 2005, 127, 1563–1575.
13. M. Mas-Torrent, C. Rovira and J. Veciana, *Adv. Mater.*, 2013, 25, 462–468.
14. J. ado-Montenegro, M. Mas-Torrent, F. Otón, N. Crivillers, J. Veciana and C. Rovira, *Chem. Commun.*, 2013, 49, 8084–8086 .
15. P. A. Derosa, S. Guda and J. M. Seminario, *J. Am. Chem. Soc.*, 2003, 125, 14240–14241.
16. C. Simão, M. Mas-Torrent, J. ado-Montenegro, F. Otón, J. Veciana and C. Rovira, *J. Am. Chem. Soc.*, 2011, 133, 13256–13259 .
17. G. de Ruiter, L. Motiei, J. Choudhury, N. Oded and M. E. van der Boom, *Angew. Chem., Int. Ed.*, 2010, 49, 4780–4783.
18. A. Kumar, M. Chhatwal, P. C. Mondal, V. Singh, A. K. Singh, D. A. Cristaldi, R. D. Gupta and A. Gulino, *Chem. Commun.*, 2014, 50, 3783–3785 .

19. M. L. Rodriguez-Mendez and J. Antonio de Saja, *J. Porphyrins Phthalocyanines*, 2009, 13, 606–615 .

20. P. Zhu, F. Lu, N. Pan, D. P. Arnold, S. Zhang and J. Jiang, *Eur. J. Inorg. Chem.*, 2004, 510–517 .

21. I. A. Akinbulu, S. Khene and T. Nyokong, *Electrochim. Acta*, 2010, 55, 7085–7093 .

22. N. Ishikawa, *J. Porphyrins Phthalocyanines*, 2001, 05, 87–101 .

23. K. M. Kadish, T. Nakanishi, A. Gürek, V. Ahsen and I. Yilmaz, *J. Phys. Chem. B*, 2001, 105, 9817–9821 .

24. M. Gonidec, E. S. Davies, J. McMaster, D. B. Amabilino and J. Veciana, *J. Am. Chem. Soc.*, 2010, 132, 1756–1757.

25. A. De Cian, M. Moussavi, J. Fischer and R. Weiss, *Inorg. Chem.*, 1985, 24, 3162–3167 .

26. A. V. Rudnev, K. Yoshida and T. Wandlowski, *Electrochim. Acta*, 2013, 87, 770–778 .

27. M. Gonidec, F. Luis, À. Vilchez, J. Esquena, D. B. Amabilino and J. Veciana, *Angew. Chem., Int. Ed.*, 2010, 49, 1623–1626.

CHAPTER 12

STABLE LUMINESCENT IRIDIUM(III) COMPLEXES WITH BIS(N-HETEROCYCLIC CARBENE) LIGANDS: PHOTO-STABILITY, EXCITED STATE PROPERTIES, VISIBLE-LIGHT-DRIVEN RADICAL CYCLIZATION AND CO_2 REDUCTION, AND CELLULAR IMAGING

Chen Yang[ab], Faisal Mehmood[a], Tsz Lung Lam[cd], Sharon Lai-Fung Chan[cd], Yuan Wu[a], Chi-Shun Yeung[a], Xiangguo Guan[a], Kai Li[ab], Clive Yik-Sham Chung[a], Cong-Ying Zhou[ab], Taotao Zou[a] and Chi-Ming Che[ab]

[a]State Key Laboratory of Synthetic Chemistry, Institute of Molecular Functional Materials, HKU-Joint Laboratory on New Materials and Department of Chemistry, The University of Hong Kong, Pokfulam Road, Hong Kong, China

[b]HKU Shenzhen Institute of Re and Innovation, Shenzhen, China

[c]The Hong Kong Polytechnic University Shenzhen Re Institute, Shenzhen, PR China

[d]Department of Applied Biology and Chemical Technology, The Hong Kong Polytechnic University, Hung Hom, Hong Kong, China

ABSTRACT

A new class of cyclometalated Ir(III) complexes supported by various bidentate C-deprotonated (C^N) and *cis*-chelating bis(N-heterocyclic carbene) (bis-NHC) ligands has been synthesized. These complexes display strong emission in deaerated solutions at room temperature with photoluminescence quantum yields up to 89% and emission lifetimes up to 96 µs. A photo-stable complex containing C-deprotonated fluorenyl-substituted C^N shows no significant decomposition even upon irradiation for over 120 h by blue LEDs (12 W). These, together with the strong absorption in the visible region and rich photo-redox properties, allow the bis-NHC Ir(III) complexes to act as good photo-catalysts for reductive C–C bond formation from $C(sp^3/sp^2)$–Br bonds cleavage using visible-light irradiation ($\lambda > 440$ nm). A water-soluble complex with a glucose-functionalized bis-NHC ligand catalysed a visible-light-driven radical cyclization for the synthesis of pyrrolidine in aqueous media. Also, the bis-NHC Ir(III) complex in combination with a cobalt catalyst can catalyse the visible-light-driven CO_2reduction with excellent turnover numbers (>2400) and selectivity (CO over H_2 in gas phase: >95%). Additionally, this series of bis-NHC Ir(III) complexes are found to localize in and stain endoplasmic reticulum (ER) of various cell lines with high selectivity, and exhibit high cytotoxicity towards cancer cells, revealing their potential uses as bioimaging and/or anti-cancer agents.

INTRODUCTION

Luminescent organometallic complexes of 3rd row transition metals, such as Ir,[1–5] Pt,[6–11] and Au[12–19] are currently receiving burgeoning interest due to their profound applications in materials science,[20–22] biology,[23,24] and organic synthesis.[25–27] In particular, the favourable emission properties of cyclometalated Ir(III) complexes have been harnessed for applications by

Citation: Chen Yang, Faisal Mehmood, Tsz Lung Lam, Sharon Lai-Fung Chan, Yuan Wu, Chi-Shun Yeung, Xiangguo Guan, Kai Li, Clive Yik-Sham Chung, Cong-Ying Zhou, Taotao Zou and Chi-Ming Che, Stable luminescent iridium(III) complexes with bis(N-heterocyclic carbene) ligands: photo-stability, excited state properties, visible-light-driven radical cyclization and CO_2 reduction, and cellular imaging, DOI: 10.1039/C5SC04458H.

many re groups with examples such as the development of high performance OLEDs by Thompson and co-workers[3–5,28–31] and the design of bioimaging and cellular probes by Lo and co-workers.[32–36] Over the past decade, as a result of extensive work most notably by MacMillan,[37,38] Yoon,[27,39–41] and Stephenson and co-workers,[42–45] luminescent cyclometalated iridium(III) complexes are widely used in photo-redox catalysis, which has been shown to have useful applications in organic synthesis.

In 2008, Yoon and co-workers[40] reported [2 + 2] enone cycloadditions, and MacMillan and co-workers[38] published the alkylation of aldehydes, both of which were catalysed by triplet metal-to-ligand-charge-transfer (^{3}MLCT) excited state of $[Ru(bpy)_3]^{2+}$ generated upon visible-light irradiation. Subsequently, Stephenson and co-workers achieved the reductive dehalogenation of activated alkyl halides catalysed by $[Ru(bpy)_3]^{2+}$,[44] and reductive dehalogenation of alkyl, alkenyl and aryl iodides by using *fac*-$Ir(ppy)_3$ as photo-redox catalyst.[45] Compared to $[Ru(bpy)_3]^{2+}$ and *fac*-$Ir(ppy)_3$, the application of luminescent platinum(II) photo-catalysts is still nascent. Recently, our group and Wu's group demonstrated that pincer Pt(II) complexes are capable of catalysing light induced C–C bond formation.[46,47]

The important features that allow luminescent transition metal complexes to act as useful photo-redox catalysts or photo-sensitizers for light induced reactions include: (1) the long lifetime of their electronic excited states, thus allowing bimolecular reaction to proceed in solution; (2) their electronic excited states as both strong reducing and oxidizing reagents with reduction potentials systematically varied by the auxiliary ligands.[41] For the photo-catalysis to have practical interest, the design of highly stable photo-redox catalysts with long-lived electronic excited states in solution is desirable.

Our endeavour to develop transition metal photo-catalysis is to use visible light for activation of small molecules such as CO_2, described in this work, and for C–X bond functionalization. A major consideration is to utilize visible light, which falls within the solar spectrum and avoids deleterious high-energy UV-initialized photochemical side reactions.[48,49] In the literature, platinum group metal complexes and semi-conductors are usually used as photo-redox catalysts for the photochemical reduction of CO_2.[50] Earlier examples of transition metal photo-catalysts used for the photochemical reduction of CO_2 include cobalt porphyrins,[51] $Re(bpy)(CO)_3Cl$[52,53] and Ir(terpy)(ppy)Cl.[54] More recently, systems comprising *fac*-$Ir(ppy)_3$ in conjunction with $[Ni(^{Pr}bimiq1)]^{2+}$,[55] Fe(porphyrin),[56][Co(TPA)Cl]Cl,[57] (TPA = tris(2-pyridylmethyl)amine) and $[Co(N5)]^{2+}$ (N5 = 2,13-dimethyl-3,6,9,12,18-pentaazabicyclo-[12.3.1]octadeca-1(18),2,12,14,16-pentaene)[58] were reported for photochemical reduction of CO_2.

The stability of photo-catalysts is an important issue for the practical application of transition metal photochemistry. Numerous studies revealed that the photochemically active excited states of $[Ru(bpy)_3]^{2+}$,[59–61] $[Ir(ppy)_2(bpy)]^+$,[60,62] and *fac*-$Ir(ppy)_3$[55,57,63] are not stable under light irradiation for a long period of time as a result of dissociation of coordinated ligand(s) presumably *via* low lying d–d excited state(s). To address the photo-stability issue, we considered the use of N-heterocyclic carbene (NHC) ligands which have been receiving burgeoning attention in coordination chemistry due to their strong σ-donor strength to develop robust metal photo-sensitizers and photo-catalysts.[64–70] Also, the *N*-substituent of NHC ligands can be used to tune both the physical and chemical properties of the resultant photo-active transition metal complexes such as their solubility in various solvents including water. NHC ligands functionalized with carboxylate, sulfonate, amine/ammonium, and alcohol motifs have been reported for the development of water-soluble transition metal catalysts for Suzuki coupling, hydrosilylation, hydrogenation, olefin metathesis and CO_2 reduction.[65,71]

Compared with Ir(III) complexes supported by bidentate acetylacetonate[28] and/or 2,2′-bypyridine[1] ligands, the photophysical and application studies of the related bis-NHC Ir(III) complexes are relatively scarce. In 2010, cationic bis-NHC Ir(III) complexes were reported by De Cola and co-workers[72] to have application in blue-light emitting electrochemical cells; subsequently, a number of bis-NHC Ir(III) complexes were reported for photophysical and biological studies.[65,67,72–74] In this work, a series of strongly luminescent Ir(III) complexes (Chart 1) containing bis-NHC ligands and visible light absorbing C-deprotonated (C^N) ligands was synthesized and their photophysical and electrochemical properties were examined. These complexes display high photo-stability and are strongly emissive with long lifetimes of up to 96 μs in solution at room temperature. The water-soluble luminescent Ir(III) complexes, containing the glucose-functionalized NHC ligand, were found to be active photo-catalysts for radical cyclization leading to the formation of 5-membered pyrrole rings in aqueous media with high substrate conversions and yields. One of the photo-stable Ir(III) complexes was utilized as a photo-sensitizer and in conjunction with a recently reported catalyst [Co(TPA)Cl]Cl to convert CO_2 into CO with a turnover number (TON) > 2400, selectivity in gas phase > 95% and yield of 5.6% (1 mL out of 18 mL of CO_2 was converted into CO at 5 μM concentration of Co complex). Some of the complexes were also demonstrated as potential bioimaging and/or anti-cancer agents.

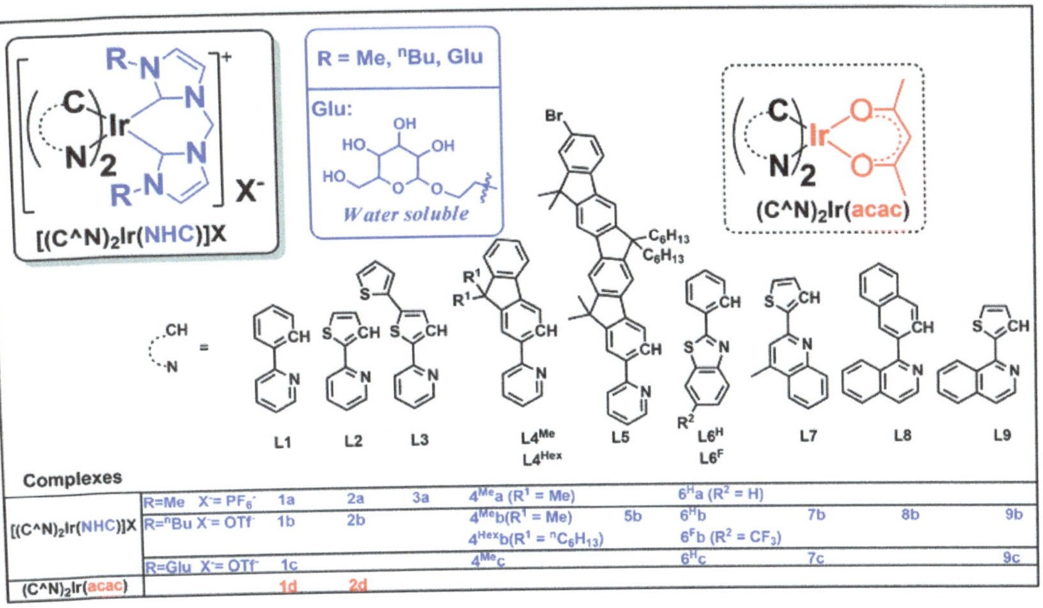

Complexes		L1	L2	L3	L4	L5	L6	L7	L8	L9
[(C^N)$_2$Ir(NHC)]X	R=Me X$^-$= PF$_6^-$	1a	2a	3a	4^{Me}a (R^1 = Me)		6^{H}a (R^2 = H)			
	R=^{n}Bu X$^-$= OTf$^-$	1b	2b		4^{Me}b(R^1 = Me) 4^{Hex}b(R^1 = $^{n}C_6H_{13}$)	5b	6^{H}b 6^{F}b (R^2 = CF_3)	7b	8b	9b
	R=Glu X$^-$= OTf$^-$	1c			4^{Me}c		6^{H}c	7c		9c
(C^N)$_2$Ir(acac)		1d	2d							

Chart 1: Iridium(III) complexes in this work. Complexes 1a,[73,74] 1b,[73] 1d[28] and 2d[31] have been reported in the literature.

RESULTS

Synthesis, characterization and photo-stability of [(C^N)$_2$Ir(NHC)]X complexes

The structures of 18 bis-NHC Ir(III) complexes synthesized in this work (**1c** and **2a**, **2b**, **3–9**), together with previously reported Ir(III) complexes **1a**,[73,74] **1b**,[73] **1d**[28] and **2d**,[31] are depicted in Chart 1. Complexes **1–9** were synthesized in good yields by refluxing [(C^N)$_2$Ir(μ-Cl)]$_2$ with bis(imidazolium) salts in the presence of silver(I) oxide in 2-methoxylethanol. Details of synthesis and characterization data of ligands and complexes are provided in the ESI.† Notably, ^{1}H NMR spectra of **7b** and **7c** show poorly resolved peaks in the aromatic region (6.7–8.2 ppm) at ambient temperature, and hence ^{1}H NMR spectra are recorded from 238 K to 300 K (Fig. S1, ESI†) in order to assure the purity of **7b** and **7c**.

Complexes **1a**–$\mathbf{4^{Me}a}$, **1b**–$\mathbf{4^{Me}b}$, $\mathbf{4^{Hex}b}$–**9b** with *N*-methyl or *N*-butyl substituent on bis-NHC ligands are soluble in most common aprotic solvents, but not in protic solvents *e.g.* methanol (MeOH) or water. Complexes **1c**, $\mathbf{4^{Me}c}$, $\mathbf{6^{H}c}$, **7c** and **9c** with glucose functionalized bis-NHC ligand are soluble in MeOH, ethanol (EtOH) and water.

The photo-stability of these Ir(III) complexes with bis-NHC ligands was examined by using $\mathbf{4^{Me}b}$ and $\mathbf{6^{H}b}$ as representative examples. $\mathbf{4^{Me}b}$ and $\mathbf{6^{H}b}$ in degassed deuterated MeCN were irradiated using blue light (12 W blue LEDs) for 5 days. The photolysis was monitored by ^{1}H NMR spectroscopy. As depicted in Fig. 1a, in the e of $\mathbf{4^{Me}b}$, less than 5% of the complex was observed to undergo photochemical decomposition after irradiation of 120 h, revealing its outstanding photo-stability. Under the same conditions, Ru(bpy)$_3$Cl$_2$, *fac*-Ir(ppy)$_3$ and [(dFCF$_3$ppy)$_2$Ir(dtbbpy)]PF$_6$(dFCF$_3$ppy = 3,5-difluoro-2-[5-(trifluoromethyl)-2-pyridinyl]phenyl; dtbbpy = 4,4′-bis(1,1-dimethylethyl)-2,2′-bipyridine) were found to decompose after 10 h irradiation as revealed by the changes of their ^{1}H NMR spectra (Fig. 1c).

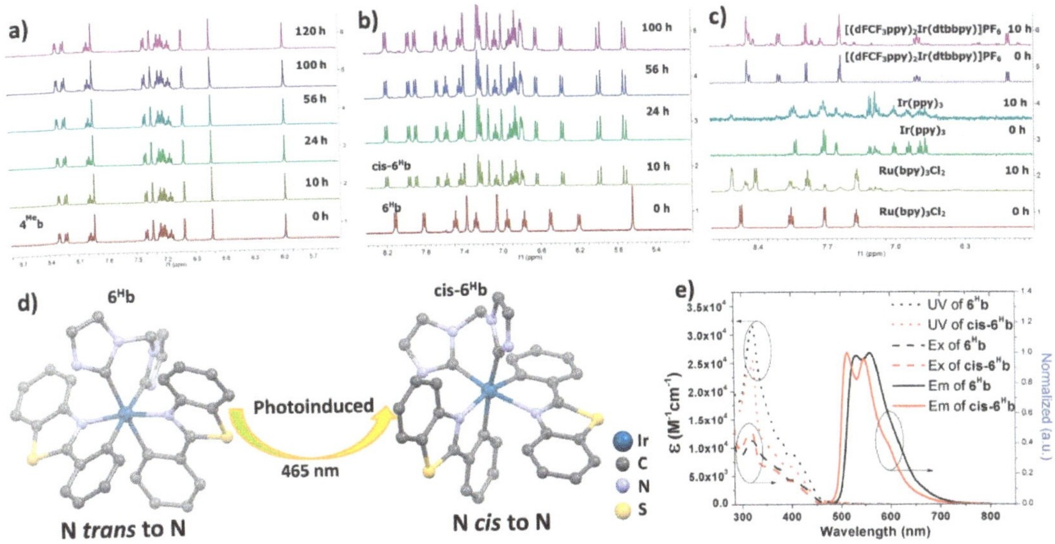

Figure 1: ^{1}H NMR spectra for (a) $\mathbf{4^{Me}b}$; (b) $\mathbf{6^{H}b}$; (c) Ru(bpy)$_3$Cl$_2$, *fac*-Ir(ppy)$_3$ and [(dFCF$_3$ppy)$_2$Ir(dtbbpy)]PF$_6$ in deuterated MeCN solution for irradiating by blue light (12 W, $_{max}$= 462 nm)[46] (all solutions are degased by nitrogen gas for 10 min); (d) crystal structure diagrams

showing the photoinduced transformation of coordination for Ir metals in 6^{H}**b** and ***cis*-6^{H}b** (butyl group on bis-NHC ligand and hydrogen atoms are omitted for clarity); (e) UV/Vis absorption (dotted line), excitation (dashed line) and emission (solid line) spectra of solutions of 6^{H}**b** and ***cis*-6^{H}b** (concentration of 2.0×10^{-5} M) in degassed DCM at 298 K.

Interestingly, irradiation of 6^{H}**b** using blue LEDs for 10 h led to a clean and quantitative conversion to a new species which did not show further changes upon subsequent irradiation for another 90 h (Fig. 1b, this species is noted as ***cis*-6^{H}b**). The yield of scale synthesis of ***cis*-6^{H}b** from the irradiation of degassed MeCN solution of 6^{H}**b** (100 mg in 4 mL MeCN) was 94%. ***cis*-6^{H}b** was found to be stable upon standing in solution in the dark for another 40 h, or exposed to air for another 20 h (Fig. S2, ESI†). To verify that the transformation of 6^{H}**b** to ***cis*-6^{H}b** was caused by visible-light irradiation, a negative control experiment was conducted by keeping 6^{H}**b** in deuterated MeCN in the dark (Fig. S2, ESI†), and no structural changes were detected by ^{1}H NMR spectroscopy.

The UV-Vis absorption, emission and excitation spectra of ***cis*-6^{H}b** measured in DCM solution at 298 K are different from those of 6^{H}**b** with hypsochromic shifts in peak maxima (Fig. 1e). The emission band for ***cis*-6^{H}b** displays vibronic spacings of 1245 cm^{-1} while that of 6^{H}**b** shows spacings of 908 cm^{-1}. Taking into consideration the ESI-MS and spectroscopic data of ***cis*-6^{H}b**, ***cis*-6^{H}b** is likely a structural isomer for 6^{H}**b**. The exact structure of ***cis*-6^{H}b** has been determined by X-ray crystallography.

X-Ray crystallography

Crystals of 4^{Me}**b** (with PF_6^- counter anion), 6^{H}**b**, ***cis*-6^{H}b** and **7b** suitable for X-ray crystallographic analysis were obtained by slow diffusion of diethyl ether into DCM (6^{H}**b**), MeCN (4^{Me}**b** and **7b**) and chloroform (***cis*-6^{H}b**) solution of these complexes, respectively. Perspective views of 6^{H}**b** and ***cis*-6^{H}b** are shown in Fig. 1d (those of 4^{Me}**b**, **7b** are shown Fig. S3 in ESI†). Selected bond lengths and angles are compiled in Table S1 (ESI†). Similar to the reported examples,[72] 6^{H}**b** adopts a distorted octahedral geometry (Fig. 1), the iridium atom is coordinated by two cyclometalated C^N ligands and one bis-NHC ligand. The two C^N ligands adopt a mutually eclipsed configuration, with the two N atoms (N1 and N2) *trans* to each other and with Ir–N bond length of 2.067(2) and 2.071(2) Å. The two substituted phenyl rings are oriented *cis* to each other with Ir–C bond length of 2.054(2) and 2.077(2) Å. The C^{NHC} atoms of bis-NHC ligand are *trans* to two C atoms from C^N ligands. The Ir–C^{NHC} distances of 2.105(2) Å and 2.128(2) Å are comparable to the literature values of Ir–C^{NHC} bonds *trans* to phenyl groups.[72] The C^{NHC}–Ir–C^{NHC} bite angles of 4^{Me}**b** (84.7(3)°), 6^{H}**b** (85.10(9)°), **7b**(85.58(6)°) and that of reported [(dfppy)$_2$Ir(NHC^{Me})]PF_6 (85.28(15)°) (dfppy = 4,6-difluoro-phenylpyridine) are consistent.[72] The structures of 4^{Me}**b** and **7b** are in line with that of 6^{H}**b**. Interestingly, for***cis*-6^{H}b** the N atoms from the two C^N ligand are *cis* to each other (Fig. 1d), indicating that 6^{H}**b** and ***cis*-6^{H}b** are structural isomers. The crystallographic refinement parameters for 4^{Me}**b**, 6^{H}**b**, ***cis*-6^{H}b**and **7b** are summarized in Table S2 (ESI†).

Electronic absorption and emission spectroscopy

Photophysical data of **1–9** including their UV/Vis absorption and emission maxima, emission lifetimes and photo-luminescent quantum yields are tabulated in Table 1. All the complexes show intense high-energy absorptions ranging from 250 to 380 nm and lower-energy absorptions at 416–494 nm (Fig. 2, 3 and S4 in ESI†). The high-energy absorptions are attributed to the $^{1}\pi \rightarrow \pi^*$ transition of the ligands, whereas the low energy absorption bands should be ascribed to the admixture of metal-to-ligand charge transfer (MLCT) transitions and ligand centered (LC) π–π^* transitions.[72] The charge transfer (CT) assignment is supported by the energy trend observed in the complexes with different C^N ligands. For example, the low-energy absorptions of **8b** and **5b** reveal significant red-shifts from those of **1b** and 4^{Me}**b** respectively. This can be rationalized by the extended π-conjugation of the C^N ligand in **8b** and **5b**, leading to the lowering of the π^*(C^N) orbitals and hence decreases in CT energy. These red-shifts in the UV-Vis absorptions, together with the large molar absorptivities in the visible region (*e.g.*, **5b**: 7.2×10^{4} M^{-1} cm^{-1}), allow the complexes to show strong absorption in the visible region. This is believed to be crucial for harvesting visible light in the solar spectrum as well as avoiding UV-initiated side reactions due to the use of UV in the photo-catalysis by the Ir(III) complexes. As revealed by the UV absorption spectra of 6^{F}**b** and 6^{H}**b** (Fig. S4b, ESI†), a bathochromic shift about 500 cm^{-1} is observed for the lower-energy absorption band(s). These could be attributed to the electron-withdrawing group of CF_3 on the cyclometalated ligand which lowers the C^N based LUMO level.

Table 1: Photophysical data of complexes **1–9**

Complex	**Medium**^{a}	**Absorption (λ_{max}/nm) ($10^{-3}\varepsilon$/ M^{-1} cm^{-1})**	**Emission λ_{em}/nm (τ/ μs)**	**Φ /%**
a **Measured in degassed CH_2Cl_2 and water (2.0×10^{-5} M) at 298 K. *b* 4^{Me}a, 4^{Me}b and 4^{Hex}b were measured at the concentration of 5×10^{-6} M. *c* 5b was measured at the concentration of 1×10^{-6} M. *d* Phosphorescence quantum yields were measured by using [Ru(bpy)$_3$](PF$_6$)$_2$ (Φ = 0.062 in MeCN) as standard.**				
1a	CH_2Cl_2	255 (26.8), 267 (24.9), 311 (10.6), 342 (6.1), 380 (3.7), 416 (1.2)	470 (2.1), 499, 534	89
1b	CH_2Cl_2	254 (36.8), 266 (35.5), 311 (15.5), 342 (9.1), 381 (5.82), 416 (2.16)	470 (2.1), 500, 534	89

1c	H_2O	254 (36.8), 266 (35.5), 311 (15.5), 342 (9.1), 381 (5.82), 416 (2.16)	469 (2.0), 498, 534	89
2a	CH_2Cl_2	252 (13.5), 290 (21.7), 332 (12.6), 369 (6.07), 405 (4.92)	530, 547, 570 (3.03)	11.1
3a	CH_2Cl_2	252 (15.5), 282 (15.1), 313 (20.5), 335 (23.1), 375 (21.9), 389 (20.9), 440 (20.9)	472, 674 (4.8), 824	0.2
4^{Me}a	$CH_2Cl_2^{b}$	260 (31.4), 297 (30.0), 316 (33.5), 334 (32.3), 360 (22.4), 378 (16.8), 421 (10.5)	524, 564 (28.6), 614	65
4^{Me}b	$CH_2Cl_2^{b}$	267 (52.2), 297 (46.8), 316 (50.9), 335 (49.0), 358 (36.3), 376 (27.3), 421 (16.2)	525, 566 (28.7), 614	75
4^{Hex}b	$CH_2Cl_2^{b}$	268 (42.1), 299 (42.3), 320 (50.4), 338 (49.2), 364 (32.6), 379 (25.6), 421 (14.6)	527, 568 (32.7), 616	82
5b	$CH_2Cl_2^{c}$	260 (54.1), 273 (42.5), 310 (32.9), 323 (34.8), 377 (67.2), 398 (97.0), 436 (71.6)	576 (96.1), 625, 683	22.6
6^{H}b	CH_2Cl_2	271 (23.0), 321 (25.0), 377 (9.57), 407 (6.81), 436 (3.38)	531, 560 (6.4)	78
6^{H}c	H_2O	268 (21.4), 321 (22.1), 377 (8.06), 407 (5.49), 436 (2.42)	536, 560 (5.0)	66
6^{F}b	CH_2Cl_2	270 (23.0), 323 (32.1), 383 (10.7), 415 (7.88), 447 (4.08)	530, 558 (3.2)	68
7b	CH_2Cl_2	296 (26.8), 333 (18.5), 388 (9.31), 437 (7.76), 467 (4.60)	582, 623 (6.2)	36
7c	H_2O	293 (25.9), 330 (17.4), 386 (9.63), 432 (6.92), 464 (3.71)	583, 622 (5.9)	32
8b	CH_2Cl_2	260 (80.1), 290 (48.5), 315 (35.4), 376 (34.5), 429 (7.52), 485 (3.47), 494 (2.88)	618, 659 (5.2)	3
9b	CH_2Cl_2	313 (24.2), 334 (23.8), 384 (12.7), 428 (10.7), 457 (8.20), 485 (3.47)	617, 667 (7.4)	13
9c	H_2O	315 (18.4), 380 (9.69), 420 (7.6), 454 (4.95), 475 (2.15)	620 (4.5), 668	9

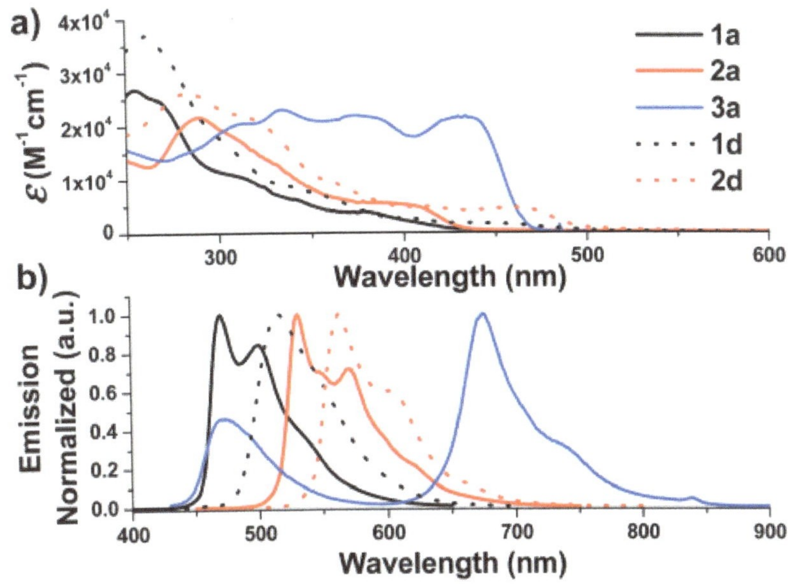

Figure 2: UV-Vis absorption (top) and emission (bottom) spectra of solutions of **1a–3a** and **1d–2d** in degassed DCM (concentration of 2.0×10^{-5} M) at 298 K.

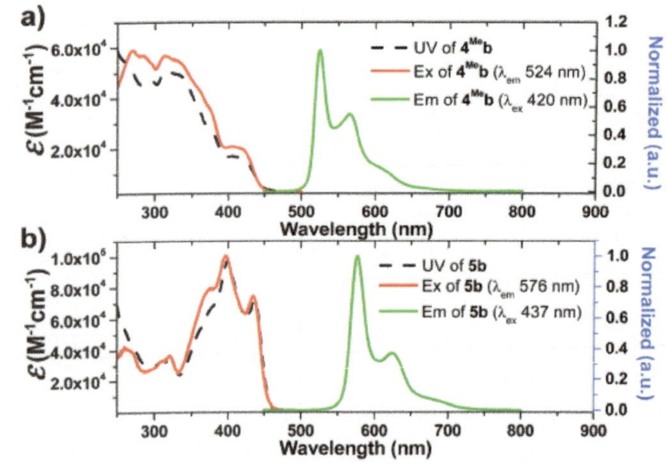

Figure 3: UV/Vis absorption (black dashed line), excitation (red solid line) and emission (green solid line) spectra of solutions of (a) **4^{Me}b** (5.0×10^{-6} M); (b) **5b** (1.0×10^{-6} M) in degassed DCM at 298 K.

On the other hand, the *N*-alkyl substituents on the bis-NHC ligand are found to have only minor effects, if any, on the UV/Vis absorption spectra of the Ir(III) complexes, as revealed by the overlaid spectra of the group of **1a**, **1b**, **1c** (Fig. S4a, ESI†). Complex **1a** absorbs weakly at the wavelength from 430 nm to 500 nm with molar absorptivity less than 500 M^{-1} cm^{-1} in contrast to the high values of 2500 M^{-1} cm^{-1} for **1d** with the same C^N luminophore. The calculated wavelength for ground state HOMO → LUMO ($S_0 \rightarrow S_1$ transition) is 369 nm and 414 nm for **1a** and **1d**, respectively (Fig. S4e and f, ESI†). These calculated values are in reasonable agreement with the corresponding experimental absorption $_{max}$ values 410 and 460 nm respectively.

Complexes **1–9** display strong phosphorescence in deaerated solution at room temperature (Fig. 2, 3 and S4 (ESI†); Table 1). All the complexes show vibronic-structured emissions, and their emission lifetimes are found to be in the microsecond regime. For example, **4^{Me}b** exhibits structured emission bands with vibrational spacings of ~1380 cm^{-1} and a long emission lifetime of 28.2 μs. Only small negative solvatochromic effects on the emission at 524 nm are found (±5 nm; Fig. S4d, ESI†). These findings, together with TD-DFT calculations, suggest that the photoluminescence of the complexes is derived from triplet metal-perturbed ligand-centred (^{3}LC) π–π* excited states. On the other hand, the structureless emission of **1d** (the acetylacetonate (acac) analogous of **1a**) at 516 nm should be ascribed to the ^{3}MLCT/LLCT emission.[2] Interestingly, complex **3a** (= 0.2%; = 4.8 μs) displays dual emission (Fig. 2) in contrast to the single emission of the acac analogue **3d**,[75]suggesting the possibility of modulation of photophysical properties of Ir(III) complexes by the NHC ligands.

Changes in chemical structure of the cyclometalated ligands also result in a profound effect on the photophysical properties of the Ir(III) complexes. For example, **5b** displays a significant red-shift in emission maximum (576 nm) when compared to **4^{Me}b** and **4^{Hex}b** (525 and 527 nm respectively). This can be rationalized by the extended π-conjugation of the C^N ligands, leading to the decrease in energy of metal-perturbed ^{3}LC emission. Moreover, for **4^{Me}a**, **4^{Me}b** and **4^{Hex}b**, their photoluminescence quantum yields in solution are affected by the substituents on the fluorenyl moiety as well as the*N*-alkyl groups of the bis-NHC ligands, *e.g.* complex **4^{Hex}b** exhibits a higher photoluminescence quantum yield than **4^{Me}a** and **4^{Me}b**. This is probably attributed to the fact that the long hexyl and *N*-butyl chains disfavour intermolecular stacking interactions among the planar C^N ligands, leading to reduced triplet–triplet annihilation and a higher photoluminescence quantum yield.

In addition, **5b** shows a significantly longer emission lifetime (96.1 μs) than **4^{Me}b** (28.7 μs). This might be due to the reduced metal character in the electronic excited state of **5b**, and hence slower triplet radiative decay. Similarly, with a smaller parentage of metal character in the frontier molecular orbitals, **6^{H}b** (6.4 μs) shows a longer emission lifetime than **6^{F}b** (3.2 μs). These long-lived triplet excited states allow the Ir(III) NHC complexes to undergo a variety of photochemical reactions, notably for visible-light-driven photo-catalytic reductive C–Br bond cleavage and CO_2 reduction which will be illustrated later.

Electrochemistry

The electrochemical data of **1b–9b** are summarized in Table 2 (all values *versus* Ag/$AgNO_3$, scan rate of 100 mV s^{-1}, 0.1 M ^{n}Bu_4NPF_6 in MeCN as supporting electrolyte) and Table S3† (values *vs.*$Cp_2Fe^{+/0}$, ESI†). These complexes display one irreversible oxidative wave at E_{pa} = 0.62–1.04 V and one irreversible reductive wave at E_{pc} = −2.52 to −1.94 V (*vs.* Ag/$AgNO_3$). The cyclic voltammograms of **1a** have been reported elsewhere[74] with the oxidation potential of +1.16 V (*vs.* Ag/$AgNO_3$) in CH_2Cl_2. By comparing with the electrochemical data of **1d** and **2d**[2], both of which have the ancillary acac ligand, the first oxidation wave of **1a** should originate from Ir(III)-centred/C^N ligand-centred oxidation. This assignment is further supported by the observation of the more positive oxidation potential of **6^{F}b** (+1.04 V) than **6^{H}b** (+0.98 V). The electron-withdrawing $-CF_3$ group of **6^{F}b** will lower the energy levels of the π(C^N) orbitals, leading to the more positive oxidation potential of **6^{F}b**. On the other hand, **4^{Me}b** and **5b** have more extended π-conjugated C^N ligands than **1b**. This will result in an increase in the energy level of π(C^N) orbitals, thus **4^{Me}b** and**5b** can undergo oxidation more readily than **1b** and hence less positive oxidation potentials are found.

Table 2: Electrochemical data[a] of bis-NHC Ir(III) complexes

Complex	E_{pa} /V	E_{pc} /V
***a* Supporting electrolyte: 0.1 M $^nBu_4NPF_6$ in MeCN and values are recorded *vs.* Ag/AgNO$_3$ (0.1 M) in MeCN; $Cp_2Fe^{+/0}$ occurs at the range of 0.05–0.08 (V) *vs.* Ag/AgNO$_3$. *b* Values refer to oxidation peak potential (E_{pa}) at 25 °C for irreversible couples at a scan rate of 100 mV s^{-1}. *c* Values refer to reduction peak potential (E_{pc}) for the irreversible reduction waves.**		
1b	0.87	−2.52
2a	0.64	−2.24
4^{Me}a	0.74	−2.41
4^{Me}b	0.79	−2.40
4^{Hex}b	0.74	−2.44
5b	0.72	−2.25
6^{H}b	0.98	−2.14
6^{F}b	1.04	−1.94
7b	0.69	−2.09
8b	0.62	−1.94
9b	0.62	−2.16

Compared with **1b**, [(dfppy)$_2$Ir(bis-NHC^{Bu})]PF$_6$ (ref. 72) displays a more anodic oxidation potential of E_{ox} = 1.04 V and a similar reduction potential of E_{re} = −2.37 V (*vs.* $Cp_2Fe^{+/0}$), which is attributed to stabilization of HOMOs as a result of the presence of highly electron-withdrawing F substitution on the C^N ligand (dfppy). For the irreversible reduction wave, a less negative reduction potential is found for **6^{H}b** (E_{pc} = −2.14 V *vs.* Ag/AgNO$_3$, Table 2, Fig. S4†), as compared to that of **1b** (E_{pc} = −2.52 V *vs.* Ag/AgNO$_3$, Table 2, Fig. 4) which has a less extended π-conjugation of the C^N ligand. The reductive wave is further anodically shifted in the e of **6^{F}b** (E_{pc} = −1.94 V *vs.* Ag/AgNO$_3$, Table 2, Fig. S5a, ESI†). Therefore, the reduction process should be localized on the C^N ligands.

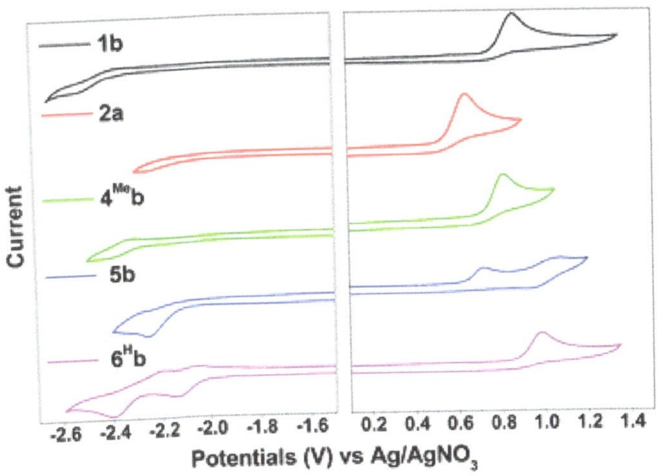

Figure 4: Cyclic voltammograms of **1b**, **2a**, **4^{Me}b**, **5b** and **6^{H}b** in MeCN with $^nBu_4NPF_6$ (0.1 M) as supporting electrolyte. Conditions: glass-carbon, working electrode, scan rate: 100 mV s^{-1}.

Excited state properties

Nano-second transient absorption and emission spectroscopic (tr-abs and tr-em) measurements were undertaken for the highly emissive bis-NHC Ir(III) complexes with long-lived electronic excited states *i.e.***4^{Me}b** (Fig. 5) and **4^{Me}c** (Fig. 6a) in MeCN and water respectively. These two complexes have the same lumophore but different functionalized NHC groups, leading to a different solubility in solvents, such as one is soluble in organic solvents and the other in water. The tr-abs and tr-em spectra of the other bis-NHC Ir(III) complexes *e.g.***5b** in MeCN (Fig. S6a, ESI†), and **6^{H}c** (Fig. S6b, ESI†) in water were also recorded and the results can be found in the ESI.†

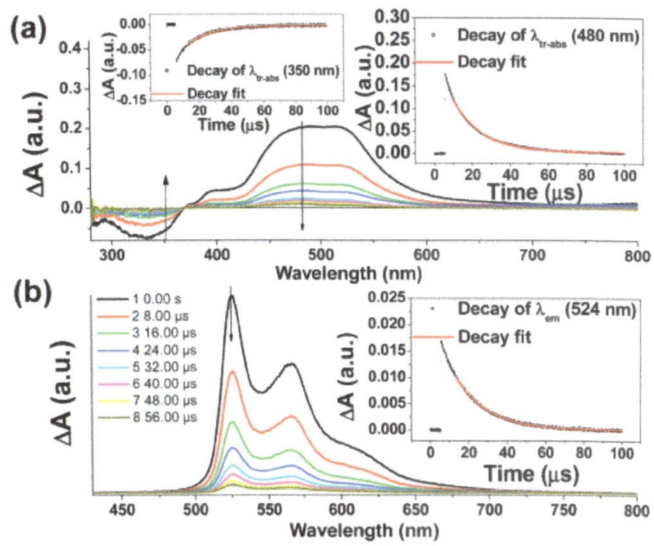

Figure 5: Time-resolved spectra of **4^{Me}b** (a) tr-abs (insets: decay of tr-abs at = 350 nm and 480 nm); (b) tr-em (decay of tr-em at = 524 nm) spectra recorded at specified times after laser pulse excitation (355 nm) in degassed MeCN at 298 K.

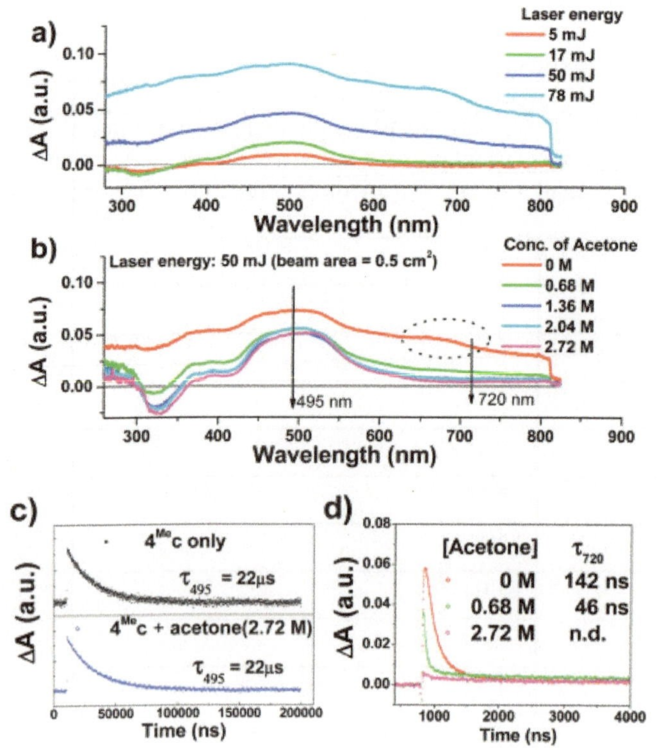

Figure 6: Transient absorption spectra recorded (a) under specified laser energy; (b) in presence of specified concentrations of acetone; of a degassed aqueous solution of **4^Me^c** (about 1 × 10^{-5} M). Kinetic studies of (c) $\lambda_{tr\text{-}abs}$ (495 nm) and (d) $\lambda_{tr\text{-}abs}$ (720 nm) in the absence/presence of specified concentrations of acetone.

The time-resolved absorption and emission spectra of **4^{Me}b** recorded at various time intervals after excitation at 355 nm are depicted in Fig. 5. The kinetic decay analysis of bleaching of ground-state of **4^{Me}b** ($_{350\,nm}$ = 15.4 μs, Fig. 5a, left inset) matches well with the growth of the absorption of triplet excited state ($_{480\,nm}$ = 15.2 μs, Fig. 5a, right inset) as well as the emission lifetime ($_{524\,nm}$ = 16.5 μs, Fig. 5b, right inset) in MeCN at 298 K.

The transient absorption spectra of **4^{Me}c** in aqueous solution recorded at different energies of laser beams (355 nm) reveal different spectral changes. As depicted in Fig. 6a, in addition to the growth of the absorption of triplet excited state of **4^{Me}c** from 380 to 700 nm, the emergence of an absorption band from 650 to 730 nm was observed at laser pulse energy ≥ 50 mJ (beam area: 0.5 cm^2). This absorption band could be quenched upon addition of acetone (Fig. 6b and S7a and b, ESI†), but no quenching of transient absorption at 495 nm and emission at 524 nm were observed in the presence of acetone (Fig. 6c and S7, ESI†). The decay rate constant monitored at 720 nm of **4^{Me}c** (Fig. 6d, absence of acetone) was much faster than that measured at 495 nm in Fig. 6c/S7c† and decay of emission at 524 nm (Fig. S7d, ESI†). In view of these different kinetic behaviours, the transient absorption from 650 to 730 nm depicted in Fig. 6b might originate from hydrated electrons e_{aq}^{-},[76,77] which were formed by the photo-induced ionization of **4^{Me}c** in aqueous solution upon excitation with high energy laser beams. This is in line with a reported photoionization of $[Pt_2(POP)_4]^{4-}$,[78] as well as the findings in the studies of solvated electron with acetone.[79,80] Similarly, complex **6^{H}c** in aqueous solution was also observed to undergo photo-ionization as revealed by the increase in transient absorption in the region of 600 to 700 nm (Fig. S8b, ESI†). For **4^{Me}b**, its transient absorption spectra monitored at high laser pulse energy in MeCN exhibit similar profiles as for lower energy (Fig. S8b and c, ESI†). However, newly generated long-lived species have been observed at high laser pulse energies after 100 μs (Fig. S8b–e, ESI†), revealing that photo-ionization of **4^{Me}b** with the generation of [**4^{Me}b**]$^+$ likely occurs. The accompanying solvated electrons are not observed in this e due to ready quenching by MeCN.[81]

Based on the electrochemical data and the determination of $E_{0\text{-}0}$ from the spectroscopic measurements, the excited state redox properties of the bis-NHC Ir(III) complexes can be estimated (Table S3, ESI†). The triplet excited states of the complexes are found to be powerful oxidants and reductants, and some of them are even more reactive towards photoredox reactions than $[Ru(bpy)_3]^{2+}$ and*fac*-Ir$(ppy)_3$. A representative example is **4^{Me}b** (E($Ir^{IV/III*}$) = −1.51 V *vs.* SCE) (here Ir^{IV} is a simple notation to denote the oxidized Ir^{III} species, the site of the oxidation can be the metal and/or C^N ligand), which is a stronger reductant than $[Ru(bpy)_3]^{2+}$(E($Ru^{III/II*}$) of −0.81 *vs.* SCE).[37,82] As a result, it is anticipated that the bis-NHC Ir(III) complexes described herein, upon photoexcitation in the visible-light region, can catalyse a number of reactions which are not feasible by the widely used $[Ru(bpy)_3]^{2+}$.

Photo-catalysis

With good photo-stability, long excited state lifetime, favourable absorption in the spectral region of blue LED and tunable photo-redox properties, these bis-NHC Ir(III) complexes have been investigated for photo-redox reactions (Chart 2), examples of which are described in the following section.

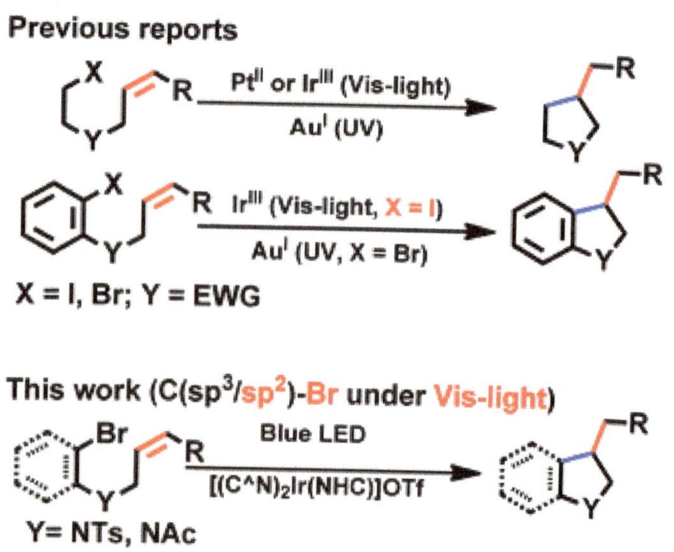

Chart 2: General photo-catalytic reactions by cyclometalated complexes. EWG = electron-withdrawing group.

We considered a recent work by Lee and co-workers[83] on visible-light-induced reductive cyclization of aromatic iodides and bromides to form indoline using $[(ppy)_2Ir(dtbbpy)]PF_6$ as photo-catalyst (PC). Yet, aryl bromides are found to be less reactive than iodides.[84] Barriault and co-workers addressed this issue by using dinuclear gold(I) complexes ($[Au_2(dppm)_2]OTf_2$) as photo-catalysts.[85]However, this gold complex only shows absorption in the high-energy UV region (< 300 nm), which may result in destructive effects on the products and/or lead to undesired side reactions.

As the present bis-NHC Ir(III) complexes show strong absorption at > 400 nm, they were used to photo-catalyse the reductive cyclization of aryl bromides using blue LEDs. Among the bis-NHC Ir(III) complexes examined, **1b**, **4^{Me}b** and **6^{H}b** displayed good photo-catalytic activity for reductive cyclization of aryl bromide (**A1**) in terms of both substrate conversion and product yield (Table 3).

Table 3: Screening bis-NHC Ir(III) complexes for photo-catalysis[a]

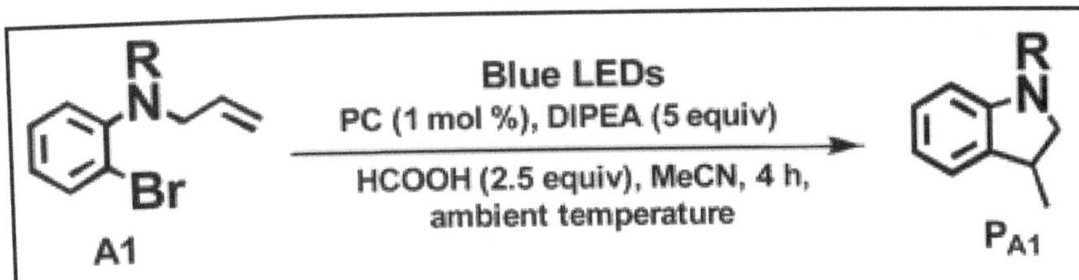

Entry[b]	PC	Conversion[c]/%	Yield[c]/%
1	**1b**	93	59
2	**2b**	1	0
3	**4^{Me}b**	90	59
4	**5b**	90	52
5	**6^{H}b**	93	59
6	**6^{F}b**	80	44
7	**7b**	49	24
8	**8b**	11	7.3
9	**9b**	0.1	0

a **Complex 3a was not tested because of the low quantum yield (0.2%, see Table 1).** ***b*** **Procedure: substrate 50 μmol, PC (1 mol%), DIPEA (5 equiv.), HCOOH (2.5 equiv.) in 4 mL MeCN solution was degassed by nitrogen, and irradiated by blue light (12 W, λ_{max} = 462 nm) at ambient temperature for 4 h.** ***c*** **Determined by ^{1}H NMR spectroscopy by adding internal standard of 5,5′-dimethyl-2,2′-bipyridine.**

In the e of another aryl bromide substrate **A2**, **4^{Me}b** showed both good substrate conversion and product yield (Table 4, entries 4–6). It was chosen as photo-sensitizer for optimization of the reaction conditions. A number of control experiments were performed, and no reaction was observed in the absence of amine or light (entries 13 and 14). Lowering the loading of PC (Table 3, entry 4), the absence of HCOOH or exposure to air (entries 11 and 12) were observed to decrease the substrate conversion of this reaction. Interestingly, using 1,8-diazabicyclo[5.4.0]undec-7-ene (DBU, entry 10, $E_{onset}^{+/0}$ = 0.60 V *vs.* $Cp_2Fe^{+/0}$), also led to excellent substrate conversion and good product yield comparable to what was obtained with tetramethylethylenediamine (TMEDA, entry 9, $E_{onset}^{+/0}$ = 0.11 V *vs.* $Cp_2Fe^{+/0}$, Fig. S9, ESI†). In contrast, the widely used photo-catalysts $[Ru(bpy)_3]Cl_2$ and *fac*-$Ir(ppy)_3$ (Table 4, entries 15 and 16) showed little or no conversion under similar conditions.

Table 4: Visible-light-induced C–C bond formation for aryl halide

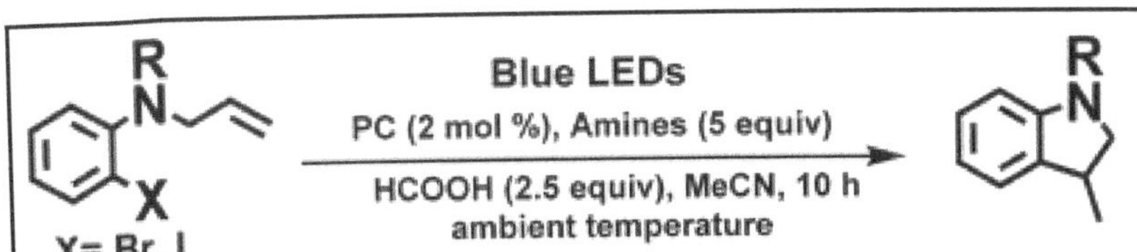

X= Br, I
A1 (R = Ac; X = Br) — PA1
A2 (R = Ts; X = Br) — PA2
A3 (R = Ac; X = I) — PA1

Entry[a]	PC (substrate)	Amines	Conversion[b]/%	Yield[b]/%

***a* Procedure: substrate 50 μmol, PC (2 mol%), amine (5 equiv.), HCOOH (2.5 equiv.) in 4 mL aqueous solution was degassed by nitrogen, and irradiated by blue light (12 W, λ_{max} = 462 nm) at ambient temperature for 10 h. *b* Determined by ^{1}H NMR spectroscopy by adding internal standard of 5,5′-dimethyl-2,2′-bipyridine. *c* Irradiated for 4 h. *d* Absence of HCOOH. *e* Presence of air (no degassing). *f* Absence of amine. *g* Absence of light.**

1	**1b (A1)**	DIPEA	97	67
2	**4^{Me}b (A1)**	DIPEA	97	64
3	**6^{H}b (A1)**	DIPEA	96	64
4	**4^{Me}b (A2)**	DIPEA	98	54
5	**1b (A2)**	DIPEA	99	<15
6	**6^{H}b (A2)**	DIPEA	84	51
7[c]	**4^{Me}b (A3)**	TEA	97	79
8	**4^{Me}b (A1)**	TEA	91	64
9	**4^{Me}b (A1)**	TMEDA	52	32
10	**4^{Me}b (A1)**	DBU	96	72
11[d]	**4^{Me}b (A1)**	DIPEA	85	54
12[e]	**4^{Me}b (A1)**	DIPEA	62	28
13[f]	**4^{Me}b (A1)**	—	0	0
14[g]	**4^{Me}b (A1)**	DIPEA	0	0
15	$Ru(bpy)_3Cl_2$ **(A1)**	DIPEA	0	0
16	*fac*-$Ir(ppy)_3$ **(A1)**	DIPEA	27	16

The radical cyclization of the alkyl bromides, **B1** and **B2**, catalysed by **1b**, **4^{Me}b** or **6^{H}b** proceeded smoothly, with reasonable to excellent substrate conversion and product yields. The yield of cyclization of **B1** was improved to up to 90% (entry 5, Table 3) when formic acid was added and **6^{H}b** was used as a photo-catalyst (PC). When 2,2,6,6-tetramethyl-1-piperidinyloxy (TEMPO) was added, the reaction was totally inhibited, indicating the involvement of a radical intermediate in the reaction (entry 12).

The photo-catalytic reaction could be initiated from the oxidative quenching of **4^{Me}b*** with aryl/alkyl halides. This is because the excited state reduction potential of **4^{Me}b*** ($E(Ir^{IV/III*})_{pc}$ = −1.51 V (*vs.* SCE), Table S3, Fig. S9†) can allow a direct one electron reduction of aryl/alkyl halides by **4^{Me}b***, leading to carbon–halogen (σ*(C–X)) bond cleavage to give alkyl radical in the e of sp^{3}carbon or radical anion intermediate for sp^2 carbon.[86] Subsequent reactions of the alkyl radical or radical anion intermediate with C(sp^2)–H bond lead to C–C bond formation. An aminium radical cation generated from the oxidation of amine by [**4^{Me}b**]$^{+}$[83,85] could serve as an electron donor to complete the reductive process. In the e of [$(ppy)_2Ir(dtbbpy)$]PF_6,[83] its triplet excited state ([$(ppy)_2Ir(dtbbpy)]^{+*}$) reacts with DIPEA *via* a reductive quenching mechanism to generate Ir(II) species ($E(Ir^{III/II})_{pc}$ = −1.51 V (*vs.* SCE)[1], Fig. S9, ESI†), which initiates the subsequent reducing catalytic reaction (Table 5).

Table 5: Visible-light-induced C–C bond formation of alkyl bromide

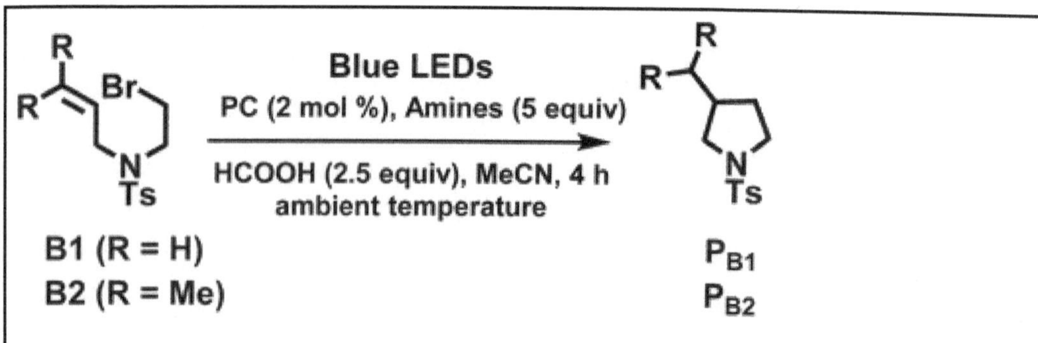

Entry[a]	**PC**	**Amines**	**Conversion[b]/%**	**Yield[b]/%**
1	$Ru(bpy)_3Cl_2$	DIPEA	19	5

***a* Entry 1–12: R = H (substrate B1); procedure: substrate 50 μmol, PC (2 mol%), amine (5 equiv.) and HCOOH (2.5 equiv.) in 4 mL MeCN solution was degassed by nitrogen, and irradiated by blue light (12 W, λ_{max} = 462 nm) at 25 °C. *b*Determined by ^{1}H NMR spectroscopy by adding an internal standard of 5,5′-dimethyl-2,2′-bipyridine. *c* Absence of HCOOH (20 equiv.). *d* Absence of light. *e* Presence of TEMPO (radical trapping reagent, 2 equiv.). *f* Entry 13: R = Me (substrate B2).**

2	*fac*-Ir(ppy)$_3$	DIPEA	90	72
3	**1b**	DIPEA	99	85
4	**4^{Me}b**	DIPEA	99	76
5	**6^{H}b**	DIPEA	99	90
6	**4^{Me}b**	TEA	99	72
7	**4^{Me}b**	TMEDA	84	55
8	**4^{Me}b**	DBU	99	64
9	**4^{Me}b**	—	0	0
10^{c}	**4^{Me}b**	DIPEA	27	15
11^{d}	**4^{Me}b**	DIPEA	0	0
12^{e}	**4^{Me}b**	DIPEA	0	0
13^{f}	**4^{Me}b**	DIPEA	99	59

Interestingly, modifying the *N*-substituent of bis-NHC ligand from an alkyl group to a glucose moiety renders photocatalyst **6^{H}c** soluble in aqueous media. At the outset, we examined the **6^{H}c**-catalysed reductive cyclization of **B1** in a mixture of H_2O/MeOH (3 : 1) with ascorbic acid as reductant, but both the substrate conversion and product yield were low. The use of diisopropylethylamine (DIPEA) as reductant instead of ascorbic acid and addition of tetrabutylammonium chloride improved the conversion and yield to 98% and 49%, respectively. Increasing the vol% of methanol in aqueous solution to 75% leads to 99% conversion and 87% product yield. To the best of our knowledge, this is the first example of visible-light-driven radical cyclization for synthesis of pyrrolidine in aqueous media (Table 6).

Table 6: Visible-light-induced radical cyclization in aqueous solution

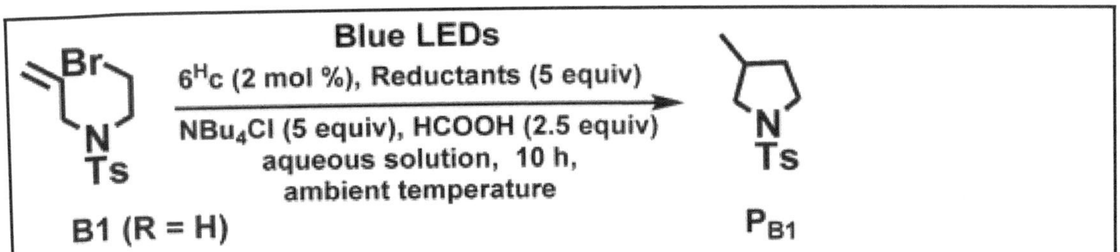

Entry^{a}	Solvents (H_2O/MeOH)^{b}	Reductant	Conversion^{c}/%	Yield^{c}/%
1	3/1	Ascorbic acid	23	10
2^{d}	3/1	Ascorbic acid	20	14
3	3/1	DIPEA	98	49
4^{d,e}	3/1	DIPEA	90	21
5^{d}	3/1	DIPEA	79	31
6	1/1	DIPEA	99	64
7	1/3	DIPEA	99	87
8	0/1	DIPEA	99	66

a **Procedure: substrate 50 μmol, 6^{H}c (2 mol%), reductant (5 equiv.), ^{n}Bu$_4$NCl (5 equiv.) in 4 mL aqueous solution was degassed by nitrogen, and irradiated by blue light (12 W, λ_{max} = 462 nm) at 25 °C. *b* Solvent system used is water/methanol in volume ratio (v/v). *c* Determined by ^{1}H NMR spectroscopy by adding internal standard of 5,5′-dimethyl-2,2′-bipyridine. *d* Absence of HCOOH. *e* Absence of ^{n}Bu$_4$NCl.**

Visible-light-driven CO_2 reduction

There is a surge of interest in developing photo-catalytic CO_2 reduction using earth abundant metal complexes as catalysts and luminescent cyclometalated Ir(III) complexes particularly *fac*-Ir(ppy)$_3$ as PC. Nevertheless, several recent reports drew attention to the instability of *fac*-Ir(ppy)$_3$, which is a challenge for achieving efficient light-driven CO_2 reduction in the long run.[55,57,63] In view of the good photo-stability of **4^{Me}b**, we investigated the visible-light-driven CO_2 reduction by utilizing **4^{Me}b** in combination with the recently reported [Co(TPA)Cl]Cl complex.[57]

A CO_2-saturated MeCN/triethylamine solution (4 : 1, v/v; 4 mL) containing catalytic amounts of **4^{Me}b** and [Co(TPA)Cl]Cl was irradiated by blue LEDs (12 W) for a specified time period, and the evolved gases were separated and identified by GC-TCD equipped with a molecular sieve column. The volume of H_2 and CO gases were calculated by using CH_4 as the internal standard.

As shown in Fig. 7a and b, the amount of CO and H_2 generated from the reaction mixture is found to show strong dependence on the concentrations of **4^{Me}b** and Co(II) catalysts. Particularly, the highest TON value of over 5000 can be accomplished at 0.5 μM of Co(II), while no generation of gases is found at low concentration of Co(II) (50 nM; Fig. S11, ESI†). Similarly, only negligible amount of product gases can be detected after irradiation for 24 h when [Ir] (0.005 mM) is lower than [Co] (0.05 mM). With the representative system containing **4^{Me}b** (0.5 mM) and Co(II) (0.005 mM), (Fig. 7c) the visible-light-driven CO_2 reduction in three parallel reaction runs gives TON (CO) > 2400 (conversion of about 18 mL of CO_2 into 1 mL of CO) with excellent selectivity in generating CO over H_2 in the gaseous phase (>95%) after reaction for 72 h. This result is better than for the system utilizing *fac*-Ir(ppy)$_3$ as PC, which reveals only TON (CO) > 900 and selectivity (CO) of 85% under similar reaction conditions.[57]

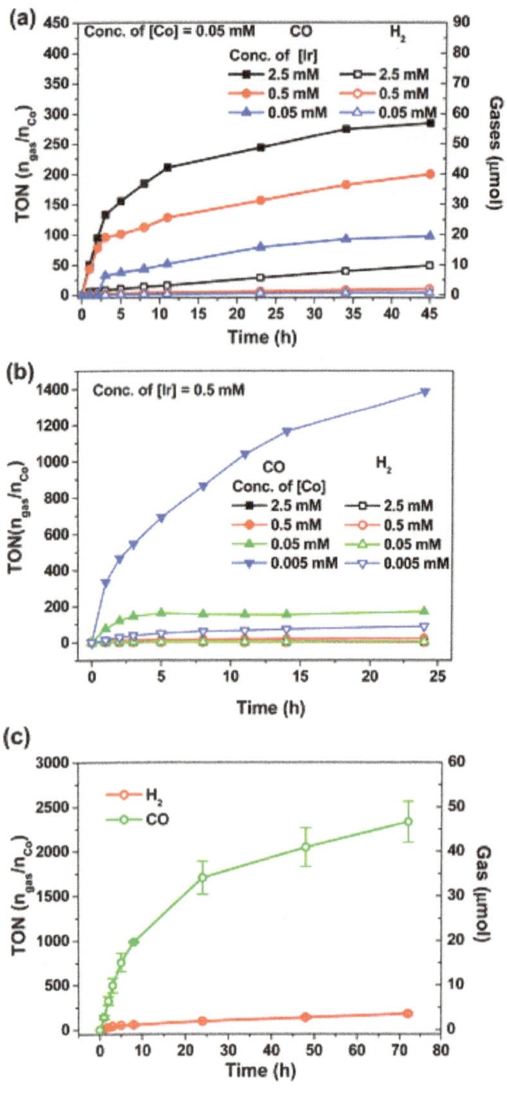

Figure 7: TON value and amounts of gases (CO and H_2) generation from CO_2 in a CO_2-saturated MeCN/TEA (4/1, v/v, 4 mL in total) solution as a function of irradiation time: concentration dependence of (a) PC **4^{Me}b** and (b) catalyst [Co(TPA)Cl]Cl; (c) a solution containing 0.005 mM [Co(TPA)Cl]Cl, 0.5 mM **4^{Me}b** was irradiated using blue LEDs (12 W) based on the averaged results from three parallel reaction runs.

In order to confirm the roles of catalysts in photo-driven CO_2 reduction, several control experiments were performed (Table S5, ESI†). Firstly, in the absence of Co(II) complex, no CO gas was observed in a CO_2-saturated MeCN/TEA (4 : 1, v/v; 4 mL) solution after irradiation for 24 h. On the other hand, in the absence of light, sacrificial amine or PC, the reaction mixture only gives negligible amounts of CO. To ascertain the catalytic role of the Co(II) complex in the reaction, mercury was added to the reaction mixture in order to exclude the possibility of CO generation from heterogeneous Co nanoparticles. To specify, 29.2 μmol of CO (0.714 mL, TON 146) could be generated from the solution with [Ir] (0.5 mM), [Co] (0.05 mM) and elementary Hg (1 mL) after irradiation for 18 h, and this result is comparable with that using the solution mixture without Hg (28.4 μmol of CO formed, TON 142).

Cellular imaging and cell viability assay

We have a long-standing interest in luminescent transition metal complexes, particularly those supported by NHC ligands that display anti-cancer activities.[9,16,87–90] The luminescence would allow direct monitoring of cellular uptake and tracking of cellular location in cancer cells by fluorescence microscopy, and such properties of transition metal NHC complexes have been demonstrated to be useful in the elucidation of the anti-cancer mechanism of action. In this work, in view of their favourable photophysical properties, the *in vitro* cytotoxicity of the bis-NHC Ir(III) complexes against HeLa cells was investigated by MTT assay. As shown in Table S6 (ESI†), the Ir(III) complexes display potent cytotoxicity, with IC_{50} values ranging from 0.5 to 56.1 μM depending on the lipophilicity and structures of the Ir(III) complexes. Those complexes with butyl groups on the bis-NHC ligand are more cytotoxic than those with glucose units.

Since complexes **1b**, **4^{Me}b**, **6^{H}a**, **6^{H}b** and **6^{H}c** demonstrate outstanding photophysical properties *i.e.* high quantum yield with long-lived electronic excited states, cellular imaging of these complexes in HeLa cells were performed. After treatment of human cervical cancer cells (HeLa) with the Ir(III) complexes for 15 min, strong green/yellow luminescence was observed in the cytoplasm (but not nucleus) of cancer cells, as revealed by fluorescence microscopy (Fig. 8). Co-localization analyses indicate that the emission of these complexes are mainly localized in the endoplasmic reticulum (ER), which is stained by red-emissive ER-specific ER-Tracker™; a high Pearson›s correlation coefficient (*R*) between the emissions of complexes and ER-Tracker™ is found (for example, **6^{H}b** shows a high*R* value of 0.80). Consistently, these complexes do not accumulate in other organelles such as lysosome or mitochondria, as shown by the poor co-localization of the emissions of the complexes with the emission of Lyso-Tracker® and Mito-Tracker® respectively (Fig. S13, ESI†). Noticeably, **1b**$^{+}$ (both the counter anions of triflate and chloride) was found in our laboratory to be specifically localized in the ER of cancer cells (Fig. S14†), but not, as reported elsewhere, in mitochondria.[73] With the specific accumulation of the complexes in ER, the cytotoxic properties of the complexes may originate from the induction of ER stress[88] and immunogenic cell death.[91]

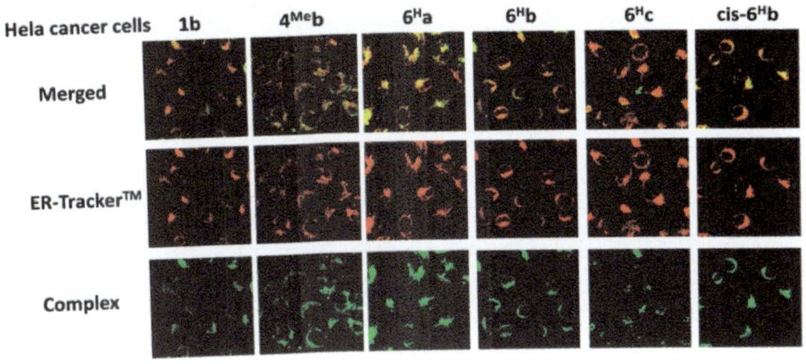

Figure 8: Fluorescence microscopy images of HeLa cancer cells incubated with the Ir(III) NHC complexes: merged (top), ER-tracker™ (middle) and complexes (bottom). Complexes were excited at 340 nm using an emission filter of 510 nm. ER-tracker™ was excited at 546 nm using an emission filter of >580 nm.

DISCUSSION

This work reveals the potential impact and usefulness of bis-NHC ligands for the development of robust metal photo-catalysts. Compared with the well-known complexes *fac*-Ir(ppy)$_3$, this series of bis-NHC Ir(III) complexes exhibits: (i) outstanding photo-stability under visible light irradiation; (ii) good photo-catalytic performance for several photo-electrochemically reactions; and (iii) long-lived emissive excited states.

Stability of NHC metal complexes

The most important feature of metal–ligand bonding between transition metals and NHCs can be rationalized as the σ-coordination (sp^2-hybridized lone pair electron) from NHCs. The contribution of both π-back-bonding into carbene p-orbital and π-donation from the carbene p-orbital might be considered as less significant for NHC metal complexes. These behaviours are similar to the coordination characteristics of phosphines, but NHCs are in general better electron-donors than phosphines. Thus, the stronger metal–ligand interaction renders NHC–metal coordination less labile than metal–phosphine bonding and the NHC complexes are more thermally stable.[92] The distinct electronic properties and coordination chemistry of NHCs can also lead to improved catalytic activity of the metal complexes, owing to the increased catalyst stability and consequently lower rates of catalyst decomposition.[68] In the previous section, the studies on the photo-stability of **4^{Me}b** and *fac*-Ir(ppy)$_3$ have revealed the outstanding stabilization contribution from bis-NHC carbene ligands. Thus this might be one reason that explains the better performance for photo-catalytic reactions using the present bis-NHC Ir(III) complexes as PC.

Role of bis-NHC ligands in tuning emission energy and lifetimes

A comparison between the photophysical data of bis-NHC Ir(III) complexes with two notable Ir(III) complexes (*fac*-Ir(ppy)$_3$ and (ppy)$_2$Ir(acac) (**1d**)) will be helpful in understanding the role of NHC on the strong absorptivities, high emission efficiencies and long lifetimes of the present bis-NHC Ir(III) complexes. On the basis of DFT calculations on **1a** as the representative example (Fig. S4, ESI†), the lowest-energy absorption bands of **1a**, **1d** and *fac*-Ir(ppy)$_3$ are ascribed to the HOMO → LUMO transitions, the energy of which is sensitive to the charge of ancillary ligands. The transition energy of *fac*-Ir(ppy)$_3$ and **1d** with the negatively charge ancillary ligands (C^N and acac) are quite close with wavelengths of 416 and 414 nm respectively. However, for **1a** with the neutral ancillary ligand (bis-NHC), the HOMO → LUMO transition is markedly hypsochromically shifted to 369 nm, which is consistent with the experimental observations. Fig. S10 (ESI†) shows the surfaces and the energies of HOMO and LUMO of **1a**, *fac*-Ir(ppy)$_3$ and **1d**. On the whole, the HOMO contains comparable components of iridium and C^N ligand, and the LUMO is mainly localized on the C^N ligand. Thus, the transition can be assigned as an admixture of metal-to-ligand charge transfer (MLCT) and ligand centered (LC) π–π* transition, which is in accordance with the assignments reported in the literature.

The different amounts of metal character in the frontier molecular orbitals of **1a**, **1d** and Ir(ppy)$_3$, as deduced by TD-DFT calculations, can account for the photophysical properties and long emission lifetimes of the Ir(III) NHC complexes. As shown in Fig. S10 (ESI†), *fac*-Ir(ppy)$_3$ and **1d** have similar energy levels in HOMO and LUMO (around −4.5 eV and −2.0 eV). By simple substitution of the negatively charged ancillary ligands (C^N and acac) in Ir(ppy)$_3$ and **1d** by neutral NHC ligand, **1a** is found to show lower energy levels of HOMO and LUMO (about −5.2 eV and −2.4 eV). This can be explained by the less electron-donating effect of the neutral NHC ligand compared to the negatively charged auxiliary ligands, as well as a certain contribution of the stabilization of dπ(Ir) by the π-acceptor orbitals of NHC ligand of **1a**.[68] As a result, the energy level of the HOMO of **1a** is 700 mV lower than for **1d**, which coincides with the experimental observation that the oxidation potential of **1a** is 410 mV more positive than that of **1d**. On the other hand, the transition energy of HOMO → LUMO in **1a** is estimated to be 0.3 eV larger than in Ir(ppy)$_3$ and **1d**, which may account for the blue shift of low-energy absorption of **1a** (369 nm *vs.* 410 nm), as compared to that of **1d** and Ir(ppy)$_3$, in UV-Vis absorption spectra. Notably, TD-DFT calculation reveals that frontier molecular orbitals of **1a** show a smaller metal character than those of **1d** (Ir character in the HOMO of **1a** and **1d** are found to be ~30 and ~40% respectively), and hence the triplet excited states of **1a** are likely to have smaller metal character. This would probably slow down the spin–orbit coupling, resulting in slower radiative and non-radiative $T_1 \rightarrow S_0$ decay for **1a**. As a result, **1a**shows a longer emission lifetime than **1d**.

Triplet excited states for photo-catalytic reactivity

In addition to the photo-stability and strong absorptivity in the visible light region, the long-lived triplet excited states, especially their propensity to lose an electron, are to a large extent believed to be crucial for the photo-catalytic properties of the complexes. The long triplet excited state lifetime of **4^{Me}b** is beneficial for bi-molecular photochemical reactions allowing the electron-transfer pathways to have sufficient time to take place prior to the decay of the excited state to the ground state.

Considering the fact that the excited state of [Ru(bpy)$_3$]$^{2+}$ (E(Ru$^{II*/I}$) = 0.77 V *vs.* SCE) has a more powerful oxidative potential than for **4^Me^b** ($E(Ir^{III*/II})$ = 0.51 V *vs.* SCE, Table S3†) and no radical cyclization products are obtained when using $[Ru(bpy)_3]^{2+}$, the photo-catalytic route *via* reductive quenching cycle is not feasible. The photo-catalytic reaction would possibly be initialized *via* an oxidative quenching cycle of excited states of the photo-catalyst. By carefully examining the transient-absorption spectra of triplet state of **4^Me^b** in MeCN, newly generated long-lived species were observed by using higher energy laser beams (Fig. S8b–g†). These long-lived species were found to be increased in the presence of substrate **A1**, and could be quenched by DIPEA. This long-lived species might be Ir(IV), which is generated by single electron transfer from [**4^Me^b**]* to **A1**. The calculated excited-state reduction potential of **4^Me^b** ($E(Ir^{IV/III*})$ = −1.51 V *vs.* SCE, Table S3†) reveals that **4^Me^b** is not a stronger photo-reductant than *fac*-$Ir(ppy)_3$ ($E(Ir^{IV/III*})$ = −1.73 V *vs.* SCE).[37,93] However, the performance of **4^Me^b** in visible-light-driven radical cyclization and CO_2 reduction prove to be better than for *fac*-$Ir(ppy)_3$. For example, the conversion of substrates **A1**/**B1** to indoline/pyrrolidine by **4^Me^b** (97%/99%) are higher than those by *fac*-$Ir(ppy)_3$ (27%/90%). Therefore, there should be other reasons for the good performance of **4^Me^b** in photo-catalysis. Plausible reasons could be: (i) the generated Ir(IV) $[\mathbf{4^{Me}b}]^+$ ($E(Ir^{IV/III})$ = 0.96 V *vs.* SCE, Table S3†) are more easily reduced by amines than [*fac*-$Ir(ppy)_3]^+$ ($E(Ir^{IV/III})$ = 0.77 V *vs.* SCE); and (ii) the higher photo-stability of **4^Me^b** compared with that of *fac*-$Ir(ppy)_3$.

Consequently, the excellent photo-stability, strong absorptivity, long lifetimes and the photo-ionization behaviours of our bis-NHC Ir(III) complexes possibly enable the photolysis of long-lived excited states of Ir(III)* to Ir(IV) more easily and thus promote the radical cyclization in a catalytic cycle.

Diversities of biological activities of Ir(III) complexes with different *N*-substituents on NHC ligands

Although *N*-substituents of NHC ligands are found to show negligible effects on the luminescent properties of the bis-NHC Ir(III) complexes, they have a determinant role on the physical properties and hence the anti-cancer activities of the complexes. For example, complexes with *N*-butyl substituents on the bis-NHC ligands generally display lower IC_{50} values than those with *N*-methyl substituents (Table S6 in ESI†). This is probably attributed to the increase in lipophilicity of the complexes, resulting in better permeability through cellular membrane and higher accumulation of the complexes in cancer cells. The fast cellular uptake of the complexes is supported by the strong luminescence observed in HeLa cells after incubation of the cells with the complexes for 15 min (Fig. 8and S12 in ESI†). On the other hand, the glucose-functionalized NHC ligands give the complexes good aqueous solubility and a more hydrophilic nature, leading to likely slower cellular uptake as well as reduced cytotoxicity toward cancer cells. It is noteworthy that the complexes accumulate in cellular ER as revealed by fluorescence microscopy images of the co-staining experiments. This can consequently induce ER stress[88] and immunogenic cell death, which probably accounts for the high cytotoxicity of the complexes. Further manipulations of the functionalities on NHC ligands of the Ir(III) complexes may realize the development of a new class of diagnostic and/or therapeutic agents.

CONCLUSIONS

A new series of cyclometalated Ir(III) complexes bearing bis-NHC ligands has been demonstrated as strongly luminescent materials and promising photo-catalysts for visible-light-driven radical cyclization and CO_2 reduction and as biological theranostic agents. Owing to the high stability of the Ir–C^{NHC} bond, these bis-NHC Ir(III) complexes show excellent photo-stability compared with widely-used PC, such as *fac*-$Ir(ppy)_3$ and $[Ru(bpy)_3]^{2+}$. With long-lived triplet excited states and rich photoredox properties, **4^Me^c** can undergo photo-ionization in aqueous media as supported by transient absorption experiments, while **4^Me^b** is found to be a more effective catalyst for radical cyclization and CO_2 reduction than *fac*-$Ir(ppy)_3$ under visible-light irradiation. Interestingly, water-soluble **6^H^c**, which contains a glucose moiety on the bis-NHC ligands, has been demonstrated as the first PC for the synthesis of pyrrolidine in aqueous media. In addition, through modulations of the chemical structures of the cyclometalated ligands and/or *N*-substituents on the bis-NHC ligands, the Ir(III) complexes have been found to show different luminescent properties as well as anti-cancer activities, indicating the potential of the complexes as theranostic agents.

AUTHOR CONTRIBUTION

Chen Yang, Faisal Mehmood, Tsz-Lung Lam, Yuan Wu, Chi-Shun Yeung, Xiangguo Guan, Kai Li, Clive Yik-Sham Chung, Cong-Ying Zhou and Taotao Zou carried out all the experiments and performed data analysis. Chen Yang, Chi-Ming Che and Sharon Lai-Fung Chan designed the experiments, analysed the data and wrote the manuscript. All authors reviewed the manuscript.

ACKNOWLEDGEMENTS

This work was supported by the National Key Basic Re Program of China (No. 2013CB834802), the University Grants Committee of the HKSAR Area of Excellence Scheme (AoE/P-03/08), and the -Croucher Foundation Funding Scheme for Joint Laboratories. C. Yang acknowledges the support of the postgraduate studentships from the University of Hong Kong and Miss Yingshuo Zhang for the synthesis of part of the substrates. S. L.-F. Chan thanks the Hong Kong Re Grants Council (PolyU 253038/15P) and the National Science Foundation of China (21401157) for financial support. Cellular imaging data were acquired using equipment maintained by the University of Hong Kong Li Ka Shing Faculty of Medicine Faculty Core Facility.

NOTES AND REFERENCES

1. M. S. Lowry, J. I. Goldsmith, J. D. Slinker, R. Rohl, J. Robert, A. Pascal, G. G. Malliaras and S. Bernhard, *Chem. Mater.*, 2005, 17, 5712–5719.
2. C. Yang, S.-L. Lai, S. L.-F. Chan, K.-H. Low, G. Cheng, K.-T. Yeung, C.-C. Kwok and C.-M. Che, *Chem.–Asian J.*, 2014, 9, 3572–3585.
3. A. B. Tamayo, B. D. Alleyne, P. I. Djurovich, S. Lamansky, I. Tsyba, N. N. Ho, R. Bau and M. E.

Thompson, *J. Am. Chem. Soc.*, 2003, 125, 7377–7387.

4. M. A. Baldo, D. F. O'Brien, Y. You, A. Shoustikov, S. Sibley, M. E. Thompson and S. R. Forrest, *Nature*, 1998, 395, 151–154.
5. M. A. Baldo, M. E. Thompson and S. R. Forrest, *Nature*, 2000, 403, 750–753.
6. S. C. F. Kui, P. K. Chow, G. Cheng, C.-C. Kwok, C. L. Kwong, K.-H. Low and C.-M. Che, *Chem. Commun.*, 2013, 49, 1497–1499 .
7. K. Li, G. Cheng, C. Ma, X. Guan, W.-M. Kwok, Y. Chen, W. Lu and C.-M. Che, *Chem. Sci.*, 2013, 4, 2630–2644 .
8. S. C. F. Kui, P. K. Chow, G. S. M. Tong, S.-L. Lai, G. Cheng, C.-C. Kwok, K.-H. Low, M. Y. Ko and C.-M. Che, *Chem.–Eur. J.*, 2013, 19, 69–73.
9. K. Li, T. Zou, Y. Chen, X. Guan and C. M. Che, *Chem.–Eur. J.*, 2015, 21, 7441–7453.
10. E. S.-H. Lam, D. P.-K. Tsang, W. H. Lam, A. Y.-Y. Tam, M.-Y. Chan, W.-T. Wong and V. W.-W. Yam, *Chem.–Eur. J.*, 2013, 19, 6385–6397.
11. Z. He, W.-Y. Wong, X. Yu, H.-S. Kwok and Z. Lin, *Inorg. Chem.*, 2006, 45, 10922–10937.
12. S.-L. Lai, L. Wang, C. Yang, M.-Y. Chan, X. Guan, C.-C. Kwok and C.-M. Che, *Adv. Funct. Mater.*, 2014, 24, 4655–4665.
13. W. X. Ni, M. Li, J. Zheng, S. Z. Zhan, Y. M. Qiu, S. W. Ng and D. Li, *Angew. Chem., Int. Ed.*, 2013, 52, 13472–13476.
14. M.-C. Tang, D. P.-K. Tsang, M. M.-Y. Chan, K. M.-C. Wong and V. W.-W. Yam, *Angew. Chem., Int. Ed.*, 2013, 52, 446–449.
15. W.-P. To, K. T. Chan, G. S. M. Tong, C. Ma, W.-M. Kwok, X. Guan, K.-H. Low and C.-M. Che, *Angew. Chem., Int. Ed.*, 2013, 52, 6648–6652.
16. T. Zou, C. T. Lum, S. S.-Y. Chui and C.-M. Che, *Angew. Chem., Int. Ed.*, 2013, 52, 2930–2933.
17. F.-F. Hung, W.-P. To, J.-J. Zhang, C. Ma, W.-Y. Wong and C.-M. Che, *Chem.–Eur. J.*, 2014, 20, 8604–8614.
18. X.-S. Xiao, W.-L. Kwong, X. Guan, C. Yang, W. Lu and C.-M. Che, *Chem.–Eur. J.*, 2013, 19, 9457–9462.
19. R. Visbal, J. M. Lopez-de-Luzuriaga, A. Laguna and M. C. Gimeno, *Dalton Trans.*, 2014, 43, 328–334 .
20. C.-M. Che, C.-C. Kwok, S. C. F. Kui, S.-L. Lai and K.-H. Low, in *Compr. Inorg. Chem. II*, ed. J. R. Poeppelmeier, Elsevier, Amsterdam, 2nd edn, 2013, pp. 607–655.
21. H. Yersin and W. J. Finkenzeller, in *Highly Efficient OLEDs with Phosphorescent Materials*, Wiley-VCH Verlag GmbH & Co. KGaA, 2008, pp. 1–97.
22. Y. Ma, H. Zhang, J. Shen and C. Che, *Synth. Met.*, 1998, 94, 245–248.
23. K. K.-W. Lo, A. W.-T. Choi and W. H.-T. Law, *Dalton Trans.*, 2012, 41, 6021–6047 .
24. F. L. Thorp-Greenwood, R. G. Balasingham and M. P. Coogan, *J. Organomet. Chem.*, 2012, 714, 12–21.
25. J. N. Demas and B. A. DeGraff, *Anal. Chem.*, 1991, 63, 829A–837A.
26. V. W.-W. Yam and K. M.-C. Wong, *Chem. Commun.*, 2011, 47, 11579–11592 .
27. T. P. Yoon, M. A. Ischay and J. Du, *Nat. Chem.*, 2010, 2, 527–532.
28. S. Lamansky, P. Djurovich, D. Murphy, F. Abdel-Razzaq, R. Kwong, I. Tsyba, M. Bortz, B. Mui, R. Bau and M. E. Thompson, *Inorg. Chem.*, 2001, 40, 1704–1711.
29. J. Li, P. I. Djurovich, B. D. Alleyne, M. Yousufuddin, N. N. Ho, C. J. Thomas, J. C. Peters, R. Bau and M. E. Thompson, *Inorg. Chem.*, 2005, 44, 1713–1727.
30. A. F. Rausch, M. E. Thompson and H. Yersin, *J. Phys. Chem. A*, 2009, 113, 5927–5932.
31. S. Lamansky, P. Djurovich, D. Murphy, F. Abdel-Razzaq, H.-E. Lee, C. Adachi, P. E. Burrows, S. R. Forrest and M. E. Thompson, *J. Am. Chem. Soc.*, 2001, 123, 4304–4312.
32. K. K.-W. Lo, B. T.-N. Chan, H.-W. Liu, K. Y. Zhang, S. P.-Y. Li and T. S.-M. Tang, *Chem. Commun.*, 2013, 49, 4271–4273 .
33. S. P.-Y. Li, T. S.-M. Tang, K. S.-M. Yiu and K. K.-W. Lo, *Chem.–Eur. J.*, 2012, 18, 13342–13354.
34. W. H.-T. Law, L. C.-C. Lee, M.-W. Louie, H.-W. Liu, T. W.-H. Ang and K. K.-W. Lo, *Inorg. Chem.*, 2013, 52, 13029–13041.
35. K. K.-W. Lo and K. K.-S. Tso, *Inorg. Chem. Front.*, 2015, 2, 510–524 .
36. K. K.-W. Lo and K. Y. Zhang, *Adv.*, 2012, 2, 12069–12083 .
37. C. K. Prier, D. A. Rankic and D. W. C. MacMillan, *Chem. Rev.*, 2013, 113, 5322–5363.
38. D. A. Nicewicz and D. W. C. MacMillan, *Science*, 2008, 322, 77–80.
39. Z. Lu and T. P. Yoon, *Angew. Chem., Int. Ed.*, 2012, 51, 10329–10332.
40. M. A. Ischay, M. E. Anzovino, J. Du and T. P. Yoon, *J. Am. Chem. Soc.*, 2008, 130, 12886–12887.
41. D. M. Schultz and T. P. Yoon, *Science*, 2014, 343, 1239176.
42. J. W. Tucker, J. D. Nguyen, J. M. R. Narayanam, S. W. Krabbe and C. R. J. Stephenson, *Chem. Commun.*, 2010, 46, 4985–4987 .
43. J. M. R. Narayanam and C. R. J. Stephenson, *Chem. Soc. Rev.*, 2011, 40, 102–113 .
44. J. M. R. Narayanam, J. W. Tucker and C. R. J. Stephenson, *J. Am. Chem. Soc.*, 2009, 131, 8756–8757.
45. J. D. Nguyen, E. M. D'Amato, J. M. R. Narayanam and C. R. J. Stephenson, *Nat. Chem.*, 2012, 4, 854–859.
46. P.-K. Chow, G. Cheng, G. S. M. Tong, W.-P. To, W.-L. Kwong, K.-H. Low, C.-C. Kwok, C. Ma and C.-M. Che, *Angew. Chem., Int. Ed.*, 2015, 54, 2084–2089.
47. J.-J. Zhong, Q.-Y. Meng, G.-X. Wang, Q. Liu, B. Chen, K. Feng, C.-H. Tung and L.-Z. Wu, *Chem.–Eur. J.*, 2013, 19, 6443–6450.
48. B. Kumar, M. Llorente, J. Froehlich, T. Dang, A. Sathrum and C. P. Kubiak, *Annu. Rev. Phys. Chem.*, 2012, 63, 541–569.

49. C. D. Windle and R. N. Perutz, *Coord. Chem. Rev.*, 2012, 256, 2562–2570.

50. C. Costentin, M. Robert and J.-M. Savéant, *Chem. Soc. Rev.*, 2013, 42, 2423–2436 .

51. D. Behar, A. T. Dhanasekaran, P. Neta, C. M. Hosten, A. D. Ejeh, P. Hambright and E. Fujita, *J. Phys. Chem. A*, 1998, 102, 2870–2877.

52. R. Ziessel, J. Hawecker and J.-M. Lehn, *Helv. Chim. Acta*, 1986, 69, 1065–1084.

53. H. Hori, F. P. A. Johnson, K. Koike, O. Ishitani and T. Ibusuki, *J. Photochem. Photobiol., A*, 1996, 96, 171–174.

54. S. Sato, T. Morikawa, T. Kajino and O. Ishitani, *Angew. Chem., Int. Ed.*, 2013, 52, 988–992.

55. V. S. Thoi, N. Kornienko, C. G. Margarit, P. Yang and C. J. Chang, *J. Am. Chem. Soc.*, 2013, 135, 14413–14424.

56. J. Bonin, M. Robert and M. Routier, *J. Am. Chem. Soc.*, 2014, 136, 16768–16771.

57. S. L.-F. Chan, T. L. Lam, C. Yang, S.-C. Yan and N. M. Cheng, *Chem. Commun.*, 2015, 51, 7799–7801 .

58. L. Chen, Z. Guo, X.-G. Wei, C. Gallenkamp, J. Bonin, E. Anxolabéhère-Mallart, K.-C. Lau, T.-C. Lau and M. Robert, *J. Am. Chem. Soc.*, 2015, 137, 10918–10921.

59. R. S. Khnayzer, V. S. Thoi, M. Nippe, A. E. King, J. W. Jurss, K. A. El Roz, J. R. Long, C. J. Chang and F. N. tellano, *Energy Environ. Sci.*, 2014, 7, 1477–1488 .

60. C.-F. Leung, S.-M. Ng, C.-C. Ko, W.-L. Man, J. Wu, L. Chen and T.-C. Lau, *Energy Environ. Sci.*, 2012, 5, 7903–7907 .

61. A. B. Tossi and H. Görner, *J. Photochem. Photobiol., B*, 1993, 17, 115–125.

62. Y.-J. Yuan, Z.-T. Yu, X.-J. Liu, J.-G. Cai, Z.-J. Guan and Z.-G. Zou, *Sci. Rep.*, 2014, 4, 4045.

63. S. Schmidbauer, A. Hohenleutner and B. König, *Beilstein J. Org. Chem.*, 2013, 9, 2088–2096.

64. W. A. Herrmann, *Angew. Chem., Int. Ed.*, 2002, 41, 1290–1309.

65. A. Azua, S. Sanz and E. Peris, *Chem.–Eur. J.*, 2011, 17, 3963–3967.

66. Y. Unger, A. Zeller, M. A. Taige and T. Strassner, *Dalton Trans.*, 2009, 4786–4794 .

67. A. G. Tennyson, E. L. Rosen, M. S. Collins, V. M. Lynch and C. W. Bielawski, *Inorg. Chem.*, 2009, 48, 6924–6933.

68. M. N. Hopkinson, C. Richter, M. Schedler and F. Glorius, *Nature*, 2014, 510, 485–496.

69. D. Martin, M. Melaimi, M. Soleilhavoup and G. Bertrand, *Organometallics*, 2011, 30, 5304–5313.

70. B. Rao, H. Tang, X. Zeng, L. Liu, M. Melaimi and G. Bertrand, *Angew. Chem., Int. Ed.*, 2015, 54, 14915–14919.

71. L.-A. Schaper, S. J. Hock, W. A. Herrmann and F. E. Kühn, *Angew. Chem., Int. Ed.*, 2013, 52, 270–289.

72. C.-H. Yang, J. Beltran, V. Lemaur, J. Cornil, D. Hartmann, W. Sarfert, R. Fröhlich, C. Bizzarri and L. De Cola, *Inorg. Chem.*, 2010, 49, 9891–9901.

73. Y. Li, C.-P. Tan, W. Zhang, L. He, L.-N. Ji and Z.-W. Mao, *Biomaterials*, 2015, 39, 95–104.

74. J. Jin, H.-W. Shin, J. H. Park, J. H. Park, E. Kim, T. K. Ahn, D. H. Ryu and S. U. Son, *Organometallics*, 2013, 32, 3954–3959.

75. D. N. Kozhevnikov, V. N. Kozhevnikov, M. Z. Shafikov, A. M. Prokhorov, D. W. Bruce and J. A. Gareth Williams, *Inorg. Chem.*, 2011, 50, 3804–3815.

76. E. J. Hart and J. W. Boag, *J. Am. Chem. Soc.*, 1962, 84, 4090–4095.

77. M. H. Elkins, H. L. Williams, A. T. Shreve and D. M. Neumark, *Science*, 2013, 342, 1496–1499.

78. K. C. Cho and C. M. Che, *Chem. Phys. Lett.*, 1986, 124, 313–315.

79. R. Y. N. Gengler, D. S. Badali, D. Zhang, K. Dimos, K. Spyrou, D. Gournis and R. J. D. Miller, *Nat. Commun.*, 2013, 4, 2560.

80. A. K. El Omar, U. Schmidhammer, P. Pernot, S. Murata and M. Mostafavi, *J. Phys. Chem. A*, 2012, 116, 11989–11996.

81. C. Xia, J. Peon and B. Kohler, *J. Chem. Phys.*, 2002, 117, 8855–8866.

82. A. Juris, V. Balzani, P. Belser and A. von Zelewsky, *Helv. Chim. Acta*, 1981, 64, 2175–2182.

83. H. Kim and C. Lee, *Angew. Chem., Int. Ed.*, 2012, 51, 12303–12306.

84. M. Majek, U. Faltermeier, B. Dick, R. Pérez-Ruiz and A. Jacobivon Wangelin, *Chem.–Eur. J.*, 2015, 21, 15496–15501.

85. G. Revol, T. McCallum, M. Morin, F. Gagosz and L. Barriault, *Angew. Chem., Int. Ed.*, 2013, 52, 13342–13345.

86. M. Arun Prasad and M. V. Sangaranarayanan, *Chem. Phys. Lett.*, 2005, 414, 55–60 .

87. R. W.-Y. Sun, A. L.-F. Chow, X.-H. Li, J. J. Yan, S. S.-Y. Chui and C.-M. Che, *Chem. Sci.*, 2011, 2, 728–736 .

88. T. Zou, C.-N. Lok, Y. M. E. Fung and C.-M. Che, *Chem. Commun.*, 2013, 49, 5423–5425 .

89. T. Zou, C. T. Lum, C.-N. Lok, W.-P. To, K.-H. Low and C.-M. Che, *Angew. Chem., Int. Ed.*, 2014, 53, 5810–5814.

90. J. L.-L. Tsai, A. O.-Y. Chan and C.-M. Che, *Chem. Commun.*, 2015, 51, 8547–8550 .

91. D. Y. Q. Wong, W. W. F. Ong and W. H. Ang, *Angew. Chem., Int. Ed.*, 2015, 54, 6483–6487.

92. C. M. Crudden and D. P. Allen, *Coord. Chem. Rev.*, 2004, 248, 2247–2273.

93. L. Flamigni, A. Barbieri, C. Sabatini, B. Ventura and F. Barigelletti, *Top. Curr. Chem.*, ed. V. Balzani and S. Campagna, Springer, Berlin/Heidelberg, 2007, vol. 281, pp. 143–203.

CHAPTER 13

SYNTHESIS, SPECTRAL CHARACTERIZATION AND BIOLOGICAL ACTIVITY STUDIES OF TRANSITION METAL COMPLEXES OF SCHIFF BASE LIGAND CONTAINING INDOLE MOIETY

Fazlur Rahaman[a] & B.H.M. Mruthyunjayaswamy[b]

[a]Department of Chemistry, C.M.R. Institute of Technology, Bengaluru 560037, Karnataka, India

[b]Department of Studies and Research in Chemistry, Gulbarga University, Gulbarga 585106, Karnataka, India

ABSTRACT

A new series of transition metal complexes of Co(II), Ni(II), Cu(II), Zn(II), Cd(II) and Hg(II), with a tridentate (oxygen nitrogen nitrogen) ligand 3,5-dimethyl-*N*′-[(1*E*)-1′-pyridin-2′-ylethylidene]-1*H*-indole-2-carbohydrazide (L) have been synthesized and characterized on the basis of analytical, magnetic susceptibility, infra red, nuclear magnetic resonance, fast atom bombardment-mass, electron paramagnetic resonance spectral studies, X-ray powder diffraction and UV-visible spectral data. Based on analytical and molar conductance data, the complexes may be formulated as $[M(L)_2]\cdot(H_2O)_n$, where M Co, Cu and Ni and $[M(L)Cl_2].(H_2O)_n$, where M=Zn, Cd and Hg. The spectral studies revealed that, an octahedral geometry for Co(II), Ni(II) and Cu(II) complexes, whereas square pyramidal geometry for Zn(II), Cd(II) and Hg(II) complexes. The ligand and its Co(II) complex have been characterized by X-ray powder diffraction studies. The ligand and its complexes were tested for their antibacterial activity against *Escherichia coli* **&** *Staphyloccocus aureus* **and antifungal activity against** *Aspergillus niger* **&** *Candida albicans*.

INTRODUCTION

Indole nucleus and its derivatives occupied a unique place in chemistry of nitrogen heterocyclic compounds, because of their varied biodynamic properties [1]. The indole nucleus is present in a number of physiologically significant compounds like serotonin, tryptophan, indole-3-acetic acid, gramine, reserpine, lysergic acid diethylamide and also in important antibiotics like mitomycin and gliotoxin [1]. The pyridine derivatives like, 2-acetylpyridine, 2-benzoylpyridine- and 2-formylpyridine thiosemicarbazones and their metal complexes have shown substantial *in vitro* activity against various human tumor lines [2]. Their Cu(II) complexes possesses good activity against animal and human tumor strains [3]. The Zn(II), Cd(II) and Hg(II) complexes of the ligands 2-acetylpyridine-4-*N*-ethylthiosemicarbazone and

Citation: Fazlur Rahaman & B.H.M. Mruthyunjayaswamy, Synthesis, spectral characterization and biological activity studies of transition metal complexes of Schiff base ligand containing indole moiety, DOI: 10.1080/2164232X.2014.889580.

2-acetylpyridine-*N*-oxide-4-*N*-ethyl thiosemicarbazone have been synthesized and tested for their antifungal activity against *Aspergillus niger*. Both the ligands and their Zn(II) complexes have exhibited good antifungal activity [4]. The 2-acetylpyridine-4-*N*-methylthiosemicarbazone ligand and its Zn(II), Cd(II), Pd(II) and Pt(II) complexes were tested for their antifungal activity against pathogens *A. niger* and*Paecilomyces variotii* using the standard drug Nystatin [5]. Elena et al. [4] reported on the synthesis and varied biological studies of 2-acetylpyridine thiosemicarbazone ligands and their metal complexes. Published reports are sparse on the arylhydrazones of 2-acetylpyridine, containing heterocyclic moiety, which enhances the inhibitory action of the molecule.

As a part of our continuing efforts to synthesize and characterize transition metal chelates using Schiff base ligand [6–9]. We now report the synthesis, characterization and biological activities of transition metal complexes containing Schiff base derived from 2-acetylpyridine and 3,5-dimethylindole-2-carbohydrazide. This ligand system coordinates with the metal ion in a tridentate manner through the carbonyl oxygen, azomethine nitrogen and pyridine ring nitrogen.

EXPERIMENTAL

All the chemicals are of analytical reagent grade. Solvents were dried and distilled before use according to standard methods [10]. The precursor 3,5-dimethylindole-2-carbohydrazide was prepared by standard method [11]. The metal chlorides were used in their hydrated form.

Synthesis of the ligand

An equimolar mixture of3,5-dimethylindole-2- carbohydrazide (1.0 mmol) and 2-acetylpyridine (1.0 mmol) in ethanol (30 mL) were refluxed in the presence of catalytic amount of glacial acetic acid (1–2 drops) for about 6 h on a water bath. The reaction mixture was cooled to room temperature, the separated Schiff base was collected by filtration, washed with hot ethanol (10 mL) dried and recrystallized from 1,4 dioxan. Yield of the Ligand: 80% m.p: 258°C. Proposed molecular formula: $C_{18}H_{18}N_4O$ (FW=306): Calcd: C, 70.57%; H, 5.92%; N, 18.29 %. Found: C, 70.52%; H, 5.89%; N, 18.26%. The structure of the Schiff base (**L**) is given in Scheme 1.

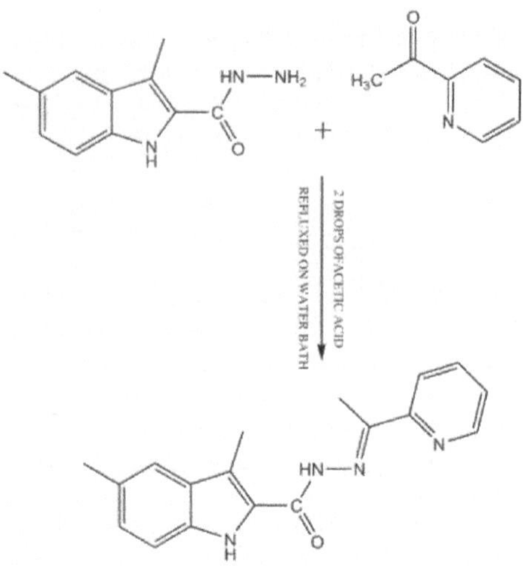

Scheme 1: Synthesis of ligand.

Synthesis of the complexes

Synthesis of cobalt(II), nickel(II) and copper(II) complexes

The metal complexes were prepared by adding stoichiometric quantities to a hot methanolic solution (30 mL) containing 1.0 mmol of ligand and methanolic solution (20 mL) of a metal chlorides (hydrated) (1.0 mmol). The mixtures were refluxed on a water bath for 6 h. The resulting solid complexes were collected by filtration and washed with 4–5 mL of hot methanol and dried in a vacuum over anhydrous calcium chloride in a desiccator (yield 65–80%).

Synthesis of zinc(II), cadmium(II) and mercury(II) complexes

The metal complexes were prepared by adding stoichiometric quantities to a hot methanolic solution (30 mL) containing 1.0 mmol of ligand and methanolic solution (20 mL) of a metal chlorides (hydrated) (1.0 mmol). The mixtures were refluxed on a water bath for 8 h. The resulting solid complexes were collected by filtration and washed with 4–5 mL of hot methanol and dried in a vacuum over anhydrous calcium chloride in a desiccator (yield 65–80%).

Physical measurements

Infrared (IR) spectra of the synthesized compounds were recorded as KBr pellets on PERKIN-ELMER Spectrum One FT-IR spectrometer. ^{1}H nuclear magnetic resonance (NMR) spectra were recorded on a Bruker Avance 400 MHz spectrometer. UV-VIS spectra of the complexes were recorded on Elico-SL 164 double beam spectrometer in the range 200–1200 nm in dimethylformamide (DMF) solution (1×10^{-3} M). Fast atom bombardment (FAB)-mass spectrum of the ligand was obtained on JEOL SX 102/DA-6000 mass spectrometer using Argon/Xenon as the gas. The accelerating voltage was 10 kV and the spectrum was recorded at room temperature using meta-dinitrobenzyl alcohol as a matrix. Elemental analyses were obtained from HERAEUS C, H, N- O rapid analyzer and metal analyses were carried out by following the standard methods. Electron paramagnetic resonance (EPR) measurements were carried out on a Bruker Biospin Gmbh spectrometer working at a microwave frequency of 9.4 GHz using DPPH as a reference with the field set at 3200 Gauss.

Antimicrobial activity

The antibacterial and antifungal activities were performed by the cup-plate method [7]. In a typical procedure, molten agar kept at 45°C was then poured into Petri dishes and allowed to solidify. Holes of 8 mm diameter were punched using a sterile cork borer and these were completely filled with test solutions (1 mgmL^{-1} in DMF). The plates were incubated for 24 h at 37°C. The diameters of the zones of inhibition for all test compounds were measured and the results compared with streptomycin of the same concentration as that of the test compound under identical conditions.

Antifungal activity of the compounds was evaluated against the *A. niger* and *Candida albicans* by the cup-plate method, cultured on potato-dextrose agar medium adapting similar procedure as described above. The plates were incubated at 37°C for 48 h. The diameters of the zone of inhibition for all the test compounds were measured and the results compared with the standard drug griseofulvin of the same concentration under identical conditions.

RESULTS AND DISCUSSION

The Schiff base is soluble in DMF, dimethyl sulfoxide (DMSO) and insoluble in common organic solvents. The purity of the ligand was checked by elemental analysis, FT-IR, ^{1}HNMR and FAB-mass. The physical characteristics of the Schiff base and its metal complexes are given in Table 1. The metal complexes are soluble in DMF and DMSO and insoluble in acetone, chloroform, ethanol, methanol and water. The elemental analysis data of the Schiff base and its metal complexes are consistent with the calculated results from the empirical formula (Figures 1 and 2). The observed molar conductance of the complexes in DMF for 10^{-3} M solutions at room temperature is consistent with the non-electrolytic nature of the complexes [12].

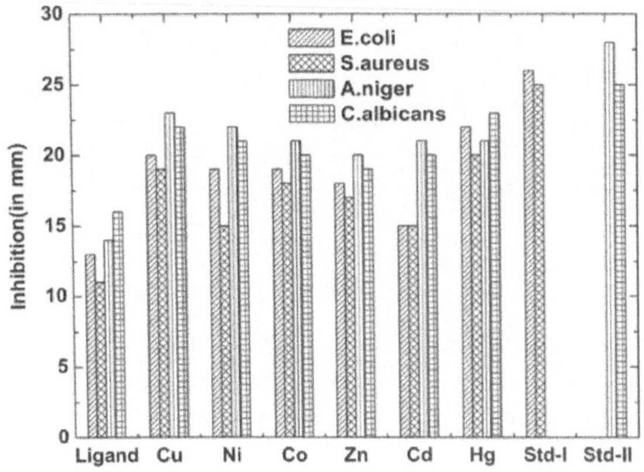

Figure 1: Antimicrobial screening data of the ligand HL (ligand containing one dissociable hydrogen) and its complexes.

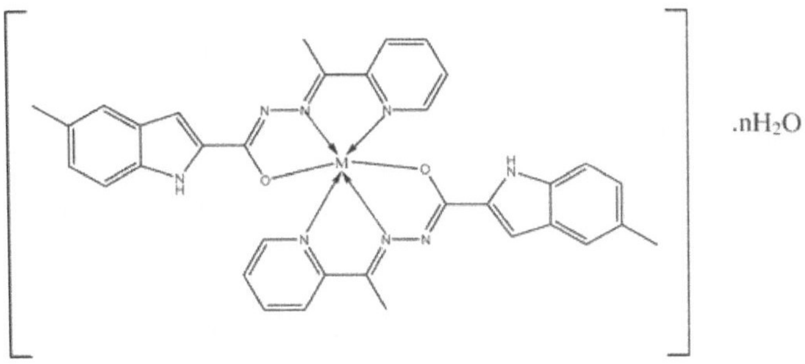

Figure 2: Suggested geometries of the complexes, where M=Co (II) Ni(II) and Cu(II) and *n*=1 for Co(II), Ni(II) and Cu(II).

Table 1: Elemental analysis, color, yield, and molar conductance of the Schiff base and its metal complexes

Compound/ complex	Color	Elemental analysis (%) calcd/(found) C	H	N	Cl	M	Magnetic moment μ_{eff}(B.M.)	λ_M (Ω^{-1} m^2 mol^{-1})	m.p. (^{0}C), Yield (%)
$C_{18}H_{18}N_4O$	Yellow	70.57 (70.52)	5.92 (5.89)	18.29 (18.26)	–	–	–	–	258 (80)
$[Co(C_{18}H_{17}N_4O)_2]\cdot(H_2O)$	Dark brown	62.88 (62.84)	5.28 (5.24)	16.30 (16.25)	–	8.57 (8.54)	4.89	27	305 (65)
$[Ni(C_{18}H_{17}N_4O)_2]\cdot(H_2O)$	Yellow	62.90 (62.86)	5.28 (5.25)	8.54 (8.49)	–	16.30 (16.26)	2.99	24	285 (65)
$[Cu(C_{18}H_{17}N_4O)_2]\cdot(H_2O)$	Yellow	62.46 (62.41)	5.24 (5.20)	16.19 (16.16)	–	9.18 (9.14)	1.99	31	305 (70)
$Zn[C_{18}H_{18}N_4O)Cl_2]\cdot(H_2O)$	Yellow	46.93 (46.89)	4.38 (4.34)	12.16 (12.11)	15.39 (15.33)	14.20 (14.17)	Diamag.	31	302 (75)
$Cd[C_{18}H_{18}N_4O)Cl_2]\cdot(H_2O)$	Orange	42.58 (42.55)	3.97 (3.92)	11.04 (10.99)	13.97 (13.94)	22.14 (22.10)	Diamag.	29	280 (75)
$[Hg(C_{18}H_{18}N_4O)Cl_2]$	Dark brown	36.28 (36.25)	3.38 (3.35)	9.40 (9.37)	11.90 (11.86)	33.66 (33.61)	Diamag.	30	310 (80)

Note: Diamag., diamagnetic.

IR spectra

The important IR frequencies exhibited by the ligand and their Co(II), Ni(II), Cu(II), Zn(II), Cd(II) and Hg(II) complexes are given inTable 2 along with their assignments. The IR spectrum of ligand showed a strong intensity band at 3265 and 3083 cm^{-1}attributable to indole ring ν_{NH}, and ν_{NH} of amide group, respectively. In the case of IR spectra of Cu(II), Co(II) and Ni(II) complexes, the band due to amide NH has disappeared, absence of this band due to ν_{NH}, indicates enolization of the amide carbonyl function during complexation and subsequent coordination of the amide carbonyl oxygen *via* deprotonation [13]. The band due the ν_{NH} of indole ring [14,15] observed in the region 3330–3210 cm^{-1} in the above complexes, indicating its non-involvement in the complexation. The band appeared at 1649 cm^{-1} in the ligand due to $\nu_{C=O}$ has been disappeared in all complexes except Zn(II), Cd(II) and Hg(II) complexes, confirm enolization of amide function during complexation which is evident by the appearance of a new band in the region 1602–1597 cm^{-1} in these complexes due to the formation of azine moiety, >C- N– N- C< further confirms the enolization of amide carbonyl function during complexation [16–18]. The band due to $\nu_{C=N}$ which was observed at 1577 cm^{-1} in ligand has shown a negative shift of 36–26 cm^{-1} in of all the complexes and appeared in the region 1551–1541 cm^{-1} indicating the involvement of azomethine nitrogen in the complexation with metal ions [19]. The appearance of a new bands in the region 3445–3409 cm^{-1} for Cu(II), Ni(II), Co(II), Zn(II), Cd(II) and Hg(II) complexes, indicate the presence of coordinated or lattice water in these complexes [20]. Pyridine ring vibrations which were observed at 1430, 1048, 620 cm^{-1} in the case of ligand have shifted to higher wave number side in all complexes and appeared in the region 1493–1464, 1070–1065 and 674–635 cm^{-1} indicating the involvement of pyridine ring nitrogen in the coordination [21]. Appearance of a new set of bands in all the complexes, due to ν_{M-O} and ν_{M-N} vibrations is the direct evidence for complexation. The bands observed in the region 440–426 and 531–478 cm^{-1} are assigned to ν_{M-O} and ν_{M-N}, respectively. The bands in the region 321–318 cm^{-1} have been assigned to ν_M– *Cl* bands in the case of Zn(II), Cd(II) and Hg(II) complexes [22]. The presence of chlorine atoms in these complexes is further confirmed by quantitative chloride estimation.

Table 2: Characteristics IR bands of the Schiff Base and its complexes (in cm^{-1})

Ligand/complex	ν_{H_2O}	$\nu_{NH/NH}$	$\nu_{C=O}$	$\nu_{C=N}$	Pyridine ring vibrations	Azine >C=N–N=C<	ν_{M-O}	ν_{M-N}	ν_{M-Cl}
$C_{18}H_{18}N_4O$	–	3265/3083	1649	1577	1430,1048, 620	–	–	–	–
$[Co(C_{18}H_{17}N_4O)_2]\cdot(H_2O)$	3432	3210	–	1545	1467,1070, 658	1597	531	426	–
$[Ni(C_{18}H_{17}N_4O)_2]\cdot(H_2O)$	3409	3220	–	1548	1464,1070, 674	1602	514	437	–
$[Cu(C_{18}H_{17}N_4O)_2]\cdot(H_2O)$	3445	3330	–	1551	1493, 1068, 647	1600	478	440	–
$[Zn(C_{18}H_{18}N_4O)Cl_2]\cdot(H_2O)$	3436	3252/3055	1620	1549	1469, 1070, 637	–	482	427	318
$[Cd(C_{18}H_{18}N_4O)Cl_2]\cdot(H_2O)$	3441	3218/3065	1615	1547	1471, 1067, 648	–	485	446	321
$[Hg(C_{18}H_{18}N_4O)Cl_2]$	–	3223/3058	1627	1541	1466, 1065, 635	–	483	428	319

Electronic spectra

The electronic spectra of Co(II), Ni(II) and Cu(II) complexes were recorded in DMF solution at 10^{-3} molar concentration. The electronic spectrum of the Co(II) complex exhibited absorption bands 10,582, 15,625 and 20,230 cm^{-1} due to the $^4T_{1g}(F) \rightarrow {}^4T_{2g}(F)(\nu_1)$, $^4T_{1g}(F) \rightarrow {}^4A_{2g}(F)(\nu_2)$ and $^4T_{1g}(F) \rightarrow {}^4T_{1g}(P)(\nu_3)$ transitions, respectively [23–25]. The Ni(II) complex displayed an electronic spectrum with transitions at 10,526, 16,129 and 24,390 cm^{-1}, respectively. These bands may be assigned to the transitions $^3A_{2g}(F) \rightarrow {}^3T_{2g}(F)(\nu_1)$, $^3A_{2g}(F) \rightarrow {}^3T_{1g}(F)(\nu_2)$ and $^3A_{2g}(F) \rightarrow {}^3T_{1g}(P)(\nu_3)$, respectively [23–25]. The Cu(II) complex displayed bands at 13,157 and 18,348 cm^{-1}, which may be assigned to the transitions $^2B_{1g} \rightarrow {}^2A_{1g}(d_x^2 - y^2 \leftarrow d_z^2$, $\nu_1)$, $^2B_{1g} \rightarrow {}^2E_g(d_x^2 - y^2 \leftarrow d_{xyyz}, \nu_2)$ and $^2B_{1g} \rightarrow {}^2B_{2g}(d_x^2 - y^2 \leftarrow d_{xy}, \nu_3)$. The third band at 34,482 cm^{-1} may be due to charge transfer [16,26] (Table 2).

^{1}H NMR spectral studies

The ^{1}H NMR spectral data of the ligand and its complex were recorded in DMSO-d6. These ^{1}H NMR spectrum of the ligand and complex were compared, it was observed that the signals of the protons of different functionalities of the ligand have been shifted toward the downfield region indicating the coordination of the ligand to the metal ion. The ^{1}H NMR spectrum of the ligand showed a fine singlet at δ 2.50 s (- N- C- CH_3) due to three protons of - CH_3 group on the azomethine carbon. Another two singlets appeared at δ 2.45 s (CH_3) and δ 2.40 s (CH_3) are due to protons of methyl groups at 3, 5-position of indole nucleus, respectively. Proton of the amide NH has resonated as a singlet at δ 11.40 s (1H), whereas proton on indole nitrogen appeared at δ 10.45 s (1H) as a distinct singlet. Three aromatic protons of the indole ring and four protons of pyridine ring have resonated as a multiplet in the region δ 7.07–8.65 m (Ar-H, 7H). In the ^{1}H NMR spectrum of Zn(II) complex, the two singlets appeared at δ 2.50 s (CH_3) and δ 2.45 s (CH_3) are due to protons of methyl groups at 3, 5-position of indole nucleus. Another singlet at δ 2.59 s (CH_3) is due to the protons of the methyl group on the carbon atom of azomethine function of the ligand. NH proton of the amide function and indole NH has resonated as a distinct singlet at δ 11.65 s (1H) and δ 10.60 s (1H), respectively. The seven aromatic protons have resonated in the region δ 7.46–8.75 m (Ar-H, 7H) as a multiplet (Table 3).

Table 3: ^{1}HNMR spectral data of the Schiff base and its Zn(II) complex

Compound	Chemical shift, δ (ppm)	Assignments
$C_{18}H_{18}N_4O$	2.50	(s, N=CCH_3)
	2.45	(s, –CH_3) 3-position of indole
	2.40	(s, –CH_3) 5-position of indole
	10.45	(s, –NH) indole-NH
	11.40	(s,–NH) amide-NH
	7.07–8.65	(M, Ar-H, 7H)Aromatic protons of indole and pyridine ring
$[Zn(C_{18}H_{18}N_4O)Cl_2]\cdot(H_2O)$	2.59	(s, N=CCH_3)
	2.52	(s, –CH_3) 3-position of indole
	2.47	(s, –CH_3) 5-position of indole
	10.60	(s, –NH) indole-NH
	11.65	(s, –NH) amide-NH
	7.46–8.75	(M, Ar-H, 7H)Aromatic protons of indole and pyridine ring

FAB-mass spectral analysis

Mass spectrum of ligand showed a molecular ion peak M$^+$ at *m/z* 306 (13%), which is equivalent to its molecular weight. The molecular ion M$^+$ by the expulsion of $C_7H_8N_3$ species gave fragment ion A_1 at *m/z* 172 (M$^+$-C_7H_8N = A_1, 100%), which is also a base peak. The fragment ion A_1 by the loss of carbon monoxide molecule gave fragment ion A_2 at *m/z* 144

($A_1-CO = A_2$, 10%). Fragment ion A_2 underwent further fragmentation by the loss methyl radical and gave fragment ion A_3 at m/z 129 ($A_2-CH_3 = A_3$, 3%). Fragment ion A_3 by the loss of C_2H species gave another fragment ion A_4 at m/z 104 ($A_3-C_2H = A_4$, 6%) (Scheme 2).

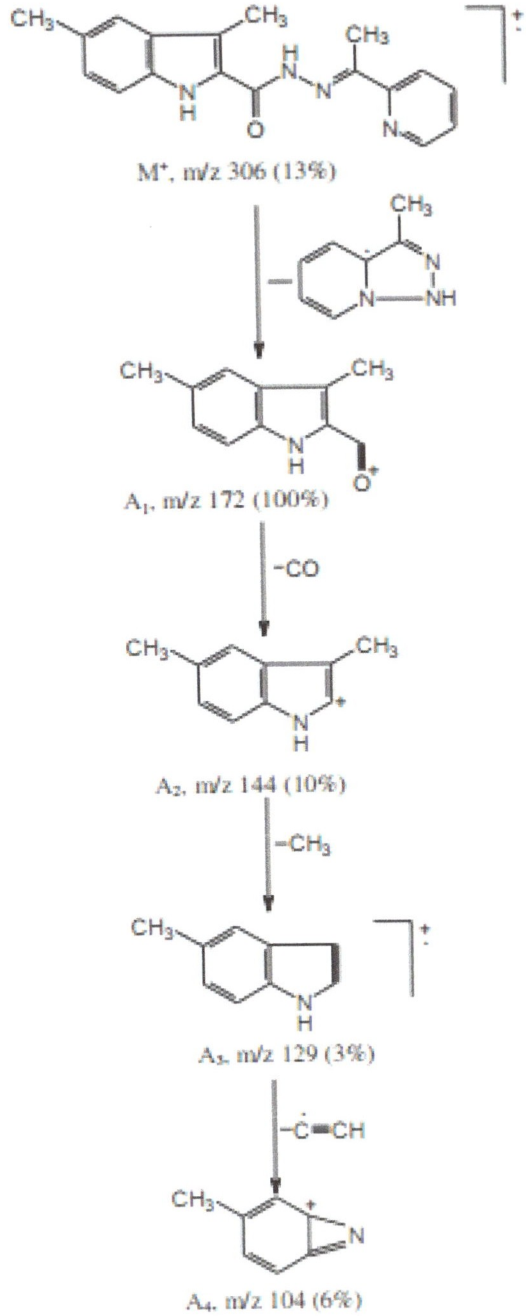

Scheme 2: Fragmentation pattern of ligand.

Magnetic susceptibilities

Magnetic susceptibilities of the complexes calculated at ambient temperature and the data presented in Table 1. The magnetic moments are in good agreement with the expected high spin configurations. The Co(II), Ni(II) and Cu(II) complexes showed magnetic moments of 4.89, 2.99 and 1.99 μB, corresponding to 3, 2 and 1 unpaired electrons respectively [23,27,28]. The Zn(II), Cd(II) and Hg(II) complexes show a diamagnetic behavior having a d^{10} system which is as per expectation.

EPR spectrum

The EPR spectrum of the Cu(II) complex was recorded at 300 K as a polycrystalline sample and in acetonitrile solution, on the X-band at the frequency of 9.4 GHz under a magnetic field strength 3200 G. The polycrystalline spectrum showed a well-resolved anisotropically broad signal. The analysis of the spectrum gives $g_\parallel$ 2.25 and $g_\perp$ 2.06 (Table 4). The trend $g_\parallel > g_\perp > 2.0023$ observed for the complex under study indicates that the unpaired electron is localized in the $d_{x^2-y^2}$ orbital of the Cu(II) ion. The observed geometric parameter calculated by the expression $G = (g_\parallel - 2)/(g_\perp - 2)$ is greater than 4 $(G = 4.17)$ evidenced the exchange interaction is negligible between metal ion in the solid complex [29–31]. Based on the spectral studies, the following geometries may be suggested for the complexes (Figure 1).

Table 4: Electronic spectral bands (in cm^{-1}) of Co(II), Ni(II) and Cu(II) complexes and EPR data of Cu(II) complex

Complex	Absorption (cm^{-1})	Band assignments	ESR data $g_{\perp}$	g_{11}	G	$g_{iso/av}$
$[Co(C_{18}H_{17}N_4O)_2].(H_2O)$	10,582	$^4T_{1g}(F) \rightarrow {}^4T_{2g}(F)\ (\nu_1)$				
	15,625	$^4T_{1g}(F) \rightarrow {}^4A_{2g}(F)\ (\nu_2)$				
	20,230	$^4T_{1g}(F) \rightarrow {}^4T_{1g}(P)\ (\nu_3)$				
$[Ni(C_{18}H_{17}N_4O)_2].(H_2O)$	10,526	$^3A_{2g}(F) \rightarrow {}^3T_{2g}(F)(\nu_1)$				
	16,129	$^3A_{2g}(F) \rightarrow {}^3T_{1g}(F)\ (\nu_2)$				
	24,390	$^3A_{2g}(F) \rightarrow {}^3T_{1g}(P)\ (\nu_3)$				
$[Cu(C_{18}H_{17}N_4O)_2].(H_2O)$	13,157	$^2B_{1g} \rightarrow {}^2A_{1g}(d^{22}_{x-y} \leftarrow d^2_z, \nu_1)$,	2.25	2.06	4.17	2.186
	18,348	$^2B_{1g} \rightarrow {}^2E_g(d^{22}_{x-y} \leftarrow d_{xyyz}, \nu_2)$				
	34,482	$^2B_{1g} \rightarrow {}^2B_{2g}(d^{22}_{x-y} \leftarrow d_{xy}, \nu_3)$				

X-ray diffraction studies

Powder X-ray diffraction was done for ligand and its cobalt complex and spectra is given in Figures 1 and 2. X-ray diffraction analysis of the ligand and its cobalt(II) complex confirm the cubic crystal system for these compounds. Ligand has unit cell dimensions *a*=*b*=*c*=16.59, $\alpha = \beta = \gamma = 90$ and $[Co(L)_2] \cdot (H_2O)$ has unit cell dimensions *a*=*b*=*c*=16.19, $\alpha = \beta = \gamma = 90$ [32,33] (Tables 5 and 6).

Table 5: Powder X-ray diffraction data of ligand HL

Peak no.	2θ	θ	Sinθ	Sin$^2\theta$	$h^2 + k^2 + l^2$	*hkl*	d Calc.	d Obser.	*a* in A°
1	5.324	2.662	0.04644	0.00216	1	1 0 0	16.58558	16.58558	16.59
2	10.445	5.2225	0.09102	0.00829	4	2 0 0	8.46264	8.46264	16.59
3	14.636	7.318	0.12738	0.01622	8	2 2 0	6.04744	6.04744	16.59
4	19.932	9.966	0.17306	0.02995	14	3 2 1	4.45096	4.45096	16.59
5	22.656	11.328	0.19643	0.03858	18	3 3 0	3.92159	3.92159	16.59
6	23.911	11.9555	0.20715	0.04291	20	4 2 0	3.71853	3.71853	16.59
7	26.832	13.416	0.23202	0.05383	25	5 0 0	3.31998	3.31998	16.59
8	36.772	18.386	0.31542	0.09949	46	6 3 1	2.44216	2.44216	16.59

Note: HL, ligand containing one dissociable hydrogen.

Table 6: Powder X-ray diffraction data of Co(II) complex

Peak no.	2θ	θ	Sinθ	Sin$^2\theta$	$h^2 + k^2 + l^2$	*hkl*	d Calc.	d Obser.	*a* in A°
1	5.4548	2.7274	0.04758	0.00226	1	1 0 0	16.18817	16.18817	16.19
2	11.82	5.91	0.10297	0.0106	5	2 1 0	7.4811	7.4811	16.19
3	12.843	6.4215	0.11184	0.01251	6	2 1 1	6.88741	6.88741	16.19
4	15.214	7.607	0.13238	0.01752	8	2 2 0	5.81897	5.81897	16.19
5	16.84	8.42	0.14643	0.02144	9	3 0 0	5.26059	5.26059	16.19
6	19.554	9.777	0.16981	0.02884	13	3 2 0	4.53614	4.53614	16.19
7	21.782	10.891	0.18894	0.0357	16	4 0 0	4.07693	4.07693	16.19
8	23.887	11.9435	0.20695	0.04283	19	3 3 1	3.72221	3.72221	16.19
9	25.051	12.5255	0.21687	0.04703	21	4 2 1	3.55183	3.55183	16.19
10	33.428	16.714	0.28759	0.08271	37	6 1 0	2.67842	2.67842	16.19

Antimicrobial activity

A cup plate method was employed for the *in vitro* study of antibacterial and antifungal effect against, *Staphylococcus aureus, Escherichia coli,* ***C.*** *albicans and A. niger*. The inhibitory effect of ligand and its complexes against these organisms are given inFigure 1. The antimicrobial screening data show that the metal chelates exhibit a higher inhibitory effect than the free ligand. The increased activity of the metal chelates can be explained based on the chelation theory [7,34–36] (Table 7).

Table 7: Antibacterial and antifungal activity of HL and its complexes

Compound	Mean of zone diameter (mm)			
	Antibacterial activity		Antifungal activity	
	E. coli	*S. aureus*	*A. niger*	*C. albicans*
Ligand	13	11	14	16
Cu	20	19	23	22
Ni	19	15	22	21
Co	19	18	21	20
Zn	18	17	20	19
Cd	15	15	21	20
Hg	22	20	21	23
Std-I	26	25	–	–
Std-II	–	–	28	25

Note: HL, ligand containing one dissociable hydrogen.

CONCLUSIONS

The present study revealed octahedral geometry for the Co(II), Ni(II), Cu(II) and square pyramidal geometry for Zn(II), Cd(II) and Hg(II) complexes, respectively. The ligand acts in a tridentate manner coordinating through carbonyl oxygen, azomethine-N and pyridine ring-N groups in an oxygen nitrogen nitrogen fashion. Moreover, the microbial data revealed that the complexes were superior to the free ligand in the inhibition of the tested bacteria and fungi. It is proposed that concentration plays a vital role in increasing the degree of inhibition the activity increased with increasing concentration of the complexes (Figures 2and 3).

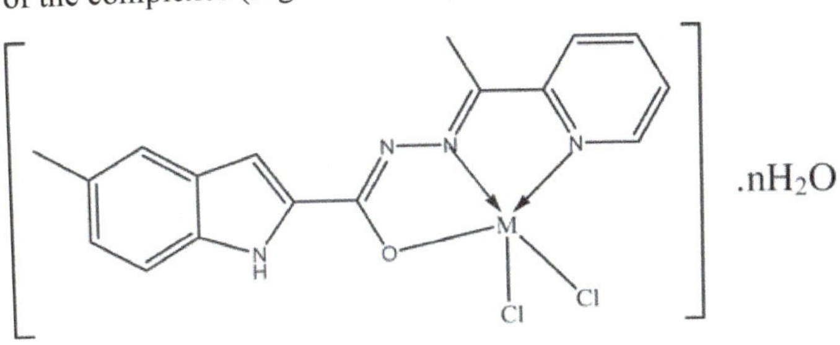

Figure 3: Suggested geometries of the complexes, M′=Zn (II), Cd(II) and Hg(II), where n=1 for Zn(II), Cd(II) and n=0 for Hg(II).

Supplementary Material

The authors are thankful to the Chairman, Department of Chemistry, Gulbarga University, Gulbarga for providing laboratory facilities. They also thank IISc Bangalore, STIC Kochi and CDRI Lucknow for ^{1}HNMR, ESR, Mass, FAB-Mass, XRD and elemental data.

REFERENCES

1. A. Bayer. *Chem. Ber.*, 13, 2254 (1880). doi: 10.1002/cber.188001302243
2. E. Liberta, D.X. West. *Biometals*, 5, 121 (1992). doi: 10.1007/BF01062223 , [PubMed],
3. (a) M.C. Millar, C.N. Stineman, J.R. Vance, D.X. West, I.H. Hall. *Anticancer Res.*, 18, 4131 (1998); (b) M.C. Millar, C.N. Stineman, J.R. Vance, D.X. West, I.H. Hall. *Appl. Organomet. Chem.*, 13, 9 (1999).
4. B. Elena, C. Rosa, C. Alfonso, D. Ricarso, M.C. Maichle, S. Joachin, X.W. Douglas. *Polyhedron*, 18, 3695–3702 (1999). doi:10.1016/S0277-5387(99)00309-5 ,
5. B. Elena, C. Rosa, C. Alfonso, D. Ricarso, L.E. Authorny, M.-M. Cacilia, M.S. Michelle, X.W. Douglas. *Eur. J. Inorg. Chem.*, 1999, 965–973 (1999).
6. B.H.M. Mruthyunjayaswamy, Y. Jadegoud, B.I. Omkar, G.P. somanath, M.K. Shreeshaila. *Trans. Met. Chem.*, 30, 234–242 (2005). doi: 10.1007/s11243-004-2997-z ,
7. B.H.M. Mruthyunjayaswamy, B.I. Omkar, Y. Jadegoud. *J. Braz. Chem. Soc.*, 16, 783–789 (2005). doi: 10.1590/S0103-50532005000500016 ,
8. F. Rahaman, B. Hiremath, S.M. Basavarajaiah, B.H.M. Jayakumarswamy, B.H.M. Mruthyunjayaswamy. *J. Indian. Chem. Soc.*, 85, 381–386 (2008).
9. F. Rahaman, O.B. Ijare, Y. Jadegoud, B.H.M. Mruthyunjayaswamy. *JCC*, 62, 1457–1467 (2009).
10. A.I. Vogel. *A Text Book of Quantitative Organic Analysis*, 3rd Edn, ELBS Longmans Green and Co. Ltd., London (1962).
11. S.P. Hiremath, P.S. Badami, M.G. Purohit. *Indian J. Chem.*, 24B, 1115 (1985).
12. W.J. Geary. *Coord. Chem. Rev.*, 7, 81 (1971). doi: 10.1016/S0010-8545(00)80009-0 ,
13. N.K. Singh, S.B. Singh. *Indian J. Chem.*, 40A, 1071–1075 (2001).
14. P.H. Shivayogi, B.H.M. Mruthyunjayaswamy, G.P. Muralidhar. *Indian J. Chem.*, 16B, 789–792 (1978).
15. J.S. Biradar, B. Sharanbasappa. *Green Chem. Lett. Rev.*, 2, 237–241 (2009). doi: 10.1080/17518250903393890 [Taylor & Francis Online],
16. A.V. Lakshmi, N.R. Sangeetha, S. Pal. *Indian J. Chem.*, 36, 844–848 (1997).
17. E.R. Blout, M. Fields, R. Karplus. *J. Am. Chem. Soc.*, 70, 194 (1948). doi: 10.1021/ja01181a056 ,
18. R.S. Baligar, V.K. Revankar. *J. Serb. Chem. Soc.*, 71, 1301–1310, JSCS–352 (2006). doi: 10.2298/JSC0612301B ,

19. A. Koji, M. Kanako, O. Masaki, O. Hishashi. *Inorg. Chem.*, 41, 4461 (2002). doi: 10.1021/ic020002f , [PubMed],

20. Rajesh Malhotra, Sudhir Kumar, Jyoti, Hari Ram Singal, Kuldip Singh Dhindsa. *Indian J. Chem.*, 39A, 421 (2000).

21. W.E. Marsh, D. Eggleston, W.E. Hall Field, D.J. Hodgson. *Inorg. Chem. Acta*, 70(2), 137 (1983). doi: 10.1016/S0020-1693(00)82792-9 ,

22. B.B. Mohapatra, S.K. Saraf. *J. Indian Chem. Soc.*, 80, 696 (2003).

23. F.A. Cotton, G. Wilkinson. *Advanced Inorganic Chemistry*. 3rd Edn, Wiley Eastern Pvt. Ltd., New Delhi (1972).

24. (a) I.M. Procter, R.J. Hathaway, P. Nicholls, *J. Chem. Soc.* A, 1678–1684 (1968); (b) K.G. Kocwin, W. Wajciechowski, *Trans. Met. Chem.*, 21, 312 (1996).

25. (a) K.C. Satpathy, B.B. Jal, R. Mishra, *Trans. Met. Chem.*, 9, 8 (1984); (b) A.C. Fabretti, G. Peyrond, G.C. Franchini. *Inorg. Chem. Acta*, 35, 49 (1979).

26. A.B.P. Lever. *Inorganic Electronic Spectroscopy*. 2nd Edn, pp. 487, Elsevier, Amsterdam (1984).

27. (a) S. Chandra, K. Gupta. *Indian J. Chem.*, 40A, 775 (2001); (b) S. Chandra, A. Gautam, M. Tyagi. *Trans. Met. Chem.* 32, 1079 (2007).

28. R.K. Agarwal, Kisher Arora, Prashant Dutt. *Polyhedron*, 13, 957 (1994). doi: 10.1016/S0277-5387(00)83016-8 ,

29. L.K. Mishra, Y. Jha, B.K. Sinha, R. Kant, Rajeswar Singh. *J. Indian. Chem. Soc.*, 76, 65 (1999).

30. (a) H. Hathaway, D.E. Billing. *Coord. Chem. Rev.*, 5, 143–207 (1970); (b) R.J. Dudley, B.J. Hathway. *J. Chem. Soc.*, Sec. A, 1725 (1970); (c) S. Chandra, K. Gupta. *Trans. Metal Chem.*, 27, 196–197 (2002); (d) K. Burger, *Coordination Chemistry Experimental Methods*, 1st Edn, Butterworth and Co. Ltd., London. (1973).

31. M.F. Fouda, M.M. Abd-Elzaher, M.M. Shadofa, F.A. El Saied, M.I. Ayad, A.S. El Tabl. *Trans. Metal Chem.*, 33, 219–228 (2008). doi: 10.1007/s11243-007-9024-0 ,

32. B.D. Cullity. *Elements of X-Ray Diffraction,* Addison-Wesley Pub. Co., Notre dame, IN (1978).

33. (a) N.F.M. Henry, H. Lipson, W.A. Wooster. *Interpretation of X-Ray Diffraction Photography*, p. 179, Macmillon, London (1959); (b) M.J. Burger. *X-Ray Crystallography*, p. 100, Willey, New York (1953); (c) M.M. Woolfson. *An Introduction to X-Ray Crystallography*, p. 125, Cambridge University Press, Cambridge (1980).

34. A.L. Barry. *The Antimicrobial Susceptibility Test, Principles and Practices*, 4th Edn, ELBS, London, 180–193 (1976).

35. S.K. Sengupta, O.P. Pandey, B.K. Srivatsava, V.K. Sharma. *Trans. Metal Chem.*, V-23, 349 (1998). doi:10.1023/A:1006986131435 ,

36. S. Chandra, M. Tyagi. *J. Serb. Soc.*, V-73, 727 (2008). doi: 10.2298/JSC0807727C ,

CHAPTER 14

RUTHENIUM(II) COMPLEXES CONTAINING FUNCTIONALISED Β-DIKETONATE LIGANDS: DEVELOPING A FERROCENE MIMIC FOR BIOSENSING APPLICATIONS

Yeng Ying Lee, D. Barney Walker, J. Justin Gooding and Barbara A. Messerle

School of Chemistry, The University of New South Wales, Sydney, NSW 2052, Australia

ABSTRACT

Three series of ruthenium complexes with the general formula $Ru(bpy)_n(\beta\text{-diketonato})_{3-n}$ (bpy = 2,2′-bipyridine, n = 0, 1, 2) were prepared and investigated using cyclic voltammetry and UV-vis spectroscopy. Variation of both the number and electronic demand of the β-diketonato ligands resulted in well-defined modulation of the potential at which oxidation of the metal centre occurred. The observed potentials were shown to be in good agreement with calculated ligand electrochemical parameters. A novel ruthenium(II) complex with electrochemical behaviour similar to that of ferrocene was identified.

INTRODUCTION

Ferrocene is well established as an electrochemically active tag for investigating and monitoring biological activity in situ.[1–9] This is largely due to the fact that, in addition to its favourable electrochemical properties, the ferrocenyl group can be readily functionalised and is considered to be relatively stable in aqueous, aerobic media. However, the ferricenium ion formed following oxidation has been shown to decompose when exposed to chloride ions,[10–13] potentially limiting the application of ferrocene as a redox label in sensors designed for long-term analyte monitoring (e.g.implantable devices).

The stable and reversible nature of the $Ru^{2+/3+}$ redox couple suggests that ruthenium complexes could be an attractive alternative to ferrocene for biological sensors and several groups have explored this possibility.[14–22] An advantage of utilising ruthenium species is that the potential at which oxidation of the metal centre occurs can be influenced by the electronic demand of the ligands occupying the primary coordination sphere. Large changes in the $E_{1/2}$ of a Ru^{2+} centre can be induced either by introducing anionic ligands[23,24] and/or by changing the number of π-acceptor coordinating species

Citation: Yeng Ying Lee, D. Barney Walker, J. Justin Gooding and Barbara A. Messerle, Ruthenium(II) complexes containing functionalised β-diketonate ligands: developing a ferrocene mimic for biosensing applications, DOI: 10.1039/C4DT01459F.

around the metal.[25–30] More subtle 'tuning' of the redox potential has previously been demonstrated by attaching either electron-withdrawing or electron-donating groups (EWG or EDG) to the peripheral sites around the coordinating ligands.[28,31,32]

Here we present the synthesis and electrochemical characterisation of three series of ruthenium complexes (of the general formulae $[Ru(bpy)_2(\beta\text{-diketonato})](PF_6)$, $[Ru(\beta\text{-diketonato})_3]$ and $[Ru(bpy)(\beta\text{-diketonato})_2]$, bpy = 2,2′-bipyridine) wherein the number and type of ligand is varied in order to tune the redox potential of the metal centre towards values suitable for biosensing applications. We take advantage of the fact that the potential at which oxidation of these ruthenium complexes occurs can be attenuated by varying the substituents on the β-diketonato ligands and identify several candidates for biosensor integration that have a very similar electrochemical profile to that of ferrocene.

RESULTS AND DISCUSSION

Synthesis

Three series of ruthenium complexes with β-diketonato ligands were prepared (Scheme 1) where the complexes of Series I each contains a single β-diketonato ligand, Series II, three β-diketonato ligands and Series III, two β-diketonato ligands.

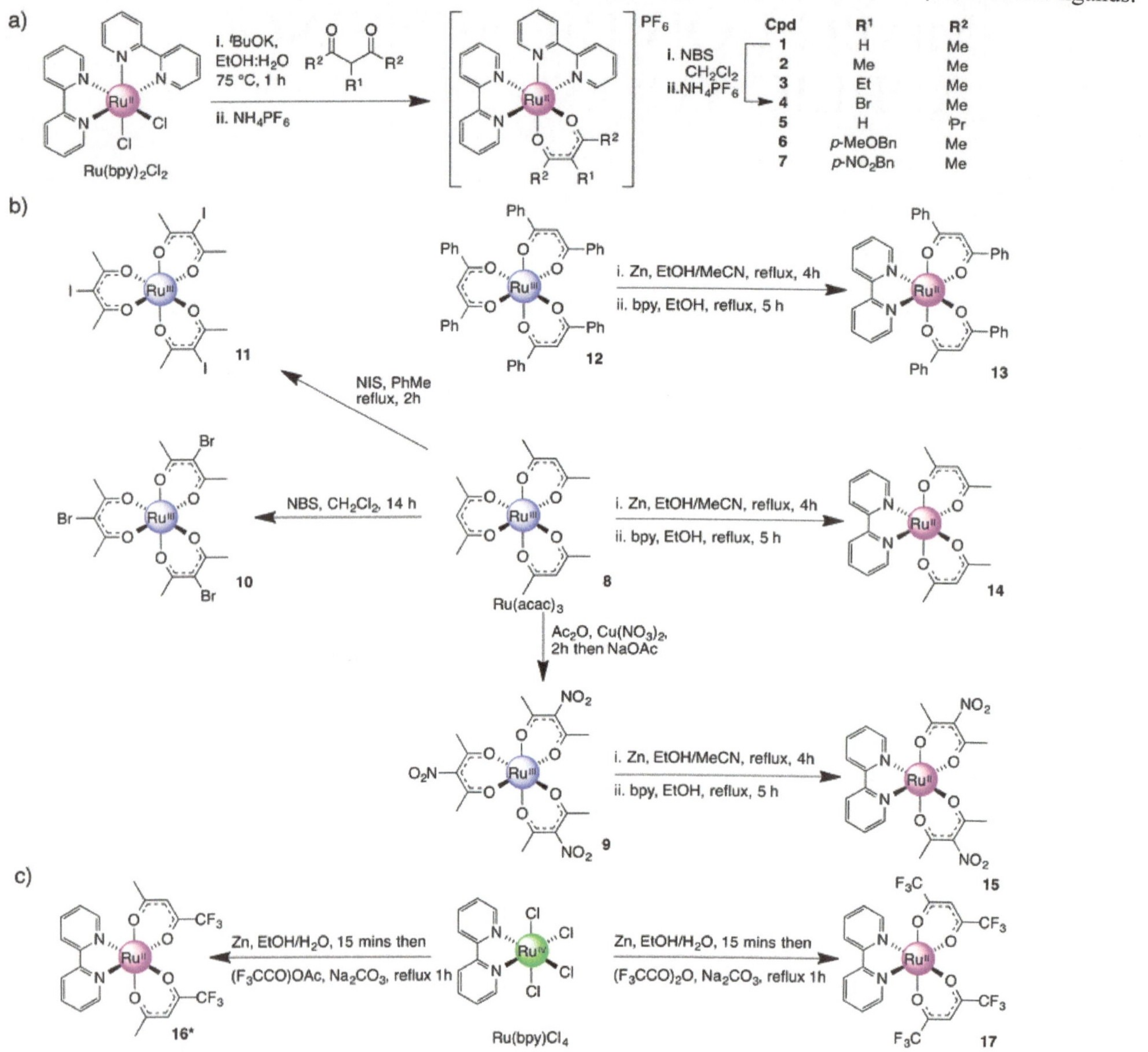

Scheme 1: Synthesis of complexes **1–17**.

Synthesis of Series I $[Ru(bpy)_2(\beta\text{-diketonato})](PF_6)$ complexes. Ruthenium complexes **1–7** (Scheme 1a) were prepared using a method adapted from a literature procedure.[33] Compounds **1**, **2**and **4** have been prepared previously (see ESI† for more information).[34] All Series I complexes (excluding **4**) were prepared by displacing chloride from $Ru(bpy)_2Cl_2$ with selected β-diketones in the presence of a stoichiometric amount of ᵗBuOK in a hot $EtOH–H_2O$ mixture. The desired product was precipitated following cooling and the addition of an aqueous solution of NH_4PF_6 to the reaction mixture. Complex **4** was prepared by treatment of **1** with N-bromosuccinimide in CH_2Cl_2.[35] Novel compounds **5**, **6** and **7** were prepared in the same way as compounds **1–4** in 60%, 53% and 27% yields respectively.

Synthesis of Series II $[Ru(\beta\text{-diketonato})_3]$ complexes. Methods for the direct functionalisation of the acetylacetonate (acac) ligand of complex **8** have been described previously[36,37] and these were utilised to expand the family of complexes investigated in this study.[32,38–42] Following the method of Collman et al.,[42] nitration of **8** with $Cu(NO_3)_2$ in acetic anhydride was achieved to give $Ru(NO_2\text{-acac})_3$ (**9**) in 70% yield (Scheme 1b). Complex **8** also underwent facile bromination with N-bromosuccinimide to give $Ru(Br\text{-acac})_3$ (**10**) in a 90% yield.[32] $Ru(I\text{-acac})_3$ (**11**) was synthesised in the same way using N-iodosuccinimide in 85% yield Complex **12** was prepared in a 25% yield by following a literature procedure.[28,31]

Synthesis of Series III [Ru(bpy)(β-diketonato)$_2$] complexes. Due to the relatively substitution-inert nature of Ru(β-diketonato)$_3$ complexes, one approach to the synthesis of the mixed β-diketonato and bpy complexes is to prepare an intermediate that allows clean ligand substitution reactions with bpy.[43] Ru(β-diketonato)$_2$(MeCN)$_2$ complexes were targeted as MeCN is sufficiently labile to be readily displaced by bpy. This method was originally applied to the synthesis of $Ru(acac)_2(bpy)$ (**14**).[44]

Complex **13** was synthesised in two steps via a bis-acetonitrile ($Ru(dbm)_2(MeCN)_2$) (dbm = dibenzoylmethane) intermediate. This intermediate was accessed by the reduction of complex **12** with activated zinc dust in refluxing EtOH followed by cooling and the addition of excess MeCN. The mixture was then refluxed for a further 2 h resulting in a colour change from bright red to orange.[45] Treatment of the intermediate ($Ru(dbm)_2(MeCN)_2$) with an equimolar amount of bpy in refluxing EtOH gave $Ru(dbm)_2(bpy)$ (**13**) as a dark green solid in 60% yield. The methyl analogue (R^2 = Me) was prepared using the same procedure (**8** was readily converted to **14** in a 60% overall yield). This method also proved successful for the preparation of nitro-analogue **15** but treatment of the tri-bromo/iodo ruthenium(III) complexes (**10, 11**) with elemental zinc resulted in dehalogenation of the ruthenium species and instead formed **14** along with other decomposition products.

A second approach to preparing Series III complexes via $Ru(bpy)(Cl)_4$ was investigated concurrently. The starting material was obtained in 85% yield by dissolving $RuCl_3{\cdot}3H_2O$ in dilute HCl, adding 1.2 equiv. of bpy and then leaving the solution for 21 days at room temperature.[46] Subsequent treatment of $Ru(bpy)(Cl)_4$ with activated zinc dust in EtOH–H_2O for 15 min followed by the addition of Na_2CO_3 and either 1,1,1-trifluoropentane-2,4-dione (tfac) or 1,1,1,5,5,5-hexafluoropentane-2,4-dione (hfac) followed by refluxing for 1 h resulted in the formation of complexes **16** and **17**in 25% and 33% yields respectively (Scheme 1c).[47] Complex **16** exists as a mixture of three geometrical isomers in 1 : 1 : 2 ratio of cis- and trans-isomers due to its asymmetric β-diketonato ligand. This was confirmed by ^{1}H NMR where the following configurations were observed: trans-CF_3-cis-H-[Ru(tfac)$_2$(bpy)], cis-CF_3-cis-H-[Ru(tfac)$_2$(bpy)] and cis-CF_3-trans-H-[Ru(tfac)$_2$(bpy)]. This is further supported by the presence of four signals in ^{19}F NMR (see Experimental section).

Whilst this route initially appeared to be a more direct way of preparing Ru(bpy)(β-diketonato)$_2$ complexes compared with preforming the Ru(β-diketonato)$_3$ species first (Scheme 1b) our observation was that the overall yields tended to be significantly lower and in some es the final product did not form at all. However this method did allow us to furnish sufficient amounts of complexes **16** and **17** for subsequent electrochemical analysis.

Cyclic voltammetry

All complexes in Series I–III were shown to undergo a redox cycle (assumed to be $Ru^{2+/3+}$) with a peak separation (ΔE) between 59 to 95 mV, and i_{pa}/i_{pc} at near unity indicating that the complexes were both electrochemically and chemically reversible. The $E_{1/2}$ of the complexes was independent of the scan rate (v) and the redox processes were diffusion-controlled as i_p vs. $v^{1/2}$ was found to be linear. Waves were assigned by comparison to those of analogous metal complexes.[48]

Series I. Ruthenium(II) complexes 1–7 [Ru(bpy)$_2$(β-diketonato)](PF$_6$). The potential at which oxidation of the ruthenium(II) complexes occurs in Series I (compounds **1** to **7**) were observed to be subtly influenced by the various substituents on the β-diketonato ligands. Complex **2** has an electron-donating methyl group at the methine position (R^1 = Me, Scheme 1a). Consequently, a cathodic shift of the redox potential of **2** by 100 mV (128 mV vs. 228 mV) was observed when compared with that of complex **1** (Table 1). A similar cathodic shift was observed for the redox potential of **3** (R^1= Et) although in this e the change is less pronounced ($\Delta E_{1/2}$ = 47 mV) in line with the reduced electron donating capacity of the ethyl groups of **3** compared with the methyl groups of **2**. These observations are in agreement with previous reports[49] and can be rationalised as follows: an electron-donating group (EDG) on the β-diketonato ligand increases the electron density around the metal centre and consequently destabilises electrons in the metal d-orbitals. Thus the metal centre is more readily oxidised resulting in the observed cathodic shift in the $E_{1/2}$.

Table 1: $E_{1/2}$ for Series I Ru(bpy)$_2$(β-diketonato) complexes

[Ru(bpy)$_2$((R^2C(O))$_2$CR^1))]PF$_6$	$E_{1/2}$ ^{a}/V
a $E_{1/2}$ ***vs.*** **FcH$^{+/0}$ in 0.01 M AgNO$_3$ in MeCN with 0.1 M NBu$_4$PF$_6$, v = 0.1 V s^{-1}.**	
1 (R^1 = H, R^2 = Me)	0.228
2 (R^1 = Me, R^2 = Me)	0.128
3 (R^1 = Et, R^2 = Me)	0.181
4 (R^1 = Br, R^2 = Me)	0.292
5 (R^1 = H, R^2 = ^{i}Pr)	0.201
6 (R^1 = p-MeOBn, R^2 = H)	0.167
7 (R^1 = p-NO$_2$Bn, R^2 = H)	0.194

By the same argument, introducing an electron-withdrawing group (EWG) should result in an anodic shift in the $E_{1/2}$ and this effect was observed for complex **4** (R^1 = Br) when compared with **1**($\Delta E_{1/2}$ = 64 mV).

As expected, complex **5** (R^2 = ^{i}Pr) and complex **1** (R^2 = Me) had almost identical electrochemical profiles ($\Delta E_{1/2}$ = 27 mV). The potentials at which complexes **6** and **7** were oxidised are quite similar (167 mV and 194 mV respectively) despite having functional groups with quite different electronic demands in the para position of the benzyl group (OMe vs. NO_2). This is not surprising considering that the aromatic ring is not in direct conjugation with the diketonate portion of the ligand bound to the ruthenium(II) ion. Consequently a modest cathodic shift ($\Delta E_{1/2}$ = 61 mV for **6**, $\Delta E_{1/2}$ = 34 mV for **7**) is observed in both es due to the slight electron donating effect of the benzylic CH_2 group.

Series II and III. Ruthenium(II) complexes 8–17, Ru (β-diketonato)$_3$*vs.* [Ru(bpy) (β-diketonato)$_2$](PF$_6$). The potentials at which all the Series II complexes (**8–12**) oxidised were all shifted cathodically compared with those of Series I due to the presence of three anionic ligands. In addition, as with Series I, the effect of a strongly EWG (R^1 = NO_2, **9**) results in a large anodic shift (671 mV) of the potential at which the complex oxidised when compared with **8**. The halogenated analogues of $Ru(acac)_3$ (**10** and **11**) were also found to have an anodically shifted $E_{1/2}$ although the effect was less significant ($\Delta E_{1/2}$ = 230 mV for Br, 231 mV for I). The potential at which complex **12** was oxidised was relatively close to that of **8** despite having six phenyl rings conjugated with the coordinating oxygen atoms (Table 2).

Table 2: $E_{1/2}$ for Series II and III complexes

Series II	$E_{1/2}$ ^{a}/V
8 (acac)	−1.06
9 (NO_2-acac)	−0.389
10 (Br-acac)	−0.830
11^b (I-acac)	−0.829
12 (dbm)	−0.906
Series III	$E_{1/2}$ ^{a}/V
13 (dbm)	−0.416
14 (acac)	−0.478
15 (NO_2-acac)	−0.054
16 (tfac)	−0.087
17 (hfac)	0.395

^{a} $E_{1/2}$ *vs.* $FcH^{+/0}$ in 0.01 M $AgNO_3$ in MeCN with 0.1 M NBu_4PF_6, ν = 0.1 V s^{-1}. ^{b} CV recorded in CH_2Cl_2 with 0.1 M NBu_4PF_6 as supporting electrolyte.

It might be expected that moving from Series II to Series III compounds, wherein one β-diketonato ligand is substituted for a bpy ligand, would result in a set of complexes with an $E_{1/2}$ closer to the desired redox window (−0.743 V to +0.067 V vs. $FcH^{+/0}$, −0.343 V to +0.467 V vs. Ag/AgCl).

The replacement of a β-diketone in **12** and **8** by bpy to form **13** and **14** raised the $E_{1/2}$ by 490 mV and 672 mV respectively. This was attributed to the fact that bpy is a π-acceptor – and the HOMO–LUMO gap of the complex could be increased and may result in a more stabilised Ru^{2+} complex. The stabilising effect of bpy works in concert with the NO_2-acac ligands used to form **15** resulting in an observed $E_{1/2}$ of −54 mV (vs. $FcH^{+/0}$). In the same way it was possible to anodically shift the potential at which the ruthenium metal centre is oxidised by replacing the acac ligands with tri- and hexafluorinated 2,4-diones. A comparison of the $E_{1/2}$s of **16** with **14** demonstrates this effect ($\Delta E_{1/2}$ = 391 mV). The potential at which **17** oxidised was anodically shifted by 482 mV compared to that of **16**. This difference corresponds to the number of trifluoromethyl substituents on the ligands: there are four in **17** compared to two in **16**. Of the Series III complexes, **15** and **16** were each shown to have an $E_{1/2}$ very close to that of ferrocene ($E_{1/2}$ = −54 mV and −87 mV respectively).

Diffusion coefficient, D_o

The diffusion coefficients, D_o, of complexes **1–17** in acetonitrile were obtained using the Randles–Sevcik equation:

$$i_p = 2.69 \times 10^5 n^{3/2} A D_o{}^{1/2} C \nu^{1/2} \quad (1)$$

where i_p = current, n = number of electrons transferred, A = electrode area, D = diffusion coefficient, C = bulk concentration of redox species and v = scan rate (Table 3).

Table 3: Diffusion coefficients for ruthenium complexes in Series I and III

Series I	$D_o/10^{-9}$ m^2 s^{-1}
1 (R^1 = H, R^2 = Me)	1.759
2 (R^1 = Me, R^2 = Me)	1.489
3 (R^1 = Et, R^2 = Me)	1.182
4 (R^1 = Br, R^2 = Me)	1.330
5 (R^1 = H, R^2 = ^{i}Pr)	1.561
6 (R^1 = *p*-MeOBn, R^2 = H)	1.495
7 (R^1 = *p*-NO_2Bn, R^2 = H)	1.346
Series III	
13 (dbm)	0.100
14 (acac)	1.354
15 (NO_2-acac)	0.645
16 (tfac)	1.745
17 (hfac)	1.369

The electrode area on the glassy carbon electrode, A, was determined using the known D_o of ferrocene in acetonitrile with 0.1 M NBu_4PF_6 as a supporting electrolyte (2.24 × 10^{-5} cm^2 s^{-1}).[50]

Complexes with bulky ligands, and those of higher charge, would be expected to have smaller D_o values as they would be expected to migrate more slowly in solution.[15,51,52] Additionally, changes in electron density around ruthenium brought on by donor/acceptor groups on β-diketonates will affect the degree to which counterions will attach to the reduced complexes.[53] The structurally simplest complex in Series I, **1**, has the largest D_o while the rest of the complexes are between 1.182 to 1.561 × 10^{-9} m^2 s^{-1}. However, the bulkiest complex of the series did not have the smallest D_o. This could be attributed to the interaction between these charged complexes of Series I with acetonitrile and the supporting electrolytes.

In Series III, the D_o for **13**, which contains the bulkiest ligand, dbm, is about one order of magnitude smaller than that of **8**. In the same way complex **17**, with two trifluoromethyl groups on each β-diketonato ligand, has a smaller D_o than **16** which only has one trifluoromethyl group on each ligand. Complex **15** has a lower D_o when compared to **16** and **17**, which can possibly be attributed to its interaction with the supporting electrolyte as the nitro group, although formally neutral, has considerable polarisation between its nitrogen and oxygen atoms.

Relationships between the nature of the ligands and $E_{1/2}$

The ruthenocycle formed by the coordination of acac-type ligands to Ru cations can be considered pseudo-aromatic. As such, the Hammett constant can be used to correlate the combined electronic effects of 'para' and 'meta' substituents on the β-diketonates with the potential at which the metal centres oxidised on the basis that such complexes can be thought of as having similar geometries and a common redox centre, for which linear free energy relationships can be established.[54–56] In this study correlation of the calculated Hammett constants with the recorded $E_{1/2}$s is limited by the fact that the contribution of the bpy ligands is not included. Consequently, a reasonably linear correlation was observed for Series II and III but not for Series I (see ESI†).

Instead a more comprehensive method for probing the relationship between the structural and electronic features of a complex and its redox potential was applied.[57] The most commonly utilised additive model was originally developed by Lever[58] and has proved to be a useful tool for investigating metal-centred redox processes. The ligand electrochemical parameter, $E_L(L)$, used in this model and based on the $Ru^{2+/3+}$ redox couple, has been widely applied to examine the correlation between ligands and $E_{1/2}$ of metal complexes. For a metal (M) centre bound to multiple varying ligands (X_x, Y_y, Z_z), $E_L(L)$ is defined as:

$$\sum E_{(L)} = xE_L(X) + yE_L(Y) + zE_L(Z). \quad (2)$$

This formula can be used to predict the $E_{1/2}$ of complexes with an octahedral geometry.[58]

The $\sum E_{(L)}$ was calculated for all complexes in Series I–III where values for $E_{(L)}(L)$ were available (Table S4). A strongly linear correlation was established between $E_{(L)}(L)$ and $E_{1/2}$ of the complexes of Series I–III suggesting that the ligand contributions are additive (Fig. 1).

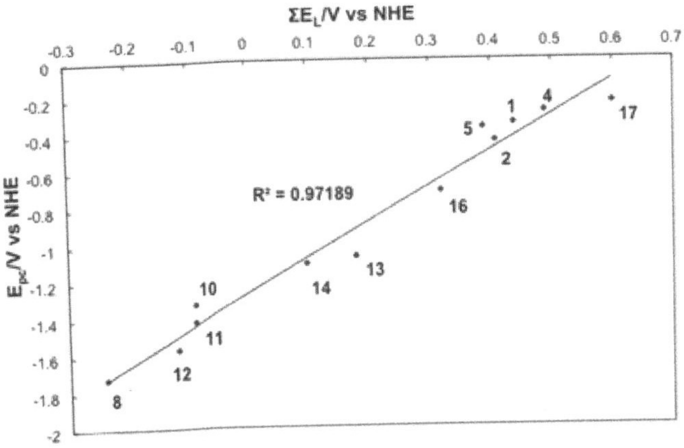

Figure 1: $\sum E_{(L)}$ vs. $E_{1/2}$ for selected complexes.

Correlation between UV-Vis absorbance and $E_{1/2}$

UV-vis spectra of complexes **1–7** (Series I) were recorded in MeCN at a concentration of 0.05 mM and the data is summarised in Table 4. In total, there are two strong absorption bands assigned to π→π* intraligand transitions in the UV region and three MLCT bands dπ (Ru^{2+})→π* (L) (L = bpy, β-diketonate) in the visible region. For complexes **1–3** the data reported here is in agreement with literature values.[48,59,60] The UV-vis spectrum for complex **4** (R^1 = Br, R^2 = Me) shows a slight hypsochromic shift in the MLCT bands (504 vs. 515 nm, 558 vs. 568 nm) when compared with those of **1**that could be indicative of an increased HOMO–LUMO gap. This is in agreement with the electrochemical data as a notably higher potential is required to oxidise the Ru(II) metal centre of **4** ($E_{1/2}$ = 292 mV) when compared with **1** ($E_{1/2}$ = 228 mV). Overall, the absorbance spectra of all Series I complexes are dominated by the influence of the anionic β-diketonato ligand and are consequently very similar to each other (see Fig. S4† for spectra).[61]

Table 4: Absorption maxima, λ_{max}, and molar extinction coefficients, ε for Series I

Complex	λ/nm (ε/10^3 M^{-1} cm^{-1})
1 (R^1 = H, R^2 = Me)	247 (28.4), 295 (53.8), 369 (11.3), 515 (9.3), 568sh (6.1)
2 (R^1 = Me, R^2 = Me)	247 (22.9), 296 (46.5), 374 (10), 521 (7.5), 586sh (5.1)
3 (R^1 = Et, R^2 = Me)	247 (26.0), 295 (5.9), 373 (11.2), 519 (10.4), 580sh (5.6)
4 (R^1 = Br, R^2 = Me)	246 (23.8), 294 (53.6), 370 (10.7), 504 (8.3), 558sh (5.5)
5 (R^1 = H, R^2 = ^iPr)	247 (25.2), 295 (47.8), 370 (10.2), 516 (7.7), 574sh (5.5)
6 (R^1 = p-MeOBn, R^2 = H)	246 (25.5), 295 (51.1), 376 (11.2), 518 (8.0), 577sh (5.7)
7 (R^1 = p-NO_2Bn, R^2 = H)	247 (20.7), 295 (42.1), 373 (8.5), 512 (5.9), 575sh (4.1)

In contrast to Series I, the UV-vis absorption spectra of complexes in series III are markedly different from each other (Table 5). The absorptions due to the four phenyl rings attached to complex **13**are clearly visible in the UV-vis spectrum at 246 nm. These additional aromatic groups may also account for the broad band observed around 300 nm which could result from a degree of overlapping with the π→π* intraligand transitions of bpy.

Table 5: Absorption maxima, λ_{max}, and molar extinction coefficients, ε for Series III complexes

Complex	λ/nm (ε/10^3 M^{-1} cm^{-1})
13 (dbm)	246 (28.8), 300 (34.8), 323 (31.2), 485 (9.0), 608 (7.5)
14 (acac)	277sh (24.5), 297 (17.5), 411 (8.8), 617 (4.9)
15 (NO_2-acac)	249 (14.8), 287 (19.7), 297 (21.0), 408 (8.5), 546 (5.14)
16 (tfac)	248 (15.0), 287sh (27.5), 294 (25.4), 425 (7.7), 559 (6.7)
17 (hfac)	243 (6.5), 289 (16.2), 337 (1.6), 509 (5.6)

The spectrum for complex **15** contains significant distortion of the band at around 300 nm where no clear maximum is visible. This could be attributed to an overlap of the nitro group absorption with the t_{2g}→π* (MLCT) band.[62] The strongly electron-withdrawing nitro group may also account for the significantly blue-shifted MLCT maxima observed at 546 nm.

Both **16** and **17** contain strongly electron-withdrawing trifluoromethyl groups on the β-diketonato ligands. In line with this, the spectrum of complex **16** contains a MLCT band at 559 nm whereas the corresponding peak for **17** appears at 509 nm (Fig. 2 and 3).

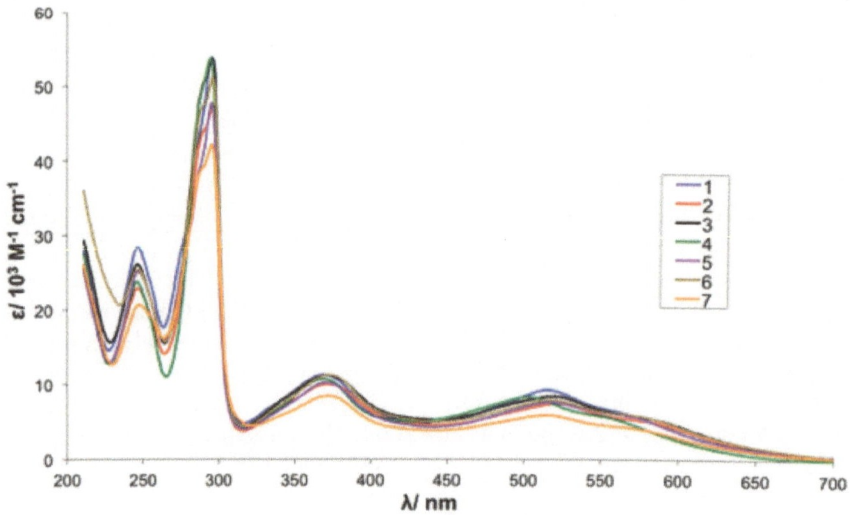

Figure 2: UV-Vis absorbance spectra for complexes **1–7**.

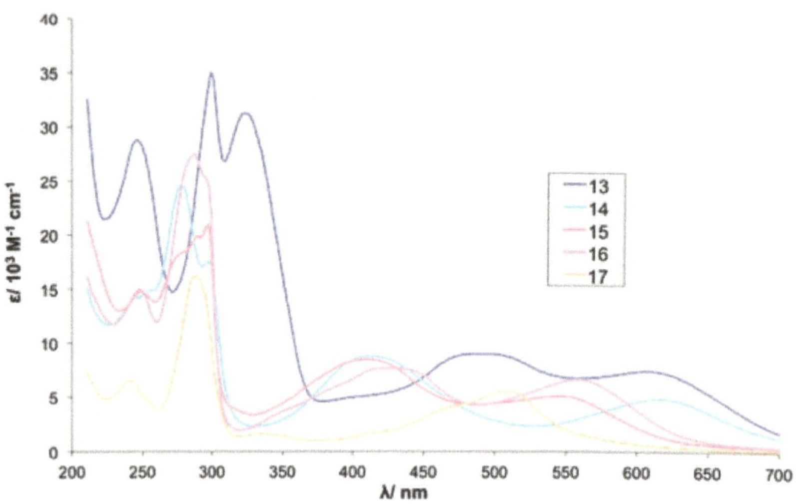

Figure 3: UV-Vis absorbance spectra for complexes **13–17**.

CONCLUSIONS

In this paper, ruthenium complexes bearing bpy and β-diketonato ligands were prepared, fully characterised and investigated using cyclic voltammetry and UV-vis spectroscopy. Three series of ruthenium complexes were investigated: Series I (**1–7**) of the general formula [Ru(bpy)$_2$(β-diketonato)](PF_6); Series II (**8–12**) of the general formula Ru(β-diketonato)$_3$ and Series III (**13–17**) of the general formula Ru(bpy)(β-diketonato)$_2$. In line with previous studies, varying the substituents tethered to the β-diketonato ligand attenuated the $E_{1/2}$ of the complexes. It was observed that EDG shifted the $E_{1/2}$ cathodically whereas EWG shifted the $E_{1/2}$ anodically. Furthermore the effects were shown to be cumulative based on the number of β-diketonato ligands bound to the metal centre. Further correlations were made between $E_{1/2}$ of complexes and the ligand electrochemical parameter ($\sum E_{(L)}$). The ligand electrochemical parameter was shown to have a linear relationship with $E_{1/2}$ for all three series.

Of the complexes prepared, the $E_{1/2}$ of complexes **7**, **13**, **14**, **15** and **16** were demonstrated to have $E_{1/2}$s within the range considered suitable for biological sensors (−0.743 V to +0.067 V vs. $FcH^{+/0}$). Furthermore, the novel complex **15** and the previously reported complex **16** were demonstrated each to have an $E_{1/2}$ very close to that of ferrocene ($E_{1/2}$ = −54 mV and −87 mV vs. $FcH^{+/0}$, respectively) suggesting that they would be also good candidates for redox labels in electroactive biosensors.

EXPERIMENTAL METHODS

All manipulations of metal complexes and air sensitive reagents were performed using either standard Schlenk techniques or in a nitrogen/argon filled Braun glove-box. Reagents were purchased from Aldrich Chemical Company Inc. or Alfa Aesar Inc. and were used without further purification unless otherwise stated. For the purposes of air sensitive manipulations and in the preparation of air sensitive complexes, pentane, hexane, dichloromethane and tetrahydrofuran were dispensed from a PuraSolv solvent purification system. Solvents were dried and distilled under an atmosphere of nitrogen using standard procedures and stored under nitrogen in glass ampoules, each fitted with a Youngs© Teflon valve prior to use. Ethanol (EtOH) was distilled from diethoxymagnesium and dimethylformamide (DMF) was first dried over molecular sieves (4 Å) and distilled. Argon (>99.999%) was obtained from Air Liquide and used as received. Nitrogen gas for Schlenk line operation comes from in-house liquid nitrogen boil-off. ^{1}H and ^{13}C NMR spectra were recorded on Bruker DPX300, DMX400, DMX500 and DMX600 spectrometers operating at 300, 400, 500 and 600 MHz (^{1}H) respectively and 75, 100, 125 and 150 MHz (^{13}C) respectively. Unless otherwise stated, spectra were recorded at 25 °C and chemical shifts (δ) are quoted in ppm. Coupling constants (J) are quoted in Hz and have uncertainties of ±0.05 Hz for ^{1}H and ±0.5 Hz for ^{13}C. ^{1}H and ^{13}C NMR chemical shifts were referenced internally to residual solvent resonances. Deuterated solvents were purchased from Cambridge Stable Isotopes and used as received. Microanalyses were carried out at the Campbell Micro-analytical Laboratory, University of Otago, New Zealand or at the Research School of Chemistry, the Australian National University, Canberra, Australia. Mass spectra were acquired using a Thermo LTQ Orbitrap XL located in the Bio-analytical Mass Spectrometry Facility (BMSF) of the Mark Wainwright Analytical Centre, UNSW or on a Micromass ZQ (ESI-MS) mass spectrometer located in the School of Chemistry, UNSW. M is defined as the molecular weight of the compound of interest or cationic fragment for cationic metal complexes.

Cyclic voltammetry was performed using an Autolab PGSTAT 12 potentiostat (Eco Chemie, Netherlands). The CV data was processed using Nova Windows software. A conventional three-electrode electrochemical cell comprised of a working electrode (glassy carbon), a counter electrode (platinum wire) and a reference electrode (aqueous or non-aqueous depending on the solvent system of the experiment) was used for all analysis. The aqueous reference electrode was a Ag/AgCl (3 M KCl) electrode (CH Instruments, Inc., TX, USA) while the non-aqueous reference electrode was a Ag/Ag^{+} electrode (CH Instruments, Inc., TX, USA) which was referenced against ferrocene. The working electrodes used were glassy carbon electrodes (CH Instruments, Inc., TX, USA). The solutions for electrochemical measurements were deoxygenated with nitrogen gas for 10 min prior to measurements and kept under a blanket of nitrogen during the course of measurements. UV-Vis spectra were recorded on Shimadzu UV-2401PC in dry MeCN (5×10^{-4} M) and reported as λ_{max}/nm (ε/M^{-1} cm^{-1}). For characterisation of previously reported compounds see the ESI.† All complexes were isolated as a racemic mixture of Δ and Λ enantiomers.

$[Ru(bpy)_2(eacac)][PF_6]$, 3

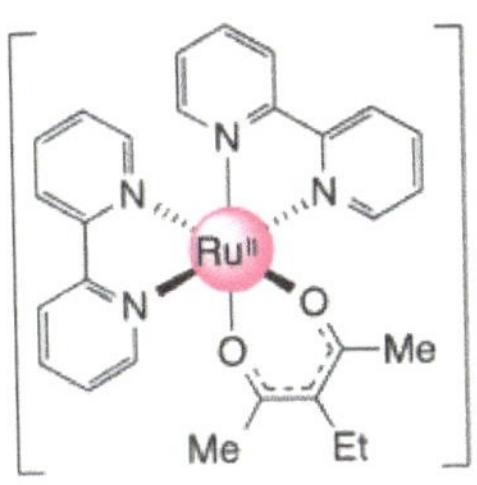

3

$Ru(bpy)_2Cl_2{\cdot}2H_2O$ (0.406 g, 0.780 mmol, 1 equiv.) was dissolved in degassed H_2O–EtOH (1 : 1) and heated to 75 °C for 30 minutes. 3-ethyl-2,4-pentanedione (eacac, 0.100 g, 0.780 mmol, 1.5 equiv.) was added to the solution followed by t-BuOK (0.131 g, 1.17 mmol, 1.5 equiv.). The mixture was then stirred at 75 °C for 1 h and cooled to room temperature before NH_4PF_6 (0.699 g, 4.29 mmol, 5.5 equiv.) was added to precipitate the product. The solid was collected and recrystallised from CH_2Cl_2–hexane to give the crude products. The dark solid was then purified using column chromatography (silica gel, CH_2Cl_2–MeCN 4 : 1). Yield: 0.05 g, 15%. ^{1}H NMR (400 MHz, DMSO-d_6): δ 0.92 (t, J = 6.8 Hz, 3H), 1.90 (s, 6H), 2.21 (q, J = 6.8 Hz, 2H), 7.21 (m, 2H), 7.66 (m, 2H), 7.74 (m, 2H), 7.83 (m, 2H), 8.16 (m, 2H), 8.62 (m, 4H), 8.75 (m, 2H) ppm. ^{13}C NMR (150.90 MHz, DMSO-d_6): δ 15.32, 23.51, 27.02, 109.61, 123.34, 123.45, 125.61, 126.37, 134.60, 136.44, 149.77, 152.73, 157.34, 158.76, 184.99 ppm. MS (ESI): m/z 541.1166 ($[M]^{+}$ required 541.1172). Anal. Calcd for $C_{27}H_{27}F_6N_4O_2PRu$: C, 47.30; H, 3.97; N, 8.17; found: C, 46.93; H, 3.94; N, 7.91.

$[Ru(bpy)_2(dmhd)][PF_6]$, 5

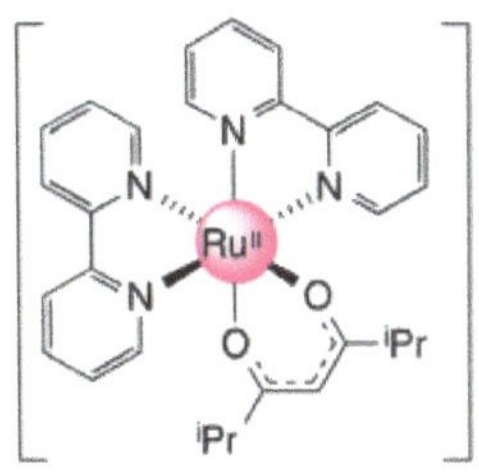

5

Prepared as for **3** with 0.100 g (0.640 mmol, 1.2 equiv.) of 2,6-dimethyl-3,5-heptanedione (dmhd). Yield: 0.37 g, 60%. ^{1}H NMR (300 MHz, DMSO-d_6): δ 0.58 (d, J = 6.8 Hz, 6H), 0.79 (d, J = 6.8 Hz, 6H), 2.25 (qq, J = 6.8, 6.8 Hz, 2H), 5.34 (s, 1H), 7.23 (ddd, J = 7.3, 7.3, 1.3 Hz, 2H), 7.73 (ddd, J = 7.3, 7.3, 1.3 Hz, 2H), 7.83–7.90 (m, 4H), 8.16 (ddd, J = 7.3, 7.3, 1.5 Hz, 2H), 8.50 (dd, J = 7.3, 1.3 Hz, 2H), 8.64 (dd, J = 7.3, 1.3 Hz, 2H), 8.75 (dd, J = 7.3, 1.5 Hz, 2H) ppm. ^{13}C NMR (150.90 MHz, DMSO-d_6): δ 19.98, 38.18, 94.27, 123.01, 123.21, 125.40, 126.03, 134.74, 136.35, 149.63, 153.11, 157.46, 158.75, 192.85 ppm. MS (ESI): m/z 569.1547 ($[M]^{+}$ required 569.1485). Anal. Calcd for $C_{29}H_{31}F_6N_4O_2PRu$: C, 48.81; H, 4.38; N, 7.85. Found: C, 48.15; H, 4.52; N, 7.70.

$[Ru(bpy)_2(mbpd)][PF_6]$, 6

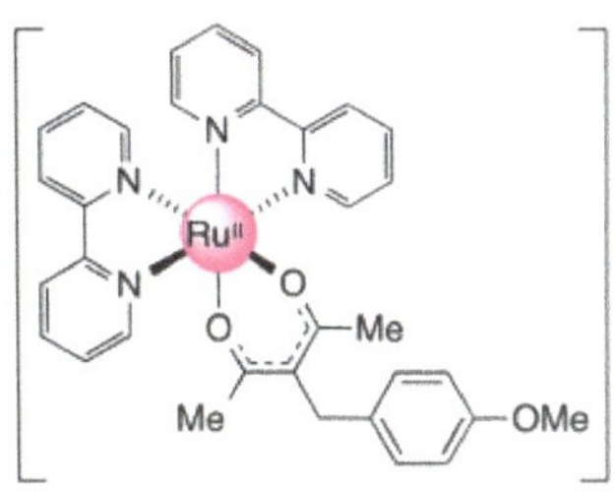

6

Prepared as for **3** with 0.111 g (0.503 mmol, 1.2 equiv.) of mbpd. Yield: 0.20 g, 53%. ^{1}H NMR (400 MHz, DMSO-d_6): δ 1.79 (s, 6H), 3.57 (s, 2H), 3.73 (s, 3H), 6.86 (m, 4H), 7.22 (m, 2H), 7.71 (m, 2H), 7.80–7.88 (m, 4H), 8.22 (m, 2H), 8.67 (m, 4H), 8.78 (m, 2H) ppm. ^{13}C NMR (100.64 MHz, DMSO-d_6): δ 27.56, 34.70, 54.99, 106.26, 113.76, 123.38,

123.44, 125.57, 126.37, 128.03, 133.26, 134.64, 136.51, 149.74, 152.79, 157.28, 157.36, 158.68, 185.82 ppm. MS (ESI): m/z 633.1420 ($[M]^+$ required 633.1434). Anal. Calcd for $C_{33}H_{31}F_6N_4O_3PRu$: C, 50.97; H, 4.02, N, 7.20. Found: C, 50.34; H, 4.04; N, 7.24.

$[Ru(bpy)_2(nbpd)][PF_6]$, 7

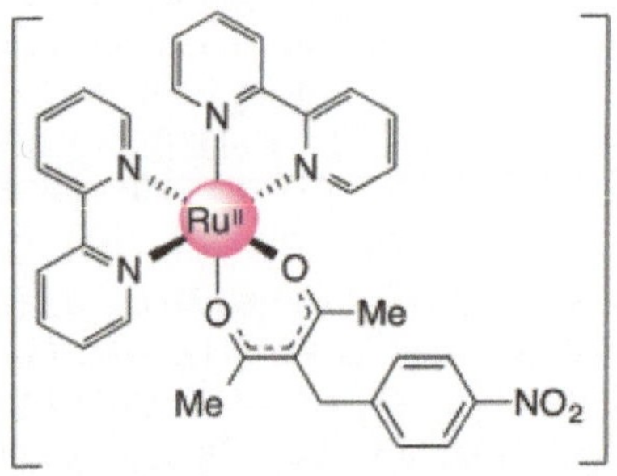

7

Prepared as for **3** with 0.113 g (0.480 mmol, 1 equiv.) of nbpd. Yield: 0.20 g, 27%. ^{1}H NMR (300 MHz, DMSO-d_6): δ 1.79 (s, 6H), 3.81 (s, 2H), 7.23 (m, 2H), 7.28 (m, 2H), 7.72 (m, 2H), 7.84–7.88 (m, 4H), 8.16 (m, 2H), 8.23 (m, 2H), 8.65 (m, 2H), 8.71 (m, 2H), 8.78 (m, 2H) ppm. ^{13}C NMR (100.64 MHz, DMSO-d_6): δ 27.74, 35.82, 105.25, 123.42, 123.46, 123.52, 125.61, 126.49, 128.39, 134.73, 136.65, 145.84, 149.77, 150.49, 152.85, 157.29, 158.68, 185.96 ppm. MS (ESI): m/z 648.1188 ($[M]^+$ required 648.1179). Anal. Calcd for $C_{32}H_{28}F_6N_5O_4PRu$: C, 48.49; H, 3.56, N, 8.84. Found: C, 47.85; H, 3.53; N, 8.80.

$Ru(acac\text{-}NO_2)_3$, 9

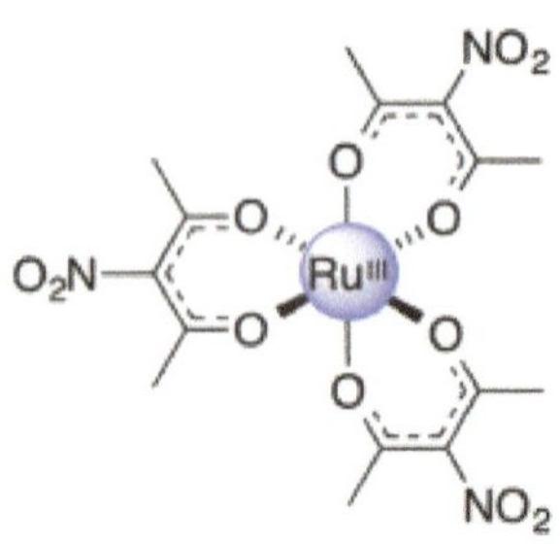

9

Acetic anhydride (30 mL) was added to finely ground $Cu(NO_3)_2{\cdot}3H_2O$ (2.35 g, 9.73 mmol) to give a light blue suspension. The contents were stirred at 0 °C for 15 min after the flask was fitted with a calcium chloride drying tube. **8** (1.20 g, 3.01 mmol) was added to the cold deep blue solution. The mixture was stirred at 0 °C for 2 h and a further 2 h at room temperature. Ice (100 g), deionised water (100 g) and anhydrous sodium acetate (2.14 g, 26.1 mmol) were added to the now reddish brown mixture. The colour immediately turned greenish blue. The solution was left to stir for 18 h. The contents were filtered to give bright red powder. The solid was washed with water to give product as a bright red powder. Yield: 1.12 g, 70%. ^{1}H NMR (400 MHz, $CDCl_3$): δ −3.55 (s, 18H) ppm. MS (ESI): m/z 556.9813 ($[M + Na]^+$ required 556.9826). Anal. Calcd for $C_{15}H_{18}N_3O_{12}Ru$: C, 33.78; H, 3.40; N, 7.88. Found: C, 33.01; H, 3.54; N, 7.65.

$Ru(NO_2\text{-}acac)_2(bpy)$, 15

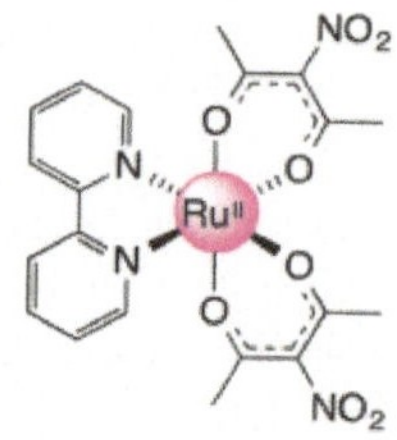

15

(a) Compound **9** (0.574 g, 1.08 mmol) was stirred in EtOH with activated zinc dust (0.5 g) for 1 h, during which time the colour changed from bright red to brown. MeCN (5 mL) was added to the brown mixture and refluxed for 4 h. The mixture was filtered through a bed of celite on which a brown layer remained. The crude product was subjected to silica gel column chromatography to first elute unreacted **9** with CH_2Cl_2 followed by EtOAc to flush the product out from the column as an orange fraction. Solvent was removed to give the title product as an orange solid. Yield: 0.46 g, 92%. ^{1}H NMR (400 MHz, DMSO-d_6): δ 2.09 (s, 6H), 2.11 (s, 6H), 2.73 (s, 6H) ppm. ^{13}C NMR (100.64 MHz, DMSO-d_6): δ 3.83, 26.59, 26.83, 128.46, 139.13, 183.15, 184.52 ppm. MS (ESI): m/z 495.0052 ($[M + Na]^+$ 495.0060).

(b) The $Ru(acac\text{-}NO_2)_2(MeCN)_2$ intermediate isolated in (a) (0.200 g, 0.424 mmol) and bpy (0.0660 g, 0.423 mmol) were added to a Schlenk flask followed by EtOH (~15 mL). The reaction was refluxed for 5 h before solvent was removed under reduced pressure to give a dark brown solid. The solid was purified by column chromatography (silica gel, MeCN–CH_2Cl_2 1 : 5) to give a dark brown solid. Yield: 0.030 g, 13%. ^{1}H NMR (500 MHz, DMSO-d_6): δ 1.76 (s, 6H), 2.28 (s, 6H), 7.52 (m, 2H), 7.90 (m, 2H), 8.59 (m, 2H), 8.68 (m, 2H) ppm. ^{13}C NMR (150.90 MHz, DMSO-d_6): 26.57, 27.11, 122.74, 125.21, 128.44, 134.66, 152.07, 159.82, 182.14, 184.49 ppm. MS (ESI): m/z 569.0209 ($[M + Na]^+$ required 569.0217). Anal. Calcd for $C_{20}H_{20}N_4O_8Ru$: C, 44.04; H 3.70; N 10.27. Found: C, 44.17; H, 3.77; N, 10.04.

ACKNOWLEDGEMENTS

Financial support from the University of New South Wales and the Australian Research Council (LP100200593) is gratefully acknowledged. We also would like to thank staff of the NMR Facility, Mark Wainwright Analytical Centre, UNSW for their assistance.

NOTES AND REFERENCES

1. N. Metzler-Nolte and M. Salmain, in Ferrocenes, John Wiley & Sons, Ltd, 2008, pp. 499–639 DOI:10.1002/9780470985663.ch13.
2. D. R. van Staveren and N. Metzler-Nolte, Chem. Rev., 2004, 104, 5931–5986 .
3. M. F. R. Fouda, M. M. Abd-Elzaher, R. A. Abdelsamaia and A. A. Labib, Appl. Organomet. Chem., 2007, 21, 613–625.

4. M. Kind and C. Woell, Prog. Surf. Sci., 2009, 84, 230–278 .

5. I. Willner and E. Katz, Angew. Chem., Int. Ed., 2000, 39, 1181–1218 .

6. Y. Wan, Y. Su, X. Zhu, G. Liu and C. Fan, Biosens. Bioelectron., 2013, 47, 1–11 .

7. S. Martic, M. Labib, P. O. Shipman and H.-B. Kraatz, Dalton Trans., 2011, 40, 7264–7290 .

8. T. Joshi, G. J. Barbante, P. S. Francis, C. F. Hogan, A. M. Bond, G. Gasser and L. Spiccia, Inorg. Chem., 2012, 51, 3302–3315 .

9. S. M. Khor, G. Liu, J. R. Peterson, S. G. Iyengar and J. J. Gooding, Electroanalysis, 2011, 23, 1797–1804.

10. R. Prins, A. R. Korwagen and A. G. T. G. Kortbeek, J. Organomet. Chem., 1972, 39, 335–344.

11. D. Ge and R. Levicky, Chem. Commun., 2010, 46, 7190–7192 .

12. R. Prins, A. R. Korswagen and A. G. T. G. Kortbeek, J. Organomet. Chem., 1972, 39, 335–344.

13. H. Ju and D. Leech, Phys. Chem. Chem. Phys., 1999, 1, 1549–1554 .

14. S. M. Zakeeruddin, D. M. Fraser, M. K. Nazeeruddin and M. Grätzel, J. Electroanal. Chem., 1992, 337, 253–283.

15. D. M. Fraser, S. M. Zakeeruddin and M. Grätzel, J. Electroanal. Chem., 1993, 359, 125–139.

16. C. Zhang, T. Haruyama, E. Kobatake and M. Aizawa, Anal. Chim. Acta, 2000, 408, 225–232.

17. S. Warren, T. McCormac and E. Dempsey, Bioelectrochemistry, 2005, 67, 23–35 .

18. T. V. Federova, A. S. Vilesov, S. A. Kurzeev, E. V. Stepanova, E. O. Landesman and O. V. Koroleva, Appl. Biochem. Microbiol., 2006, 42, 550–557 .

19. E. V. Ivanova, V. S. Sergeeva, J. Oni, C. Kurzawa, A. D. Ryabov and W. Schuhmann, Bioelectrochemistry, 2003, 60, 65–71.

20. I. L. Medintz, D. Farrell, K. Susumu, S. A. Trammell, J. R. Deschamps, F. M. Brunel, P. E. Dawson and H. Mattoussi, Anal. Chem., 2009, 81, 4831–4839 .

21. M.-Y. Wei, S.-D. Wen, X.-Q. Yang and L.-H. Guo, Biosens. Bioelectron., 2009, 24, 2909–2914 .

22. N. A. Yusof, Asian J. Chem., 2011, 23, 1153–1157 .

23. W.-W. Yang, Y.-W. Zhong, S. Yoshikawa, J.-Y. Shao, S. Masaoka, K. Sakai, J. Yao and M. Haga, Inorg. Chem., 2012, 51, 890–899 .

24. R. Cerón-Camacho, S. Hernández, R. L. Lagadec and A. D. Ryabov, Chem. Commun., 2011, 47, 2823–2825 .

25. H. Weizman and Y. Tor, J. Am. Chem. Soc., 2002, 124, 1568–1569 .

26. M. Vrábel, M. Hocek, L. Havran, M. Fojta, I. Votruba, B. Klepetářá, R. Pohl, L. Rulíšek, L. Zendlová, P. Hobza, I. Shih, E. Mabery and R. Mackman, Eur. J. Inorg. Chem., 2007, 1752–1769 .

27. J. J. Rack, E. S. Krider and T. J. Meade, J. Am. Chem. Soc., 2000, 122, 6287–6288.

28. A. Endo, M. Kajitani, M. Mukaida, K. Shimizu and G. P. Sato, Inorg. Chim. Acta, 1988, 150, 25–34.

29. N. L. Frank and T. J. Meade, Inorg. Chem., 2003, 42, 1039–1044 .

30. M. Haga, T. Matsumura-Inoue, K. Shimizu and G. P. Satô, J. Chem. Soc., Dalton Trans., 1989, 371–373 .

31. L. Wolf, E. Butter and H. Weinelt, Z. Anorg. Allg. Chem., 1960, 306, 87–93.

32. A. Endo, K. Shimizu and G. P. Satô, Chem. Lett., 1985, 1985, 581–584 .

33. S. Munery, J. Jaud and J. Bonvoisin, Inorg. Chem. Commun., 2008, 11, 975–977 .

34. M. S. Chan and A. C. Wahl, J. Phys. Chem., 1985, 89, 5829–5831.

35. S. Munery, J. Jaud and J. Bonvoisin, Inorg. Chem. Commun., 2008, 11, 975–977 .

36. J.-H. Olivier, J. Harrowfield and R. Ziessel, Chem. Commun., 2011, 47, 11179–11188 Search .

37. J. P. Collman, Angew. Chem., Int. Ed., 1965, 4, 132–138 .

38. J. P. Collman, R. A. Moss, H. Maltz and C. C. Heindel, J. Am. Chem. Soc., 1961, 83, 531–534.

39. R. W. Kluiber, J. Am. Chem. Soc., 1960, 82, 4839–4842.

40. P. R. Singh and R. Sahai, Aust. J. Chem., 1967, 20, 639–648.

41. P. R. Singh and R. Sahai, Aust. J. Chem., 1967, 20, 649–655.

42. J. P. Collman, R. L. Marshall, W. L. I. Young and S. D. Goldby, Inorg. Chem., 1962, 1, 704–710.

43. I. R. Baird, S. J. Rettig, B. R. James and K. A. Skov, Can. J. Chem., 1999, 77, 1821–1833.

44. A. K. Gupta and R. K. Poddar, Indian J. Chem. Sect. A: Inorg., Bio-inorg., Phys., Theor. Anal. Chem., 2000, 39, 457–460 Search .

45. T. Kobayashi, Y. Nishina, K. Shimizu and G. P. Satô, Chem. Lett., 1988, 1137–1140.

46. R. A. Krause, Inorg. Chim. Acta, 1977, 22, 209–213.

47. A. M. El-Hendawy, A. H. Al-Kubaisi and H. A. Al-Madfa, Polyhedron, 1997, 16, 3039–3045.

48. A. M. El-Hendawy, A. H. Al-Kubaisi and H. A. Al-Madfa, Polyhedron, 1997, 16, 3039–3045.

49. E. Eskelinen, M. Haukka, T.-J. J. Kinnunen and T. A. Pakkanen, J. Electroanal. Chem., 2003, 556, 103–108.

50. N. G. Tsierkezos, J. Solution Chem., 2007, 36, 289–302.

51. A. Endo, Y. Hoshino, K. Hirakata, Y. Takeuchi, K. Shimizu, Y. Furushima, H. Ikeuchi and G. P. Satô, Bull. Chem. Soc. Jpn., 1989, 62, 709–716.

52. H. Ikeuchi, K. Naganuma, M. Ichikawa, H. Ozawa, T. Ino, M. Sato, H. Yonezawa, S. Mukaida, A. Yamamoto and T. Hashimoto, J. Solution Chem., 2007, 36, 1243–1259.

53. A. Endo, Bull. Chem. Soc. Jpn., 1983, 56, 2733–2738.

54. C. Hansch, A. Leo and R. W. Taft, Chem. Rev., 1991, 91, 165–195.

55. G. S. Patterson and R. H. Holm, Inorg. Chem., 1972, 11, 2285–2288.

56. S. J. Slattery, W. D. Bare, D. L. Jameson and K. A. Goldsby, J. Chem. Soc., Dalton Trans., 1999, 1999, 1347–1352 .

57. A. J. L. Pombeiro, Eur. J. Inorg. Chem., 2007, 1473–1482.

58. A. B. P. Lever, Inorg. Chem., 1990, 29, 1271–1285.

59. A. Wu, J. Masland, R. D. Swartz, W. Kaminsky and J. M. Mayer, Inorg. Chem., 2007, 46, 11190–11201 .

60. G. M. Bryant, J. E. Fergusson and H. K. J. Powell, Aust. J. Chem., 1971, 24, 257–273.

61. Y. Takahashi, H. Arakawa, H. Sugihara, K. Hara, A. Islam, R. Katoh, Y. Tachibana and M. Yanagida, Inorg. Chim. Acta, 2000, 310, 169–174.

62. K. Kuroda, K. Yoshitani, K. Kunigita, Y. Kamiiba and K. Watanabe, Bull. Chem. Soc. Jpn., 1976, 49, 2445–2450.

CHAPTER 15

INVESTIGATION INTO THE USE OF ELECTROCHEMICAL EXTRACTION TO DRAW GOLD FROM REFRACTORY ORES

Bagdaulet Kenzhalyevich Kenzhaliyev, Zamzagul Duisenovna Dosymbaeva, Renata Ramilevna Iskhakova and Essen Nurgalievich Suleimenov

Kazakh-British Technical University, Almaty, Republic of Kazakhstan

ABSTRACT

This paper studies the possibility ofextracting gold by electrochemical leaching through the use of a sodiumthiosulfate solution. In this regard, the requirements for environmentallysound technologies play a huge and vital role. Low extraction ratios and metallosses increase the cost of this process significantly and therefore it is necessaryto reduce or prevent these disadvantages. Dissolution of gold at variouselectrolyte concentrations has been studied. The influence of current densityon the leaching of gold within the solution has been investigated. Theinfluence of the ratio S:L on gold extraction during electroleaching has beenstudied and the specific rate of gold leaching with different variations of theparameters has been defined. The sorption behavior of gold thiosulfatecomplexes on the anion exchanger AM-2B at various concentrations of sodiumthiosulfate taken from model solutions has been studied. Research into thecomparative study relating to the sorption of gold compounds from complexproductive solutions used in electrochemical leaching has been conducted.

INTRODUCTION

The majority of nonferrous metal ores deposits are developed over many decades, at the same time stocks of semi rich and relatively free milling ores are annually reduced, with the so-called refractory ores being sent on to concentration factories (refining factory) for further processing. These ores are complex in nature having high commercial value for both minerals of nonferrous metals (copper, zinc, lead) as well as for noble metals: Gold, silver, the platinum group metals and rare elements. Technologies used in the processing of these ores include, along with the release of concentrates of nonferrous metals, the extraction of noble metals.

Extraction throughout the world is ever increasing with regards to raw materials with low contents of valuable components with complex material composition, increasing

Citation: Bagdaulet Kenzhalyevich Kenzhaliyev, Zamzagul Duisenovna Dosymbaeva, Renata Ramilevna Iskhakova and Essen Nurgalievich Suleimenov, Investigation into the Use of Electrochemical Extraction to Draw Gold from Refractory Ores, DOI : 10.3844/ajassp.2015.857.864.

the requirements during processing. This means that more advanced technological schemes, as well as highly efficient and cost-effective methods of extraction of nonferrous and noble metals are required (Parga et al., 2012) and exactly through greater involvement and technical innovation in the extraction of complex refractory gold and goldbearing ores perhaps, in the future, we can provide the main growth of gold production in the world (Adams, 2005; Vignes, 2011). In the present paper the possibility of extracting gold by electrochemical leaching using a solution of sodium thiosulfate has been studied. In this regard, the requirements for environmentally sound technologies play a significant role. Low extraction ratios and metal losses leads to higher prices of ongoing processes. Therefore it is necessary to prevent these disadvantages.

LITERATURE REVIEW

Kholmogorov et al. (2002a) concluded that thiocyanate is an "effective lixiviant for gold in acidic conditions, increasing the rate of dissolution, which is comparable with the results obtained with thiourea". According to (Kholmogorov et al., 2002b), more than 95% of thiocyanate complexes of gold on carbon adsorbents and ion exchangers have been extracted from weak acidic solutions. Modern alternatives to cyanide leaching (McNulty, 2001) are ammonium thiosulfate (Breuer and Jeffrey, 2000; Aylmore, 2001) and thiocarbamide leaching (Ubaldini et al., 1998). Despite the proven effectiveness of thiourea as a gold leaching reagent (Lee et al., 1997; Yao et al., 1992; Tremblay et al., 1996) certain deficiencies of this method have been identified: Increased costs compared to cyanide; high reagent consumption; gold extraction requires many steps.

Thiosulphates of alkali metals are biodegradable and practically safe (Abbruzzese et al., 1995; Feng and Van Deventer, 2002), which permits their use in the geotechnologies of underground and heap leaching, they are cheaper than cyanide and they are more effective in the case of refractory carbon and copper-containing ores. Thiosulfate leaching has obvious advantages, when considering both technological and environmental parameters (Aylmore and Muir, 2001; Zipperian and Raghavan, 1988). The problem of reagents does not exist. This, along with the resources savings, is the main advantage of using thiosulfate technology which remains safe during all stages of technological conversion as well as ensuring staff health and safety. Currently, there are almost no examples of actual industrial use of thiosulfate leaching in the gold industry primarily due economic reasons.

Currently, the primary way of increasing the competitiveness and profitability of gold production is a combination of traditional mining technology with new technologies (Qingshuang et al., 1995; Balaz et al., 2000), heap and electrosorption leaching, electrochemical extraction (Zhou et al., 2009), hydrometallurgical repartitioning of solutions (Grosse et al., 2003) etc. However, capital costs for gold production remain high and the cost price of gold is constantly growing. There are a large number of studies on the use of electro processing in the extraction of metals in solutions, however, the reactions when transferring metals in solutions of different chemical composition by electric current-electrochemical leaching (Kenzhaliev, 2014) occupy a special place in the studies of electrochemical processes.

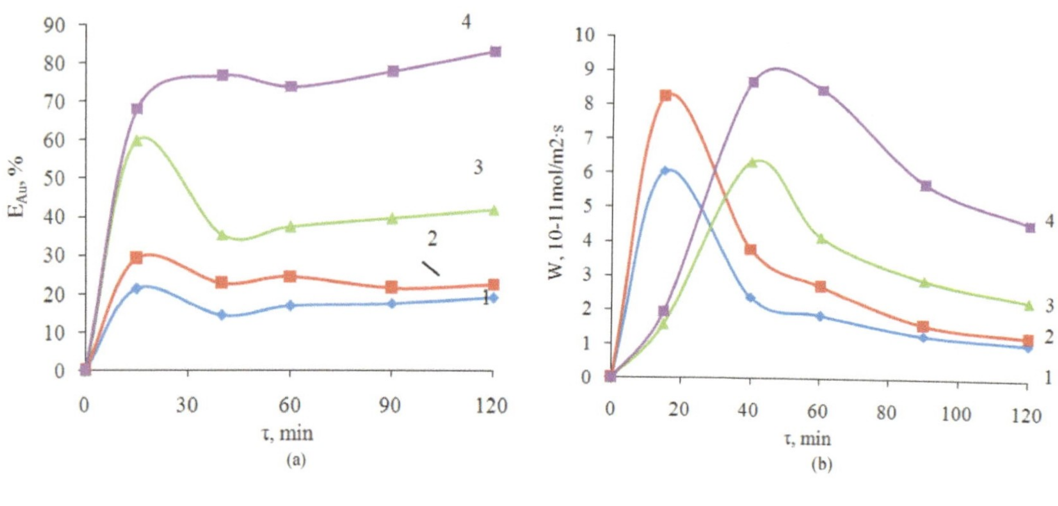

Figure 1: The degree of gold extraction versus the concentration of the electrolyte (a) and the specific rate of gold leaching (b) Concentration levels of the electrolyte: 1-0.1, 2-0.5, 3-1.0 and 4-2.0 M.

MATERIALS AND METHODS

Studies into the effects that the parameters of the electroleaching process have in gold extraction were carried out in a specially designed and constructed thermostatic reaction cell. The cell has a threeelectrode system. As the working electrode sulfur graphite was used, as the auxiliary electrode graphite electrodes were used with a silver chloride electrode serving as a reference electrode. Electroleaching was performed with sodium thiosulfate in an alkaline medium under various experimental parameters: Temperature (25; 40; 50; 60°C), current density (100, 200, 300 and 400 A/m^2), the concentration of the electrolyte (0.1; 0.5; 1.0 and 2.0 M) the ratio S:L (1:3; 1:4; 1:6; 1:10). Study duration-2 h. During the experiment the samples were taken for analysis as well as the pH of the medium and electroconductivity-Eh and concentration of Dissolved Oxygen (DO_2) have been measured by using a SensIon 156 portable analyzer, made by "HACH" (USA).

RESULTS

Defining the parameters when electrochemical leaching a gold solution. During electrochemical leaching the dissolution of metals at various concentrations of the electrolyte was observed. Dependence of the degree of gold extraction into the solution (a) and the specific rate of the gold leaching (b) are shown in Fig. 1.

The dependence of the change of oxygen concentration from the electrolyte concentration is shown in Fig. 2. With a thiosulfate concentration of 0.1 M, maximum gold extraction is 21.4%, the leaching of metals continues and the volume of the oxygen decreases, this is indirectly confirmed by the graph of the electrical conductivity from the electrolyte concentration shown in Fig. 3. Up to the 30 min mark of the process, the electric conductivity and the oxygen concentration are both decreasing. After 30 min, the electric conductivity begins to increase up to a value of 59.15 μS/cm by shifting additional components (leaching of metals) into the solution.

As can be seen in the figures, when the electrical conductivity of the solution decreases slightly and then increases it is likely due to changes in the concentration of the solution which is probably a result of the solution restructuring.

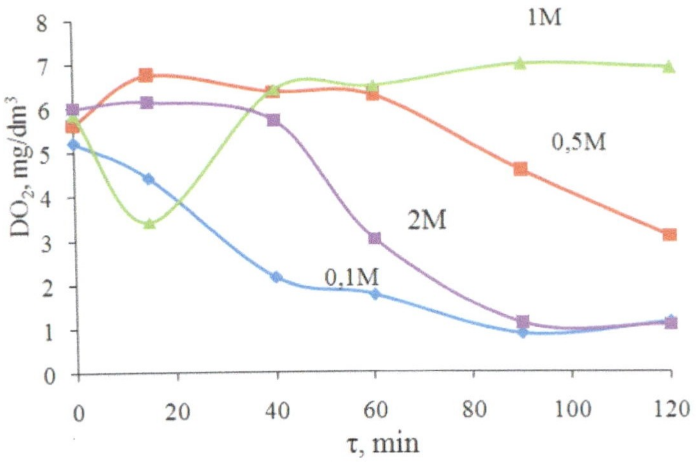

Figure 2: The oxygen concentration variation versus the electrolyte concentration Concentration levels of the electrolyte: 1-0.1, 2-0.5, 3-1.0 and 4-2.0 M

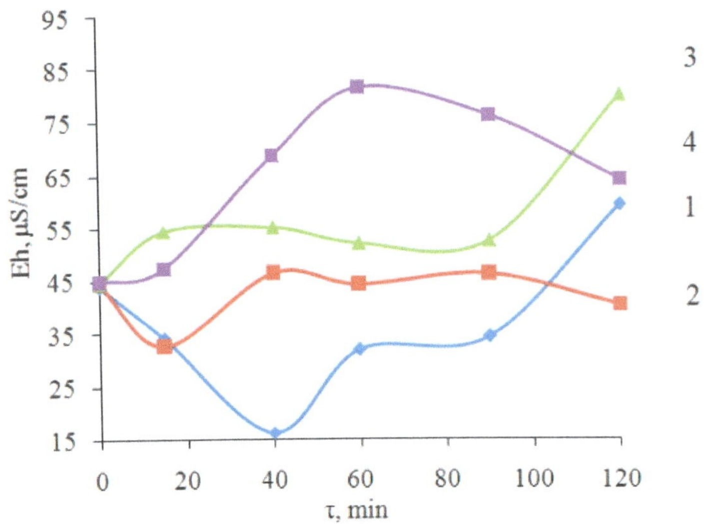

Figure 3: The electric conductivity variation versus the electrolyte concentration Electrolyte concentration: 1- 0.1, 2-0.5, 3-1.0 and 4-2.0 M.

Further studies have been conducted for determining the optimum current density using sodium thiosulfate with a concentration of 2.0 M with different current densities: 100, 200, 300 and 400 A/m² (Fig. 4a). The results showed that the specific rate of gold leaching during the process increases sharply, namely within the first 40 min of the experiment, followed by a gradual decrease in the reaction rate as the process continues (Fig. 4b). Here the characteristics of dependences is the same with all current densities with the highest specific rate of gold leaching achieved with a density of 200 A/m2 ($8.65\cdot10^{-11}$ mol/m² ·s). Results of the research into the extent of gold extraction from the concentrate depending on the S:L ratio are shown in Fig. 5. As shown in Fig. 5b, the specific rates increased up to the 40 min mark and then declined, this is directly connected with the decreasing metal content in the initial product as well as with decreasing dissolved oxygen content, which is consumed in the oxidation process of metals. The influence that temperature has on the dissolution rate of gold and silver from the concentrate in the temperature ranges from 25 to 60°C has been studied. A solution with a concentration of 2.0 M $Na_2S_2O_3{\cdot}5H_2O$ was used. Results for the gold leaching are shown in Fig. 6 as an example.

The dependence of the change in the specific rate from leaching time is shown in Fig. 6b. The effectiveness of sorption extraction of thiosulfate gold complexes have been evaluated according to the values of the extraction degree (E, %). Sorption isotherms have been obtained by method of variation in the molar ratio. The next step was to study the effect of the ratio in the volume of the anion exchanger to the volume of the contacting solution (V_{resin}:$V_{solution}$, S:L) on the overall sorption process. The effectiveness of sorption extraction on thiosulfate complexes of gold have been evaluated based on the values of the degree of sorption (extraction) (E, %) (Fig. 8).

In the next stage of the work a comparative study into the sorption of gold compounds from complex productive solutions of electrochemical leaching has been conducted. The kinetic curves of sorption from productive solutions of electrochemical leaching are shown in Fig. 9. The rates of sorption shown in Table 1 for solutions of electrochemical leaching correspond to the results on a comparative study of the sorption behavior of gold compounds and accompanying metals from technological solutions.

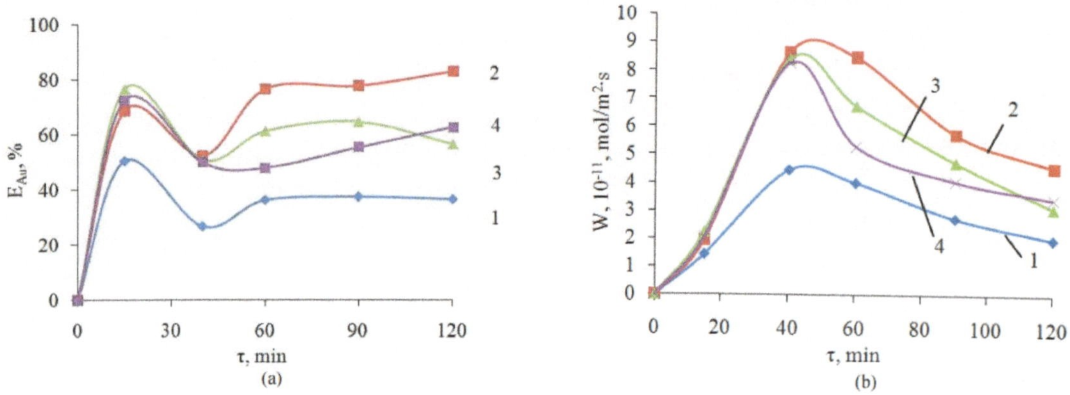

Figure 4: The degree of gold extraction versus current density (a) and the specific rate of gold leaching (b) Current density: 1-100, 2- 200, 3-300 and 4-400 A/m².

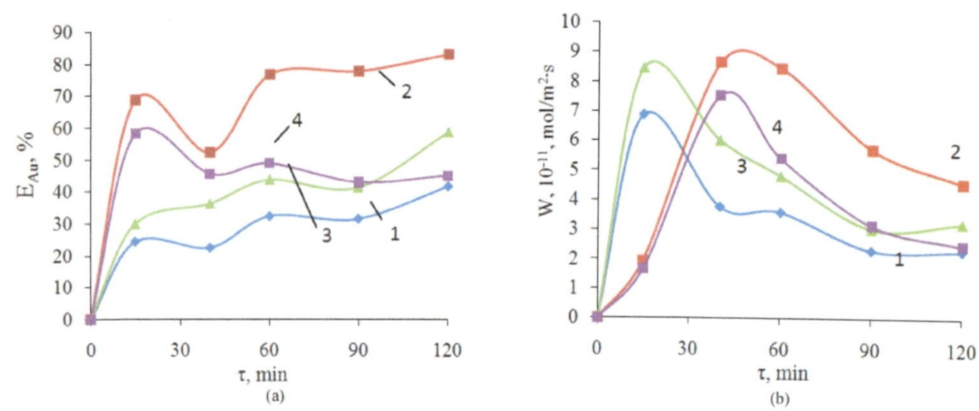

Figure 5: The degree of gold extraction by ratio of S:L (a) versus the specific rate of gold leaching by ratio of S:L (b) Ratioof S:L : 1- 1:3; 2-1:4; 3-1:6; 4-1:10; i = 200 A/m2 ; t = 25°C.

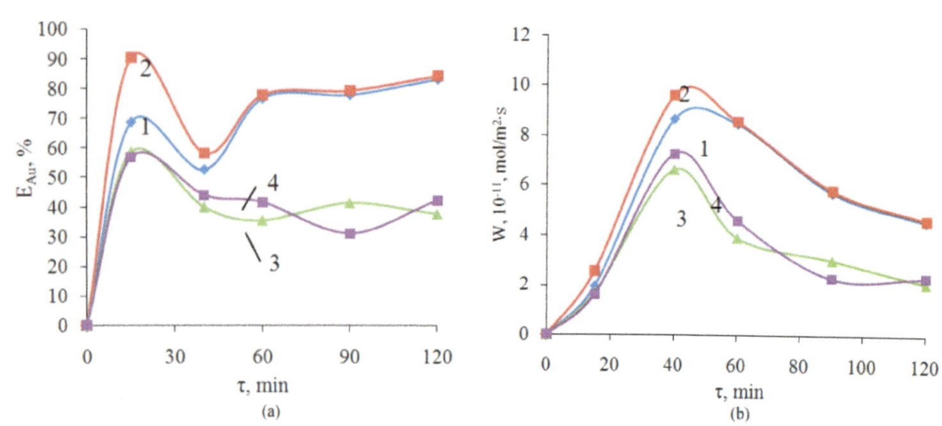

Figure 6: The dependence of the degree of gold extraction by temperature (a) and the specific rate of gold leaching by temperature (b) Temperature: 1-25, 2-40, 3-50 and 4-60°C; S:L: = 1:4; i = 200 A/m².

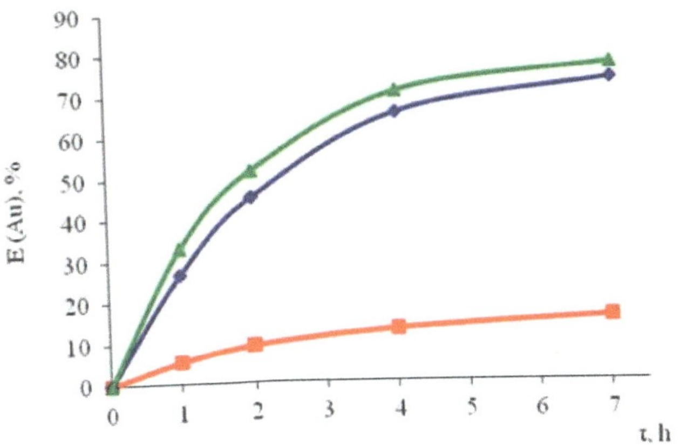

Figure 7: The dependence of the degree of gold sorption (a) and silver (b) by concentration of sodium thiosulfate Concentration of $Na_2S_2O_3$: 1-0.1 mol/dm³, 2-0.2 mol/dm³, 3-0.5 mol/dm3 ; t = 25°C.

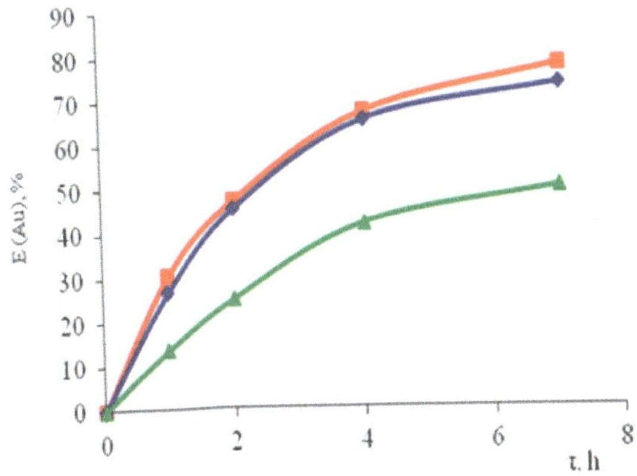

Figure 8: The effect of the ratio S:L on the degree of sorption of gold from model solutions S:L ratio: 1-1:200, 2-1:500, 3- 1:800; concentration of $Na_2S_2O_3$ -0.2 mol/dm3 ; t = 25°C.

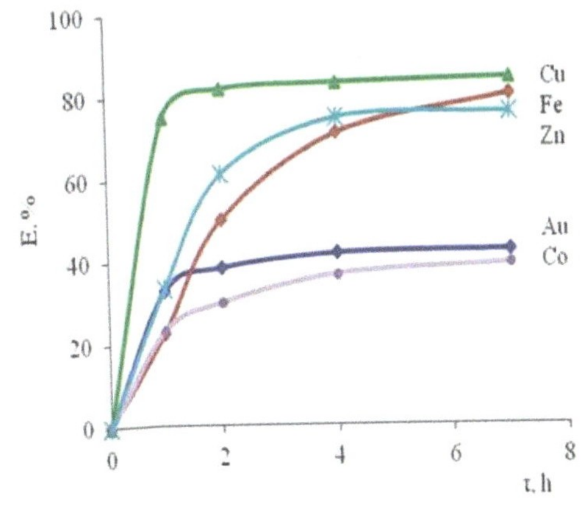

Figure 9: The degree of sorption extraction of metals from the electrochemical leaching solution S:L = 1:500, t = 25°C.

Table 1: The rates of sorption of compounds of gold, silver and non-ferrous metals from electrochemical leaching solutions

	The solution after electrochemical leaching			
Time, min	60	120	240	420
$W_{Au} \cdot 10^{-10}$, mg·eqv/s	3,5468	1,9851	1,0550	0,6102
$W_{Cu} \cdot 10^{-10}$, mg·eqv/s	3633,40	1984,06	1008,50	583,31
$W_{Fe} \cdot 10^{-10}$, mg·eqv	0,7587	0,8333	0,5907	0,3817
$W_{Co} \cdot 10^{-10}$, mg·eqv/s	0,3300	0,2124	0,1297	0,0793
$W_{Zn} \cdot 10^{-10}$, mg·eqv/s	319,9815	293,1660	179,3192	104,0458

DISCUSSION

As can be clearly seen, during the growth of gold extraction the concentration of oxygen is significantly reduced. At a concentration level of 0.1 M gold extraction increases up to 14.34%, at 0.5 M up to 22.78%, at 1.0 M up to 35.33% and at a concentration of2 M up to 76.67% in the first 30 min of leaching and then gradually increasing to 82% (Fig. 1a) with the oxygen concentration beginning to decrease after 40 min.

The specific rate at concentrations of 0.1 M and 0.5 M is increasing up to the 20 min mark of the leaching process and then reduces. At concentrations of 1.0 M and 2.0 M it is increasing up to 40 min and then reduces which characterizes the complexity of the process (Fig. 1b). In carrying out leaching over a 15 min period with 0.1 M and 2.0 M thiosulphate sodium the specific rates of gold leaching were $6.04 \cdot 10^{-11}$ mol/m² ·s and $8.44 \cdot 10^{-11}$ mol/m² ·s respectively. With leaching using 0.5 M and 1.0 M thiosulphate sodium concentrations the specific rates of reactions were: $8.22 \cdot 10^{-11}$ mol/m² ·s and $6.32 \cdot 10^{-11}$ mol/m² ·s when leaching is carried out over a 40 min period.

At a concentration of 1.0 M the oxygen concentration is reduced during the first 20 min then increases up to 6.98 mg/dm³ . As seen on the curve, reprecipitation of gold on the electrode occurs. At a thiosulfate concentration of 0.5 M, the concentration of oxygen in the system has little connection with the extraction of gold in the system. At a concentration of 0.1 M, extraction of gold in the solution is low with the oxygen concentration dropping to a value of 0.87 mg/dm³ . With gold leaching by thiosulfate with a density of 100 and 300 A/m² , values of the specific rate are $4.44 \cdot 10^{-11}$ mol/m² ·s and $8.41 \cdot 10^{-11}$ mol/m² ·s respectively. At 400 A/m² the specific rate of reactions is $8.26 \cdot 10^{-11}$ mol/m² ·s.

Research results showed that the specific rate of the process for all values of current densities at the initial time ($24 \cdot 10^2$ s) increases, then over time (from $24 \cdot 10^2$ -to $54 \cdot 10^2$ s) this indicator gets sharply reduced, which is connected with the reduction of metal content in the initial product sample as well as with the decreasing dissolved oxygen content, which has been consumed in the process of metal oxidation. Moreover there is a reduction in oxygen solubility with decreasing of pH levels in the solution. The specific rate of gold dissolution for all studied variants of electrochemical leaching had its maximum value in the first 24. 10^2 s of the leaching process. Influence of the S:L ratio on the extraction of gold during electroleaching has been studied.

With a thiosulfate concentration of 0.1 M, maximum gold extraction is 21.4%, the leaching of metals continues and the volume of the oxygen decreases, this is indirectly confirmed by the graph of the electrical conductivity from the electrolyte concentration shown in Fig. 3. Up to the 30 min mark of the process, the electric conductivity and the oxygen concentration are both decreasing. After 30 min, the electric conductivity begins to increase up to a value of 59.15μS/cm by shifting additional components (leaching of metals) into the solution. As can be seen in the figures, when the electrical conductivity of the solution decreases slightly and then increases it is likely due to changes in the concentration of the solution which is probably a result of the solution restructuring.

As can be seen (Fig. 5a) the gold extraction increased during the first 20 min of leaching and then reduced. Maximum extraction was achieved with S:L = 1:4 and is 86.67% efficient. With S:L = 1:3 and S:L = 1:6 there was an increase in the extent of gold extraction as the oxygen concentration decreased and with a decrease in the extent of gold extraction, the oxygen concentration increased. This may indicate that the oxygen concentration in the system has little connection with the gold recovery system. At S:L = 1:4 with an increase in the extent of gold extraction, the oxygen concentration at the during the initial period increased, then decreased down to 3.41 mg/dm^3 .

As shown in Fig. 5b, the specific rates increased up to the 40 min mark and then declined, this is directly connected with the decreasing metal content in the initial product as well as with decreasing dissolved oxygen content, which is consumed in the oxidation process of metals. It has been established that increasing the solution's temperature up to 60°C does not affect the efficiency of gold extraction in the solution. With temperatures of 25, 40, 50, 60°C, the maximum extraction level is 82.99, 90.47, 58.53 and 56.67% respectively.

Research results show that the specific rate of the process for all temperatures at the initial time ($24 \cdot 10^2$ s) increases, then over time (from $24 \cdot 10^2$ -to $54 \cdot 10^2$ s) this indicator is sharply reduced, which is connected with the reduction in the metal content of the initial product as well as with decreasing dissolved oxygen content, which has been consumed in the process of metal oxidation (Fig. 6b). With this the maximum specific rate is reached and the results are as follows: At 40°C a rate of $9.62 \cdot 10^{-11} mol/m^2 \cdot s$. At 50 and 60°C the values of the specific rate are 6.6 $mol/m^2 \cdot s$ and 7.25 $mol/m^2 \cdot s$ respectively. The nature of the dependence is explained firstly by a gradual increase in the degree of saturation within the solutions, with gold and other impurity elements, that takes place during the course of the leaching process and secondly by the simultaneous reduction of the gold content in the samples. The study of the sorption of gold from solutions by electrochemical leaching. In the modern gold extracting industry sorption method extraction of gold and silver is one of the most common hydrometallurgical operations. Ion exchange resins or activated carbons are used as sorbents (Kholmogorov et al., 2002b). It is known that the process of dissolution of gold and silver can be written as the following equation:

$$4Au + 8S_2O_3^{2-} + O_2 + 2H_2O \rightarrow 4\left[Au\left(S_2O_3\right)_2\right]^{3-} + 4OH^-$$

The main parameter of this process is the concentration of thiosulfate ions and, as a result of dissolution, sustained thiosulfate complexes of gold arise (I). The same complexes also are formed in the model solutions (Salarirad and Behnamfard, 2011). The sorption of gold from the individual model thiosulfate solutions by the anion exchanger AM-2B in OH^- form has been investigated. At the initial stage, the sorption behavior of thiosulfate complexes of gold on the anion exchanger AM-2B with various concentrations of sodium thiosulfate: 0.1, 0.2 and 0.5 M were studied (Fig. 7). The results showed that sorption, in a solution of 0.1 M sodium thiosulfate, is inefficient because a low degree of sorption (15%) is achieved. The best results have been obtained in the experiments with a thiosulfate concentration of 0.2 M: E_{Au} = 73.15% and 0.5 M: E_{Au} = 76.88% (Fig. 7).

It can be seen that further increases in concentration above 0.2 M has little effect on the degree of sorption, as a result of this all further experiments were performed with solutions of thiosulfate containing 0.2 mol/dm^3 . The presented curves show that the maximum degree of gold extraction is 73.15%, which is achieved at the ratio of $V_{resin}:V_{solution}$ = 1:500. It is established that with a decrease of the volume of solution, the degree of sorption in turn gets reduced by a factor of 2 times, while an increase of the volume of the solution has an insignificant effect on the degree of sorption of gold (Fig. 8).

In addition to gold a large number of accompanying metals, which have high sorption susceptibility, are removed from the solution, which is the reason for the sharp decrease in the working capacity of the anion exchanger AM-2B on the target metal. In all experiments, the presence of impurity metals adversely affects the gold sorption process, reducing its degree of sorption by several times. The extent of sorption of gold reached 42%. From the presented dependencies it can be seen that with increasing degrees of sorption of impurity metals, the degree of sorption of gold reduces. The sorption behavior of copper in solutions of electrochemical leaching is similar to behavior observed in thiosulfate solutions: During the first hour of experiments copper gets almost completely sorbed-80%. A high degree of sorption is achieved by other related elements: 75% of zinc is sorbed, 80% of iron and 35% of cobalt.

The results of the study showed that the specific rates in the process of sorption of metals from electrochemical leaching increases sharply during the initial period of time, namely, during the first 60 min of the experiment and, further to this there is a gradual decrease in the reaction rate. The maximum specific rate of sorption of gold from technological solutions of electrochemical leaching is achieved at an electric current of 200 A. The nature of the dependence is explained firstly by the gradual increase in the degree of saturation of the anion exchanger with gold and nonferrous metals, which takes place during the process of sorption and secondly by the simultaneous decrease in the metal content of the solutions.

It has been shown that the presence of a large number of accompanying (associated) elements have a high sorption ability-this is the primary reason for the sharp decrease in the working capacity of the anion exchanger AM-2B on the target metals as it exerts a strong influence on the process of sorption of noble metals from complex productive solutions during electrochemical leaching. In order to increase the maximum potential degree of sorption of gold and silver the pre-treatment (preliminary cleaning) of productive solutions, to reduce the quantity of impurity components, is recommended.

CONCLUSION

The purpose of this study is to give a preliminary assessment of the advisability of thiosulfate leaching in the extraction of gold from ores on a laboratory scale. The experimental work has allowed for a study into the effects that arise from: The concentration of the leaching agent, impress current and the ratio S:L on the process of gold extraction. The leaching of gold in a solution in the presence of a carbon disulfide electrode takes place with high intensity. Furthermore the gold can then be extracted from the solution by sorption. It is found that in solutions of electrochemical leaching with a sodium thiosulfate concentration of 2.0 mg/dm^3 , a current density of 200 A/m^2 , S:L = 1:4 and a temperature of 40°C the extent of the gold extraction reaches values of up to 80%, as in the case of conventional cyanidation.

As a result of studies, the use of sodium thiosulfate as a solvent in the process of electrochemical leaching in order to increase the process selectivity and increase the degree of extraction of metals can be recommended for the extraction of metals from refractory materials. One of the reasons for the low extraction ratio of noble metals from solutions is the presence of impurity metals in the leaching solutions. It was discovered that the efficiency of the sorption process of noble metals from electrochemical leaching solutions (sodium thiosulfate) is higher than from aqueous chemical leaching solutions. As a result, the selective extraction of metals and the absence of copper and iron directly increase the affinity of gold to the anion exchanger AM-2B.

ACKNOWLEDGMENT

The authors gratefully acknowledge the financial support of the research by the Science Committee of Ministry of Education and Science of the Republic of Kazakhstan.

AUTHOR'S CONTRIBUTIONS

Bagdaulet Kenzalyevich Kenzhaliyev: Designed the research plan, participated in the results analysis, contributed to the reviewing of the article critically. Zamzagul Duisenovna Dosymbaeva and Renata Ramilevna Iskhakova: Designed and performed experiments, contributed in the development of the conceptual framework. Essen Nurgalievich Suleimenov: Organized the study, data collection of the study sample, analysis and writing of the manuscript.

REFERENCES

1. Abbruzzese, C., P. Fornaria, R. Massiddaa, F. Vegliòb and S. Ubaldini, 1995. Thiosulphate leaching for gold hydrometallurgy. Hydrometallurgy, 39: 265-276. DOI: 10.1016/0304-386X(95)00035-F
2. Adams, M.D., 2005. Advances in Gold Ore Processing. 1st Edn., Elsevier, Amsterdam, ISBN-10: 0080459080, pp: 1076.
3. Aylmore, M.G. and D.M. Muir, 2001. Thiosulfate leaching of gold-a review. Miner. Eng., 14: 135-174. DOI: 10.1016/S0892-6875(00)00172-2
4. Aylmore, M.G., 2001. Treatment of a refractory goldcopper sulfide concentrate by copper ammoniacal thiosulfate leaching. Miner. Eng., 14: 615-637.
5. Balaz, P., J. Ficeriova, E. Boldizarova, M. Haber and S. Jelen et al., 2000. Thiosulphate leaching of gold from a mechanochemically pretreated complex sulphide concentrate. Proceedings of the 21 International Mineral Processing Congress, (MPC' 00), Rome, pp: A6-74-A6-81.
6. Breuer, P.L. and M.I. Jeffrey, 2000. Thiosulfate leaching kinetics of gold in the presence of copper and ammonia. Miner. Eng., 13: 1071-1081. DOI: 10.1016/S0892-6875(00)00091-1
7. Feng, D. and J.S. Van Deventer, 2002. Leaching behaviour of sulphides in ammoniacal thiosulphate systems. Hydrometallurgy, 63: 189-200. DOI: 10.1016/S0304-386X(01)00225-0
8. Grosse, A.C., G.W. Dicinoski, M.J. Shaw and P.R. Haddad, 2003. Leaching and recovery of gold using ammoniacal thiosulfate leach liquors (a review). Hydrometallurgy, 692: 1-21. DOI: 10.1016/S0304-386X(02)00169-X
9. Kenzhaliev, B., 2014. Electrochemical method for extracting non-ferrous and precious metals from refractory materials using combined reactions. Proceedings of the 15th Topical Meeting of the International Society of Electrochemistry. Interfacial Electrochemistry at Atomic, Molecular and Nanoscale Domains (Niagara Falls, Canada).
10. Kholmogorov, A.G., O.N. Kononova, G.L. Pashkov and Y.S. Kononov, 2002a. Thiocyanate solutions in gold technology. Hydrometallurgy, 64: 43-48. DOI: 10.1016/S0304-386X(02)00005-1
11. Kholmogorov, A.G., O.N. Kononova, G.L. Pashkov and Y.S. Kononov, 2002b. Sorption recovery of gold thiosulphate complexes. Chinese J. Chem. Eng., 10: 123-127.
12. Lee, H.Y., S.G. Kim and J.K. Oh, 1997. Cementation behavior of gold and silver onto Zn, Al and Fe powders from acid thiourea solutions. Canadian Metallurgical Q., 36: 149-155. DOI: 10.1016/S0008-4433(97)00004-9
13. McNulty, T., 2001. Cyanide substitutes. Mining Magazine, 184: 256-261.
14. Parga, J.R., J.L. Valenzuela and J.A. Díaz, 2012. New Technology for Recovery of Gold and Silver by Pressure Cyanidation Leaching and Electrocoagulation. INTECH Open Access Publisher, ISBN-10: 9533078987.
15. Qingshuang, S., W. Dequan and C. Jinju, 1995. New method of extracting gold and silver from high silver-gold concentrates. Trans. Nonferrous Metals Society China, 5: 50-53.
16. Salarirad, M.M. and A. Behnamfard, 2011. Fouling effect of different flotation and dewatering reagents on activated carbon and sorption kinetics of gold. Hydrometallurgy, 109: 23-28. DOI: 10.1016/j.hydromet.2011.05.002
17. Tremblay, L., G. Deschenes, E. Ghali, J. McMullen and M. Lanouette, 1996. Gold recovery from a sulphide bearing gold ore by percolation leaching with thiourea. Int. J. Mineral Process., 48: 225-244. DOI: 10.1016/S0301-7516(96)00029-4
18. Ubaldini, S., P. Fornari, R. Massidda and C. Abbruzzese, 1998. An innovative thiourea gold leaching process. Hydrometallurgy, 48: 113-124. DOI: 10.1016/S0304-386X(97)00076-5
19. Vignes, A., 2011. Extractive Metallurgy 3: Processing Operations and Routes. 1st Edn., John Wiley and Sons, Inc., Hoboken, New Jersey, ISBN-10: 1118617029, pp: 352.

20. Yao, S.Z., F.J. He and L.H. Nie, 1992. Piezoelectric determination of traces of thiourea. Analytica Chimica Acta, 268: 311-314. DOI: 10.1016/0003-2670(92)85226-V

21. Zhou, J., M. Zhou, P. Zhu and L. Kunfang, 2009. Method for leaching gold mine by electrochemistry oxidation method.

22. Zipperian, D. and S. Raghavan, 1988. Gold and silver extraction by ammoniacal thiosulfate leaching from a rhyolite ore. Hydrometallurgy, 19: 361-375. DOI: 10.1016/0304-386X(88)90041-2

CHAPTER 16

EFFECTIVE UTILIZATION FOR DATA OF NATURAL ENVIRONMENT CORROSION OF MATERIALS

Haitao Wang[1], Leijiang Yao[2], Xiaofeng Li[1], En-Hou Han[1] and Xiaoyan Tong[2]

[1] Key Laboratory of Nuclear Materials and Safety Assessment, Institute of Metal Research, Chinese Academy of Sciences, Shenyang, 110016, China

[2] Science and Technology on UVA Laboratory, Northwester Polytechnical University, Xi'an, 710065, China

ABSTRACT

An analysis architecture is built to make full use of natural environment corrosion of materials data. All environmental factors, such as average temperature, average relative humidity, rainfall hours, sunshine duration, chloride ion settlement, etc., in nine atmospheric test stations, including Beijing, Wuhan, Jiangjin, and Wanning, and seawater temperature, salinity, dissolved oxygen, pH values, etc. in four marine test stations, including Qingdao, Zhoushan, Xiamen, and Yulin, are modelled monthly using a statistical analysis technique. Normal distribution, logarithmic normal distribution, Weibull distribution, and uniform distribution of the samples are checked by the Shapiro-Wilk and Kolmogorov-Smirnov methods. Subsequently, the probability distribution function, the expected population value and variance, and the confidence interval are estimated. Corrosion data for twenty-two kinds of carbon steels, low alloy steels, and stainless steels, including A3, 3C, 16Mn, 10CrMoAl, 1Cr18Ni9Ti, in atmospheric and marine test stations are analyzed. The main environmental factors for atmospheric and marine corrosion of each type of steel are determined by grey relational analysis. Subsequently, a predictive model incorporating the main corrosion environmental factors, exposure time, and corrosion rate is built by the BP artificial neural network. After the evolution of corrosion pit depth is predicted by the artificial neural network, a fracture mechanics calculation is used to evaluate the residual life of a structure with corrosion pit defects.

INTRODUCTION

Corrosion is very harmful to the useful lifetime of materials. It can bring enormous economic losses and severe dangers to society. Natural environment corrosion is one of the most common corrosion phenomena. It is very complex and has a vast set of factors responsible for its existence. Natural environment corrosion includes mainly atmospheric corrosion and marine corrosion.

Atmospheric corrosion is defined as the corrosion of materials exposed to the air and its pollutants. Electrochemical corrosion and chemical corrosion can occur in the atmospheric environment. However, electrochemical corrosion is more important than chemical corrosion in atmospheric corrosion (Morcillo et al., 2013). Metal absorbs a layer of ultra-thin water film on its surface that forms an electrolyte under atmospheric

Citation: Haitao Wang, Leijiang Yao, Xiaofeng Li, En-Hou Han and Xiaoyan Tong, Effective Utilization for Data of Natural Environment Corrosion of Materials, DOI: http://doi.org/10.5334/dsj-2015-018.

exposure conditions after a certain critical humidity level has been reached (Melchers, 2013; Wall et al., 2005). In the presence of this thin-film electrolyte, atmospheric corrosion proceeds by balancing anodic and cathodic reactions. The anodic oxidation reaction involves dissolution of the metal while the cathodic reaction is commonly an oxygen reduction reaction. Under thin-film corrosion conditions, oxygen from the atmosphere is readily supplied to the electrolyte. Seawater is the electrolyte solution with very high salinity. It is one of the most corrosive of all the natural corrosion agents. Thus, electrochemical corrosion of metals is stronger in a marine environment (Traverso et al., 2014; Shifler, 2005).

A large amount of data concerning the natural environment corrosion of materials has accumulated over the last 16 years through research supported by the Major Program of the National Natural Science Foundation of China (Xiao et al., 2002; Zhu et al., 2002). In this work, analysis architecture is built to make full use of these data. This architecture analyzes the inherent laws of data and obtains information on the influences of natural environment factors and the evolutionary rule of corrosion damage. This provides corresponding support and help for research on natural environment corrosion.

STATISTICAL ANALYSIS MODELLING FOR ENVIRONMENTAL FACTORS

The framework for the module of statistical analysis modelling for environmental factors is shown in Figure 1. The module builds the atmospheric environment and the marine environment database. All environmental factors for different atmospheric and marine test stations are modelled monthly by a statistical analysis technique. The Shapiro-Wilk method (W method) and the Kolmogorov-Smirnov method (K-S method) are used to check the normal distribution, the logarithmic normal distribution, the Weibull distribution, and the uniform distribution for the sample. Subsequently, the probability distribution function, the expected value and variance of population, and the confidence interval are estimated.

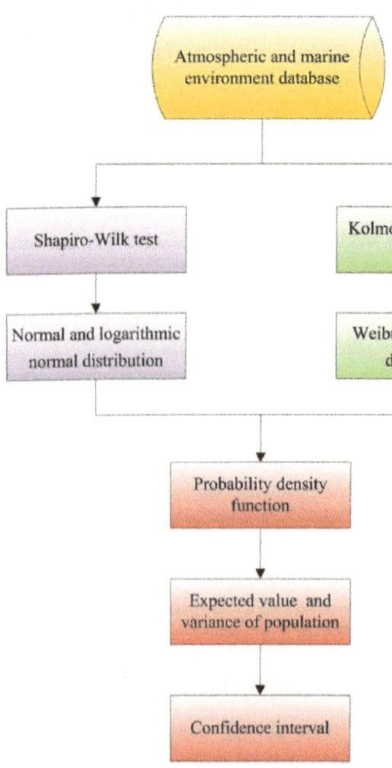

Figure 1: Framework for the module of statistical analysis modelling for environmental factors.

The atmospheric environment database involves seventeen kinds of atmospheric environmental factors: average temperature, average relative humidity, rainfall, rainfall hours, sunshine duration, SO_2 settlement, NO_2 settlement, H_2S settlement, HCl settlement, SO_3 settlement, Cl^- settlement, NH_3 settlement, dustfall of water-solubility, dustfall of water-insolubility, pH value of rainwater, and the SO_4 and Cl^- content of rainwater at the nine atmospheric test stations, Beijing, Wuhan, Jiangjin, Wanning, Qingdao, Guangzhou, Shenyang, Qionghai, and Hailar. The marine environment database involves nine kinds of marine environmental factors: seawater temperature, salinity, dissolved oxygen, pH value, temperature of the marine atmosphere, humidity, rainfall, sunshine, and wind speed at the four marine test stations, Qingdao, Zhoushan, Xiamen, and Yulin. In the statistical analysis module, the normal distribution and the logarithmic normal distribution for the sample are checked by the W method. The W test is a kind of small sample test ($3 \leq n \leq 50$) that can only check the normal distribution (Jurečková et al., 2007; Srivastava et al., 1987). It can deal with the small sample events of atmospheric and marine environments. The W test is shown in Eq. (1)

$$W = \frac{\left(\sum_{i=1}^{k} a_i \left(x_{n+1-i} - x_i\right)\right)^2}{\sum_{i=1}^{n} (x_i - \mu)^2} \tag{1}$$

where n is the sample number, k = n/2 when n is an even number or k = (n-1)/2 when n is an odd number, μ is the mean of the sample, x_i is the value of the ith sample, and a_i is the ith constant. If $W > W_\alpha$, the sample follows normal distribution, W_α is the quantile of the W test, and α is the level of significance. In order to check the logarithmic normal distribution, the sample needs a logarithm conversion before Eq. (1) is used.

The probability density functions of the normal and logarithmic normal distributions are shown in Eq. (2)

$$f(x) = \frac{1}{\sqrt{2\pi}\sigma} e^{-\frac{(x-\mu)^2}{2\sigma^2}} \qquad f(x) = \frac{1}{\sqrt{2\pi}\sigma x} e^{-\frac{(\ln x-\mu)^2}{2\sigma^2}} \tag{2}$$

where σ is the standard deviation of the sample. For normal distribution, the estimate of expected value of population E(x) = μ and the estimate of variance of population $D(x) = \sigma^2$. For the logarithmic normal distribution, the estimate of expected value of population$E(X) = \exp(\mu + {}^2/2)$ and the estimate of variance of population $D(X) = (\exp({}^2) - 1)\exp(2\mu + {}^2)$.

The K-S test is a type of universal test. It is used to check normal and nonnormal distributions (Vlček et al., 2009; Schröer et al., 1995). However, its reliability in checking the normal distribution is lower than that of the W test for a small sample. In the statistical analysis module, the Weibull and uniform distributions of the sample are checked by the K-S method. Suppose that $F_0(x)$ is the probability distribution function of the theoretical distribution and $F_n(x)$ is the cumulative frequency function of the current sample,

$$D = \max\left| F_n(x) - F_0(x) \right| \tag{3}$$

If $D < D_\alpha$, the sample follows the current theoretical distribution, D_α is the critical value of the W test, and α is the level of significance.

fracture toughness. Thus d_2 is the critical corrosion pit depth. The residual life of the structure is equal to t_2-t_1.

CONCLUSIONS

This paper introduces the overall architecture for the analysis of natural environment corrosion. Atmospheric and marine environmental data are analyzed by a statistical method. The W method's reliability in checking the normal distribution for a small sample is higher than that of the K-S method; but the W method is a special method to check normal distribution. It does not check any other distributions. The K-S test is a kind of universal test. It can check any distribution as long as the distribution function is known. Although the expected values and variances of the population can be estimated by the distribution function, a more reliable estimate is given by the confidence interval at different confidence levels. For atmospheric and marine corrosion data, the main corrosion environmental factors are confirmed by grey relational analysis. These factors are the input variables for the BP artificial neural network. This can simplify the BP artificial neural network modelling and improve the reliability of the model. After the relationship between corrosion pit depth and time is predicted by the BP network, the residual life of the structure can be obtained by a fracture mechanics calculation on the critical corrosion pit depth.

ACKNOWLEDGMENTS

This work was supported by the National Science and Technology Major Project (No. 2011ZX06004-009).

REFERENCES

1. Cai, J.P., Cottis, R.A., & Lyon, S.B. (1999) Phenomenological modelling of atmospheric corrosion using an artificial neural network. *Corrosion Science 41*, pp 2001–2030.
2. Carpenter, W.C. & Hoffman, M.E. (1995) Training Backprop Neural Networks. *Journal of AI Expert 10*, pp 30–33.
3. Deng, J.L. (1989) Grey information space. *The Journal of Grey System 1*, pp 103–117.
4. Huwang, L.C. & Hwang, J.T.G. (2002) Prediction and confidence intervals for nonlinear measurement error models without identifiability information. *Statistics & Probability Letters 58*, pp 355–362.
5. Jurečková, J. & Picek, J. (2007) Shapiro–Wilk-type test of normality under nuisance regression and scale. *Computational Statistics & Data Analysis 51*, pp 5184–5191.
6. Melchers, R.E. (2013) Long-term corrosion of cast irons and steel in marine and atmospheric environments. *Corrosion Science 68*, pp 186–194.
7. Morcillo, M., Chico, B., Díaz, I., Cano, H., & Fuente, D. (2013) Atmospheric corrosion data of weathering steels. A review. *Corrosion Science* 77, pp 6–24.
8. Schröer, G. & Trenkler, D. (1995) Exact and randomization distributions of Kolmogorov-Smirnov tests two or three samples. *Computational Statistics & Data Analysis 20*, pp 185–202.
9. Shifler, D.A. (2005) Understanding material interactions in marine environments to promote extended structural life. *Corrosion Science 47*, pp 2335–2352.
10. Srivastava, M.S. & Hui, T.K. (1987) On assessing multivariate normality based on Shapiro-Wilk W statistic. *Statistics & Probability Letters 5*, pp 15–18.
11. Traverso, P. & Canepa, E. (2014) A review of studies on corrosion of metals and alloys in deep-sea environment. *Ocean Engineering 87*, pp 10–15.
12. Vlček, O. & Huth, R. (2009) Is daily precipitation Gamma-distributed?: Adverse effects of an incorrect use of the Kolmogorov–Smirnov test. *Atmospheric Research 93*, pp 759–766.
13. Wall, F.D., Martinez, M.A., Missert, N.A., Copeland, R.G., & Kilgo, A.C. (2005) Characterizing corrosion behavior under atmospheric conditions using electrochemical techniques. *Corrosion Science 47*, pp 17–32.
14. Wang, H.T., Han, E.H., & Ke, W. (2006) Artificial neural network modeling for atmospheric corrosion of carbon steel and low alloy steel. *Corrosion Science and Protection Technology 18*, pp 144–147.
15. Wang, H.T., Han, E.H., & Ke, W. (2006) Gray model and gray relation analysis for atmospheric corrosion of carbon steel and low alloy steel. *Corrosion Science and Protection Technology 18*, pp 278–280.
16. Xiao, Y.D., Liang, C.F., Zheng, Q.F., Hu, X.J., Wang, L., Zhang, E.P., et al. (2002) "Nine five" data collection of atmospheric corrosion of materials, The Major Program of the National Natural Science Foundation of China (No. 59899141).
17. Zhang, X.Z. (1994) Application of probability weighted moments to estimate parameters in Weibull distribution. *Marine Forecast 11*, pp 55–61.
18. Zhu, X.K. & Joyce, J.A. (2012) Review of fracture toughness (G, K, J, CTOD, CTOA) testing and standardization. *Engineering Fracture Mechanics 85*, pp 1–46.
19. Zhu, X.R., Lin, L.Y., Zhang, S.P., Liu, D.Y., Lin, Z.J, Jin, W.X, et al. (2002) "Nine five" data collection of marine corrosion of materials, The Major Program of the National Natural Science Foundation of China (No. 59899142).

INDEX